T0289920

# Amide Bond Activation: Concepts and Reactions

# Amide Bond Activation: Concepts and Reactions

Edited by
Caleb Murphy

www.willfordpress.com

Published by Willford Press,
118-35 Queens Blvd., Suite 400,
Forest Hills, NY 11375, USA



ISBN: 978-1-64728-520-3

**Cataloging-in-Publication Data**

Amide bond activation : concepts and reactions / edited by Caleb Murphy.
p. cm.
Includes bibliographical references and index.
ISBN 978-1-64728-520-3
1. Amides. 2. Chemical bonds. 3. Amides--Synthesis. 4. Organic compounds--Synthesis. I. Murphy, Caleb.
QD341.A7 A45 2023
547.04--dc23

For information on all Willford Press publications
visit our website at www.willfordpress.com

# Contents

# Preface

Amides refer to functional groups wherein a carbonyl carbon atom is connected by a single bond to a nitrogen atom, and either a hydrogen or a carbon atom. An amide bond refers to a type of peptide bond. The condensation of an amine and a carboxylic acid is the most common process for the creation of this type of bond. The high stability of amide bonds is due to their tendency to create a resonating structure which gives a double bond character to the amide CO-N bond. It also leads to a decrease in the rotation around the peptide bond. The creation of amide bonds is a fundamental reaction in chemistry that is used in the synthesis of polymers, peptides and food additives. Inert amide bonds can be activated through acyl, acyl addition, radical and decarbonylative pathways. This book aims to shed light on the concepts and reactions of amide bond activation. It presents researches and studies performed by experts across the globe. The extensive content of this book provides the readers with a thorough understanding of the subject.

Various studies have approached the subject by analyzing it with a single perspective, but the present book provides diverse methodologies and techniques to address this field. This book contains theories and applications needed for understanding the subject from different perspectives. The aim is to keep the readers informed about the progresses in the field; therefore, the contributions were carefully examined to compile novel researches by specialists from across the globe.

Indeed, the job of the editor is the most crucial and challenging in compiling all chapters into a single book. In the end, I would extend my sincere thanks to the chapter authors for their profound work. I am also thankful for the support provided by my family and colleagues during the compilation of this book.

**Editor**

# 1

# 1,4-Disubstituted 1,2,3-Triazoles as Amide Bond Surrogates for the Stabilisation of Linear Peptides with Biological Activity

**Lisa-Maria Rečnik [1,2,3], Wolfgang Kandioller [2] and Thomas L. Mindt [1,2,3,*]**

1 Ludwig Boltzmann Institute Applied Diagnostics, General Hospital Vienna, 1090 Vienna, Austria; lisa.recnik@lbiad.lbg.ac.at
2 Institute of Inorganic Chemistry, Faculty of Chemistry, University of Vienna, 1090 Vienna, Austria; wolfgang.kandioller@univie.ac.at
3 Department of Biomedical Imaging and Image Guided Therapy, Division of Nuclear Medicine, Medical University of Vienna, 1090 Vienna, Austria
* Correspondence: Thomas.Mindt@lbiad.lbg.ac.at

**Abstract:** Peptides represent an important class of biologically active molecules with high potential for the development of diagnostic and therapeutic agents due to their structural diversity, favourable pharmacokinetic properties, and synthetic availability. However, the widespread use of peptides and conjugates thereof in clinical applications can be hampered by their low stability in vivo due to rapid degradation by endogenous proteases. A promising approach to circumvent this potential limitation includes the substitution of metabolically labile amide bonds in the peptide backbone by stable isosteric amide bond mimetics. In this review, we focus on the incorporation of 1,4-disubstituted 1,2,3-triazoles as amide bond surrogates in linear peptides with the aim to increase their stability without impacting their biological function(s). We highlight the properties of this heterocycle as a *trans*-amide bond surrogate and summarise approaches for the synthesis of triazole-containing peptidomimetics via the Cu(I)-catalysed azide-alkyne cycloaddition (CuAAC). The impacts of the incorporation of triazoles in the backbone of diverse peptides on their biological properties such as, e.g., blood serum stability and affinity as well as selectivity towards their respective molecular target(s) are discussed.

**Keywords:** 1,4-disubstituted 1,2,3-triazoles; CuAAC; peptidomimetics; amide bond surrogate; metabolic stabilisation

---

## 1. Introduction

Peptides represent one of the major classes of biomolecules used as diagnostics and therapeutics. [1] Since the introduction of solid-phase peptide synthesis (SPPS) by Merrifield [2], which allows for a quick, easy and automatable synthesis of peptides, their pharmaceutical impact grew enormously due to their high specificity and low off-target side-effects. However, applications of peptides in medicine can be limited due to their low in vivo stability [3]. They are susceptible to rapid degradation by proteases via hydrolysis of the amide bonds in their backbone structure [4]. This leads to a poor oral bioavailability due to the presence of proteases in the digestive system and a low metabolic stability in vivo derived from the degradation in the blood plasma mediated by soluble and membrane-bound peptidases [5,6].

Therefore, substantial effort has been put into the development of stabilisation techniques for metabolically labile peptides. Different approaches have been studied involving structural variations of the peptide sequence. For example, modification of the C- or N-terminus of a peptide, cyclisation of linear peptides or incorporation of D-amino acids and other unnatural amino acids in the sequence have been shown to have a positive impact on peptide stability in blood serum [3,5,7]. Another approach is the coadministration of protease inhibitors to impede fast degradation and enhance in vivo blood serum stability [5,8].

In order to prevent degradation by proteases, amide bonds susceptible to cleavage can be substituted with isosteric amide bond surrogates. [9] Those peptidomimetics resemble native peptides but contain synthetic non-peptidic structural features to improve the therapeutic or diagnostic properties of the molecule. The most common peptide bond mimetics include esters, *N*-methylated amide bonds, reduced amide bonds, semicarbazides, retro-inversed peptide bonds, peptoids and alkenes (Figure 1) [9,10]. Introduction of heteroatoms such as sulfur, fluorine or boron lead to further variation [11–15]. The incorporation of heterocycles in the peptide backbone is also reported. [16,17] The first example of an aromatic heterocyclic amide bond mimetic in literature is the substitution of the amide bond with 2-imidazolidine [18]. Other well-studied heterocyclic surrogates are pyrazoles, tetrazoles and 1,2,4-triazoles and many more examples can be found in the literature [19–21]. One of the most recently emerged amide bond mimetics in peptides is the 1,2,3-triazole scaffold [22].

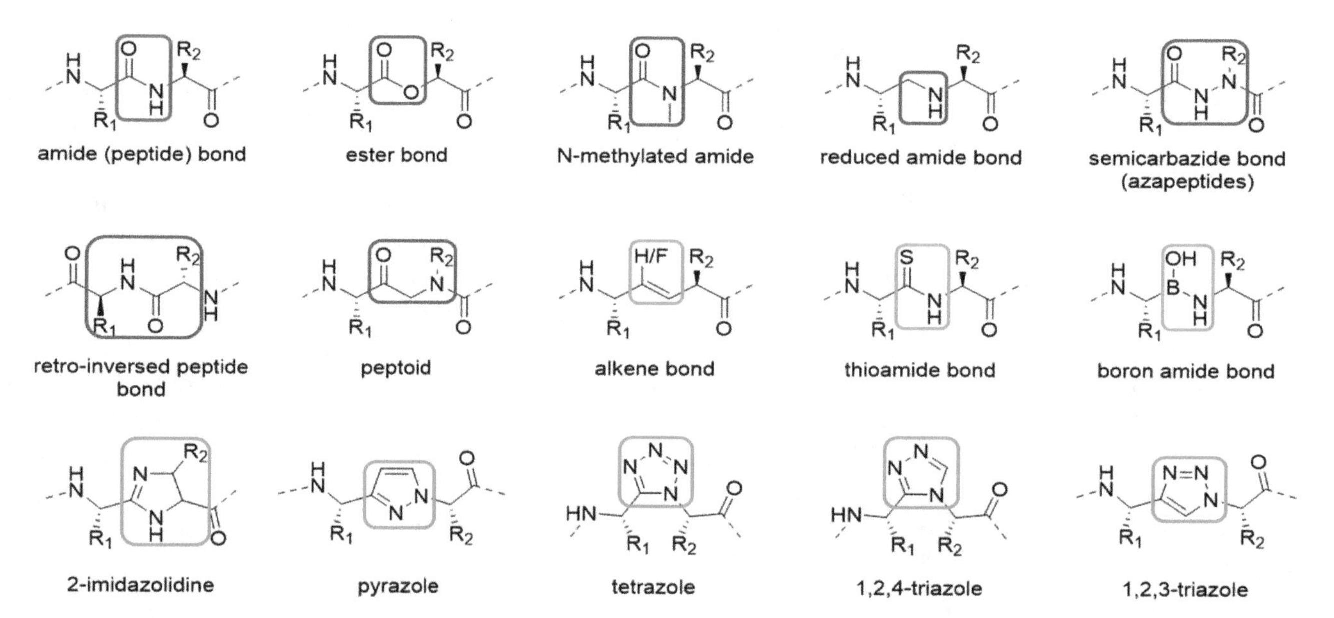


**Figure 1.** Various reported mimetics of a peptide bond (top left, red) including common motifs (blue), motifs containing heteroatoms (light blue) or heterocycles (green).

In medicinal chemistry, the 1,2,3-triazole motif is frequently used for the replacement of amide bonds and their application in small molecules has already been highlighted elsewhere [23,24]. In 2003, the first report of a cyclic peptidomimetic was published where a 1,4-substituted 1,2,3-triazole was used as an isostere for the amide bond to obtain molecules which self-assemble into nanotubes [25]. In the following years, several groups investigated the structural consequences of the incorporation of one or more triazoles on the secondary structure of resulting peptidomimetics such as $\alpha$-helices and $\beta$-sheets [26–33]. Another focus of attention was the synthesis and biological evaluation of cyclic peptidomimetics with one or more triazoles in the backbone [34–43]. In most cases, the formation of the triazole ring was used as a strategy for the (macro)cyclisation of the peptidomimetic. In a similar manner, CuAAC was also applied for the synthesis of large peptides by ligating smaller fragments [44]. A handful of studies investigated the conformation of a dipeptide with the central amide bond being substituted by a triazole [45–47].

In contrast to a metabolically labile amide bond, 1,2,3-triazoles are not cleaved by proteases. Hence the substitution of a peptide bond by a triazole ($\Psi$[Tz], a so-called amide-to-triazole switch) represents a promising strategy to increase the peptide's in vivo stability. This property is not only important for the therapeutic efficacy of a drug but also for diagnostic means [48]. A number of examples of triazolo-peptidomimetics included in this review concern their application as radiopharmaceuticals. In nuclear medicine, small radiolabelled peptides are the agents of choice for tumour-targeting vectors in diagnostic imaging of tumour entities. For example, the peptide can be conjugated via its N- or C-terminus or a sidechain to a chelator that can stably complex a metallic radionuclide such as lutetium-177 or gallium-68 [48]. The radiopharmaceutical is applied intravenously and then distributes in the entire body. It specifically binds to the cancer cells due to high receptor affinity and after internalisation, it gets accumulated inside the tumour. This leads to an increased concentration of radioactivity inside the target, which is then detected by either single photon emission computed tomography (SPECT) or positron emission tomography (PET) depending on the employed radionuclide [48,49]. In the field of SPECT- and PET-imaging, peptides are preferred over antibodies as tumour-targeting vectors because of their favourable pharmacokinetic properties, their synthetic availability and their low cost [50]. In general, their rapid distribution, fast tissue penetration and quick blood clearance enable imaging with higher resolution and after shorter incubation times compared to antibodies. Furthermore, healthy tissue is less exposed to radiation which makes radiolabelled peptides a safer option for treatment of patients [50,51]. The abundance of regulatory and other peptides, which display high affinity for receptors that are often massively overexpressed in cancer cells, constitutes a vast pool of possibilities for the development of diagnostic tools. However, one limitation is their short biological half-life in blood plasma due to proteolysis. The number of approved radiolabelled peptides for imaging or therapy is still low partly due to low in vivo stability. Currently a great effort is ongoing to improve their stability whilst maintaining (or even increasing) the affinity to the receptor [48]. Besides their use as radiopharmaceuticals, other applications of triazolo-peptidomimetics include endogenous peptides involved in signalling, protease inhibitors and inhibitors of protein-protein interaction and will be discussed in detail in the following.

In this review, we highlight the properties of 1,2,3-triazoles as *trans*-amide bond isosteres and the synthesis of triazolo-peptidomimetics. We discuss the application of the amide-to-triazole switch in peptides with the aim to improve their stability. We limit our review to the modification of linear, high affinity peptides which exhibit a known biological function and where one or more amide bonds are replaced with a 1,4-disubstituted 1,2,3-triazole. The use of 1,5-disubstituted 1,2,3-triazoles as surrogates for *cis*-amide bonds is not within the scope of this review, but can be found summarised elsewhere [42,52–56]. This review is organised by peptides and listed in chronological order.

## 2. 1,2,3-Triazole as Amide Bond Isostere

1,4-Disubstituted 1,2,3-triazoles are excellent isosteres for the *trans*-amide bond as they mimic the electronic properties such as dipole moment and hydrogen-bond properties of the amide bond in contrast to many other surrogates (Figure 2) [22]. The dipole moment of the triazole moiety is slightly higher than in amide bonds (~4.5 Debye vs. ~3.5 Debye) which allows for the polarisation of the proton at C-4 so that it can serve as H-bond donor similar to the NH in amides [57,58]. The lone pairs at *N*-2 and *N*-3 serve as weak H-bond acceptors, hence well mimicking the amide bond. [59,60] Because of the high dipole moment, the triazole unit could align with that of other amide bonds to stabilise the secondary structure of peptides [61]. Furthermore, the aromatic heterocycle is able to form intra- or intermolecular $\pi$-interactions [62]. When comparing the structure and the size of the two units, the triazole surrogate leads to a slightly longer distance (4.9–5.1 Å) between the two neighbouring amino acid residues than the amide bond (3.8–3.9 Å), which is thus more similar to the distance encountered in peptides incorporating β-amino acids (4.9 Å) [22,23,63]. The heterocycle reproduces the planarity of the amide bond, although it also displays a higher rigidity due to the aromatic structure. All these similar features make this heterocyclic motif a good amide bond mimetic, which is highly stable to enzymatic cleavage by proteases, acid or base hydrolysis and reductive or oxidative conditions [64]. However, due to this high chemical stability, it is still unknown how triazoles are metabolised and further investigations are needed to determine their biological fate [61].

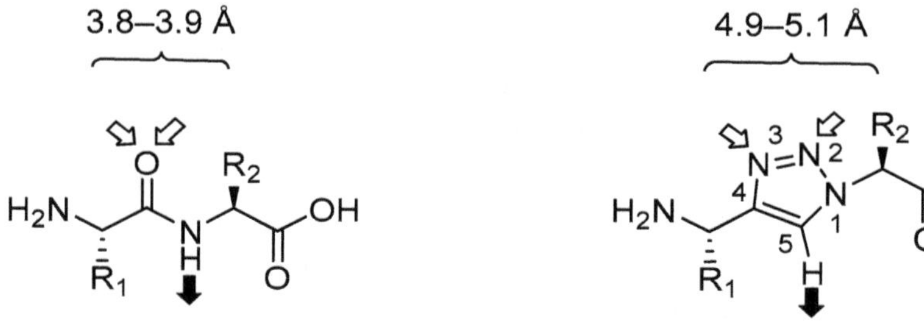

3.8–3.9 Å

4.9–5.1 Å

**Figure 2.** Comparison of an amide bond (left) and an 1,4-disubstituted 1,2,3-triazole (right). Distance between two residual side chains given. White arrows indicate H-bond receptors, black arrow H-bond donors. Modified from Valverde et al. [22].

## 3. Synthesis of Triazolo-Peptidomimetics

In recent years, the regioselective synthesis of 1,2,3-triazoles became easily accessible. In 2002, the groups of Meldal and Sharpless reported independently the Cu(I)-catalysed azide-alkyne cycloaddition reaction (CuAAC) which leads to the regioselective formation of 1,4-disubstituted 1,2,3-triazoles under mild conditions [65,66]. This reaction is classified as a click chemistry reaction as it is high yielding, wide in scope, simple to perform, produces no (or easily removable) side-products and can be conducted in a wide range of solvents [67]. Due to this outstanding versatility, this reaction quickly had a high impact in many areas of chemistry. Applications of the CuAAC range from organic and medicinal chemistry to material science and biological chemistry [68–72]. This reaction is also a versatile tool for peptide chemists, either for cyclisation reactions (head-to-tail, head/tail-to-sidechain or sidechain-to-sidechain) or cyclodimerisations, as disulfide bridge replacement, for ligation of peptide fragments or bioconjugation to other (bio)molecules [73,74]. Meldal reported the use of CuAAC on solid phase which further enhanced its use in peptide synthesis [65]. The introduction of the triazole motif into the backbone of the peptide can be easily performed in solution or on solid phase as depicted in Figure 3.

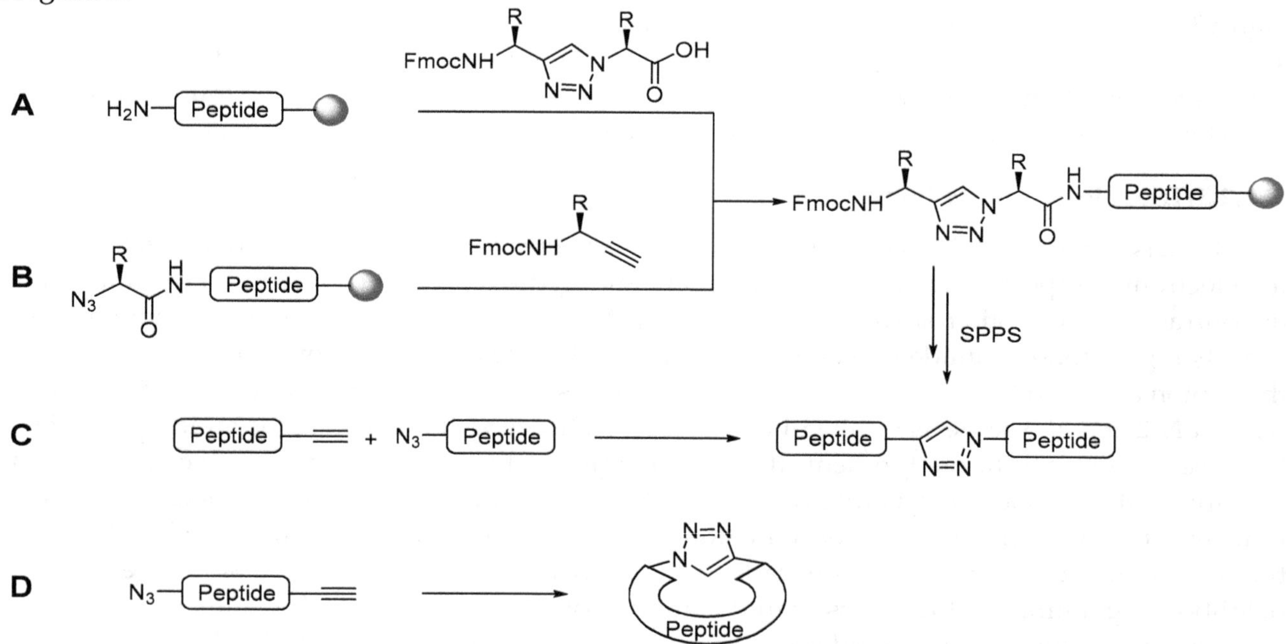


**Figure 3.** Schematic representation of the different strategies for the synthesis of triazolo-peptidomimetics: **A**: Formation of triazole-containing dipeptide in solution prior to coupling onto solid-phase synthesis. **B**: Formation of triazole on solid-phase. **C**: Triazole formation as ligation step of two peptide fragments. **D**: Triazole formation as cyclisation strategy. Modified from Valverde et al. [22].

The CuAAC can be performed in solution between an *N*-protected α-amino alkyne and an α-azido carboxylic acid leading to a triazole-containing dipeptoid building block which can then be employed for SPPS (pathway A). Alternatively, the CuAAC can be carried out directly on solid phase by transforming the immobilised, deprotected N-terminal amine on-resin into an azide and subsequently reacting it with a protected α-amino alkyne together with a Cu(I) salt (or a source thereof) for the CuAAC (pathway B). CuAAC can also be envisaged for the ligation of two larger peptide fragments with an azide- and an alkyne-terminus, respectively (pathway C). In case of cyclic triazole-containing peptides, the triazole formation may be used as the cyclisation step after the synthesis of the linear peptide sequence bearing an azide and an alkyne unit (pathway D).

In solid phase synthesis, the triazole moiety is incorporated into the peptide backbone in two steps. First, the free amine of the elongated peptide chain is transformed into an azide via a diazotransfer reaction. Imidazole-1-sulfonyl azide hydrochloride (ISA·HCl) is a relatively safe and stable diazotransfer reagent [75] and the reaction can be carried out at room temperature in solvents compatible with solid phase synthesis (e.g., DMF). Alternatively, chiral α-azido acids are also commercially available and can be used directly for SPPS. The second step consists of the CuAAC where the terminal azide is reacted with an Fmoc-protected α-amino alkyne under Cu(I) catalysis employing different sources of the metal (i.e., CuI, $CuBr{\cdot}Me_2S$, $Cu(CH_3CN)]PF_6$ or $CuSO_4{\cdot}5H_2O$ with sodium ascorbate) [76–79]. α-Amino alkynes are not commercially available yet, however, their synthesis can be achieved in a few steps from Fmoc-protected amino acids (Figure 4). The most common and convenient method involves the formation of the corresponding α-amino aldehyde via the Weinreb amide and the subsequent Seyferth-Gilbert homologation using the Bestman-Ohira reagent [80]. Besides, the Corey-Fuchs reaction for the transformation of aldehydes to alkynes can be employed for the synthesis of Boc-protected α-amino alkynes [81]. However, it is known that this approach may lead to partial racemisation of the α-amino alkyne [52,56,82]. The use of α-amino alkynes as a mixture of enantiomers results in the formation of diastereomers when used for peptide synthesis and additional measures (i.e., purification of the alkynes using chiral HPLC or separation of the resulting diastereomeric peptide products by HPLC) have to be taken into account. An alternative synthesis of α-amino alkynes involves the condensation of Ellman's auxiliary onto aldehydes and subsequent conjugate addition of an organometallic reagent to the resulting sulfinamide (Figure 4) [56]. However, this method is limited to amino acids without acid-sensitive protecting groups in their side chains.

Reaction times for the CuAAC on solid phase are generally longer than for Fmoc-SPPS peptide couplings. It is usually performed at room temperature overnight and no heating or microwave reactor is necessary. CuAAC on solid phase results in quantitative conversion which can be followed by the Punna-Finn test, a colorimetric test developed for the detection of aliphatic azides similar to the Kaiser test [83]. The CuAAC can also be used to ligate several peptide fragments as an alternative cyclisation approach for peptides.

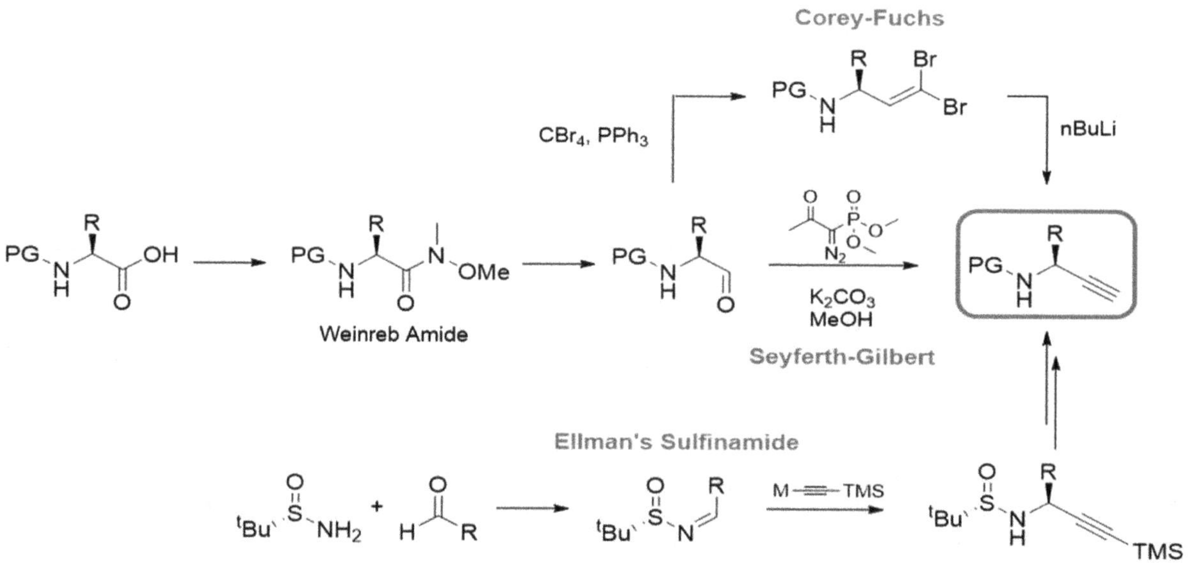

**Figure 4.** Different approaches for the synthesis of $\alpha$-amino alkynes: The most common synthesis is starting from an amino aldehyde via a Seyferth-Gilbert homologation using the Bestmann-Ohira reagent. Another possibility from the same starting material is the Corey-Fuchs reaction. An alternative approach is the condensation of an achiral aldehyde and Ellman's auxilliary followed by the conjugate addition of an organometallic alkyne reagent to the sulfinamide. M = Metal (e.g., Li or MgX), PG = protecting group, R = amino acid side chain, TMS = trimethylsilyl.

## 4. Applications

### *4.1. PACE4 Inhibitors*

Paired basic amino acid cleaving enzyme 4 (PACE4, also known as proprotein convertase subtilisin/kexin type 6 PCSK6) belongs to the family of pro-protein convertases and plays an important role in tumour aggressiveness. A patent from Dory and co-workers describes the development of more stable and selective PACE4 inhibitors for treating a variety of different cancers based on the octapeptide multi-Leu peptide [84]. Various modifications were introduced in this peptide including the replacement of amide bonds with a 1,2,3-triazole and in silico analyses were performed to select potent inhibitors. Amongst them were the two triazole-containing peptidomimetics **P1** and **P2** with a triazole in positions Leu[2]-Leu[3] and Leu[1]-Leu[2], respectively (Table 1). However, the synthesis of these two compounds is not described. Half-life ($t_{1/2}$) in blood plasma and the inhibitor constant ($K_i$) was determined. Interestingly, compound **P1** displayed a two-fold lower half-life in blood plasma and a decreased potency ($t_{1/2}$ = 1.0 h, $K_i$ = 600 nM) compared to the unmodified peptide ($t_{1/2}$ = 2.1 h, 38 nM). In contrast, peptidomimetic **P2** maintained the potency of the parent compound ($K_i$ = 37 nM) and increased the plasmatic half-life by 100% ($t_{1/2}$ = 4.0 h). Nevertheless, these two peptide analogues were not further pursued as other strategies such as the incorporation of D-amino acids, $\beta$-amino acids or modification at the N-terminus were found to be more effective.

**Table 1.** Sequences of Multi-Leu and **P1–2** and their reported biological properties [84].

| Compound | Sequence | $t_{1/2}$ [h] [b] | $K_i$ [nM] |
|---|---|---|---|
| **Multi-Leu** [a] | Ac-Leu-Leu-Leu-Leu-Arg-Val-Lys-Arg-$NH_2$ | 2.1 ± 0.2 | 38 |
| **P1** | Ac-Leu-Leu**$\Psi$[Tz]**Leu-Leu-Arg-Val-Lys-Arg-$NH_2$ | 1.0 ± 0.2 | 600 |
| **P2** | Ac-Leu**$\Psi$[Tz]**Leu-Leu-Leu-Arg-Val-Lys-Arg-$NH_2$ | 4.0 ± 0.5 | 37 |

[a] Reference compound. [b] Determined in blood plasma.

### 4.2. *Leu-Enkephalin*

The Dory group reported the synthesis of triazole-containing peptidomimetics based on Leu-enkephalin (Leu-Enk), an endogenous ligand for the delta opioid receptor (DOPr) [76]. Due to the rapid degradation of this pentapeptide, the group investigated the use of metabolically stable 1,2,3-triazoles to obtain analogues with increased biological half-lives. They designed four derivatives **Enk1–4** by systematically replacing each amide bond with a triazole (see Table 2). For the synthesis of these peptidomimetics, the triazole-bearing dipeptide was prepared in solution via CuAAC using CuI and 2,6-lutidine and the building blocks were then further used in the solid-phase synthesis of the peptidomimetics (Figure 3A). The binding affinities of all analogues were determined in a competitive binding assay on GH3 cell membrane extracts (rat pituitary tumour cell lines) containing DOPr and using the selective DOPr agonist [$^3$H]-deltorphin II as competitive ligand. Compound **Enk1** with a triazole at position Phe$^4$-Leu$^5$ showed the highest affinity with an $K_i$ = 89 nM, although it was 15-fold lower than the endogenous Leu-enkephalin ($K_i$ = 6.3 nM) [85]. Derivative **Enk2** with a triazole at position Gly$^3$-Phe$^4$ displayed a 60-fold lower affinity ($K_i$ = 460 nM) compared to the reference compound and the other analogues lost their affinity for DOPr ($K_i$ > 1000 nM). Unfortunately, experiments regarding their stability are not reported. This study shows that the 1,2,3-triazole is not in all cases a universal bioisostere for a *trans*-amide bond as their incorporation into the peptide can result in diminished or abolished biological activity.

**Table 2.** Sequences of Leu-Enk and **Enk1–4** and their reported $K_i$ values [76].

| Compound | Sequence | $K_i$ [nM] [b] |
|---|---|---|
| **Leu-Enk** [a] | Tyr-Gly-Gly-Phe-Leu | 6.3 ± 0.9 |
| **Enk1** | Try-Gly-Gly-Phe**Ψ[Tz]**Leu | 89 ± 12 |
| **Enk2** | Try-Gly-Gly**Ψ[Tz]**Phe-Leu | 460 ± 250 |
| **Enk3** | Try-Gly**Ψ[Tz]**Gly-Phe-Leu | >1000 |
| **Enk4** | Try**Ψ[Tz]**Gly-Gly-Phe-Leu | >1000 |

[a] Reference compound. [b] $K_i$ values were determined in competitive binding assays on GH3/DOPr cell membrane extracts with [$^3$H]-deltorphin II as selective agonist. $K_i$ values are expressed as the means ± standard error of means (SEM) of three to four independent experiments.

### 4.3. *Bombesin*

The amide-to-triazole switch was reported on the minimal binding sequence of bombesin BBN(7-14), an agonistic ligand of the gastrin-releasing peptide receptor (GRPR), which is overexpressed in a variety of tumours (i.e., prostate, breast, lung and pancreatic cancer) [77]. Valverde et al. chose this model peptide as an example for a radiolabelled tumour-targeting vector to demonstrate the potential of this novel peptide stabilisation method. The sequence was functionalised with a $PEG_4$ spacer and the universal chelator 1,4,7,10-tetraazacyclododecane-1,4,7,10-tetraacetic acid (DOTA) at the N-terminus which was necessary for the radiolabelling with lutetium-177, a clinically established radionuclide. In addition, the methionine residue in position 14 was changed to norleucine (Nle) to avoid side products arising from the oxidation of methionine during the radiolabelling step [86]. Afterwards, a triazole scan of [Nle$^{14}$]BBN(7-14) was performed by creating a library of nine different triazolo-peptidomimetics **BBN2–10** (Table 3) in which every amide bond was substituted with a triazole one at a time. The CuAAC was performed on solid phase according to Figure 3B by reacting the resin-bound azido-peptide with the corresponding $\alpha$-amino alkyne, $[Cu(CH_3CN)]PF_6$, tris[(1-benzyl-1*H*-1,2,3-triazol-4-yl)methyl]amine (TBTA) and *N*,*N*-diisopropylethylamine (DIPEA) to achieve the formation of the 1,4-substituted 1,2,3-triazole at the desired position. Seven out of the nine peptidomimetics showed improved in vitro serum stability ranging from $t_{1/2}$ = 8 h for **BBN8** to $t_{1/2}$ = >100 h for **BBN4** (in comparison to $t_{1/2}$ = 5 h for the reference peptide **BBN1**) thus enhancing the stability up to 20-fold. The amide-to-triazole switch in the two analogues **BBN2** and **BBN10** preserved the serum stability ($t_{1/2}$ = 5 h and 6 h, respectively). The lipophilicity of all compounds was evaluated by determining the $logD_{7.4}$ value. This is an important characteristic for radiopharmaceuticals as it impacts pharmacokinetic properties such as the renal excretion of the drug. Substitution of an amide bond with a 1,2,3-triazole did not significantly alter the lipophilicity of the novel peptide analogues at pH 7.4. When tested for their biological activity, the cell uptake and the dissociation constant ($K_D$) were evaluated using GRPR-overexpressing PC3 cells (human prostate cancer cells). The internalisation rate of **BBN2, 5** and **7** were comparable to the parental compound (29.1%, 28.3% and 24.5%, respectively compared to 27.7% for the unmodified reference peptide **BBN1**). The single digit nanomolar affinity of the reference compound **BBN1** to GRPR ($K_D$ = 2.0 nM) was also maintained in **BBN2, 5** and **7** with $K_D$ values of 3.0, 3.1 and 5.9 nM, respectively. For compound **BBN6**, the amide-to-triazole switch was slightly detrimental resulting in an internalisation rate of only 8.4% and a $K_D$ value of 48.6 nM. The other derivatives had abolished biological function on GRPR. Thus, compound **BBN5** with the triazole situated between $Gly^{11}$ and $His^{12}$ was chosen for in vivo evaluation as the most promising candidate with a 3.5-fold increase in in vitro serum stability and conserved biological activity. It displayed in vivo specificity toward GRPR and a 2-fold higher tumour uptake in athymic nude mice bearing PC3 xenografts due to its higher in vivo stability.

**Table 3.** Sequences of **BBN1–14** and their reported biological properties [77,87].

| Comp. | Sequence | $t_{1/2}$ [h] [b] | Uptake after 4 h [%] [c] | $K_D$ [nM] [d] |
|---|---|---|---|---|
| BBN1 [a] | $[^{177}Lu]$Lu-DOTA-$PEG_4$-Gln-Trp-Ala-Val-Gly-His-Leu-Nle-$NH_2$ | 5 | 27.7 | 2.0 ± 0.6 |
| BBN2 | $[^{177}Lu]$Lu-DOTA-$PEG_4$-Gln-Trp-Ala-Val-Gly-His-Leu-Nle**Ψ[Tz]**H | 6 | 29.1 | 3.0 ± 0.5 |
| BBN3 | $[^{177}Lu]$Lu-DOTA-$PEG_4$-Gln-Trp-Ala-Val-Gly-His-Leu**Ψ[Tz]**Nle-$NH_2$ | 60 | 0.2 | n.d. |
| BBN4 | $[^{177}Lu]$Lu-DOTA-$PEG_4$-Gln-Trp-Ala-Val-Gly-His**Ψ[Tz]**Leu-Nle-$NH_2$ | >100 | n.o. | n.d. |
| BBN5 | $[^{177}Lu]$Lu-DOTA-$PEG_4$-Gln-Trp-Ala-Val-Gly**Ψ[Tz]**His-Leu-Nle-$NH_2$ | 17 | 28.3 | 3.1 ± 1.0 |
| BBN6 | $[^{177}Lu]$Lu-DOTA-$PEG_4$-Gln-Trp-Ala-Val**Ψ[Tz]**Gly-His-Leu-Nle-$NH_2$ | 25 | 8.4 | 48.6 ± 11.5 |
| BBN7 | $[^{177}Lu]$Lu-DOTA-$PEG_4$-Gln-Trp-Ala**Ψ[Tz]**Val-Gly-His-Leu-Nle-$NH_2$ | 16 | 24.5 | 5.9 ± 1.8 |
| BBN8 | $[^{177}Lu]$Lu-DOTA-$PEG_4$-Gln-Trp**Ψ[Tz]**Ala-Val-Gly-His-Leu-Nle-$NH_2$ | 8 | n.o. | n.d. |
| BBN9 | $[^{177}Lu]$Lu-DOTA-$PEG_4$-Gln**Ψ[Tz]**Trp-Ala-Val-Gly-His-Leu-Nle-$NH_2$ | 14 | n.o. | n.d. |
| BBN10 | $[^{177}Lu]$Lu-DOTA-$PEG_4$**Ψ[Tz]**Gln-Trp-Ala-Val-Gly-His-Leu-Nle-$NH_2$ | 5 | 0.5 | n.d. |
| BBN11 | $[^{177}Lu]$Lu-DOTA-$PEG_4$-Gln-Trp-Ala**Ψ[Tz]**Val-Gly**Ψ[Tz]**His-Leu-Nle-$NH_2$ | 27 | 21.7 ± 0.2 | 25.6 ± 6.9 |
| BBN12 | $[^{177}Lu]$Lu-DOTA-$PEG_4$-Gln-Trp-Ala-Val**Ψ[Tz]**Gly**Ψ[Tz]**His-Leu-Nle-$NH_2$ | 40 | 3.5 ± 0.6 | >1000 |
| BBN13 | $[^{177}Lu]$Lu-DOTA-$PEG_4$-Gln-Trp-Ala**Ψ[Tz]**Val**Ψ[Tz]**Gly-His-Leu-Nle-$NH_2$ | 66 | 0.1 ± 0.1 | >1000 |
| BBN14 | $[^{177}Lu]$Lu-DOTA-$PEG_4$-Gln-Trp-Ala**Ψ[Tz]**Val**Ψ[Tz]**Gly**Ψ[Tz]**His-Leu-Nle-$NH_2$ | 61 | 0.3 ± 0.1 | >1000 |

[a] Reference compound. [b] Determined in blood serum at 37 °C. [c] Ratio of specific receptor-bound and cell-internalised compound expressed in % of administered dose. Expressed as the means ± SEM of at least two independent experiments. n.o.: not observed [d] Determined by receptor saturation binding assay on PC3 cells expressing GRPr. Expressed as the means ± SEM of at least two independent experiments. n.d.: not determined.

Valverde et al. also explored the synthesis of triazolo-peptidomimetics of $[Nle^{14}]$BBN(7-14) with multiple triazoles in the backbone [87]. Therefore, triazoles were strategically placed in the most promising positions as identified by the previous triazole scan. The three possible analogues **BBN11–13** with two triazoles in one molecule at positions $Ala^9$-$Val^{10}$, $Val^{10}$-$Gly^{11}$ or $Gly^{11}$-$His^{12}$ were synthesised as well as the peptidomimetic **BBN14** containing all three substitutions. Even though the half-lives in blood serum were increased 5- to 13-fold (27–66 h) compared to unmodified compound **BBN1**, the biological activity was lost (**BBN12–14**) or significantly decreased likely due to a constraint conformation of the peptidomimetic ($K_D$ = 25.6 nM for **BBN11**, Table 3).

Another bombesin derivative studied by Valverde et al. was the antagonist JMV594 which differs from BBN(7-14) only in the last two amino acids at the C-terminus and the addition of a D-Phe at the N-terminus (Figure 5) [88,89].

Due to the high homology of the two peptides, the three most promising positions identified in the screening of $[Nle^{14}]$BBN(7-14) were substituted with a triazole in JMV594. The obtained peptides were conjugated to $PEG_4$ and DOTA for lutetium-177 chelation at their N-terminus and the resulting derivatives **BBN15–18** were tested for cell binding and receptor affinities (Table 4). The analogues **BBN17** and **BBN18** lost their biological activity and **BBN16** with the triazole at the C-terminus displayed a lower affinity ($K_D$ = 8.1 nM) and cell binding (13%) compared to the reference compound **BBN15** ($K_D$ = 2.7 nM, cell binding 30%). Surprisingly, in vitro blood serum stability tests showed that analogue **BBN16** as well as the unmodified peptide **BBN15**, have remarkably long half-lives in blood serum compared to agonistic [Nle$^{14}$]BBN(7-14) (see above). 65% and 75% of the peptide and peptidomimetic were still intact after 48 h, respectively. Despite the high homology between the two bombesin derivatives, the successful results from the [Nle$^{14}$]BBN(7-14) triazole screening could not be transferred to JMV594 analogues. These results underline the different mode of action of the two GRPR-binding molecules (agonistic vs. antagonistic peptides). In general, backbone modifications that are tolerated by one peptide cannot be applied automatically to a structurally related peptide.

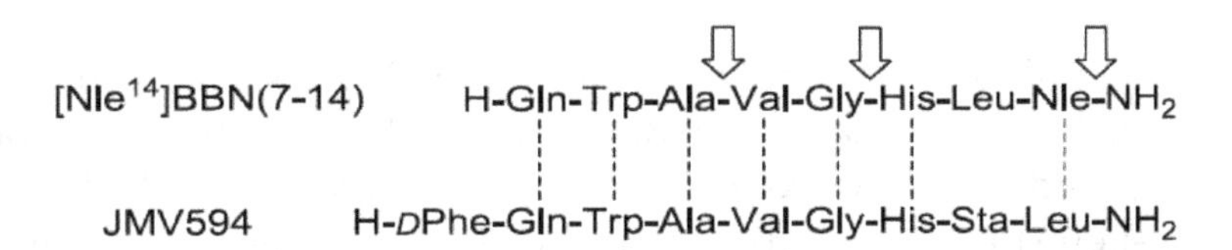


**Figure 5.** Comparison of the amino acid sequences of bombesin derivative [Nle$^{14}$]BBN(7-14) and antagonist JMV594. Dashed lines illustrate the structural similarity of the amino acid sequences, and arrows indicate positions where amide bonds can be replaced by 1,2,3-triazoles in the agonist [Nle$^{14}$]BBN(7–14) without loss of the biological properties of the vector. Reproduced with permission from Valverde et al. [89].

**Table 4.** Sequences of **BBN15–18** and their reported biological properties [89].

| Comp. | Sequence [b] | Stability after 48 h [%] [c,d] | Uptake after 4 h [%] [d,e] | $K_D$ [nM] [f] |
|---|---|---|---|---|
| **BBN15** [a] | [$^{177}$Lu]Lu-DOTA-PEG$_4$-D-Phe-Gln-Trp-Ala-Val-Gly-His-Sta-Leu | 65 | 30 | 2.7 |
| **BBN16** | [$^{177}$Lu]Lu-DOTA-PEG$_4$-D-Phe-Gln-Trp-Ala-Val-Gly-His-Sta-Leu**Ψ[Tz]**H | 75 | 13 | 8.1 |
| **BBN17** | [$^{177}$Lu]Lu-DOTA-PEG$_4$-D-Phe-Gln-Trp-Ala-Val-Gly**Ψ[Tz]**His-Sta-Leu | n.d. | n.d. | >1000 |
| **BBN18** | [$^{177}$Lu]Lu-DOTA-PEG$_4$-D-Phe-Gln-Trp-Ala**Ψ[Tz]**Val-Gly-His-Sta-Leu | n.d. | n.d. | >1000 |

[a] Reference compound. [b] Sta: statine: (3*S*,4*S*)-4-amino-3-hydroxy-6-methylheptanoic acid. [c] Determined in blood serum at 37 °C. Expressed as % of intact peptide. [d] n.d.: not determined [e] Ratio of specific receptor-bound and cell-internalised compound expressed in % of administered dose. Expressed as the means of three independent experiments. [f] Determined by receptor saturation binding assay on PC3 cells expressing GRPr. Expressed as the means ± SEM of at least two independent experiments.

### *4.4. Kisspeptin*

Kisspeptin is an endogenous peptide involved in the modulation of the gonadotropic axis, a hormonal system including luteinising hormone (LH) and follicle-stimulating hormone (FSH) which regulates fertility in mammals and human. The downregulation of this system is the cause of severe reproductive disorders resulting in partial or complete infertility in humans [90,91]. It also occurs naturally in seasonal breeders such as sheep during the anoestrous phase and its manipulation can lead to a prolonged breeding season and improved livestock management [92]. It has been shown that injection of the truncated decapeptide analogue KP10 leads to an upregulation of the gonadotropic axis although the effect is only short-lasting due to quick degradation of the peptide by proteases. Beltramo et al. examined the stabilisation of the decapeptide **KP1**, the ewe analogue of KP10, via an amide-to-triazole switch [78]. First, they *N*-acetylated the endogenous decapeptide **KP1** which led to a highly improved resistance towards proteolysis as a small fraction (2.6%) of peptide was still present after 6 h of incubation in blood serum (**KP2**, Table 5). The amount of the reference compound **KP1** was below the detection limit at this time point. The potency of the novel compounds was determined in a calcium mobilisation assay using HEK-293 cells (human embryonic kidney cells) transfected with the kisspeptin receptor KISS1R. It was shown that *N*-acetylation of **KP1** also increased drug potency from an $EC_{50}$ of 2.5 nM to 0.07 nM (**KP2**). In a next step, the two main proteolysis sites at position Phe$^6$-Gly$^7$ and Gly$^7$-Leu$^8$ were replaced by triazoles leading to two monotriazolo analogues **KP3** and **KP4** as well as bistriazolo-peptidomimetic **KP5**. The formation of these triazoles was accomplished on solid phase by the addition of Fmoc-protected $\alpha$-amino alkynes, CuBr·$Me_2S$ and DIPEA to the immobilised azidopeptide (Figure 3B).

**Table 5.** Sequences of **KP1–5** and their reported biological properties [78].

| Comp. | Sequence | Stability after 6 h [%] [b,c] | $EC_{50}$ [nM] [d] |
|---|---|---|---|
| **KP1** [a] | Tyr-Asn-Trp-Asn-Ser-Phe-Gly-Leu-Arg-Tyr-$NH_2$ | n.d. | 2.5 ± 2.2 |
| **KP2** | Ac-Tyr-Asn-Trp-Asn-Ser-Phe-Gly-Leu-Arg-Tyr-$NH_2$ | 2.6 ± 0.4 | 0.07 ± 0.06 |
| **KP3** | Ac-Tyr-Asn-Trp-Asn-Ser-Phe-Gly**Ψ[Tz]**Leu-Arg-Tyr-$NH_2$ | 40.8 ± 2.9 | 0.07 ± 0.06 |
| **KP4** | Ac-Tyr-Asn-Trp-Asn-Ser-Phe**Ψ[Tz]**Gly-Leu-Arg-Tyr-$NH_2$ | 61.4 ± 6.4 | 0.6 ± 0.05 |
| **KP5** | Ac-Tyr-Asn-Trp-Asn-Ser-Phe**Ψ[Tz]**Gly**Ψ[Tz]**Leu-Arg-Tyr-$NH_2$ | 50.7 ± 3.2 | 120 ± 87 |

[a] Reference compound. [b] Determined in blood serum at 39 °C. [c] n.d.: not detectable [d] Determined by a calcium mobilization assay on HEK-293 cells transfected with KISS1R.

The in vitro serum stabilities of triazole-containing analogues **KP3–5** increased 15- to 25-fold compared to **KP2** as after 6 h of incubation 40.8, 61.4 and 50.7% of the peptidomimetics were still intact, respectively (compared to 2.6% of **KP2**). Regarding biological activity, **KP3** with the additional triazole at position $Gly^7$-$Leu^8$ maintained the potency of **KP2**. However, the amide-to-triazole switch in **KP4** was less favourable yielding in an $EC_{50}$ of 0.6 nM and multiple triazole substitution in **KP5** was detrimental to the potency ($EC_{50}$ = 120 nM). Compounds **KP2** and **KP3** were evaluated in vivo for their capacity to increase the plasma concentration of LH and FSH in anoestrous ewes. After intravenous injection, **KP2** did not change the LH concentration compared to **KP1** but compound **KP3** resulted in a clear increase (peak concentration: 5.4 ng $mL^{-1}$ compared to 3.3 ng $mL^{-1}$ for **KP1**). The duration of LH release after administration of **KP3** was also doubled in comparison to **KP1**, which in turn resulted in a higher amount of the hormone secreted over time (Figure 6). In vivo dose-response studies of **KP3** illustrated solely a modest increase in the maximal concentration, the duration of action and the total amount of LH secretion upon application of higher **KP3** concentrations. Similar trends were observed for secretion of FSH. Overall, these results were less significant than anticipated from the in vitro data probably due to fast renal excretion. These modifications, in combination with further optimisations, hold the potential to improve the treatment of reduced or complete infertility associated with the gonadotropic axis.

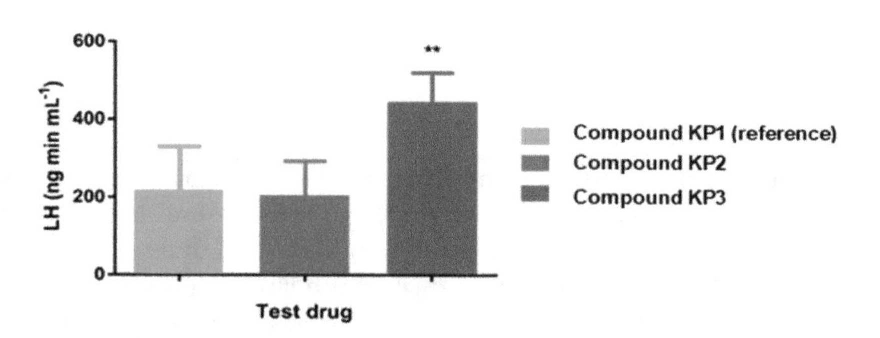


**Figure 6.** Total amount of luteinising hormone (LH) secreted in vivo in ewes after application of 5 nmol **KP1** (reference), **KP2** or **KP3**. Values are the mean ± SEM. ** for $p < 0.01$. Reproduced with permission from Beltramo et al. [78].

### *4.5. Neurotensin*

Mascarin et al. applied the amide-to-triazole switch approach to the stabilisation of the binding sequence of neurotensin, NT(8-13), a regulatory peptide which specifically binds to neurotensin receptors (NTR) [93]. The overexpression of subtype NTR1 in breast, pancreas, prostate cancer and many more makes this molecule an interesting candidate as a tumour targeting vector for the development of new therapeutic and diagnostic agents [94,95]. However, the major drawback is its low in vivo stability with a half-life of only a few minutes. Hence, the group was investigating the stabilisation of this peptide via a triazole scan. The binding sequence NT(8-13) was conjugated N-terminally to a $PEG_4$ spacer and DOTA, which was employed for the radiolabelling with lutetium-177. A first generation of triazolo-peptidomimetics was obtained by the stepwise substitution of each amide bond with a 1,4-disubstituted 1,2,3-triazole except for the amide bond between $Arg^9$-$Pro^{10}$, which does not allow for an amide-to-triazole switch due to the secondary amine of Pro. A second generation was obtained by replacing isoleucine in position 12 of the most promising triazolo-peptidomimetics with *tert*-leucine (Tle), another strategy reported for the stabilisation of this sequence. The same synthetic protocol as described above for bombesin analogues was employed (Figure 3B) [77]. All compounds were then evaluated in vitro for their stability in blood serum (Figure 7) as well as their cell internalisation into NTR1-expressing HT29 cells (human colon carcinoma) and affinity toward the NTR1 (Table 6). Substitution of amide bonds between the C-terminus and $Pro^{10}$ (compounds **NT2–5**) resulted in a loss of biological function albeit an up to 4-fold increase in vitro stability was achieved ($t_{1/2}$ = 13.0–164.0 min compared to $t_{1/2}$ = 39.4 min for reference compound **NT1**). This came as no surprise, as this region is involved in several H-bond interactions with the receptor and therefore sensitive to structural modifications. Exchanging the amide bond at the N-terminus $PEG_4$-$Arg^8$ or $Arg^8$-$Arg^9$ (compound **NT6–7**) slightly improved in vitro stability to half-lives of 64.9 min and 46.9 min, respectively. Biological evaluation of **NT6** and **NT7** showed that the activity was maintained in vitro with a cellular uptake of 6.4% and 9.4%, respectively, and $K_D$ values of 8.8 nM and 4.5 nM (for comparison, reference compound **NT1**: 7.3% internalisation and $K_D$ 3.7 nM). Earlier studies reported the remarkably improved stability of NT(8–13) when $Ile^{12}$ is replaced with a Tle residue [96,97]. Modification of the two most promising triazolo-peptidomimetics **NT6** and **NT7** with the $Ile^{12}$-to-$Tle^{12}$ switch resulted in derivatives **NT9** and **NT10** with a 100-fold increase of the in vitro stability (>95% of compound still intact after 4 h; half-lives not determined) compared to the reference compound **NT8** (0.9% after 4 h). However, only derivative **NT9** retained a submicromolar affinity toward NTR1 ($K_D$ = 214 nM). In vivo evaluation of the most promising peptide conjugates in athymic nude mice bearing HT-29 xenografts revealed that the replacement of the $Arg^8$-$Arg^9$ amide bond by a triazole resulted in two-fold tumour uptake for both, the $Ile^{12}$ derivative **NT6** and the $Tle^{12}$ derivative **NT9**, compared to their respective reference compounds without a triazole. This can be ascribed to the significantly improved stability of the triazolo-peptidomimetics. Compound **NT7** on the other hand, showed a high tumour-to-background ratio, a property which is particularly important for potential imaging applications.

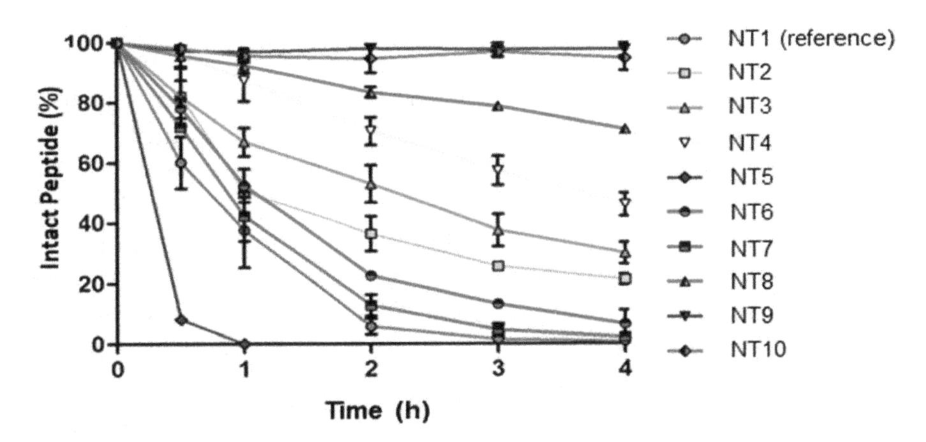


**Figure 7.** In vitro blood serum stabilities of reference compound **NT1** (green) and monotriazolo-peptidomimetics **NT2–10** at 37 °C. Values are expressed as percentage of intact peptide. Error bars indicate the standard deviation of mean values (n ≥ 2–3). Reproduced with permission from Mascarin et al. [93].

Subsequent investigations focused on the incorporation of multiple triazoles in the backbone of **NT(8–13)** [98]. Building on the results of the triazole scan discussed above, the triazolo-peptidomimetics **NT11** and **NT12** with a simultaneous substitution at $PEG_4$-$Arg^8$ and $Arg^8$-$Arg^9$ for the $Ile^{12}$ and $Tle^{12}$ analogue were synthesised and tested in vitro. Multiple triazoles could not improve the stability of the peptidomimetics further. Surprisingly, the biological activity of **NT11** was maintained ($k_D$ = 4.6 nM, 10.8% cell internalisation) but unexpectedly the stability was found to be reduced ($t_{1/2}$ = 13 min) compared to the unmodified peptide conjugate **NT1** ($t_{1/2}$ = 39.4 min). One explanation might be that the introduction of two consecutive triazoles in this peptide sequence results in a confirmation of the peptide, which is prone to enzymatic degradation. In the case of the Tle[12] derivative **NT12**, the in vitro plasma stability was maintained (>97% intact after 4 h), but the receptor affinity was abolished.

**Table 6.** Sequences of **NT1–12** and their reported biological properties [93,98].

| Comp. | Sequence | Stability after 4 h [%] [b] ($t_{1/2}$ in min) [c] | Uptake after 4 h [%] [d] | $K_D$ [nM] [e] |
|---|---|---|---|---|
| **NT1** [a] | [$^{177}$Lu]Lu-DOTA-PEG$_4$-Arg-Arg-Pro-Tyr-Ile-Leu | 0.9 ± 0.3 (39.4) | 7.3. ± 0.4 | 3.7 ± 0.8 |
| **NT2** | [$^{177}$Lu]Lu-DOTA-PEG$_4$-Arg-Arg-Pro-Tyr-Ile-Leu**Ψ[Tz]**H | 21.3 ± 1.8 (69.7) | n.o. | n.d. |
| **NT3** | [$^{177}$Lu]Lu-DOTA-PEG$_4$-Arg-Arg-Pro-Tyr-Ile**Ψ[Tz]**Leu | 30.0 ± 3.6 (72.0) | n.o. | n.d. |
| **NT4** | [$^{177}$Lu]Lu-DOTA-PEG$_4$-Arg-Arg-Pro-Tyr**Ψ[Tz]**Ile-Leu | 46.1. ± 3.8 (164.0) | n.o. | n.d. |
| **NT5** | [$^{177}$Lu]Lu-DOTA-PEG$_4$-Arg-Arg-Pro**Ψ[Tz]**Tyr-Ile-Leu | 0 (13.0) | n.o. | n.d. |
| **NT6** | [$^{177}$Lu]Lu-DOTA-PEG$_4$-Arg**Ψ[Tz]**Arg-Pro-Tyr-Ile-Leu | 6.5 ± 4.6 (64.9) | 6.4 ± 1.2 | 8.8 ± 1.7 |
| **NT7** | [$^{177}$Lu]Lu-DOTA-PEG$_4$**Ψ[Tz]**Arg-Arg-Pro-Tyr-Ile-Leu | 2.2. ± 1.2 (46.9) | 9.4 ± 0.5 | 4.5 ± 0.8 |
| **NT8** [a] | [$^{177}$Lu]Lu-DOTA-PEG$_4$-Arg-Arg-Pro-Tyr-Tle-Leu | 70.6. ± 1.4 (n.d.) | 1.3 ± 0.2 | 507 ± 114 |
| **NT9** | [$^{177}$Lu]Lu-DOTA-PEG$_4$-Arg**Ψ[Tz]**Arg-Pro-Tyr-Tle-Leu | 97.7 ± 2.3 (n.d.) | 2.1 ± 0.1 | 214 ± 45 |
| **NT10** | [$^{177}$Lu]Lu-DOTA-PEG$_4$**Ψ[Tz]**Arg-Arg-Pro-Tyr-Tle-Leu | 94.7 ± 4.2 (n.d.) | 1.2 ± 0.2 | >1000 |
| **NT11** | [$^{177}$Lu]Lu-DOTA-PEG$_4$**Ψ[Tz]**Arg**Ψ[Tz]**Arg-Pro-Tyr-Ile-Leu | 0.2 ± 0.2 (13) | 10.8 ± 0.4 | 4.6 ± 2.3 |
| **NT12** | [$^{177}$Lu]Lu-DOTA-PEG$_4$**Ψ[Tz]**Arg**Ψ[Tz]**Arg-Pro-Tyr-Tle-Leu | 97.2 ± 3.1 (n.d.) | 2.2 ± 0.1 | >1000 |

[a] Reference compound. [b] Determined in blood serum at 37 °C. [c] n.d.: not determined [d] Ratio of specific receptor-bound and cell-internalised compound expressed in % of administered dose. Expressed as the means of at least two independent experiments. [e] Determined by receptor saturation binding assay on HT-29 cells expressing NTR1. Expressed as the means ± SEM of at least two independent experiments.

### *4.6. Cathepsin K & S Inhibitors*

The Lalmanach group explored the stabilisation of peptidic inhibitors for cathepsin K and S [99]. These two proteins are cysteine proteases normally present in lysosomes and share a high homology. When found in the extracellular matrix, these enzymes are often overexpressed and dysregulated, making them innovative targets for new therapies for a range of medical conditions. Cathepsin K is considered as a target involved in osteoporosis whereas cathepsin S has been identified to play a role in autoimmune diseases and neuropathic pain [100,101]. One strategy to identify inhibitors for proteases is to turn their substrates into a non-cleavable inhibitory molecule by substituting the hydrolysed amide bond. For this purpose, the use of metabolically stable 1,2,3-triazoles as amide bond mimics at the previously identified scissile bonds in two known cathepsin K and S inhibitors were investigated. They also compared the resulting triazolo-peptidomimetics to their azapeptide counterparts in which the $\alpha$-carbon next to the labile amide bond was replaced by a nitrogen (Figure 1). The triazolo-peptidomimetics were synthesised on solid phase following the protocol of Beltramo et al. (Figure 3B) [78]. The substrates were tested for their ability to inhibit enzyme activity on a range of different proteases (including human cysteine proteases, a matrix metalloproteinase, an aspartyl protease and several serine-proteases) and $K_i$ values for cathepsin K and S were determined using fluorescently-labelled substrates (Table 7). Triazole-containing compound **CatS1** was based on the previously reported sequence of a tridecapeptide. [102] It was shown that it inhibits cathepsin S, but also cathepsin K and L without affecting any other protease. This peptidomimetic was not cleaved by cathepsin S, K or L as shown by HPLC analyses of the different incubation mixtures. It was further demonstrated that **CatS1** reversibly and competitively inhibits cathepsins S, K and L in the micromolar range ($K_i$ = 15, 10 and 30 μM, respectively). In contrast to other cathepsins present in the lysosome, cathepsin S is not only operating at acidic pH but also remains stable and active at neutral pH. Indeed, $K_i$ values obtained at pH 5.5 and pH 7.4 were comparable for cathepsin S ($K_i$ = 15 vs. 42 μM) but not for cathepsin K or L (data not shown here). However, the corresponding azapeptide **CatS2** was the more potent and specific inhibitor of cathepsin S and K with a $K_i$ value in the nanomolar range.

**Table 7.** Sequences of **CatS1–2** and **CatK1–2** and their reported biological properties [99].

| Comp. | Sequence [a] | $K_i$ for CatS, pH 5.5 [nM] [b] | $K_i$ for CatS, pH 7.4 [nM] [b] | $K_i$ for CatK, pH 5.5 [nM] [b] | $K_i$ for CatL, pH 5.5 [nM] [b] |
|---|---|---|---|---|---|
| **CatS1** | Ac-Gly-Arg-Trp-His-Pro-Met-Gly**Ψ[Tz]**Ala-Pro-Trp-Glu-D-Ala-D-Arg-$NH_2$ | 15,000 ± 5000 | 42,000 ± 8000 | 10,000 ± 3600 | 30,000 ± 3200 |
| **CatS2** | Ac-Gly-Arg-Trp-His-Pro-Met-**aza**Gly-Ala-Pro-Trp-Glu-D-Ala-D-Arg-$NH_2$ | 26 ± 5 | 17 ± 3 | 3 ± 0.8 | 5 ± 0.5 |
| **CatK1** | Abz-Arg-Pro-Pro-Gly**Ψ[Tz]**Phe-Ser-Pro-Phe-Arg-Tyr(3-$NO_2$)-$NH_2$ | n.o. | n.o. | 800 ± 330 | 22,000 ± 2000 |
| **CatK2** | Abz-Arg-Pro-Pro-**aza**Gly-Phe-Ser-Pro-Phe-Arg-Tyr(3-$NO_2$)-$NH_2$ | n.o. | n.o. | 9 ± 0.7 | 2000 ± 400 |

[a] Abz: *o*-Aminobenzoic acid [b] $K_i$ values were determined by measuring residual peptidase activity after incubation with the peptidomimetics at 37 °C using Z-Leu-Arg-AMC (Cathepsin S) or Z-Phe-Arg-AMC (Cathepsin K and L) as substrate. Data expressed as the means ± SEM of three individual experiments. n.o.: not observed.

Another set of investigated peptidomimetics was based on a bradykinin-derived substrate, previously reported to be degraded by cathepsin K [103]. Triazolo-peptidomimetic **CatK1** did not inhibit cathepsin S and other proteases but was able to inhibit cathepsin K and L in a reversible manner with a $K_i$ of 0.8 and 22 μM, respectively. However, azapeptide **CatK2** displayed a much higher affinity in the nanomolar range. Molecular modelling studies suggest that although the triazole exhibits similar geometric, steric and electronic features as the *trans*-amide bond, its constrained coplanar structure reduces the interaction of the peptidomimetics with the enzyme's active pocket. Hence, in this case the insertion of a 1,2,3-triazole turned out to be less effective than the substitution of amide bonds with a semicarbazide due to significantly lower affinities (100- to 1000-fold).

### 4.7. Caspase-3 Inhibitors

Proteases are ubiquitous enzymes executing various tasks in organisms and their activity is tightly regulated by different mechanisms. A lot of effort has been invested in studying their function and localisation. For example, Verhelst and co-workers investigated the possibility to develop a general strategy for the design of chemical probes that covalently interact with proteases [104]. Their idea was to design inhibitors based on known substrates by substituting the labile peptide bond with a metabolically stable surrogate and attach a photo-crosslinker to covalently inhibit the targeted enzyme. They studied caspase-3 as a model protease and developed several peptidomimetics based on the known pentapeptide substrate Asp-Glu-Val-Asp-Ala (Figure 8). An alkyne function was localised as a bioorthogonal tag at the N-terminus of the peptide to enable the installation of a fluorophore via CuAAC. In addition, a glycine residue with a benzophenone moiety at the α-carbon serving as photoactivatable crosslinker was incorporated at either the C-terminus (**CasC1, CasC3, CasC4**) or the N-terminus (**CasC2, CasC5, CasC6**) of the peptide. The two peptide analogues **CasC1** and **CasC2** were obtained by an amide-to-triazole switch at the previously identified scissile bond $Asp^4$-$Ala^5$ of the parent peptide. The other probes **CasC3–6** contained reduced amide bonds in the same position and either benzophenone or diazirine as substituent of the glycine residue at the C- or N-terminus (Figure 8).

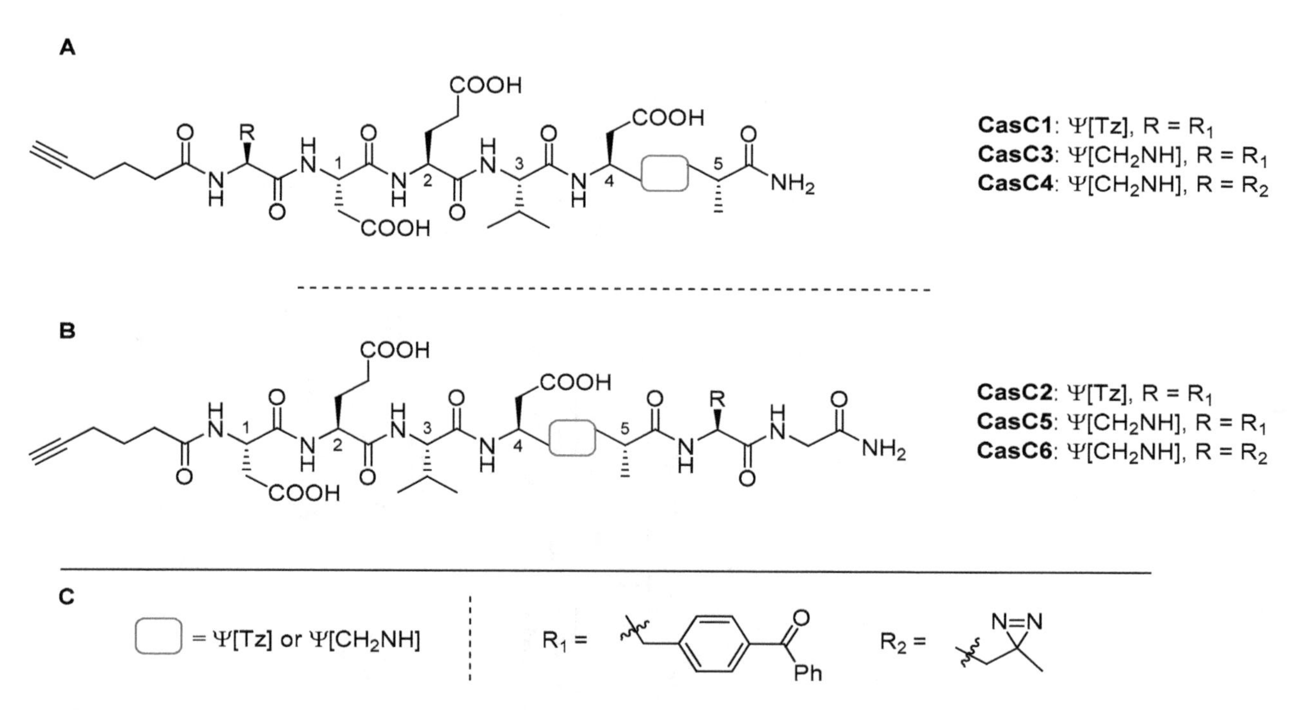


**Figure 8.** **A:** Structures of **CasC1, CasC3, CasC4** with the photo-activable crosslinker R at the N-terminus; **B**: structures of **CasC2, CasC5, CasC6** with the photo-activable crosslinker R at the C-terminus; **C**: Structures of backbone modification (red box, either Ψ[Tz] or Ψ[$CH_2NH$]) at $Asp^4$-$Ala^5$ and of the photo-cleavable linkers.

The synthesis of the triazolo-peptidomimetics was performed on solid phase, including the triazole incorporation via CuAAC (Figure 3B). For this step, immobilised azidopeptide was incubated with the corresponding protected $\alpha$-amino alkyne, CuBr, Na-ascorbate, 2,6-lutidine and DIPEA. With the chemical probes in hand, they first performed competitive activity-based protein profiling (ABPP) to investigate the inhibitory characteristics of the novel compounds. For this purpose, caspase-3 was incubated with the probes under UV irradiation and residual caspase-3 activity was measured using a fluorescent enzyme-specific probe and analysed by SDS-PAGE. Analogues **CasC3–6** with a reduced amide bond at $Asp^4$-$Ala^5$ resulted in complete inhibition of the protease whereas triazole-containing derivative **CasC1** did not show any inhibitory effect. The second triazolo-peptidomimetic **CasC2** displayed slight inhibition at 10 μM but even increasing the substrate concentration to 100 μM did not lead to full inhibition of caspase-3. Similar to the findings for cathepsin K & S inhibitors described above, triazoles were the less efficient amide bond surrogate for caspase-3 inhibitors probably due to the increased rigidity of the triazole compared to the amide bond. Further studies are necessary to determine if this is also valid for other proteases inhibitors.

### 4.8. NRP-1/VEGF$_{165}$ Inhibitors

Neuropilin-1 (NRP-1) is a protein implicated in angiogenesis and therefore also plays a role in the vascularisation of tumours [105]. It has been shown that it is heavily overexpressed in a variety of tumours and hence, related to tumour malignancy and poor prognosis [106,107]. The interaction of NRP-1 with vascular endothelial growth factor 165 (VEGF$_{165}$) is an interesting target for novel cancer therapies and several inhibitors of NRP-1/VEGF$_{165}$ have been proposed as potential drugs such as the heptapeptide **A7R**. [108] Based on these findings, the group of Misicka has developed branched pentapeptides of the type Lys(Har)-Xaa-Xaa-Arg (Har = homoarginine) with submicromolar IC$_{50}$ values, for example 0.3 μM for Xaa-Xaa= Pro-Ala (**NV1**). However, they observed enzymatic cleavage of the peptides, which led to a low blood plasma stability in vitro [109]. To overcome the peptide's instability, the authors turned their attention to 1,4-disubstituted 1,2,3-triazoles as a stable amide bond surrogate. They designed and synthesised a series of peptides where they connected the C-terminal Arg residue and the N-terminal branched Lys(Har) residue (either L-Lys or D-Lys) with different peptidomimetic spacers. The spacers consisted of a short sequence of one to three amino acids in which at least one amide bond was substituted with a triazole (Table 8) [79]. In total, seven spacers varying in length and the number of triazole bonds were investigated (not all of them are shown here). Synthesis of the triazolo-peptidomimetics was achieved on solid support (Figure 3B) with $CuSO_4 \cdot 5H_2O$ and sodium ascorbate as catalysts in the CuAAC. The inhibitory activity of all derivatives towards binding of VEGF$_{165}$ to NRP-1 was measured at a concentration of 10 μM using an enzyme-linked immunosorbent assay (ELISA) and IC$_{50}$ values were determined for the most active compounds. None of the triazolo-peptidomimetics described performed better than reference **A7R**. Two compounds **NV2** and **NV3** of the series displayed inhibition comparable to A7R (58% and 52% inhibition at 10 μM, respectively, compared to 61% for A7R). Both peptidomimetics contain the GlyΨ[Tz]GlyΨ[Tz] motif as a linker and differ in the chirality of the terminal Lys residue (L vs. D, respectively). However, the in vitro plasma stability results revealed that these two analogues are significantly more stable. After 48 h, 70% of compound **NV2** and >90% of compound **NV3** were still intact (half-lives not determined), whereas the previously reported compound **NV1** possess a half-life of only 39 min [109]. Further modification of **NV2** at the N-terminus was attempted to optimise the inhibition of NRP-1/VEGF$_{165}$ interaction but efforts remained unsuccessful. Molecular dynamics simulations of compound **NV2** bound to NRP-1 confirmed a relatively instable binding mode, which is in line with lower inhibitory activity compared to **A7R**. The simulations revealed that even when the C-terminus is tightly bound to the receptor, the triazole-containing spacer and the N-terminus still showed significant mobility which prevents stable inhibitor-protein interaction. This suggests that improved inhibition could be achieved by introducing more rigid structures and/or groups that enable further interaction with the receptor. In conclusion, **NV2** and **NV3** demonstrated a remarkably improved in vitro stability due to the triazole insertion in the peptide's backbone although no enhanced inhibitory activity compared to the reference compound **A7R** or **NV1** was observed. These results indicate that the compounds **NV2** and **NV3** could serve as lead structures in the development of more active NRP-1/VEGF$_{165}$ inhibitors.

**Table 8.** Sequences of **A7R** and **NV1–3** and their reported biological properties [79].

| Compound | Sequence [b] | Inhibition at 10 μM [%] [c,d] | $IC_{50}$ [μM] [d] |
|---|---|---|---|
| **A7R** [a] | Ala-Thr-Trp-Lys-Pro-Pro-Arg | 61.0 ± 0.4 | 5.86 |
| **NV1** | Lys(Har)-Pro-Ala-Arg [a] | n.d. | 0.3 |
| **NV2** | Lys(Har)-Gly**Ψ[Tz]**Gly**Ψ[Tz]**Arg | 58.1 ± 2.1 | 8.39 |
| **NV3** | D-Lys(Har)-Gly**Ψ[Tz]**Gly**Ψ[Tz]**Arg | 52.6 ± 1.3 | 10.22 |

[a] Reference compound. [b] Har: Homoarginine [c] Expressed as percentage of inhibition of VEGF$_{165}$ binding to NRP-1. Values represent means ± standard deviation (SD). n.d.: not determined. [d] Determined in an enzyme-linked immunosorbent assay.

### *4.9. Minigastrins*

Minigastrin is a tridecapeptide binding to the cholecystokinin-2-receptor (CCK2R), a receptor whose overexpression is linked to various cancer types such as pancreatic, thyroid or neuroendocrine tumours [110]. Minigastrin analogues have been extensively explored as potential tumour-targeting vectors for CCK2R-positive cancer cells and a screening test resulted in the truncated DOTA-conjugated octapeptide MG11 as a promising candidate with a high receptor affinity but poor plasma stability [111]. Recently, Grob et al. reported the successful application of the amide-to-triazole switch methodology to MG11 in order to increase its stability [82]. The authors focussed on the use of MG11 as a radiotracer. For easier handling, they replaced the methionine residue in position 15 with norleucine **[Nle$^{15}$]MG11.** [86] They then performed a triazole scan, where every peptide bond was substituted individually with a 1,4-disubstituted-1,2,3-triazole. Synthesis of these triazolo-peptidomimetics **MGN1–8** was performed on solid phase according to procedures described by Valverde et al. (Figure 3B) [77]. In vitro blood plasma stability tests revealed that six out of eight compounds showed enhanced stability and in five cases even more than 50% of the molecules were still being completely intact after 24 h of incubation (Table 9). It was calculated that the half-life of compound **MGN4** with a triazole at position Trp$^{14}$-Nle$^{15}$ is 90-fold higher ($t_{1/2}$ = 349.8 h) compared to the reference compound **[Nle$^{15}$]MG11** ($t_{1/2}$ = 3.9 h). Insertion of a triazole at position Gly$^{13}$-Trp$^{14}$ (**MGN5**) displayed maintained stability ($t_{1/2}$ = 3.8 h) and only substitution at Tyr$^{12}$-Gly$^{13}$ (**MGN6**) resulted in a slight decrease of the proteolytic stability of the peptide ($t_{1/2}$ = 2.6 h). The measurement of logD$_{7.4}$ values of all compounds confirmed earlier findings of the same group, that the lipophilicity was not significantly altered by the introduction of a triazole into the peptide backbone. [77] All compounds were evaluated in vitro for their internalisation into CCK2R-expressing A431 cells (human epidermoid carcinoma) and affinity toward CCK2R (Table 9). It was shown that compounds **MGN1–3** with a triazole incorporated within the last three amino acids of the C-terminus almost completely abolished their biological activity. This comes as no surprise as this region is the minimal binding sequence and known to be very sensitive even to minor modifications. Compounds **MGN4, MGN5** and **MGN7** maintained their biological activity (i.e., 33.1% of internalisation after 4 h and $IC_{50}$ = 25.4 nM for **MGN4**) when compared to reference compound **[Nle$^{15}$]MG11** (32.2% internalisation, $IC_{50}$ = 15.4 nM). **MGN6** and **MGN8** were reported to have increased biological activity. Interestingly, compound **MGN6** which displayed the lowest in vitro serum stability almost doubled its internalisation rate to 54.3% and enhanced the $IC_{50}$ value to 1.7 nM. This is the first example of an amide-to-triazole switch leading to higher receptor affinity of the resulting peptidomimetic. Compounds with maintained or improved biological activity were also tested in vivo in athymic nude mice with CCK2R-positive tumour xenografts (Figure 9). Not surprisingly, compound **MGN6** also performed best in mice with a 2.6-fold increased tumour uptake and a slower tumour washout. In addition, the analogue also showed a 3-fold increased tumour-to-kidney ratio, a property important for the translation of radiotracers to the clinic as the dose to radiation-sensitive kidneys needs to be minimised. Despite its slightly reduced in vitro plasma stability, derivative **MGN6** performed better than other minigastrin derivatives because of its increased affinity and tumour uptake. Further in silico modelling studies led to the hypothesis that an additional cation-$\pi$ interaction between the 1,2,3-triazole in **MGN6** and $Arg^{356}$ in the binding pocket of the receptor are responsible for the improved binding properties of compound **MGN6** to the receptor. Thus, this is the first example of the application of an amide-to-triazole switch which resulted in a derivative with improved pharmacodynamic properties such as receptor affinity and cell internalisation.

**Table 9.** Sequences of **[Nle$^{15}$]MG11** and **MGN1–15** and their reported biological properties [82,112].

| Compound | Sequence | $t_{1/2}$ [h] [b,d] | Uptake after 4 h [%] [b,e] | $IC_{50}$ [nM] [c,f] |
|---|---|---|---|---|
| [Nle$^{15}$]MG11 [a] | Lu-DOTA-D-Glu-Ala-Tyr-Gly-Trp-Nle-Asp-Phe-$NH_2$ | 3.9 (3.8–4.1) | 32.2 ± 3.2 | 15.4 (11.0–21.1) |
| MGN1 | Lu-DOTA-D-Glu-Ala-Tyr-Gly-Trp-Nle-Asp-Phe**Ψ[Tz]**H | 34.9 (31.3–39.3) | 2.5 ± 2.9 | 200.5 (164–245) |
| MGN2 | Lu-DOTA-D-Glu-Ala-Tyr-Gly-Trp-Nle-Asp**Ψ[Tz]**Phe-$NH_2$ | 35.9 (32.2–40.3) | 0.1 ± 0.08 | > 50,000 |
| MGN3 | Lu-DOTA-D-Glu-Ala-Tyr-Gly-Trp-Nle**Ψ[Tz]**Asp-Phe-$NH_2$ | 114.3 (96.5–139.7) | 0.2 ± 0.13 | > 50,000 |
| MGN4 | Lu-DOTA-D-Glu-Ala-Tyr-Gly-Trp**Ψ[Tz]**Nle-Asp-Phe-$NH_2$ | 349.8 (263–520) | 33.1 ± 1.9 | 25.4 (18.3–34.6) |
| MGN5 | Lu-DOTA-D-Glu-Ala-Tyr-Gly**Ψ[Tz]**Trp-Nle-Asp-Phe-$NH_2$ | 3.8 (3.6–4.0) | 41.7 ± 3.9 | 15.6 (12.3–19.7) |
| MGN6 | Lu-DOTA-D-Glu-Ala-Tyr**Ψ[Tz]**Gly-Trp-Nle-Asp-Phe-$NH_2$ | 2.6 (2.5–2.7) | 54.3 ± 5.1 | 1.7 (1.3–2.3) |
| MGN7 | Lu-DOTA-D-Glu-Ala**Ψ[Tz]**Tyr-Gly-Trp-Nle-Asp-Phe-$NH_2$ | 51.4 (46.2–57.7) | 29.6 ±2.7 | 20.9 (17.0–25.7) |
| MGN8 | Lu-DOTA-D-Glu**Ψ[Tz]**Ala-Tyr-Gly-Trp-Nle-Asp-Phe-$NH_2$ | 7.7 (7.3–8.1) | 39.6 ± 2.7 | 8.0 (6.3–10.9) |
| MGN9 | Lu-DOTA-D-Glu-Ala-Tyr**Ψ[Tz]**Gly-Trp**Ψ[Tz]**Nle-Asp-Phe-$NH_2$ | 279.5 (206–431) | 28.2 ± 3.0 | 65.8 (53.2–80.9) |
| MGN10 | Lu-DOTA-D-Glu-Ala-Tyr**Ψ[Tz]**Gly**Ψ[Tz]**Trp-Nle-Asp-Phe-$NH_2$ | 1.9 (1.85–2.04) | 49.6 ± 3.0 | 12.4 (10.9–14.0) |
| MGN11 | Lu-DOTA-D-Glu-Ala**Ψ[Tz]**Tyr-Gly-Trp**Ψ[Tz]**Nle-Asp-Phe-$NH_2$ | 386.1 (249–842) | 31.1 ± 3.9 | 91.0 (76.3–108.3) |
| MGN12 | Lu-DOTA-D-Glu-Ala**Ψ[Tz]**Tyr**Ψ[Tz]**Gly-Trp-Nle-Asp-Phe-$NH_2$ | 4.1 (3.8–4.4) | 48.3 ± 2.2 | 5.3 (4.5–6.1) |
| MGN13 | Lu-DOTA-D-Glu**Ψ[Tz]**Ala-Tyr**Ψ[Tz]**Gly-Trp-Nle-Asp-Phe-$NH_2$ | 14.7 (14.0–15.5) | 58.4 ± 3.5 | 5.8 (5.3–6.4) |
| MGN14 | Lu-DOTA-D-Glu**Ψ[Tz]**Ala**Ψ[Tz]**Tyr-Gly-Trp-Nle-Asp-Phe-$NH_2$ | 6.4 (5.8–7.2) | 40.9 ± 2.5 | 15.6 (12.6–19.1) |
| MGN15 | Lu-DOTA-D-Glu**Ψ[Tz]**Ala**Ψ[Tz]**Tyr**Ψ[Tz]**Gly-Trp-Nle-Asp-Phe-$NH_2$ | 8.1 (7.7–8.6) | 47.1 ± 1.6 | 7.3 (6.0–8.7) |

[a] Reference compound. [b] Results obtained with [$^{177}$Lu]Lu-labelled compounds. [c] Results obtained with nonradioactive $^{nat}$Lu-labelled compounds. [d] Determined in blood serum at 37 °C. Expressed as means with 95% confidence interval of nonlinear regression of at least two independent experiments. [e] Ratio of specific receptor-bound and cell-internalised compound expressed in % of administered dose. Expressed as the means ± SD of at least three independent experiments. [f] Determined by receptor saturation binding assay on A431 cells transfected with CCK2R. Expressed as means with 95% confidence interval of nonlinear regression of at least three independent experiments.

Based on these results, Grob et al. studied the incorporation of multiple triazoles in the peptide backbone of **[Nle$^{15}$]MG11** for potential additive or synergistic effects in order to further improve the properties of the peptide [112]. Therefore, the authors were strategically combining positions for amide-to-triazole switches, which led to increased stability or affinity in the case of **[Nle$^{15}$]MG11** (Table 9). In total, seven novel multi-triazole containing peptide analogues **MGN9–15** were synthesised and tested for their in vitro and in vivo properties. Except for **MGN10**, they all displayed maintained or improved half-lives in blood serum compared to **[Nle$^{15}$]MG11** ranging from 4.1 to 386.1 h. In five out of seven cases, the in vitro stability fell in between or was above the half-lives of the corresponding monotriazolo-peptidomimetics. Incorporation of multiple triazoles into the backbone of **[Nle$^{15}$]MG11** was slightly detrimental to cell internalisation for **MGN9** and **MGN11** (28.2% and 31.1%, respectively), but beneficial for all other tested compounds, with **MGN13** having the highest internalisation rate of 58.4%. A similar trend was observed for the affinities, where all compounds except **MGN9** and **MGN11** displayed maintained or improved $IC_{50}$ values, ranging from 5.3 to 15.6 nM. All examples except **MGN11** were included in in vivo studies and displayed higher tumour uptake than the reference compound **[Nle$^{15}$]MG11**. The most promising candidates were **MGN13** and **MGN15** containing further triazole substitutions at position D-Glu$^{10}$-Ala$^{11}$ and/or Ala$^{11}$-Tyr$^{12}$ in addition to the modification at Tyr$^{12}$-Gly$^{13}$. In vivo, the multitriazolo-peptidomimetics showed a further 1.5-fold increase in tumour uptake and a 2- to 3-fold increase in the tumour-to-kidney ratio in athymic nude mice with CCK2R-positive tumour xenografts compared to mono-triazole compound **MGN6**. These results confirm that the amide-to-triazole switch is a promising strategy to obtain peptidomimetics with improved tumour-targeting properties. In summary, **[Nle$^{15}$]MG11** is highly tolerant towards backbone modifications in its N-terminal region whilst preserving or improving receptor affinity. This is the first example of multiple triazole-to-amide switches in peptides leading to improved serum stability and affinity to the receptor. In particular, **MGN15** is the first reported compound where three amide bonds in the backbone were substituted with a triazole and biological function was still maintained. However, no apparent trend for synergistic or additive effects of multiple triazoles in the backbone of peptides can currently be defined to allow the prediction of the biological behaviour of multitriazolo-petidomimetics based on those of the corresponding monosubstituted triazolo-peptidomimetics.

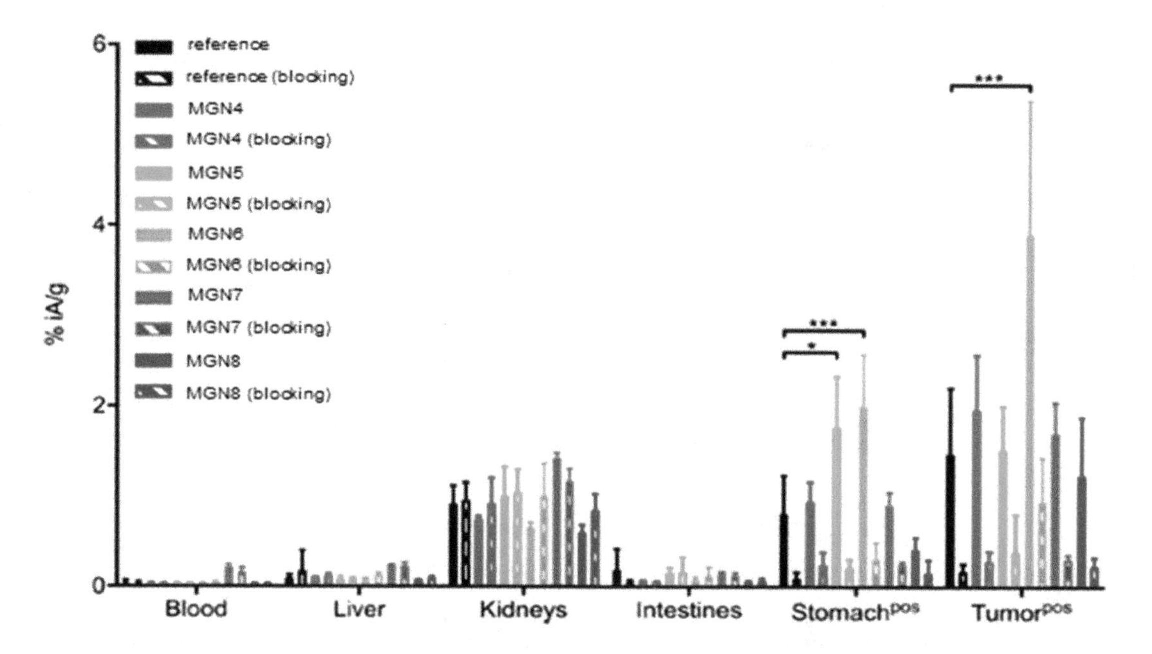


**Figure 9.** Biodistribution of $^{177}$Lu-labelled reference compound and **MG4–8** in mice bearing A431-CCK2R tumour xenografts 4 h post-injection of the radiotracer. Percentage of injected activity per gram of organ/tissue (%iA/g) without (plain columns) and with (crosshatched columns) co-injection of excess minigastrin (blocking experiments to verify receptor specificity). * for $p < 0.033$ and *** for $p < 0.001$. Reproduced with permission from Grob et al. [82].

## *4.10. Angiotensin*

Angiotensin II (**AII**) is an octapeptide and serves as ligand for the angiotensin receptor subtype 1 and 2 (AT1R and AT2R). It has been shown that AT2R represents a novel target for the development of treatments for various cancers including melanoma and breast cancer [113,114]. Nevertheless, there is still an imminent need to develop ligands with high receptor subtype-specificity (Table 10). One potent and selective derivative of **AII** is the previously reported **[Tyr$^6$]AII**, in which His$^6$ is replaced by Tyr [114]. This modification yielded in a ligand with nanomolar affinity for AT2R and an 18.000-fold selectivity for AT2R over AT1R. However, its use is limited by a low metabolic stability. Therefore, Tzakos and co-workers intended to develop more stable analogues by applying the amide-to-triazole switch to **[Tyr$^6$]AII.** [115] They introduced a 1,2,3-triazole at four strategic positions in the peptide backbone yielding triazolo-peptidomimetics **AII1–4** (Table 10). The triazoles were incorporated following the protocol from Valverde et al. (Figure 3B) [77].

**Table 10.** Sequences of **AII** and **AII1–4** and their reported biological properties [115].

| Compound | Sequence | $IC_{50}$ for AT2R [nM] [b] | AT2R/AT1R Selectivity [c] |
|---|---|---|---|
| **AII** [a] | Asp-Arg-Val-Tyr-Ile-His-Pro-Phe | 0.12 ± 0.01 | 13.7 |
| **[Tyr[6]]AII** | Asp-Arg-Val-Tyr-Ile-Tyr-Pro-Phe | 4.0 [d] | 18,000 [d] |
| **AII1** | Asp-Arg-Val-Tyr-Ile**Ψ[Tz]**Tyr-Pro-Phe | 1990 ± 88 | >5 |
| **AII2** | Asp-Arg-Val-Tyr**Ψ[Tz]**Ile-Tyr-Pro-Phe | 2.8 ± 0.08 | >3611 |
| **AII3** | Asp-Arg-Val**Ψ[Tz]**Tyr-Ile-Tyr-Pro-Phe | 155 ± 6 | >64 |
| **AII4** | Asp-Arg**Ψ[Tz]**Val-Tyr-Ile-Tyr-Pro-Phe | 7.5 ± 0.05 | >1336 |

[a] Reference compound. [b] Determined in a competition binding assay on HEK-293 cells transfected with AT2R with [$^{125}$I,Sar$^1$,Ile$^8$]AII as competitor. Expressed as means ± SD of three independent experiments. [c] Ratio of $IC_{50}$ AT2R/AT1R. [d] Values taken from Magnani et al. [114].

These novel triazolo-peptidomimetics were tested for their plasma stability. After 24 h, only approx. 10% of the compounds were found degraded, compared to 46% of the reference compound [Tyr$^6$]AII. Thus, the replacement of amide bonds in the backbone of the peptide of investigation significantly enhanced its stability in blood plasma. The binding affinity and selectivity of the peptidomimetics and reference compound **AII** was assessed in a competitive binding assays using HEK-293 cells (human embryonic kidney cells) transfected with AT1R or AT2R and [$^{125}$I,Sar$^1$,Ile$^8$]AII as competitive ligand (Figure 10). The affinities of all compounds were reduced when compared to the endogenous ligand **AII**. Interestingly, the experiments revealed that two of the novel derivatives retained high selectivity for AT2R over AT1R (3.600- and 1.300-fold for compound **AII2** and **AII4**, respectively) whereas the two other peptidomimetics only showed minimal selectivity (64- and 5-fold for compound **AII3** and **AII1**, respectively). It should be noted that the AT2R selectivities represent conservative estimations as the analogues failed to displace AT1R binding sufficiently. Further high-resolution 2D NMR studies confirmed that compounds with maintained receptor subtype selectivity, **AII2** and **AII4**, shared structural features in solution which indicates similar conformation. They clearly clustered differently from compounds **AII1** and **AII3** which exhibit poor receptor subtype selectively. This could imply that the introduction of a triazole moiety at specific positions favours a structural microenvironment that is responsible for the observed higher affinities of **AII2** and **AII4** to AT2R. Thus, these results illustrate that the amide-to-triazole switch can also provide peptide analogues of enhanced metabolic stability and maintained receptor subtype selectivity.

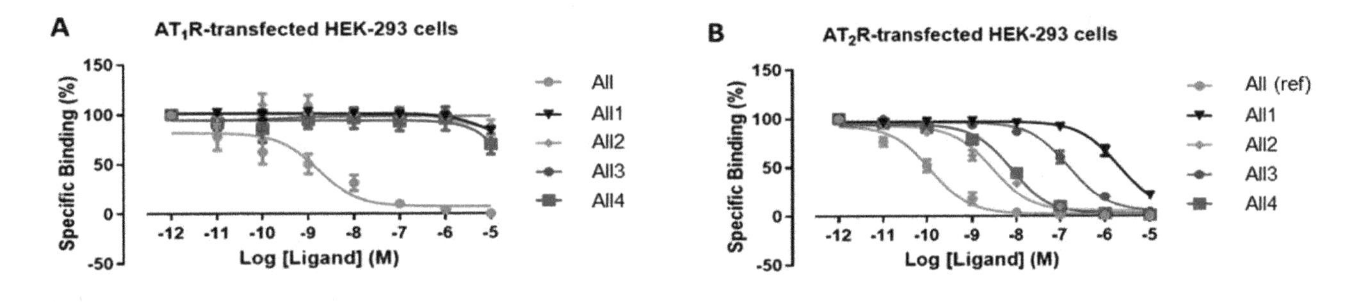


**Figure 10.** Competition-binding experiment for AII (reference compound) and the four peptidomimetic analogues **AII1–4** against [$^{125}$I,Sar$^1$,Ile$^8$]AII on AT1R-transfected HEK-293 cells (**A**) and AT2R-transfected HEK-293 cells (**B**). Reproduced with permission from Vrettos et al. [115].

## 5. Conclusions

In this review we discussed the use of 1,4-disubstituted 1,2,3-triazoles as bioisosteres for a *trans*-amide bond in peptidomimetics in order to enhance the parent peptide's metabolic stability whilst retaining or improving its biological activity.

In the majority of the reported cases, the substitution of an amide bond with a 1,4-disubstituted 1,2,3-triazole led to peptidomimetics with increased in vitro blood serum stability. These results underline the importance and thus growing interest in triazoles as promising amide bond surrogates. However, a second important factor that needs to be considered is the impact of such structural modifications on the peptides' biological activity, for example, the affinity or selectivity towards its respective receptor. Another important parameter in drug development is the lipophilicity of the compound. It has been shown that the amide-to-triazole switch leads to peptidomimetics with maintained lipophilicity. In the studied cases, single or multiple substitutions of (up to three) amide bonds with triazoles did not change the $logD_{7.4}$ values of peptides consisting of at least six amino acids [77,82,93,112]. Even though 1,4-disubstituted 1,2,3-triazoles are good bioisosteres of *trans*-amide bonds by mimicking electronic and steric features, a few essential differences exist. First, the heterocycle slightly increases the distance between the two neighbouring amino acid residues when compared to a peptide bond. Triazoles also possess a higher rigidity and planarity which can result in changes of the conformation of the peptide that can potentially influence the binding to its biological target. Furthermore, the introduced heterocycle may enable new $\pi$-interactions between the peptidic ligand and the binding site of its respective receptor, which in turn can influence its biological properties. In general, the affinity and selectivity a novel triazolo-peptidomimetic towards their receptor do not seem to be predictable. Ideally a triazole-scan, similar to established Ala-scans, is performed with the peptide of interest to evaluate all positions within a peptide sequence in order to identify the most promising one(s) which tolerate(s) the structural modification. The amide-to-triazole switch can lead to a higher success rate when compared to other established stabilisation methods such as in the case of BBN(7-14), where four novel peptidomimetics with conserved biological activity were identified whereas the reduction of amide bonds or *N*-methylation recognised only two novel promising compounds each [44,116,117]. On the other hand, in the case of protease inhibitors for caspase-3 or cathepsin K & S, the amide-to-triazole switch was not successful and more promising results were obtained with reduced amide bonds (for caspase-3) or the corresponding azapeptides (for cathepsin K & S) [99,104]. In the same way, additive or synergistic effects of combining multiple triazoles in one peptide appear to be difficult to anticipate based on the data of mono-substituted triazolo-peptidomimetics. In addition, the combination of this stabilisation strategy with other backbone modifications such as reduced amide bonds or *N*-methylated amino acids are still outstanding and leaves ample possibilities to explore in the future. In conclusion, an amide-to-triazole switch is often in favour of higher in vitro blood serum stabilities, but the biological properties of the resulting peptidomimetics need to be evaluated individually in order to develop triazolo-peptidomimetics successfully as potential drug candidates.

**Acknowledgments:** Open Access Funding was provided by the University of Vienna, Austria.

## References

1. Lau, J.L.; Dunn, M.K. Therapeutic peptides: Historical perspectives, current development trends, and future directions. *Bioorg. Med. Chem.* **2018**, *26*, 2700–2707. [CrossRef] [PubMed]
2. Merrifield, R.B. Solid Phase Peptide Synthesis. I. The Synthesis of a Tetrapeptide. *J. Am. Chem. Soc.* **1963**, *85*, 2149–2154. [CrossRef]
3. Di, L. Strategic Approaches to Optimizing Peptide ADME Properties. *AAPS J.* **2015**, *17*, 134–143. [CrossRef] [PubMed]
4. Humphrey, M.J. The Oral Bioavailability of Peptides and Related Drugs. In *Delivery Systems for Peptide Drugs*; Davis, S.S., Illum, L., Tomlinson, E., Eds.; NATO ASI Series; Springer US: Boston, MA, USA, 1986; pp. 139–151. ISBN 978-1-4757-9960-6.
5. Werle, M.; Bernkop-Schnürch, A. Strategies to improve plasma half life time of peptide and protein drugs. *Amino Acids* **2006**, *30*, 351–367. [CrossRef] [PubMed]
6. Yao, J.-F.; Yang, H.; Zhao, Y.-Z.; Xue, M. Metabolism of Peptide Drugs and Strategies to Improve their Metabolic Stability. *Curr. Drug Metab.* **2018**, *19*, 892–901. [CrossRef]

7. Goodman, M.; Felix, A. *Synthesis of Peptides and Peptidomimetics*; Georg Thieme Verlag: Stuttgart, Germany, 2002; ISBN 978-1-58890-023-4.
8. Nock, B.A.; Maina, T.; Krenning, E.P.; Jong, M. de "To Serve and Protect": Enzyme Inhibitors as Radiopeptide Escorts Promote Tumor Targeting. *J. Nucl. Med.* **2014**, *55*, 121–127. [CrossRef]
9. Choudhary, A.; Raines, R.T. An Evaluation of Peptide-Bond Isosteres. *ChemBioChem* **2011**, *12*, 1801–1807. [CrossRef]
10. Gentilucci, L.; De Marco, R.; Cerisoli, L. Chemical Modifications Designed to Improve Peptide Stability: Incorporation of Non-Natural Amino Acids, Pseudo-Peptide Bonds, and Cyclization. *Curr. Pharm. Des.* **2010**, *16*, 3185–3203. [CrossRef]
11. Cour, T.F.M.L.; Hansen, H.A.S.; Clausen, K.; Lawesson, S.-O. The geometry of the thiopeptide unit. *Int. J. Pept. Protein Res.* **1983**, *22*, 509–512. [CrossRef]
12. Chen, X.; Mietlicki-Baase, E.G.; Barrett, T.M.; McGrath, L.E.; Koch-Laskowski, K.; Ferrie, J.J.; Hayes, M.R.; Petersson, E.J. Thioamide Substitution Selectively Modulates Proteolysis and Receptor Activity of Therapeutic Peptide Hormones. *J. Am. Chem. Soc.* **2017**, *139*, 16688–16695. [CrossRef]
13. Allmendinger, T.; Furet, P.; Hungerbühler, E. Fluoroolefin dipeptide isosteres—I.: The synthesis of Glyψ(CF=CH)Gly and racemic Pheψy(CF=CH)Gly. *Tetrahedron Lett.* **1990**, *31*, 7297–7300. [CrossRef]
14. Malde, A.K.; Khedkar, S.A.; Coutinho, E.C. The B(OH)−NH Analog Is a Surrogate for the Amide Bond (CO−NH) in Peptides: An ab Initio Study. *J. Chem. Theory Comput.* **2007**, *3*, 619–627. [CrossRef] [PubMed]
15. Mathieu, S.; Trinquier, G. The –BF–NH– Link as a Peptide-Bond Surrogate. *J. Phys. Chem. B* **2012**, *116*, 8863–8872. [CrossRef] [PubMed]
16. Freidinger, R.M.; Veber, D.F.; Perlow, D.S.; Brooks; Saperstein, R. Bioactive conformation of luteinizing hormone-releasing hormone: Evidence from a conformationally constrained analog. *Science* **1980**, *210*, 656–658. [CrossRef] [PubMed]
17. DiMaio, J.; Belleau, B. Synthesis of chiral piperazin-2-ones as model peptidomimetics. *J. Chem. Soc. Perkin 1* **1989**, *9*, 1687–1689. [CrossRef]
18. Jones, R.C.F.; Ward, G.J. Amide bond isosteres: Imidazolines in pseudopeptide chemistry. *Tetrahedron Lett.* **1988**, *29*, 3853–3856. [CrossRef]
19. De Luca, L.; Falorni, M.; Giacomelli, G.; Porcheddu, A. New pyrazole containing bicarboxylic α-amino acids: Mimics of the cis amide bond. *Tetrahedron Lett.* **1999**, *40*, 8701–8704. [CrossRef]
20. May, B.C.H.; Abell, A.D. The synthesis and crystal structure of alpha-keto tetrazole-based dipeptide mimics. *Tetrahedron Lett.* **2001**, *42*, 5641–5644. [CrossRef]
21. Hitotsuyanagi, Y.; Motegi, S.; Fukaya, H.; Takeya, K. A cis Amide Bond Surrogate Incorporating 1,2,4-Triazole. *J. Org. Chem.* **2002**, *67*, 3266–3271. [CrossRef]
22. Valverde, I.E.; Mindt, T.L. 1,2,3-Triazoles as Amide-bond Surrogates in Peptidomimetics. *Chimia* **2013**, *67*, 262–266. [CrossRef]
23. Hou, J.; Liu, X.; Shen, J.; Zhao, G.; Wang, P.G. The impact of click chemistry in medicinal chemistry. *Expert Opin. Drug Discov.* **2012**, *7*, 489–501. [CrossRef] [PubMed]
24. Bonandi, E.; Christodoulou, M.S.; Fumagalli, G.; Perdicchia, D.; Rastelli, G.; Passarella, D. The 1,2,3-triazole ring as a bioisostere in medicinal chemistry. *Drug Discov. Today* **2017**, *22*, 1572–1581. [CrossRef] [PubMed]
25. Horne, W.S.; Stout, C.D.; Ghadiri, M.R. A Heterocyclic Peptide Nanotube. *J. Am. Chem. Soc.* **2003**, *125*, 9372–9376. [CrossRef]
26. Horne, W.S.; Yadav, M.K.; Stout, C.D.; Ghadiri, M.R. Heterocyclic Peptide Backbone Modifications in an α-Helical Coiled Coil. *J. Am. Chem. Soc.* **2004**, *126*, 15366–15367. [CrossRef] [PubMed]
27. Angelo, N.G.; Arora, P.S. Nonpeptidic Foldamers from Amino Acids: Synthesis and Characterization of 1,3-Substituted Triazole Oligomers. *J. Am. Chem. Soc.* **2005**, *127*, 17134–17135. [CrossRef] [PubMed]
28. van Maarseveen, J.H.; Horne, W.S.; Ghadiri, M.R. Efficient Route to C2 Symmetric Heterocyclic Backbone Modified Cyclic Peptides. *Org. Lett.* **2005**, *7*, 4503–4506. [CrossRef]
29. Angelo, N.G.; Arora, P.S. Solution- and Solid-Phase Synthesis of Triazole Oligomers That Display Protein-Like Functionality. *J. Org. Chem.* **2007**, *72*, 7963–7967. [CrossRef]
30. Boddaert, T.; Solà, J.; Helliwell, M.; Clayden, J. Chemical communication: Conductors and insulators of screw-sense preference between helical oligo(aminoisobutyric acid) domains. *Chem. Commun.* **2012**, *48*, 3397–3399. [CrossRef]

31. Salah, K.B.H.; Legrand, B.; Das, S.; Martinez, J.; Inguimbert, N. Straightforward strategy to substitute amide bonds by 1,2,3-triazoles in peptaibols analogs using Aibψ[Tz]-Xaa dipeptides. *Pept. Sci.* **2015**, *104*, 611–621. [CrossRef]
32. Salah, K.B.H.; Das, S.; Ruiz, N.; Andreu, V.; Martinez, J.; Wenger, E.; Amblard, M.; Didierjean, C.; Legrand, B.; Inguimbert, N. How are 1,2,3-triazoles accommodated in helical secondary structures? *Org. Biomol. Chem.* **2018**, *16*, 3576–3583. [CrossRef]
33. Schröder, D.C.; Kracker, O.; Fröhr, T.; Góra, J.; Jewginski, M.; Nieß, A.; Antes, I.; Latajka, R.; Marion, A.; Sewald, N. 1,4-Disubstituted 1H-1,2,3-Triazole Containing Peptidotriazolamers: A New Class of Peptidomimetics With Interesting Foldamer Properties. *Front. Chem.* **2019**, *7*. [CrossRef] [PubMed]
34. Bock, V.D.; Perciaccante, R.; Jansen, T.P.; Hiemstra, H.; van Maarseveen, J.H. Click Chemistry as a Route to Cyclic Tetrapeptide Analogues: Synthesis of cyclo-[Pro-Val-ψ(triazole)-Pro-Tyr]. *Org. Lett.* **2006**, *8*, 919–922. [CrossRef] [PubMed]
35. Bock, V.D.; Speijer, D.; Hiemstra, H.; Maarseveen, J.H. van 1,2,3-Triazoles as peptide bond isosteres: Synthesis and biological evaluation of cyclotetrapeptide mimics. *Org. Biomol. Chem.* **2007**, *5*, 971–975. [CrossRef] [PubMed]
36. Springer, J.; de Cuba, K.R.; Calvet-Vitale, S.; Geenevasen, J.A.J.; Hermkens, P.H.H.; Hiemstra, H.; van Maarseveen, J.H. Backbone Amide Linker Strategy for the Synthesis of 1,4-Triazole-Containing Cyclic Tetra- and Pentapeptides. *Eur. J. Org. Chem.* **2008**, *2008*, 2592–2600. [CrossRef]
37. Liu, Y.; Zhang, L.; Wan, J.; Li, Y.; Xu, Y.; Pan, Y. Design and synthesis of cyclo[-Arg-Gly-Asp-Ψ(triazole)-Gly-Xaa-] peptide analogues by click chemistry. *Tetrahedron* **2008**, *64*, 10728–10734. [CrossRef]
38. Horne, W.S.; Olsen, C.A.; Beierle, J.M.; Montero, A.; Ghadiri, M.R. Probing the Bioactive Conformation of an Archetypal Natural Product HDAC Inhibitor with Conformationally Homogeneous Triazole-Modified Cyclic Tetrapeptides. *Angew. Chem. Int. Ed.* **2009**, *48*, 4718–4724. [CrossRef]
39. Beierle, J.M.; Horne, W.S.; van Maarseveen, J.H.; Waser, B.; Reubi, J.C.; Ghadiri, M.R. Conformationally Homogeneous Heterocyclic Pseudotetrapeptides as Three-Dimensional Scaffolds for Rational Drug Design: Receptor-Selective Somatostatin Analogues. *Angew. Chem. Int. Ed.* **2009**, *48*, 4725–4729. [CrossRef]
40. Singh, E.K.; Nazarova, L.A.; Lapera, S.A.; Alexander, L.D.; McAlpine, S.R. Histone deacetylase inhibitors: Synthesis of cyclic tetrapeptides and their triazole analogs. *Tetrahedron Lett.* **2010**, *51*, 4357–4360. [CrossRef]
41. Davis, M.R.; Singh, E.K.; Wahyudi, H.; Alexander, L.D.; Kunicki, J.B.; Nazarova, L.A.; Fairweather, K.A.; Giltrap, A.M.; Jolliffe, K.A.; McAlpine, S.R. Synthesis of sansalvamide A peptidomimetics: Triazole, oxazole, thiazole, and pseudoproline containing compounds. *Tetrahedron* **2012**, *68*, 1029–1051. [CrossRef]
42. Tischler, M.; Nasu, D.; Empting, M.; Schmelz, S.; Heinz, D.W.; Rottmann, P.; Kolmar, H.; Buntkowsky, G.; Tietze, D.; Avrutina, O. Braces for the Peptide Backbone: Insights into Structure–Activity Relationships of Protease Inhibitor Mimics with Locked Amide Conformations. *Angew. Chem. Int. Ed.* **2012**, *51*, 3708–3712. [CrossRef]
43. Oueis, E.; Jaspars, M.; Westwood, N.J.; Naismith, J.H. Enzymatic Macrocyclization of 1,2,3-Triazole Peptide Mimetics. *Angew. Chem. Int. Ed.* **2016**, *55*, 5842–5845. [CrossRef] [PubMed]
44. Valverde, I.E.; Lecaille, F.; Lalmanach, G.; Aucagne, V.; Delmas, A.F. Synthesis of a Biologically Active Triazole-Containing Analogue of Cystatin A Through Successive Peptidomimetic Alkyne–Azide Ligations. *Angew. Chem. Int. Ed.* **2012**, *51*, 718–722. [CrossRef] [PubMed]
45. Paul, A.; Bittermann, H.; Gmeiner, P. Triazolopeptides: Chirospecific synthesis and cis/trans prolyl ratios of structural isomers. *Tetrahedron* **2006**, *62*, 8919–8927. [CrossRef]
46. Hartwig, S.; Hecht, S. Polypseudopeptides with Variable Stereochemistry: Synthesis via Click-Chemistry, Postfunctionalization, and Conformational Behavior in Solution. *Macromolecules* **2010**, *43*, 242–248. [CrossRef]
47. Kann, N.; Johansson, J.R.; Beke-Somfai, T. Conformational properties of 1,4- and 1,5-substituted 1,2,3-triazole amino acids–building units for peptidic foldamers. *Org. Biomol. Chem.* **2015**, *13*, 2776–2785. [CrossRef]
48. Charron, C.L.; Hickey, J.L.; Nsiama, T.K.; Cruickshank, D.R.; Turnbull, W.L.; Luyt, L.G. Molecular imaging probes derived from natural peptides. *Nat. Prod. Rep.* **2016**, *33*, 761–800. [CrossRef]
49. Fani, M.; Maecke, H.R.; Okarvi, S.M. Radiolabeled Peptides: Valuable Tools for the Detection and Treatment of Cancer. *Theranostics* **2012**, *2*, 481–501. [CrossRef]
50. Reubi, J.C.; Maecke, H.R. Peptide-Based Probes for Cancer Imaging. *J. Nucl. Med.* **2008**, *49*, 1735–1738. [CrossRef]

51. Reilly, R.M.; Sandhu, J.; Alvarez-Diez, T.M.; Gallinger, S.; Kirsh, J.; Stern, H. Problems of Delivery of Monoclonal Antibodies. *Clin. Pharmacokinet.* **1995**, *28*, 126–142. [CrossRef]
52. Tam, A.; Arnold, U.; Soellner, M.B.; Raines, R.T. Protein Prosthesis: 1,5-Disubstituted[1,2,3]triazoles as cis-Peptide Bond Surrogates. *J. Am. Chem. Soc.* **2007**, *129*, 12670–12671. [CrossRef]
53. Pokorski, J.K.; Miller Jenkins, L.M.; Feng, H.; Durell, S.R.; Bai, Y.; Appella, D.H. Introduction of a Triazole Amino Acid into a Peptoid Oligomer Induces Turn Formation in Aqueous Solution. *Org. Lett.* **2007**, *9*, 2381–2383. [CrossRef] [PubMed]
54. Ahsanullah; Schmieder, P.; Kühne, R.; Rademann, J. Metal-Free, Regioselective Triazole Ligations that Deliver Locked cis Peptide Mimetics. *Angew. Chem. Int. Ed.* **2009**, *48*, 5042–5045. [CrossRef] [PubMed]
55. Tietze, D.; Tischler, M.; Voigt, S.; Imhof, D.; Ohlenschläger, O.; Görlach, M.; Buntkowsky, G. Development of a Functional cis-Prolyl Bond Biomimetic and Mechanistic Implications for Nickel Superoxide Dismutase. *Chem.–Eur. J.* **2010**, *16*, 7572–7578. [CrossRef] [PubMed]
56. Kracker, O.; Góra, J.; Krzciuk-Gula, J.; Marion, A.; Neumann, B.; Stammler, H.-G.; Nieß, A.; Antes, I.; Latajka, R.; Sewald, N. 1,5-Disubstituted 1,2,3-Triazole-Containing Peptidotriazolamers: Design Principles for a Class of Versatile Peptidomimetics. *Chem.–Eur. J.* **2018**, *24*, 953–961. [CrossRef]
57. Palmer, M.H.; Findlay, R.H.; Gaskell, A.J. Electronic charge distribution and moments of five- and six-membered heterocycles. *J. Chem. Soc. Perkin Trans. 2* **1974**, 420–428. [CrossRef]
58. Bates, W.W.; Hobbs, M.E. The Dipole Moments of Some Acid Amides and the Structure of the Amide Group 1. *J. Am. Chem. Soc.* **1951**, *73*, 2151–2156. [CrossRef]
59. Bourne, Y.; Kolb, H.C.; Radić, Z.; Sharpless, K.B.; Taylor, P.; Marchot, P. Freeze-frame inhibitor captures acetylcholinesterase in a unique conformation. *Proc. Natl. Acad. Sci. USA* **2004**, *101*, 1449–1454. [CrossRef]
60. Brik, A.; Alexandratos, J.; Lin, Y.-C.; Elder, J.H.; Olson, A.J.; Wlodawer, A.; Goodsell, D.S.; Wong, C.-H. 1,2,3-Triazole as a Peptide Surrogate in the Rapid Synthesis of HIV-1 Protease Inhibitors. *ChemBioChem* **2005**, *6*, 1167–1169. [CrossRef]
61. Massarotti, A.; Aprile, S.; Mercalli, V.; Del Grosso, E.; Grosa, G.; Sorba, G.; Tron, G.C. Are 1,4- and 1,5-Disubstituted 1,2,3-Triazoles Good Pharmacophoric Groups? *ChemMedChem* **2014**, *9*, 2497–2508. [CrossRef]
62. Schulze, B.; Schubert, U.S. Beyond click chemistry–supramolecular interactions of 1,2,3-triazoles. *Chem. Soc. Rev.* **2014**, *43*, 2522–2571. [CrossRef]
63. Seebach, D.; Matthews, J.L. β-Peptides: A surprise at every turn. *Chem. Commun.* **1997**, 2015–2022. [CrossRef]
64. Prasher, P.; Sharma, M. Tailored therapeutics based on 1,2,3-1H-triazoles: A mini review. *MedChemComm* **2019**, *10*, 1302–1328. [CrossRef] [PubMed]
65. Tornøe, C.W.; Christensen, C.; Meldal, M. Peptidotriazoles on Solid Phase: [1,2,3]-Triazoles by Regiospecific Copper(I)-Catalyzed 1,3-Dipolar Cycloadditions of Terminal Alkynes to Azides. *J. Org. Chem.* **2002**, *67*, 3057–3064. [CrossRef]
66. Rostovtsev, V.V.; Green, L.G.; Fokin, V.V.; Sharpless, K.B. A Stepwise Huisgen Cycloaddition Process: Copper(I)-Catalyzed Regioselective "Ligation" of Azides and Terminal Alkynes. *Angew. Chem. Int. Ed.* **2002**, *41*, 2596–2599. [CrossRef]
67. Kolb, H.C.; Finn, M.G.; Sharpless, K.B. Click Chemistry: Diverse Chemical Function from a Few Good Reactions. *Angew. Chem. Int. Ed.* **2001**, *40*, 2004–2021. [CrossRef]
68. Meghani, N.M.; Amin, H.H.; Lee, B.-J. Mechanistic applications of click chemistry for pharmaceutical drug discovery and drug delivery. *Drug Discov. Today* **2017**, *22*, 1604–1619. [CrossRef]
69. Jiang, X.; Hao, X.; Jing, L.; Wu, G.; Kang, D.; Liu, X.; Zhan, P. Recent applications of click chemistry in drug discovery. *Expert Opin. Drug Discov.* **2019**, *14*, 779–789. [CrossRef]
70. Rani, A.; Singh, G.; Singh, A.; Maqbool, U.; Kaur, G.; Singh, J. CuAAC-ensembled 1,2,3-triazole-linked isosteres as pharmacophores in drug discovery: Review. *RSC Adv.* **2020**, *10*, 5610–5635. [CrossRef]
71. Xi, W.; Scott, T.F.; Kloxin, C.J.; Bowman, C.N. Click Chemistry in Materials Science. *Adv. Funct. Mater.* **2014**, *24*, 2572–2590. [CrossRef]
72. Döhler, D.; Michael, P.; Binder, W.H. CuAAC-Based Click Chemistry in Self-Healing Polymers. *Acc. Chem. Res.* **2017**, *50*, 2610–2620. [CrossRef]
73. Li, H.; Aneja, R.; Chaiken, I. Click Chemistry in Peptide-Based Drug Design. *Molecules* **2013**, *18*, 9797–9817. [CrossRef] [PubMed]
74. Ahmad Fuaad, A.A.H.; Azmi, F.; Skwarczynski, M.; Toth, I. Peptide Conjugation via CuAAC 'Click' Chemistry. *Molecules* **2013**, *18*, 13148–13174. [CrossRef] [PubMed]

75. Goddard-Borger, E.D.; Stick, R.V. An Efficient, Inexpensive, and Shelf-Stable Diazotransfer Reagent: Imidazole-1-sulfonyl Azide Hydrochloride. *Org. Lett.* **2007**, *9*, 3797–3800. [CrossRef] [PubMed]
76. Proteau-Gagné, A.; Rochon, K.; Roy, M.; Albert, P.-J.; Guérin, B.; Gendron, L.; Dory, Y.L. Systematic replacement of amides by 1,4-disubstituted[1,2,3]triazoles in Leu-enkephalin and the impact on the delta opioid receptor activity. *Bioorg. Med. Chem. Lett.* **2013**, *23*, 5267–5269. [CrossRef] [PubMed]
77. Valverde, I.E.; Bauman, A.; Kluba, C.A.; Vomstein, S.; Walter, M.A.; Mindt, T.L. 1,2,3-Triazoles as Amide Bond Mimics: Triazole Scan Yields Protease-Resistant Peptidomimetics for Tumor Targeting. *Angew. Chem. Int. Ed.* **2013**, *52*, 8957–8960. [CrossRef]
78. Beltramo, M.; Robert, V.; Galibert, M.; Madinier, J.-B.; Marceau, P.; Dardente, H.; Decourt, C.; De Roux, N.; Lomet, D.; Delmas, A.F.; et al. Rational Design of Triazololipopeptides Analogs of Kisspeptin Inducing a Long-Lasting Increase of Gonadotropins. *J. Med. Chem.* **2015**, *58*, 3459–3470. [CrossRef]
79. Fedorczyk, B.; Lipiński, P.F.J.; Puszko, A.K.; Tymecka, D.; Wilenska, B.; Dudka, W.; Perret, G.Y.; Wieczorek, R.; Misicka, A. Triazolopeptides Inhibiting the Interaction between Neuropilin-1 and Vascular Endothelial Growth Factor-165. *Molecules* **2019**, *24*, 1756. [CrossRef]
80. Dickson, H.D.; Smith, S.C.; Hinkle, K.W. A convenient scalable one-pot conversion of esters and Weinreb amides to terminal alkynes. *Tetrahedron Lett.* **2004**, *45*, 5597–5599. [CrossRef]
81. Reginato, G.; Mordini, A.; Messina, F.; Degl'Innocenti, A.; Poli, G. A new stereoselective synthesis of chiral $\gamma$-functionalized (E)-allylic amines. *Tetrahedron* **1996**, *52*, 10985–10996. [CrossRef]
82. Grob, N.M.; Häussinger, D.; Deupi, X.; Schibli, R.; Behe, M.; Mindt, T.L. Triazolo-Peptidomimetics: Novel Radiolabeled Minigastrin Analogs for Improved Tumor Targeting. *J. Med. Chem.* **2020**, *63*, 4484–4495. [CrossRef]
83. Punna, S.; Finn, M.G. A Convenient Colorimetric Test for Aliphatic Azides. *Synlett* **2004**, 99–100. [CrossRef]
84. Day, R.; Neugebauer, W.A.; Dory, Y. Stable Peptide-Based Pace4 Inhibitors. WO2013/029180A1, 7 March 2013.
85. Proteau-Gagné, A.; Bournival, V.; Rochon, K.; Dory, Y.L.; Gendron, L. Exploring the Backbone of Enkephalins To Adjust Their Pharmacological Profile for the $\delta$-Opioid Receptor. *ACS Chem. Neurosci.* **2010**, *1*, 757–769. [CrossRef]
86. Vigna, S.R.; Giraud, A.S.; Reeve, J.R.; Walsh, J.H. Biological activity of oxidized and reduced iodinated bombesins. *Peptides* **1988**, *9*, 923–926. [CrossRef]
87. Valverde, I.E.; Vomstein, S.; Fischer, C.A.; Mascarin, A.; Mindt, T.L. Probing the Backbone Function of Tumor Targeting Peptides by an Amide-to-Triazole Substitution Strategy. *J. Med. Chem.* **2015**, *58*, 7475–7484. [CrossRef]
88. Llinares, M.; Devin, C.; Chaloin, O.; Azay, J.; Noel-Artis, A.M.; Bernad, N.; Fehrentz, J.A.; Martinez, J. Syntheses and biological activities of potent bombesin receptor antagonists. *J. Pept. Res.* **1999**, *53*, 275–283. [CrossRef]
89. Valverde, I.E.; Huxol, E.; Mindt, T.L. Radiolabeled antagonistic bombesin peptidomimetics for tumor targeting. *J. Label. Compd. Radiopharm.* **2014**, *57*, 275–278. [CrossRef]
90. de Roux, N.; Genin, E.; Carel, J.-C.; Matsuda, F.; Chaussain, J.-L.; Milgrom, E. Hypogonadotropic hypogonadism due to loss of function of the KiSS1-derived peptide receptor GPR54. *Proc. Natl. Acad. Sci. USA* **2003**, *100*, 10972–10976. [CrossRef]
91. Seminara, S.B.; Messager, S.; Chatzidaki, E.E.; Thresher, R.R.; Acierno, J.S.; Shagoury, J.K.; Bo-Abbas, Y.; Kuohung, W.; Schwinof, K.M.; Hendrick, A.G.; et al. The GPR54 Gene as a Regulator of Puberty. *N. Engl. J. Med.* **2003**, *349*, 1614–1627. [CrossRef]
92. Caraty, A.; Smith, J.T.; Lomet, D.; Ben Saïd, S.; Morrissey, A.; Cognie, J.; Doughton, B.; Baril, G.; Briant, C.; Clarke, I.J. Kisspeptin Synchronizes Preovulatory Surges in Cyclical Ewes and Causes Ovulation in Seasonally Acyclic Ewes. *Endocrinology* **2007**, *148*, 5258–5267. [CrossRef]
93. Mascarin, A.; Valverde, I.E.; Vomstein, S.; Mindt, T.L. 1,2,3-Triazole Stabilized Neurotensin-Based Radiopeptidomimetics for Improved Tumor Targeting. *Bioconjug. Chem.* **2015**, *26*, 2143–2152. [CrossRef]
94. Souazé, F.; Dupouy, S.; Viardot-Foucault, V.; Bruyneel, E.; Attoub, S.; Gespach, C.; Gompel, A.; Forgez, P. Expression of Neurotensin and NT1 Receptor in Human Breast Cancer: A Potential Role in Tumor Progression. *Cancer Res.* **2006**, *66*, 6243–6249. [CrossRef] [PubMed]
95. Reubi, J.C.; Waser, B.; Friess, H.; Büchler, M.; Laissue, J. Neurotensin receptors: A new marker for human ductal pancreatic adenocarcinoma. *Gut* **1998**, *42*, 546–550. [CrossRef] [PubMed]

96. Bergmann, R.; Scheunemann, M.; Heichert, C.; Mäding, P.; Wittrisch, H.; Kretzschmar, M.; Rodig, H.; Tourwé, D.; Iterbeke, K.; Chavatte, K.; et al. Biodistribution and catabolism of 18F-labeled neurotensin(8–13) analogs. *Nucl. Med. Biol.* **2002**, *29*, 61–72. [CrossRef]
97. Bruehlmeier, M.; Garayoa, E.G.; Blanc, A.; Holzer, B.; Gergely, S.; Tourwé, D.; Schubiger, P.A.; Bläuenstein, P. Stabilization of neurotensin analogues: Effect on peptide catabolism, biodistribution and tumor binding. *Nucl. Med. Biol.* **2002**, *29*, 321–327. [CrossRef]
98. Mascarin, A.; Valverde, I.E.; Mindt, T.L. Radiolabeled analogs of neurotensin (8–13) containing multiple 1,2,3-triazoles as stable amide bond mimics in the backbone. *MedChemComm* **2016**, *7*, 1640–1646. [CrossRef]
99. Galibert, M.; Wartenberg, M.; Lecaille, F.; Saidi, A.; Mavel, S.; Joulin-Giet, A.; Korkmaz, B.; Brömme, D.; Aucagne, V.; Delmas, A.F.; et al. Substrate-derived triazolo- and azapeptides as inhibitors of cathepsins K and S. *Eur. J. Med. Chem.* **2018**, *144*, 201–210. [CrossRef]
100. Lecaille, F.; Brömme, D.; Lalmanach, G. Biochemical properties and regulation of cathepsin K activity. *Biochimie* **2008**, *90*, 208–226. [CrossRef]
101. Wilkinson, R.D.A.; Williams, R.; Scott, C.J.; Burden, R.E. Cathepsin S: Therapeutic, diagnostic, and prognostic potential. *Biol. Chem.* **2015**, *396*, 867–882. [CrossRef]
102. Lützner, N.; Kalbacher, H. Quantifying Cathepsin S Activity in Antigen Presenting Cells Using a Novel Specific Substrate. *J. Biol. Chem.* **2008**, *283*, 36185–36194. [CrossRef]
103. Lecaille, F.; Vandier, C.; Godat, E.; Hervé-Grépinet, V.; Brömme, D.; Lalmanach, G. Modulation of hypotensive effects of kinins by cathepsin K. *Arch. Biochem. Biophys.* **2007**, *459*, 129–136. [CrossRef]
104. Van Kersavond, T.; Konopatzki, R.; Chakrabarty, S.; Blank-Landeshammer, B.; Sickmann, A.; Verhelst, S.H.L. Short Peptides with Uncleavable Peptide Bond Mimetics as Photoactivatable Caspase-3 Inhibitors. *Molecules* **2019**, *24*, 206. [CrossRef]
105. Jubb, A.M.; Strickland, L.A.; Liu, S.D.; Mak, J.; Schmidt, M.; Koeppen, H. Neuropilin-1 expression in cancer and development. *J. Pathol.* **2012**, *226*, 50–60. [CrossRef]
106. Soker, S.; Takashima, S.; Miao, H.Q.; Neufeld, G.; Klagsbrun, M. Neuropilin-1 Is Expressed by Endothelial and Tumor Cells as an Isoform-Specific Receptor for Vascular Endothelial Growth Factor. *Cell* **1998**, *92*, 735–745. [CrossRef]
107. Naik, A.; Al-Zeheimi, N.; Bakheit, C.S.; Al Riyami, M.; Al Jarrah, A.; Al Moundhri, M.S.; Al Habsi, Z.; Basheer, M.; Adham, S.A. Neuropilin-1 Associated Molecules in the Blood Distinguish Poor Prognosis Breast Cancer: A Cross-Sectional Study. *Sci. Rep.* **2017**, *7*, 3301. [CrossRef]
108. Binétruy-Tournaire, R.; Demangel, C.; Malavaud, B.; Vassy, R.; Rouyre, S.; Kraemer, M.; Plouët, J.; Derbin, C.; Perret, G.; Mazié, J.C. Identification of a peptide blocking vascular endothelial growth factor (VEGF)-mediated angiogenesis. *EMBO J.* **2000**, *19*, 1525–1533. [CrossRef]
109. Tymecka, D.; Puszko, A.K.; Lipiński, P.F.J.; Fedorczyk, B.; Wilenska, B.; Sura, K.; Perret, G.Y.; Misicka, A. Branched pentapeptides as potent inhibitors of the vascular endothelial growth factor 165 binding to Neuropilin-1: Design, synthesis and biological activity. *Eur. J. Med. Chem.* **2018**, *158*, 453–462. [CrossRef]
110. Reubi, J.C.; Schaer, J.-C.; Waser, B. Cholecystokinin(CCK)-A and CCK-B/Gastrin Receptors in Human Tumors. *Cancer Res.* **1997**, *57*, 1377–1386.
111. Laverman, P.; Joosten, L.; Eek, A.; Roosenburg, S.; Peitl, P.K.; Maina, T.; Mäcke, H.; Aloj, L.; von Guggenberg, E.; Sosabowski, J.K.; et al. Comparative biodistribution of 12 111In-labelled gastrin/CCK2 receptor-targeting peptides. *Eur. J. Nucl. Med. Mol. Imaging* **2011**, *38*, 1410–1416. [CrossRef]
112. Grob, N.M.; Schmid, S.; Schibli, R.; Behe, M.; Mindt, T.L. Design of Radiolabeled Analogs of Minigastrin by Multiple Amide-to-Triazole Substitutions. *J. Med. Chem.* **2020**, *63*, 4496–4505. [CrossRef]
113. Renziehausen, A.; Wang, H.; Rao, B.; Weir, L.; Nigro, C.L.; Lattanzio, L.; Merlano, M.; Vega-Rioja, A.; del Carmen Fernandez-Carranco, M.; Hajji, N.; et al. The renin angiotensin system (RAS) mediates bifunctional growth regulation in melanoma and is a novel target for therapeutic intervention. *Oncogene* **2019**, *38*, 2320–2336. [CrossRef]
114. Magnani, F.; Pappas, C.G.; Crook, T.; Magafa, V.; Cordopatis, P.; Ishiguro, S.; Ohta, N.; Selent, J.; Bosnyak, S.; Jones, E.S.; et al. Electronic Sculpting of Ligand-GPCR Subtype Selectivity: The Case of Angiotensin II. *ACS Chem. Biol.* **2014**, *9*, 1420–1425. [CrossRef] [PubMed]
115. Vrettos, E.I.; Valverde, I.E.; Mascarin, A.; Pallier, P.N.; Cerofolini, L.; Fragai, M.; Parigi, G.; Hirmiz, B.; Bekas, N.; Grob, N.M.; et al. Single peptide backbone surrogate mutations to regulate angiotensin GPCR subtype selectivity. *Chem.–Eur. J.* **2020**. [CrossRef] [PubMed]

116. Coy, D.H.; Heinz-Erian, P.; Jiang, N.Y.; Sasaki, Y.; Taylor, J.; Moreau, J.P.; Wolfrey, W.T.; Gardner, J.D.; Jensen, R.T. Probing peptide backbone function in bombesin. A reduced peptide bond analogue with potent and specific receptor antagonist activity. *J. Biol. Chem.* **1988**, *263*, 5056–5060.
117. Horwell, D.C.; Howson, W.; Naylor, D.; Osborne, S.; Pinnock, R.D.; Ratcliffe, G.S.; Suman-Chauhan, N. Alanine scan and N-methyl amide derivatives of Ac-bombesin[7-14]. Development of a proposed binding conformation at the neuromedin B (NMB) and gastrin releasing peptide (GRP) receptors. *Int. J. Pept. Protein Res.* **1996**, *48*, 522–531. [CrossRef] [PubMed]

2

# Bonding Performance of a Hydrophilic Amide Monomer Containing Adhesive to Occlusal and Cervical Dentin

**Eri Seitoku [1], Shuhei Hoshika [1], Takatsumi Ikeda [1], Shigeaki Abe [2,*], Toru Tanaka [1] and Hidehiko Sano [1]**

[1] Faculty of Dental Medicine, Hokkaido University, Sapporo 060-8586, Japan; seitoku3045@den.hokudai.ac.jp (E.S.); starplatinum777@hotmail.com (S.H.); iketaka@den.hokudai.ac.jp (T.I.); tanto@den.hokudai.ac.jp (T.T.); sano@den.hokudai.ac.jp (H.S.)

[2] Graduate School of Biomedical Sciences, Nagasaki University, Nagasaki 852-8102, Japan

* Correspondence: sabe_den@nagasaki-u.ac.jp

**Abstract:** This study aimed to evaluate the bonding performance of a new one-step self-etching adhesive system containing a novel hydrophilic amide monomer. Clearfil Universal Bond Quick (CUB) and Clearfil Megabond 2 (CMB) were used as the one-step and two-step adhesive systems, respectively. Flat dentin surfaces of human premolars were exposed using #600 SiC (silicon carbide) and bonded with the respective adhesives of each system. The teeth were sectioned to obtain beams (1 mm × 1 mm) after 24 h of water storage. The mean bond strength and standard deviations (MPa) on an occlusal surface were as follows: CUB: 45.9 ± 19.7 and CMB: 67.9 ± 25.3. The values for cervical ones were CUB: 56.0 ± 20.3 and CMB: 67.6 ± 16.0, respectively. In both conditions, the microtensile bond strength (μTBS) value was lower than that of CMB. As seen during the microscopic observation, no adhesive failure was observed after μTBS testing because CUB formed a firm and tight adhesive interface.

**Keywords:** hydrophilic amide monomer; microtensile bond strength; failure modes; scanning electron microscope

---

## 1. Introduction

In recent years, it has been recommended that restoration should be made with minimum removal of tooth structure based on the concept of "Minimal Intervention" [1,2]. From one point of view, composite resin restoration might be the most valuable method, because the restoration could be esthetic, handy and tooth conservative. In composite resin restoration, the most crucial item would be the adhesive system.

In the first stage of dental adhesives to dentin, it was performed through three steps (etching, priming and bonding). Owing to the demand for simpler as well as more user-friendly and less technique sensitive adhesives, manufacturers have developed new adhesives used in two- or one-step procedures [3]. Regarding the one-step self-etch adhesives (1-SEAs), some researchers reported their comparable bonding performance with that of two-step self-etch adhesives (2-SEAs) [4–6]; however, other researchers reported their inferior performance [7,8].

A novel 1-SEA, Clearfil Universal Quick (Kuraray Noritake Co., Tokyo, Japan), containing a newly developed hydrophilic amide monomer has become commercially available. This hydrophilic amide monomer would have a higher setting ability; consequently, the formed polymer would be resistant to hydrolysis. Furthermore, as the monomer has a higher hydrophilic potential than 2-hydroxyethyl methacrylate (HEMA), it has good wettability to tooth structure [9,10]. Given the higher wettability of this monomer, the manufacturer insists that immediate air-drying after application of the adhesive should be performed. If done, it could be advantageous for good adhesion, especially in the cervical region, because the shorter manipulation time could avoid irritation from adhesion-reducing factors such as bleeding from the gingiva, gingival crevicular fluids, and moisture in the oral cavity, among others [11]. Furthermore, this 1-SEA can also be used for pretreatment in abutment construction using composite resin and the pretreatment of resin cements in the luting of prostheses [12]. In these cases, an adhesive interface would be formed on the cervical dentin surface, approximately perpendicular to the long axis of the tooth as the cervical force occurred [13]. If the adhesion at this region were to be broken, it would result in marginal leakage; consequently, secondary caries would occur. Hence, it is clinically crucial to form reliable adhesion at the cervical region for the successful use of universal type 1-SEA. Therefore, this study aimed to evaluate the bonding performance of the novel hydrophilic amide monomer containing 1-SEA on cervical dentin.

## 2. Materials and Methods

### 2.1. Materials Used

Basic information on the materials used herein can be found in Table 1. The 1-SEA and 2-SEA, Clearfil Universal Bond Quick (CUB) and Clearfil Megabond 2 (CMB) (Kuraray Noritake Dental Co. Tokyo, Japan), respectively, were used in this study.

**Table 1.** Materials used in this study.

| Products | Manufacturer | Code | pH | Contents |
|---|---|---|---|---|
| Clearfil Universal Bond Quick C (1-stepself-etch) | Kuraray Noritake Dental Co. | CUB | 2.3 (adhesive) | MDP, Bis-GMA, HEMA, Hydrophilic amide monomers, Silane coupling agent, Colloidal silica, Sodium fluoride, dl-Camphorquinone, accelerator, Ethanol, Water |
| Clearfil MegaBond 2 C (2-stepself-etch) | - | CMB | <2.5 (primer) | Primer, MDP, HEMA, Hydrophilic dimethacrylate, Photo-initiator, Water Bond, MDP, Bis-GMA, HEMA, Hydrophilic dimethacrylate, Photo-Inisiator, Silaneted colloidal silica |
| Clearfil AP-X (composite resin) | - | - | - | Bis-GMA, TEGDMA, Silaneted Barium glass filer, Silaneted colloidal silica, dl-Camphorquinone |

MDP: 10-Methacryloxydecyl dihydrogen phosphate. Bis-GMA: 2-bisphenol A diglycidyl methacrylate. HEMA: 2-hydroxyethyl methacrylate. TEGDMA: Tryethyleneglycol dimethacrylate.

### 2.2. Adhesive Specimen Preparation

Details regarding adhesive specimen preparation are presented in Figure 1. The collected teeth were stored following the approval of the committee of Ethics at Hokkaido University (#2013-7). Extracted teeth were stored following the guideline [14] until the experiment. The teeth were stored at 4 °C in 0.5% chloramine-T aqueous solution and used within 4 months after extraction. Each tooth was cut into crown and root. Regarding the coronal part, flat dentin surfaces were obtained by removing the coronal enamel of each tooth in a trimmer (Model Trimmer; Morita, Tokyo, Japan). Subsequently, the dentin surfaces were ground with #600 SiC (silicon carbide) paper for 60 s under water-cooling to produce a standardized smear layer prior to bonding (occlusal surface) [15]. For the root part, the cut surface was then ground with 600-grit SiC paper for 60 s. The contained occlusal and cervical surfaces were applied using CUB and CMB, as described in Figure 1 (application time, 10 s). After the application of the two adhesives, CLEARFIL AP-X (Kuraray Noritake Dental, Tokyo, Japan) were built on the surface using a light emitting diode light-curing device (Pencure 2000; Morita, Tokyo, Japan) for 40 s. The number of specimens for each of the four groups (two adherends × two adhesives) was three. The details were described in [16].

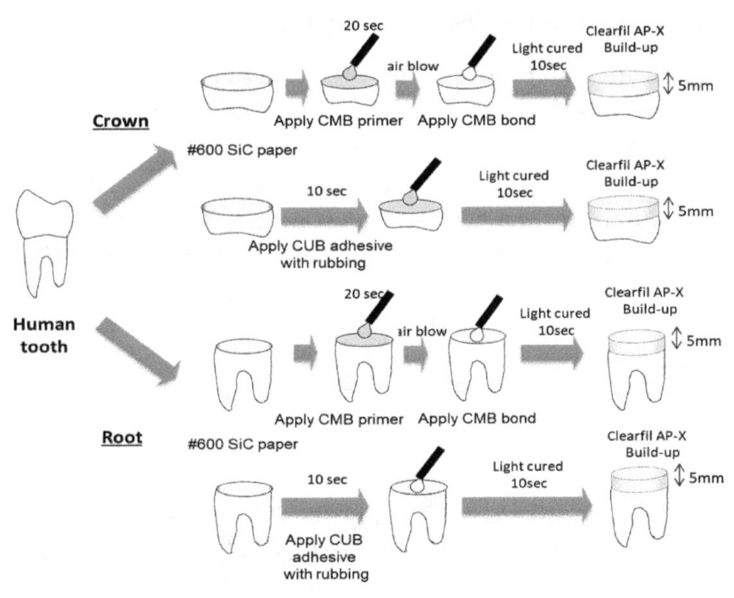


**Figure 1.** Procedure for specimen preparation.

## 2.3. *Microtensile Strength Testing*

The procedure of evaluation of μTBS is displayed in Figure 2. For both substrates, coronal and root, the dentin was prepared and bonded following the manner described in the current article. Bond strength testing was exactly following the procedure described in the previous reports [17]. The load at failure was recorded in Newtons and divided by the bonding surface area in square mm to calculate the bond strength in MPa [14,18]. Statistical analyses for the four groups were conducted using a two-way analysis of variance (ANOVA) test ($p$ = 0.05)

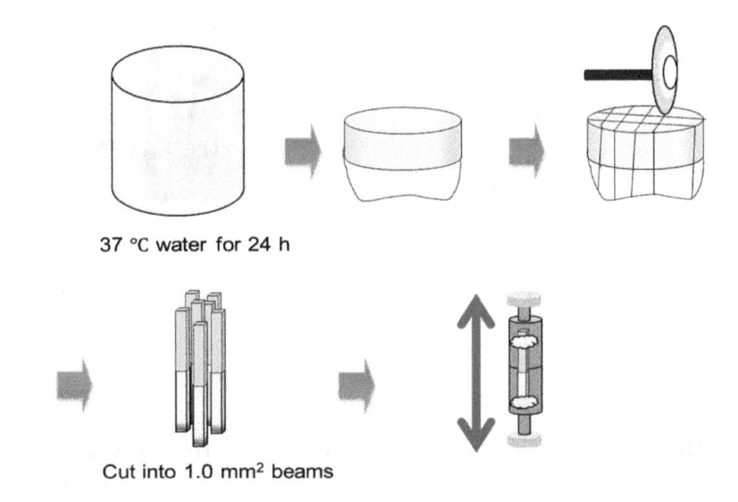


**Figure 2.** Procedure for microtensile bond strength testing.

## 2.4. *Failure Mode*

After the tensile test, dentin-side specimens were stored in a dry environment. Following desiccation, the failure surfaces of the specimens were observed through a digital microscope (VHX-5000; Keyence, Osaka, Japan) at a magnification of ×20 and recorded as either a "cohesive failure in composite resin," a "mixed failure at the interface and cohesive in dentin, " or a "cohesive failure in dentin."

*2.5. Scanning Electron Microscope (SEM) Observation of the Adhesive Interface*

Specimens were stored in water (37 °C) for 24 h. Then, the specimens were cut in the plane perpendicular to the adhesive interfaces (1.5 mm in thickness) using a low-speed diamond saw under water cooling. Subsequently, the specimens containing the adhesive interface were polished with SiC paper under water, and then followed by diamond paste in wet conditions. The polished surface was treated with 1 mol/L phosphoric acid and 5% NaOCl aqueous solution. The surfaces of the specimens were Pt-Pd coated for SEM observation (E-1030 Ion Sputter, HITACHI, Tokyo, Japan). Coated dentin surfaces were observed with field emission scanning electron microscope (FE-SEM, S-4000: HITACH, Tokyo, Japan) at an accelerated voltage of 10kV. The details were described in [19].

## 3. Results

*3.1. μTBS*

The results of μTBS testing are shown in Figure 3. In this study, no specimen failed before μTBS testing. The μTBS of CMB on the occlusal surface (CMB−occlusal) and CUB on the occlusal surface (CUB−occlusal) were 67.9 ± 25.3 MPa ($n$ = 48) and 45.9 ± 19.7 MPa ($n$ = 43), respectively. Furthermore, the μTBS of CMB on the cervical surface (CMB−cervical) and CUB on the cervical surface (CUB−cervical) were 67.6 ± 16.0 MPa ($n$ = 35) and 56.0 ± 20.3 MPa ($n$ = 37), respectively. As a result of the two-way ANOVA, CMB showed a significantly higher μTBS than CUB irrespective of adherends ($p < 0.05$). No significant difference was observed between the two adherends ($p > 0.05$).

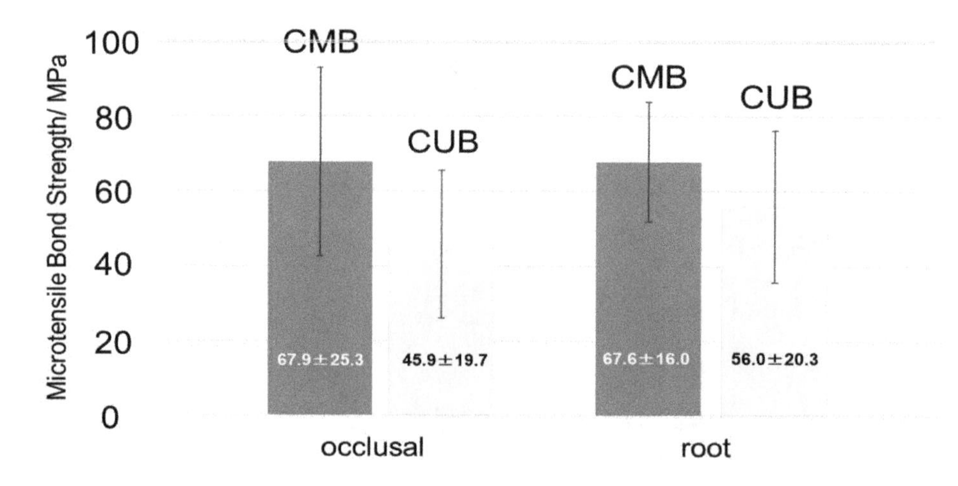


**Figure 3.** Microtensile bond strength (μTBS). CMB (blue), CUB (yellow); ($p > 0.05$).

*3.2. Failure Modes*

The failure modes after μTBS testing are indicated in Figure 4. For all the groups, "mixed failure at interface and cohesive in dentin" was the dominant mode. No specimen failed only at the adhesive interface in any of the groups.

*3.3. SEM Observation of the Adhesive Interface*

Interfacial SEM images of CMB−occlusal, CUB−occlusal, CMB−cervical and CUB−cervical are shown in Figure 5. No droplet, which acts as a sign of phase-separation was observed in the adhesive layer in both CMB and CUB specimens. Furthermore, no gap formation between the adhesive and the dentin was seen in the four groups. In the cases of CMB−occlusal and CMB−cervical, numerous resin tags that were formed in dentinal tubules could be seen. Conversely, few resin tags were seen in CUB−occlusal and CUB−cervical.

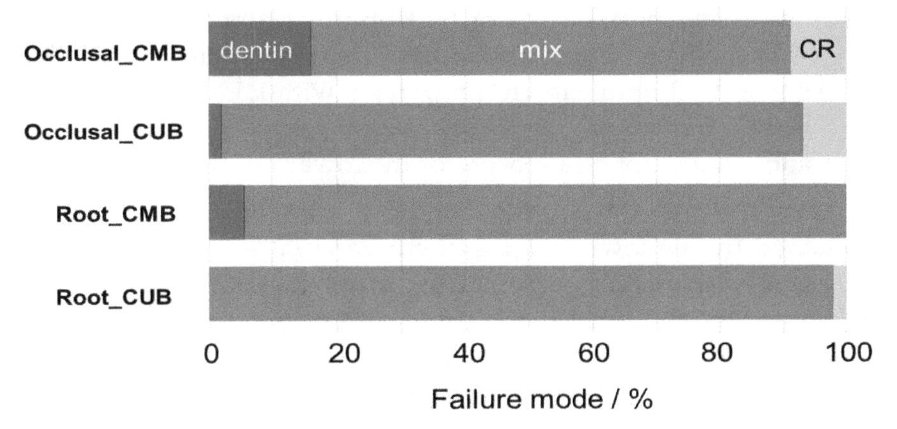


**Figure 4.** Failure modes. Cohesive failure in dentin (red), mixed failure (blue), cohesive failure in resin (yellow).

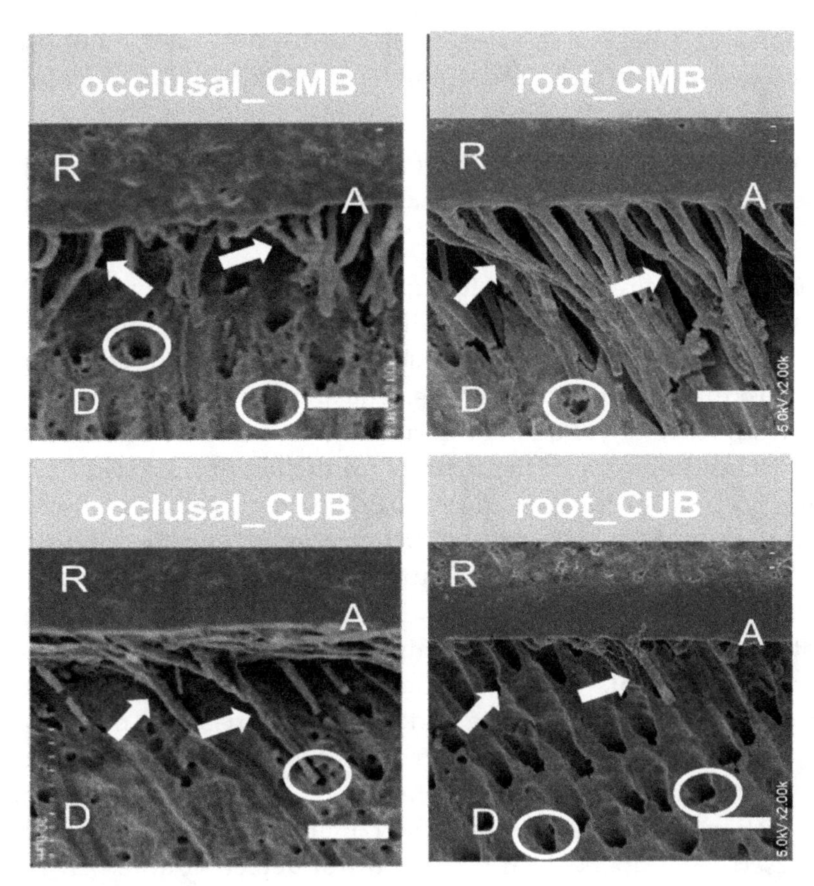


**Figure 5.** SEM images of adhesive interface. A: adhesive layer, R: composite resin, D: dentin, white arrows: resin tag, white circle: dental tubules. Scale bar: 30 μm.

## 4. Discussion

In this study, CUB indicated a lower μTBS, than that of CMB, regardless of whether adherends were cervical or occlusal ($p < 0.05$). Generally, one reason for the inferior bonding performance of 1-SEAs compared to that of 2-SEAs is a defect in the adhesive interface caused by the phase separation of the hydrophilic and the hydrophobic part contained in 1-SEAs [20]. Van Landuyt et al. revealed that some HEMA-free 1-SEA systems contained a hydrophobic monomer mixture, with which residual water could cause phase separation and blister formation in the adhesive layer [21]. The large blisters in the adhesive layer were generated by the phase separation. Consequently, the blisters acted as a weak point in the layer and adversely affected bond strength. In this study, no defect such as a "bubble" was observed in the adhesive layer of CUB. For hydrophilic properties, CUB contains hydrophilic monomers such HEMA and a hydrophilic amide monomer is contained in CUB. This hydrophilic amide monomer has a higher hydrophilicity in comparison with HEMA. As hydrophilic parts such as water were solved in the two monomers, phase separation could not occur in CUB. Thus, the defect caused by phase separation could not be a reason for the lower bond strength of CUB.

Commonly, it has been reported that 1-SEAs indicate a lower strength of adhesive than 2-SEAs. This fact can be explained by the lower degree of conversion, because 1-SEAs contain water and solvent(s) in their adhesive solution [22]. As for CUB, the polymerization properties of the hydrophilic amide monomer are considerably higher than HEMA. Therefore, the strength of the set adhesive is expected to be higher than that of conventional 1-SEAs. Kuno et al. reported that they prepared a hydrophilic 1-step self-etching adhesive which has the same composition as CUB and contained HEMA instead of hydrophilic amide monomer (denoted as CUB−H) [23]. Then they compared with ultimate tensile strength (UTS) of polymerized adhesives. The μTBS to dentin of both of them was also determined. CUB showed significantly higher UTS and μTBS than CUB−H. However, it seems difficult for the strengths of 1-SEAs to become superior to those of the bonding agents contained in 2-SEAs. Future studies should focus on the strengths of the cured adhesive of CUB in comparison to conventional 1-SEAs and the bond of 2-SEAs.

Regarding one of the major shortcomings of 1-SEAs, some researchers have reported that the bonding effectiveness of particularly (ultra-) mild self-etch adhesives might be impaired by thick smear layers [24,25].

In this study, numerous resin tags were formed in CMB−occlusal and CMB−cervical; however, few resin tags were seen in CUB−occlusal and CUB−cervical. This might be because the application time of CUB (10 s) was shorter than that of the primer of CMB (20 s). Both of the two adhesive systems used herein were (ultra-) mild self-etch adhesives. The shorter application time of CUB might have resulted in insufficient smear removal on the intertubular dentin surface and smear-plug in dentinal tubules. Hence, this fact might be one of the reasons for the lower μTBS for CUB in this study. In the present study, a smear layer was created using #600 SiC paper, although a thicker smear layer could be formed by regular diamond bars clinically. Furthermore, the application time recommended by the manufacturer is "0 (no waiting)" for CUB. Thus, many more remnants of the smear layer might be inhibited by satisfactory bonds in clinical use. Therefore, to avoid this, clinicians should ensure that a "rubbing motion" is performed, and it might be better to finish cavities and preparations using ultrafine diamond bars [25]. Future studies should focus on the influence of such factors, namely, the rubbing motion, application time and smear thickness, on the bonding performance of CUB. Moreover, the presence of defects in the tooth surface is also an important viewpoint. Grassi et al. investigated the effect of defects for endodontic sealers on bond strength of the dentin postinterface [26]. When the transverse section area of the cement layer contained defects at 12% or more, the interfacial shear strength decreased ca. 30% compared with a nondefect sample. In addition, Pettini et al. revealed that in vivo and in vitro cytogenetic and genotoxic effects by dental composite materials indicated different results [27]. Such an investigation should be required for a new material, especially.

In this study, differences in adherends (occlusal or cervical) did not affect the μTBS of the adhesives. There was a report that the parallel direction of dentinal tubules to the dentin surface resulted in a higher level of bonding because of a reduction in the number of dentinal tubules and an increase in the area of intertubular dentin. The report explained that a higher amount of collagen fiber in the intertubular dentin would form sufficient hybrid layers with monomers in the adhesive system; hence, it could result in a higher level of bonding on the adherend [28]. In this study, the teeth used were cut into two segments (crown and root) at the enamel−cement junction. It might be expected that the direction of the dentinal tubule would become parallel to the adhesive surface. However, the direction of tubules observed in the SEM images of the interface was not parallel and was not so different from the occlusal surface. Furthermore, no clear difference in the number and thickness of the tubules between occlusal and cervical dentin was observed. This fact might be the reason why the difference in the adherends (occlusal or cervical) did not affect the μTBS of the two adhesives herein.

Generally speaking, hydrophobic adhesives often generate voids and/or blister like structures at dentine–adhesive interface, because dentine contains so much water (12%). The structural defects derive inferior wettability of the adhesives to a hydroscopic dentine adherend. Though the application time of 1-SEA containing novel hydrophilic amide monomer employed in the present study was shorter (10 s) than that of 2-SEA self-etching primer (20 s), no voids, blister like-structures or gap formation could be seen at the interface with SEM observation. Furthermore, interfacial failure was not observed, and mixed failure (at interface and in dentin cohesive) was dominant for both the occlusal and cervical surface. Hence, the 1-SEA containing the novel hydrophilic amide monomer employed in the present study could quickly permeate the dentin. Therefore, CUB achieved good adhesion, regardless of whether adherends were cervical or occlusal. If the new amide monomer added to CUB rapidly permeates the dentin and enamel, it might eliminate the waiting time that could reduce the risk of contamination. It might also reduce technique sensitivity and application time. Hence, fine clinical prognosis might be expected when CUB is used clinically under intraoral conditions. Future studies should focus on the clinical outcomes of the newly introduced 1-SEA containing novel hydrophilic amide monomer [29].

## 5. Conclusions

In this study, the bonding performance of a newly introduced 1-SEA containing a novel hydrophilic amide monomer was evaluated. Under the limited conditions of this study, the conclusion obtained about the novel 1-SEA was as follows:

The novel 1-SEA could quickly permeate the dentin, though the μTBS value was lower than that of the 2-SEA. As the shorter application time could reduce the risk of contamination, fine clinical prognosis might be expected when the novel 1-SEA was used clinically under intraoral conditions. On the other hand, the shorter application time could result in insufficient smear removal. So, future studies should focus on the influence of these factors: rubbing motion, application time and smear thickness, on the bonding performance of the novel 1-SEA.

**Author Contributions:** Conceptualization, E.S. and T.I.; methodology, S.H. and T.I.; validation, E.S. and S.A.; formal analysis, E.S. and T.T. investigation, E.S. and S.A.; writing—original draft preparation, E.S.; writing—review and editing, T.I.; visualization, S.H. and T.T.; supervision, H.S.; funding acquisition, S.A. All authors have read and agreed to the published version of the manuscript.

**Acknowledgments:** This work was partially supported by MEXT, Japan.

## References

1. Assembly, F.G. FDI policy statement: Minimal intervention in the management of dental caries. *J. Minim. Interv. Dent.* **2012**, *62*, 223–243.
2. Ericson, D.; Kidd, E.; McComb, D.; Mjor, I.; Noack, M.J. Minimally invasive dentistry-Concepts and Techniques in Cariology. *Oral Health Prev. Dent.* **2003**, *1*, 59–72.
3. Van Meerbeek, B.; De Munk, J.; Yoshida, Y.; Inoue, S.; Vargas, M.; Vijay, P.; Van Landuyt, K.L.; Lambrechts, P.; Vanherle, G. Buoncore memorial lecture. Adhesion to enamel and dentin: Current status and future challenges. *Oper. Dent.* **2003**, *28*, 215–235. [PubMed]
4. Katsumata, A.; Saikaew, P.; Ting, S.; Katsumata, T.; Hoshika, T.; Sano, H.; Nishitani, Y. Microtensile bond strength bonded to dentin of a newly universal adhesive. *J. Oral Tissue Eng.* **2017**, *15*, 18–24.
5. Kawashima, S.; Shinkai, K.; Masaya Suzuki, M.; Suzuki, S. Bond Strength Comparison of One-Step/Two-Step Self-Etch Adhesives to Cavity Floor Dentin after 2.5 Years Storage in Water. *J. Interdiscip. Med. Dent. Sci.* **2017**, *5*, 1000213. [CrossRef]

6. Chan, K.M.; Tay, F.R.; King, N.M.; Imazato, S.; Pashley, D.H. Bonding of mild self-etching primers/adhesives to dentin with thick smear layers. *Am. J. Dent.* **2003**, *16*, 340–346.
7. Van Meerbeek, B.; Yoshihara, K.; Yoshida, Y.; Mine, A.; De Munk, J.; Van Landuyt, K.L. State of the art of self-etch adhesives. *Dent. Mater.* **2011**, *27*, 17–28. [CrossRef]
8. Hosaka, K.; Nakajima, M.; Takahashi, M.; Itoh, S.; Ikeda, M.; Tagami, J.; Pashley, D.H. Relationship between mechanical properties of one-step self-etch adhesives and water sorption. *Dent. Mater.* **2010**, *26*, 360–367. [CrossRef]
9. Hanabusa, M.; Kimura, S.; Hori, A.; Yamamoto, T. Effect of irradiation source on the dentin bond strength of a one-bottle universal adhesive containing an amide monomer. *J. Adhes. Sci. Technol.* **2019**, *33*, 2265–2280. [CrossRef]
10. Van Landuyt, K.L.; Snauwaert, J.; Peumans, M.; De Munck, J.; Lambrechts, P.; Van Meerbeek, B. The role of HEMA in one-step self-etch adhesives. *Dent. Mater.* **2008**, *24*, 1412–1419. [CrossRef]
11. Trzcionka, A.; Narbutaite, R.; Pranckeviciene, A.; Maskeliūnas, R.; Damaševičius, R.; Narvydas, G.; Połap, D.; Mocny-Pachońska, K.; Wozniak, M.; Tanasiewicz, M. In vitro analysis of quality of dental adhesive bond systems applied in various conditions. *Coatings* **2020**, *10*, 891. [CrossRef]
12. Ahmed, M.H.; Yoshihara, K.; Mercelis, B.; Van Landuyt, K.; Peumans, M.; Van Meerbeek, B. Quick bonding using a universal adhesive. *Clin. Oral Investig.* **2020**, *24*, 2837–2851. [CrossRef]
13. Jakupovic, S.; Anic, I.; Ajanovic, M.; Korac, S.; Konjhodzic, A.; Dzankovic, A.; Vukovic, A. Biomechanics of cervical tooth region and noncarious cervical lesions of different morphology; three-dimensional finite element analysis. *Eur. J. Dent.* **2016**, *10*, 413–418. [CrossRef] [PubMed]
14. Sano, H.; Almas Chowdhury, A.F.M.; Saikaew, P.; Matsumoto, M.; Hoshika, S.; Yamauti, M. The microtensile bond strength test: Its historical background and application to bond testing. *Jpn. Dent. Sci. Rev.* **2020**, *56*, 24–31. [CrossRef]
15. Pashley, D.H.; Tao, L.; Boyd, L.; King, G.E.; Homer, J.A. Scanning electron microscopy of the substructure of smear layer in human dentin. *Arch. Oral Biol.* **1988**, *33*, 265–270. [CrossRef]
16. Chowdhury, A.F.M.A.; Saikaew, P.; Matsumoto, M.; Sano, H.; Carvalho, R.M. Gradual dehydration affects the mechanical properties and bonding outcome of adhesives to dentin. *Dent. Mater. J.* **2019**, *38*, 361–367. [CrossRef] [PubMed]
17. Chowdhury, A.F.M.A.; Saikaeew, P.; Alam, A.; Sun, J.; Carvalho, R.M.; Sano, H. Effects of double application of contemporary self-etch adhesives on their bonding performance to dentin with clinically relevant smear layers. *J. Adhes. Dent.* **2019**, *21*, 59–66.
18. Armstrong, S.; Breschi, L.; Özcan, M.; Pfefferkorn, F.; Ferrari, M.; Van Meerbeek, B. Academy of Dental Materials guidance on in vitrotesting of dental composite bonding effectivenessto dentin/enamel using micro-tensile bondstrength (μTBS) approach. *Dent. Matter.* **2017**, *2633*, 133–143. [CrossRef]
19. Shibata, S.; Vieira, L.C.C.; Baratieri, L.N.; Fu, I.; Hoshika, S.; Matsuda, Y.; Sano, H. Evaluation of microtensile bond strength of self-etching adhesives on normal and caries-affected dentin. *Dent. Mater. J.* **2016**, *35*, 166–173.
20. Van Landuyt, K.; De Munk, J.; Snauwaert, J.; Coutinho, E.; Poitevin, A.; Yoshida, Y.; Inoue, S.; Peumans, M.; Suzuki, K.; Lambrechts, P. Monomer-solvent phase separation in one-step self-etch adhesives. *J. Dent. Res.* **2005**, *84*, 183–188. [CrossRef]
21. Shinkai, K.; Suzuki, S.; Katoh, Y. Effect of air-blowing variables on bond strength of all-in-one adhesives to bovine dentin. *Dent. Mater. J.* **2006**, *25*, 664–668. [CrossRef]
22. Ikeda, T.; De Munk, J.; Shirai, K.; Hikita, K.; Inoue, S.; Sano, H.; Lambrechts, P.; Van Meerbeek, B. Effect of air-drying and solvent evaporation on the strength of HEMA-rich versus HEMA-free one-step adhesives. *Dent. Mater.* **2008**, *24*, 1316–1323. [CrossRef] [PubMed]
23. Kuno, Y.; Hosaka, K.; Nakajima, M.; Ikeda, M.; Klein, C.A., Jr.; Foxton, R.M.; Tagami, J. Incorporation of a hydrophilic amide monomer into a one-step self-etch adhesive to increase dentin bond strength: Effect of application time. *Dent. Matter. J.* **2019**, *38*, 892–899. [CrossRef]
24. Koibuchi, H.; Yasuda, N.; Nakabayashi, N. Bonding to dentin with a self-etching primer: The effect of smear layers. *Dent. Mater.* **2001**, *17*, 122–126. [CrossRef]
25. Emis, R.B.; De Munk, J.; Cardoso, M.V.; Coutinho, E.; Van Landuyt, K.I.; Poitevin, A.; Lambrechts, P.; Van Meerbeek, B. Bond strength of self-etch adhesives to dentin prepared with three different diamond burs. *Dent. Mater.* **2008**, *24*, 978–985. [CrossRef] [PubMed]

26. Grassi, F.R.; Pappalettere, C.; Di Comite, M.; Corsalini, M.; Mori, G.; Ballini, A.; Crincoli, V.; Pettini, F.; Rapone, B.; Boccaccio, A. Effect of different irrigating solutions and endodontic sealers on bond strength of the dentin-post interface with and without defects. *Inter. J. Med. Sci.* **2012**, *9*, 642–654. [CrossRef] [PubMed]
27. Pettini, F.; Savino, M.; Corsalini, M.; Cantore, S.; Ballini, A. Cytogenetic genotoxic investigation in peripheral blood lymphocytes of subjects with dental composite restorative filling materials. *J. Biol. Regul. Homeost. Agents* **2015**, *29*, 229–233.
28. Ogata, M.; Okuda, M.; Nakajima, M.; Pereira, P.N.R.; Sano, H.; Tagami, J. Influence of the direction of tubules on bond strength to dentin. *Oper. Dent.* **2001**, *26*, 27–35.
29. Van Meerbeek, B.; Peumans, M.; Poitevin, A.; Mine, A.; Van Ende, A.; Neves, A.; De Munk, J. Relationship between bond-strength tests and clinical outcomes. *Dent. Matter.* **2010**, *26*, e100–e121. [CrossRef]

3

# Synthesis of Gemcitabine-Threonine Amide Prodrug Effective on Pancreatic Cancer Cells with Improved Pharmacokinetic Properties

**Sungwoo Hong [1,2,†], Zhenghuan Fang [3,†], Hoi-Yun Jung [1,2], Jin-Ha Yoon [4], Soon-Sun Hong [3,*] and Han-Joo Maeng [4,*]**

[1] Center for Catalytic Hydrocarbon Functionalization Institute for Basic Science (IBS), Daejeon 34141, Korea; hongorg@kaist.ac.kr (S.H.); spurs9@kaist.ac.kr (H.-Y.J.)
[2] Department and of Chemistry, Korea Advanced Institute of Science and Technology (KAIST), Daejeon 34141, Korea
[3] Department of Biomedical Sciences, College of Medicine, Inha University, Incheon 22212, Korea; junghwan110@gmail.com
[4] College of Pharmacy, Gachon University, Incheon 21936, Korea; jinha89@daum.net
* Correspondence: hongs@inha.ac.kr (S.-S.H.); hjmaeng@gachon.ac.kr (H.-J.M.)

† These authors are equally contributed to this work.

**Abstract:** To investigate the amino acid transporter-based prodrug anticancer strategy further, several amino acid-conjugated amide gemcitabine prodrugs were synthesized to target amino acid transporters in pancreatic cancer cells. The structures of the synthesized amino acid-conjugated prodrugs were confirmed by $^{1}$H-NMR and LC-MS. The pancreatic cancer cells, AsPC1, BxPC-3, PANC-1 and MIAPaCa-2, appeared to overexpress the amino acid transporter LAT-1 by conventional RT-PCR. Among the six amino acid derivatives of gemcitabine, threonine derivative of gemcitabine (Gem-Thr) was more effective than free gemcitabine in the pancreatic cancer cells, BxPC-3 and MIAPaCa-2, respectively, in terms of anti-cancer effects. Furthermore, Gem-Thr was metabolically stable in PBS (pH 7.4), rat plasma and liver microsomal fractions. When Gem-Thr was administered to rats at 4 mg/kg i.v., Gem-Thr was found to be successfully converted to gemcitabine via amide bond cleavage. Moreover, the Gem-Thr showed the increased systemic exposure of formed gemcitabine by 1.83-fold, compared to free gemcitabine treatment, due to the significantly decreased total clearance (0.60 vs. 4.23 mL/min/kg), indicating that the amide prodrug approach improves the metabolic stability of gemcitabine in vivo. Taken together, the amino acid transporter-targeting gemcitabine prodrug, Gem-Thr, was found to be effective on pancreatic cancer cells and to offer an efficient potential means of treating pancreatic cancer with significantly better pharmacokinetic characteristics than gemcitabine.

**Keywords:** amino acid transporters; amide bond; gemcitabine prodrug; metabolic stability; pancreatic cancer cells; pharmacokinetics

---

## 1. Introduction

Gemcitabine (2′-2′-difluorodeoxycytidine, dFdC) is a standard anticancer agent that is used to treat all stages of pancreatic adenocarcinoma [1], and combinations of gemcitabine with other anti-cancer agents, such as erlotinib [2], nab-paclitaxel [3,4] and others have been applied to pancreatic cancer patients. Because gemcitabine has serious toxic effects on the gastrointestinal track [5], and low oral bioavailability (BA), it is administered intravenously [6]. However, gemcitabine is a metabolically unstable drug and has a half-life of only 9 min in man, probably due to its rapid conversion to 2′-2′-difluorodeoxyuridine (dFdU) by cytidine deaminase (CD), which is abundant in blood and tissues [7]. This poor metabolic stability and poor compliance has created a clinical need for a gemcitabine prodrug. Several types of prodrugs have been reported, such as valproate amide [8,9], amino acid ester [10,11], aminoacyl amide derivative [12], and L-carnitine amide (OCTN2 targeting) [13] prodrugs.

To maintain the growth and survival of cancer cells during neoplastic transformation, amino acid production is massively increased and as a result, several uptake transporters (e.g., amino acid transporters) are overexpressed. For example, L-type amino acid transporter 1 (LAT-1), a representative amino acid transporter, is overexpressed in various types of cancer cells [14], and interestingly, LAT-1 expression has been consistently reported to be elevated in patients with pancreatic cancer and in transplanted Colo357 cells (a pancreatic cancer cell line) [15,16]. Accordingly, LAT-1 has been suggested to be a versatile target for the development of transporter-based drugs [14].

Our research group have been investigating the merits of strategies based on the amino acid transporter-based prodrug approach to overcome the multidrug resistance (MDR) triggered by various efflux transporters like P-glycoprotein (P-gp, MDR1) in cancer cells [17–19]. For example, the addition of amino acids, such as, L-valine (Val) or tyrosine (Tyr), to lapatinib by amide bond formation successfully enhanced the anti-cancer effect of lapatinib in human breast cancer cells (MDA-MB-231 and MCF7) and lung cancer cells (A549), without plasma stability issues, which demonstrated the involvements of amino acid transporters [18]. More recently, the val-modified amide prodrug of doxorubicin (Dox-Val) improved cellular uptake efficiency in MCF7 cells [19]. However, in a rat pharmacokinetic study that demonstrated the conversion of Dox-val to Dox by amide bond cleavage in vivo, the pharmacokinetic properties of the amino acid prodrug examined were limited [19].

Based on the observation that amino acid transporters (e.g., LAT-1) are overexpressed in various cancer cells for their survival, we speculated that amino acid conjugated prodrug of gemcitabine might be an effective way to improve gemcitabine uptake by overexpressing amino acid transporters in pancreatic cancer cells as well as overall physicochemical property. In the present study, we successfully synthesized a threonine-gemcitabine (Gem-Thr) prodrug and investigated its anti-cancer efficacy in pancreatic cancer cells and its in vitro/in vivo metabolic stabilities. In addition, we examined LAT-1 (an amino acid transporter) mRNA/protein expressions in pancreatic cancer cells.

## 2. Results and Discussion

### 2.1. Synthesis of Gemcitabine Prodrugs with Amino Acids

To confirm the validity of the amino acid transporter-based prodrug approach for gemcitabine, we prepared a series of amino acid conjugated gemcitabine prodrugs. As summarized in Figure 1, gemcitabine-amino acid prodrugs **3** were obtained from gemcitabine (**1**) using a two-step sequence, that is by amide bond formation between gemcitabine (**1**) and *N*-Boc-L-amino acid employing 1-hydroxy-1H-benzotriazole (HOBt), 1-ethyl-3-(3-dimethylaminopropyl)carbodiimide and 4-methylmorpholine in DMF/DMSO (3:1), and subsequent Boc deprotection of the resulting compound **2** using hydrochloric acid. In order to avoid potential problems associated with epimerization when using carbodiimides as coupling reagents, HOBt was used as an additive because it enhances the reactivities of activated ester intermediates by facilitating amine approach via hydrogen bonding. To confirm the formation of amide bond, an intermediate of representative compound (Gem-Thr-Boc) was dissolved in DMSO-$d_6$. In gemcitabine part, the chemical shift of CH(5) in oxopyrimidine core showed at $\delta$ 7.26 ($J_{5,6}$ = 7.6 Hz) as a doublet by *ortho* coupling with CH(6) at $\delta$ 8.27. The characteristic amide NH peak was detected at $\delta$ 10.99 as a broad singlet, and Boc group of amino acid was appeared at $\delta$ 1.39. The side chain of threonine $CH_3$ was appeared at $\delta$ 1.39 having a coupling constant $J$ = 6.3 Hz and OH group was resonated at $\delta$ 4.89. In addition, the MS/MS spectrum showed $[M + H]^+$ at $m/z$ 465.2, which confirmed the chemical bond between gemcitabine and *N*-Boc protected threonine. The *N*-Boc of amino acid was deprotected in acidic conditions to afford the final Gem-Thr product that was confirmed by $^{1}$H NMR in $CD_3OD$. The peak of Boc group disappeared and concurrently the side chain of threonine $CH_3$ was detected as a doublet peak at $\delta$ 7.26 ($J$ = 6.4 Hz). Furthermore, CH(5)- and CH(6)-peaks of oxopyrimidine remained at $\delta$ 6.11 and $\delta$ 7.85 coupled as a doublet ($J_{5,6}$ = 7.6).

**Figure 1.** Synthetic scheme of the procedure used to synthesize gemcitabine-amino acid prodrugs. Reagent and conditions were as follows; (i) *N*-Boc-L-amino acid, 4-methylmorpholine, HOBt, EDCI·HCl, DMF/DMSO = 3: 1, 55 °C, 17 h; (ii) 4N HCl in dioxane, dry $CH_2Cl_2$, r.t. 30 min.

*2.2. Expression of LAT-1 in Pancreatic Cancer Cell Lines*

LAT-1 mRNA and protein levels were detected in the pancreatic cancer cell lines AsPC1, BxPC-3, PANC-1 and MIAPaCa-2. LAT-1 expression was greatest in BxPC-3 (Figure 2). The following studies on anti-cancer effects were performed using BxPC-3 and MIAPaCa-2 cells. Therefore, this transcriptional activation of LAT-1 might expect to facilitate transport capability of amino acid linked gemcitabine.

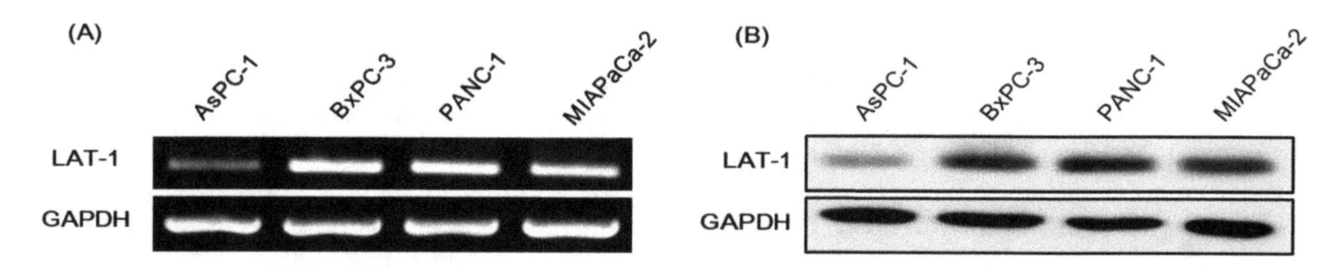


**Figure 2.** mRNA (**A**) and protein (**B**) expressions of LAT-1 in various pancreatic cancer cell lines (AsPC-1, BxPC-3, PANC-1, and MIAPaCa-2). Reverse transcription–PCR and Western blot were performed using gel electrophoresis after purification of DNA from cancer cells.

*2.3. Anticancer Effects of Prodrugs with Amino Acids in Cancer Cells*

We evaluated the cytotoxic effects of gemcitabine derivatives on various cancer cell lines including PDAC (pancreatic ductal adenocarcinoma), and compared results with those of free gemcitabine (Figure 3). In A549 (lung cancer cells) and MDA-MB-231 (breast cancer cells) neither gemcitabine nor gemcitabine derivatives had any significant anti-cancer effect versus vehicle treated controls. However, gemcitabine had significant anti-cancer effects on BxPC-3, MIAPaCa-2 (pancreatic cancer cells) and B16 (melanoma cells). The anti-cancer effects of Gem-Tyr, Gem-Val, Gem-Met, Gem-Ile and Gem-Leu were similar to those of gemcitabine in pancreatic cancer cells, whereas Gem-Thr had the most potent anticancer effect, and this was slightly superior to that of gemcitabine in BxPC-3 cells (Gem-Thr, 44.7% vs. gemcitabine, 54.1% of cell viability, $p$ = 0.0464), which was found to overexpress amino acid transporters, including LAT-1 (Figure 3). The introduction of threonine to gemcitabine is likely to be recognized by the LAT-1, which has an important role to influx the drug into the cancer cells. However, addition of some amino acid moieties to gemcitabine did not exert the enhanced cytotoxic effects on the pancreatic cancer cells. Similarly, a valine ester prodrug (Val-SN-38) showed a comparable cytotoxic effect compared to SN-38, an active metabolite of irinotecan, despite an increased intracellular accumulation in MCF cells with amino acid transporters being overexpressed [17].

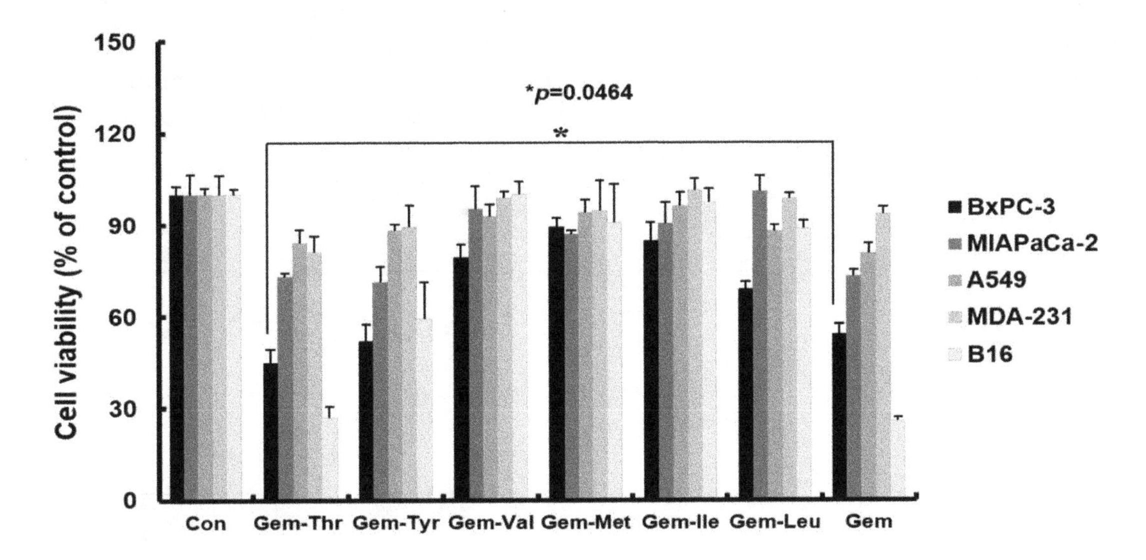


**Figure 3.** Growth-inhibitory effects of gemcitabine and of its amino-acid-conjugated derivatives on BxPC-3, MIAPaCa-2, A549, MDA-MB-231 B16 cells after exposure to 1 μM concentrations for 48 h, as estimated by MTT assay (mean ± SD, $n = 3$). Experiments were performed three times independently. * $p < 0.05$, compared with free gemcitabine group.

Additionally, consistent with MTT results, either gemcitabine or Gem-Thr induced apoptosis significantly when we measured cleaved PARP as well as TUNEL positive cells after a 24 h treatment (Figure 4).

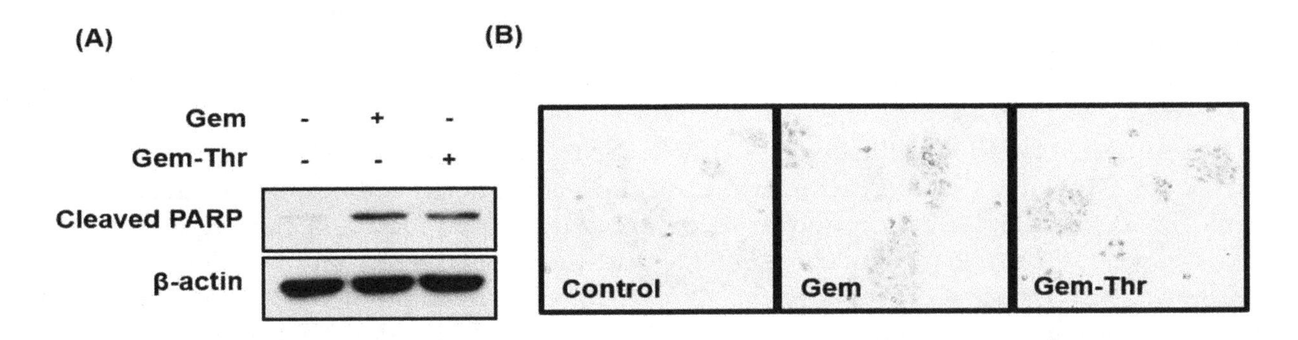


**Figure 4.** Apoptosis of pancreatic cancer cells by Gemcitabine derivative. After the treatment of gemcitabine and Gem-Thr to BxPC-3 cells for 24 h, cleaved PARP was detected by western blotting (**A**) and TUNEL assay (**B**) was performed.

### *2.4. In Vitro Plasma Stability of Gem-Thr*

To recognize a substrate by amino acid transporters in cancer cells, the stability of the amide prodrug with amino acid moiety, Gem-Thr, should be first possessed [18]. Thus, the stability of the prodrug was investigated in PBS, rat plasma and liver microsomes for up to 8 h (Figure 5). Gem-Thr was metabolically stable in PBS (pH 7.4), whereas 20% of Gem-Thr was metabolized in rat plasma after 8 h, indicating Gem-Thr can be metabolized to gemcitabine by enzymes in plasma. Surprisingly, Gem-Thr was metabolically stable in rat liver microsomal fractions, which contain high levels of cytochrome P450 (CYP) enzymes, indicating the amide bond between gemcitabine and threonine cannot be readily broken via phase I metabolism or enzymes expressed in liver microsomes. This concurs with the results of a previous study on val-lapatinib and tyr-lapatinib, in which amide linkage was found to result in appropriate metabolic stability in vitro [18]. Our results indicate the amino acid moiety of Gem-Thr is likely both stable and recognized by the amino acid transporters in cancer cells.

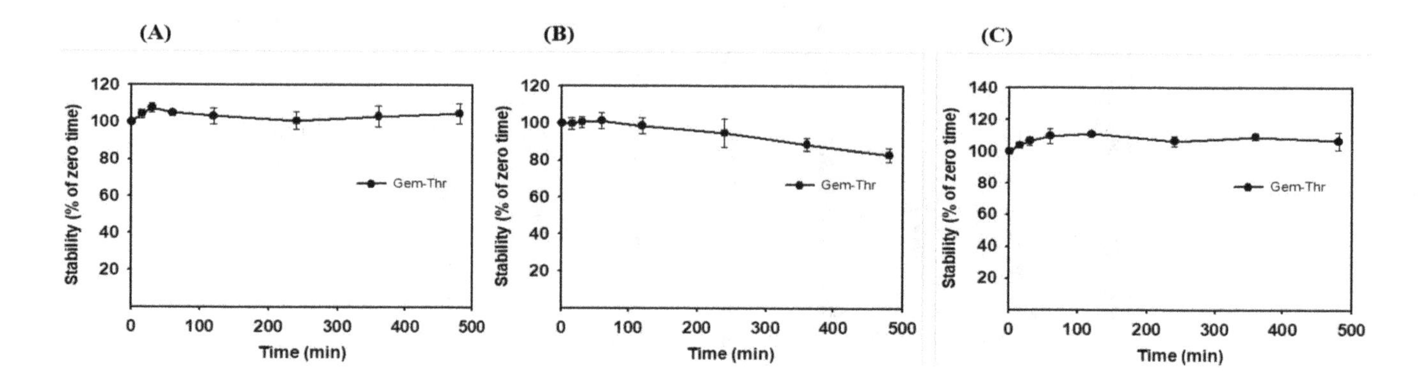


**Figure 5.** In vitro metabolic stabilities of Gem-Thr in PBS (**A**), plasma (**B**) and a liver microsome fraction (**C**) (mean ± SD, $n$ = 3).

*2.5. Comparison of Systemic Pharmacokinetics with Free Gemcitabine*

The pharmacokinetic parameters of Gem-Thr and free gemcitabine after intravenous administration in rats are summarized in Table 1. Concentrations of Gem-Thr and gemcitabine and their summed concentrations in plasma were measured and compared to those of animals administered free gemcitabine (Figure 6). After gemcitabine administration at 4 mg/kg, the oral systemic exposure (i.e., AUC) and the measure of drug elimination form the body, CL, were found to be 948.38 ± 52.04 μg·min/mL and 4.23 ± 0.23 mL/min/kg, respectively, which concurred with previously reported values [20]. The volume of distribution at steady state ($V_{SS}$) and average time for a drug molecule to reside in the body, MRT, values for gemcitabine were 2483.64 ± 867.19 mL/kg and 582.06 ± 177.90 min (Table 1). After administration of Gem-Thr at 4 mg/kg i.v., the conversion of Gem-Thr to gemcitabine was found to be probably due to amide bond cleavage, with similar systemic exposure (i.e., AUC) between Gem-Thr and the formed gemcitabine. More importantly, Gem-Thr increased systemic exposure (i.e., the AUC of gemcitabine) by 1.83-fold versus free gemcitabine, and this was attributed to a significantly lower total CL value (0.60 vs. 4.23 mL/min/kg) (Table 1). Furthermore, this suggests the amide prodrug approach improves the metabolic stability of gemcitabine in vivo. Other pharmacokinetic parameters, such as, $V_{SS}$ and MRT, were also found to be different for Gem-Thr and gemcitabine (Table 1). For example, MRT of gemcitabine for Gem-Thr was greater than that for gemcitabine, indicating that formed gemcitabine from Gem-Thr remains in the systemic circulation longer than free gemcitabine. In a previous study, an amide prodrug of gemcitabine releasing gemcitabine and valproic acid was found to enhance gemcitabine stability by blocking its deamination to uridine, and to prolong systemic exposure as compared with gemcitabine alone [8]. Furthermore, the AUC of sum of Gem-Thr and the gemcitabine (metabolite) was significantly higher (3.62-fold higher) than gemcitabine alone (Table 1). In our previous study, although we first demonstrated the successful conversion and systemic circulation of the active metabolite (doxorubicin) probably occurred due to amide bond cleavage, amide linked doxorubicin/valine failed to improve pharmacokinetic properties including systemic CL [19]. On the other hand, Gem-Thr showed enhanced gemcitabine stability in vivo and an increased AUC and decreased CL versus gemcitabine. This enhanced systemic stability of Gem-Thr is likely to improve gemcitabine uptake by cancer cells and to target amino acid transporters like LAT-1 transporters.

**Table 1.** Comparison of pharmacokinetic parameters of Gem-Thr, formed gemcitabine and sum of two species after intravenous administration of Gem-Thr at a dose of 4 mg/kg with those of free gemcitabine alone at a dose of 4 mg/kg in rats ($n$ = 3–4).

| Pharmacokinetic Parameters | Gem-Thr (4 mg/kg) | | | Free Gemcitabine (4 mg/kg) |
|---|---|---|---|---|
| | Gem-Thr | Gemcitabine | Sum | |
| AUC (μg·min/mL) | 1713.85 ± 1082.40 | 1739.88 ± 282.00 * | 3437.92 ± 1180.56 | 948.38 ± 52.04 |
| Terminal $t_{1/2}$ (min) | 236.18 ± 50.94 | 666.83 ± 271.49 | 537.23 ± 227.78 | 532.68 ± 177.90 |
| CL (mL/min/kg) | 2.85 ± 1.33 | 0.60 ± 0.10 * | 1.26 ± 0.39 | 4.23 ± 0.23 |
| $V_{ss}$ (mL/kg) | 662.35 ± 281.40 | 545.57 ± 263.01 * | 770.96 ± 435.31 | 2483.64 ± 867.19 |
| MRT (min) | 237.18 ± 20.87 | 907.18 ± 391.68 * | 577.36 ± 212.90 | 582.06 ± 177.90 |

Administration of dose for Gem-thr was 4 mg/kg, which is equivalent to 2.892 mg/kg for gemcitabine. Data presents as mean ± standard deviation (SD). * Significantly different from free gemcitabine group ($p < 0.05$).

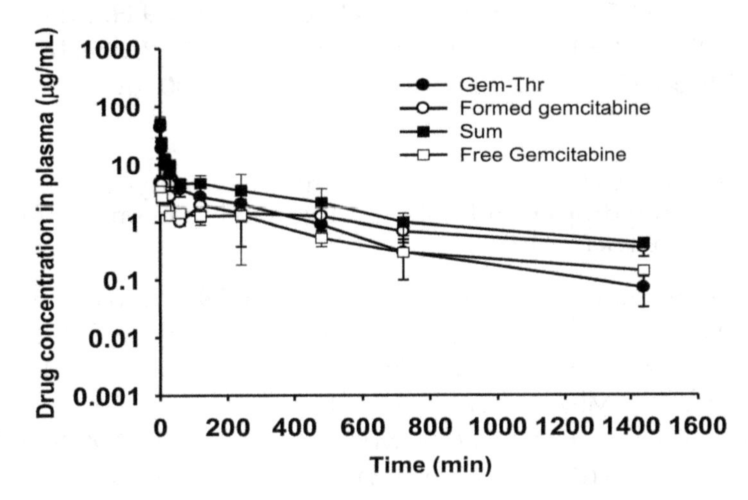


**Figure 6.** Plasma concentration-time profiles of Gem-Thr (prodrug), formed gemcitabine (metabolite), the sum of Gem-Thr and formed gemcitabine and free gemcitabine. The intravenous pharmacokinetics of Gem-Thr was compared to that of free gemcitabine (4 mg/kg, $n$ = 4).

## 3. Materials and Methods

### 3.1. Materials

Commercial grade reagents and solvents were used without further purification. Thin layer chromatography (TLC) was performed using precoated silica gel 60 $F^{254}$ plates and visualized using anisaldehyde solution, heat, and UV light (254 nm). Flash column chromatography was undertaken on silica gel (400–630 mesh). $^1$H NMR was recorded at 400 MHz and chemical shifts are quoted in parts per million (ppm) versus an appropriate solvent peak or 2.50 ppm for DMSO-$d_6$. The following abbreviations were used to describe peak splitting patterns: br = broad, s = singlet, d = doublet, t = triplet, q = quartet, m = multiplet, dd = doublet of doublets, td = triplet of doublets, ddd = doublet of doublets of doublets. Coupling constants, $J$, are reported in hertz (Hz). HPLC was conducted using an Agilent HPLC unit equipped with an Agilent Poroshell 120 EC-C18 reverse phase column (4.6 × 50 mm, 2.7 Micron) and mass spectroscopy was performed using a quadrupole LC/MS unit.

### 3.2. Synthesis and Characterization of DOX-Val

#### 3.2.1. General Procedure for Preparing Gemcitabine Derivatives

To a solution of gemcitabine (1.0 equiv), 1-ethyl-3-(3-dimethylaminopropyl)-carbodiimide hydrochloride (1.3 equiv), 4-methylmorpholine (1 equiv), and 1-hydroxybenzotriazole (1 equiv) in DMF/DMSO (5 mL, 3:1) was added dropwise $N$-Boc protected amino acid (1.1 equiv) at room temperature in a $N_2$ atmosphere. The reaction mixture was then stirred in an oil bath at 55 °C for 17 h, cooled to room temperature and quenched by adding brine (5 mL). The mixture was then extracted using ethyl acetate (3 × 10 mL) and the combined organic layer was washed with 20% LiCl solution, saturated $NaHCO_3$ aqueous solution, and brine, dried over $MgSO_4$, and concentrated under reduced pressure. The residue was purified by silica gel column chromatography (DCM/methanol = 15:1) to afford the desired product **2**. To a solution of the mixture of aforementioned intermediate in anhydrous DCM (1 mL) was added 4N HCl in dioxane (1 mL). The mixture was then stirred for 12 h at room temperature, solvent was removed, and the residue was purified by silica gel flash column chromatography (DCM/methanol = 3:1) to afford the desired product **3**.

### 3.2.2. (*S*)-2-Amino-*N*-(1-((2*R*,4*R*,5*R*)-3,3-difluoro-4-hydroxy-5-(hydroxymethyl)tetrahydrofuran-2-yl)-2-oxo-1,2-dihydropyrimidin-4-yl)-3-methylbutanamide (Gem-Val)

From the reaction of gemcitabine (27 mg, 0.104 mmol), 21 mg (56% for 2 steps) was obtained. $^1$H NMR (400 MHz, Methanol-$d_4$) δ 7.81 (d, *J* = 7.6 Hz, 1H), 6.21 (t, *J* = 7.9 Hz, 1H), 6.04 (d, *J* = 7.6 Hz, 1H), 4.51 (d, *J* = 7.5 Hz, 1H), 4.25 (td, *J* = 12.1, 8.3 Hz, 1H), 3.94 (dd, *J* = 12.7, 2.5 Hz, 1H), 3.88 (dt, *J* = 8.4, 2.9 Hz, 1H), 3.78 (dd, *J* = 12.5, 3.2 Hz, 1H), 2.14 (h, *J* = 6.8 Hz, 1H), 1.00 (dd, *J* = 6.8, 5.5 Hz, 6H); MS/MS *m*/*z* 363.1 $(M + 1)^+$.

### 3.2.3. (2*S*,3*R*)-2-Amino-*N*-(1-((2*R*,4*R*,5*R*)-3,3-difluoro-4-hydroxy-5-(hydroxylmethyl)tetrahydrofuran-2-yl)-2-oxo-1,2-dihydropyrimidin-4-yl)-3-hydroxybutanamide2-(4-((pyridin-3-ylmethyl)amino)quinazolin-2-yl)phenol (Gem-Thr)

From the reaction of gemcitabine (51 mg, 0.194 mmol), 24 mg (34%) was obtained. $^1$H NMR (400 MHz, Methanol-$d_4$) δ 7.85 (d, *J* = 7.6 Hz, 1H), 6.25–6.16 (m, 1H), 6.11 (d, *J* = 7.6 Hz, 1H), 4.67 (d, *J* = 3.9 Hz, 1H), 4.32–4.19 (m, 2H), 3.98–3.92 (m, 1H), 3.89 (dt, *J* = 8.4, 2.9 Hz, 1H), 3.79 (dd, *J* = 12.6, 3.3 Hz, 1H), 1.21 (d, *J* = 6.4 Hz, 3H); MS/MS *m*/*z* 365.1 $(M + 1)^+$.

(Gem-Thr-Boc intermediate) $^1$H NMR (400 MHz, DMSO-$d_6$) δ 10.99 (s, 1H), 8.27 (d, *J* = 7.6 Hz, 1H), 7.26 (d, *J* = 7.6 Hz, 1H), 6.49 (d, *J* = 8.7 Hz, 1H), 6.32 (d, *J* = 6.4 Hz, 1H), 6.18 (t, *J* = 7.4 Hz, 1H), 5.31 (t, *J* = 5.5 Hz, 1H), 4.89 (s, 1H), 4.25–4.11 (m, 2H), 4.08–4.00 (m, 1H), 3.89 (dt, *J* = 8.5, 3.0 Hz, 1H), 3.85–3.77 (m, 1H), 3.65 (ddd, *J* = 12.8, 6.0, 3.2 Hz, 1H), 1.39 (s, 8H), 1.09 (d, *J* = 6.3 Hz, 3H); MS/MS *m*/*z* 465.1 $(M + 1)^+$.

### 3.2.4. (*S*)-2-Amino-*N*-(1-((2*R*,4*R*,5*R*)-3,3-difluoro-4-hydroxy-5-(hydroxymethyl)tetrahydrofuran-2-yl)-2-oxo-1,2-dihydropyrimidin-4-yl)-3-(4-hydroxyphenyl)propanamide (Gem-Tyr)

From the reaction of gemcitabine (22 mg, 0.083 mmol), 27 mg (76%) was obtained. $^1$H NMR (400 MHz, Methanol-$d_4$) δ 7.77 (d, *J* = 7.6 Hz, 1H), 7.08 (d, *J* = 8.5 Hz, 2H), 6.68 (d, *J* = 8.5 Hz, 2H), 6.25–6.13 (m, 1H), 5.93 (d, *J* = 7.6 Hz, 1H), 4.23 (td, *J* = 12.1, 8.2 Hz, 1H), 3.96–3.90 (m, 1H), 3.87 (dt, *J* = 8.4, 2.9 Hz, 1H), 3.77 (dd, *J* = 12.6, 3.2 Hz, 1H), 3.09 (dd, *J* = 13.9, 6.3 Hz, 1H), 2.88 (dd, *J* = 13.8, 8.2 Hz, 1H); MS/MS *m*/*z* 427.1 $(M + 1)^+$.

### 3.2.5. (*S*)-2-Amino-*N*-(1-((2*R*,4*R*,5*R*)-3,3-difluoro-4-hydroxy-5-(hydroxymethyl)tetrahydrofuran-2-yl)-2-oxo-1,2-dihydropyrimidin-4-yl)-4-(methylthio)butanamide (Gem-Met)

From the reaction of gemcitabine (38 mg, 0.146 mmol), 23 mg (40%) was obtained. $^1$H NMR (400 MHz, Methanol-$d_4$) δ 7.83 (d, *J* = 7.7 Hz, 1H), 6.34–6.13 (m, 1H), 5.99 (d, *J* = 7.6 Hz, 1H), 4.83–4.73 (m, 1H), 4.25 (td, *J* = 12.1, 8.2 Hz, 1H), 3.98–3.91 (m, 1H), 3.89 (dt, *J* = 8.4, 2.8 Hz, 1H), 3.78 (dd, *J* = 12.5, 3.2 Hz, 1H), 2.55 (ddd, *J* = 8.4, 6.4, 3.7 Hz, 2H), 2.23–2.11 (m, 1H), 2.09 (s, 3H), 2.07–1.89 (m, 1H); MS/MS *m*/*z* 395.1 $(M + 1)^+$.

### 3.2.6. (2*S*,3*S*)-2-Amino-*N*-(1-((2*R*,4*R*,5*R*)-3,3-difluoro-4-hydroxy-5-(hydroxymethyl)tetrahydrofuran-2-yl)-2-oxo-1,2-dihydropyrimidin-4-yl)-3-methylpentanamide (Gem-Ile)

From the reaction of gemcitabine (29 mg, 0.110 mmol), 20 mg (50%) was obtained. $^1$H NMR (400 MHz, Methanol-$d_4$) δ 7.81 (d, *J* = 7.6 Hz, 1H), 6.28–6.14 (m, 1H), 6.02 (d, *J* = 7.6 Hz, 1H), 4.54 (d, *J* = 8.1

Hz, 1H), 4.25 (td, $J$ = 12.0, 8.2 Hz, 1H), 3.98–3.89 (m, 1H), 3.88 (dt, $J$ = 8.4, 2.9 Hz, 1H), 3.78 (dd, $J$ = 12.5, 3.3 Hz, 1H), 1.97–1.83 (m, 1H), 1.58 (ddd, $J$ = 13.6, 7.6, 3.4 Hz, 1H), 1.32–1.15 (m, 1H), 0.99 (d, $J$ = 6.8 Hz, 3H), 0.92 (t, $J$ = 7.4 Hz, 3H); MS/MS $m/z$ 377.2 $(M + 1)^+$.

#### 3.2.7. (*S*)-2-Amino-*N*-(1-((2*R*,4*R*,5*R*)-3,3-difluoro-4-hydroxy-5-(hydroxymethyl)tetrahydrofuran-2-yl)-2-oxo-1,2-dihydropyrimidin-4-yl)-4-methylpentanamide (Gem-Leu)

From the reaction of gemcitabine (19 mg, 0.073 mmol), 21 mg (77%) was obtained. $^1$H NMR (400 MHz, Methanol-$d_4$) δ 7.81 (d, $J$ = 7.6 Hz, 1H), 6.26–6.15 (m, 1H), 5.97 (d, $J$ = 7.6 Hz, 1H), 4.75–4.68 (m, 1H), 4.25 (td, $J$ = 12.1, 8.2 Hz, 1H), 3.98–3.91 (m, 1H), 3.88 (dt, $J$ = 8.3, 2.8 Hz, 1H), 3.78 (dd, $J$ = 12.5, 3.2 Hz, 1H), 1.77–1.56 (m, 3H), 0.98 (d, $J$ = 6.1 Hz, 3H), 0.95 (d, $J$ = 6.1 Hz, 3H); MS/MS $m/z$ 377.2 $(M + 1)^+$.

### 3.3. *Characterization of Gemcitabine Prodrugs with Amino Acid*

#### 3.3.1. Cell Culture

Human pancreatic cancer cells (MIAPaCa-2, BxPC-3 and AsPC-1) [21], lung cancer cell lines (A549) [22], human breast cancer cell lines (MDA-MB-231) [23] and melanoma cell lines (B16) [24] were purchased from the American Type Culture Collection (Manassas, VA, USA). MIAPaCa-2 and B16-F10 cells were cultured in Dulbecco's modified Eagle's medium (DMEM) supplemented with 10% heat-inactivated fetal bovine serum (FBS) and 1% penicillin/streptomycin. Aspc-1, BxPC-3 MDA-MB-231 and A549 cells were cultured in Roswell Park Memorial Institute 1640 (RPMI-1640) medium supplemented with 10% FBS and 1% penicillin/streptomycin. FBS and all other reagents used for cell culture were purchased from Invitrogen (Carlsbad, CA, USA). Cultures were maintained at 37 °C in 95% air /5% $CO_2$ humidified atmosphere.

#### 3.3.2. Reverse Transcription-PCR

Total LAT-1 RNA was isolated using Trizol reagent and subjected to reverse transcription-PCR (Promega Corp.). The PCR primers used were LAT1, 5′-CCTCTGGGCCTGTTCTCTTG-3′ (forward) and 5′-CTTGAGGCATGTCCACCTCC-3′ (reverse). PCR reaction of forward and reverse genes was performed using the Ex Taq DNA Polymerase Recombinant (TaKaRa, Tokyo, Japan), with final concentrations of 1X PCR buffer, 2.5 mM of dNTP mixture, 2.5 mM of $MgCl_2$, 10 pmol of each primer in a total reaction volume of 25 μL containing 1 μL of cDNA. Individual PCR amplification cycle of forward or reverse genes was performed with an initial denaturation step at 94 °C for 3 min, followed by 35 cycles (94 °C for 30 s; 55 °C for 60 s; 72 °C for 60 s), and finally with an elongation step at 72 °C for 5 min. The DNA products were resolved using gel electrophoresis (1.5% agarose gel).

#### 3.3.3. Western Blot Assays

BxPC-3 cells were washed with DPBS and lysed with RIPA buffer (Biosesang, Seongnam, Korea) containing 150 mM NaCl, 1% Triton X-100, 1% sodium deoxycholate, 0.1% SDS, 50 mM Tris-HCl (pH 7.5), 2 mM EDTA (pH 8.0), and Xpert protease inhibitor and phosphatase inhibitor Cocktail (GenDEPOT, Barker, TX, USA). Proteins were separated by 8% or 15% SDS-PAGE (sodium dodecylsulfate-polyacrylamide gel electrophoresis) and transferred to polyvinylidene fluoride (PVDF) membranes. Protein transfer was confirmed using a Ponceau S staining solution (AMRESCO, Solon, OH, USA), and the blots were then immunostained with appropriate primary antibodies (1:1000) and secondary antibodies (1:5000) conjugated to horseradish peroxidase (HRP). Antibody binding was detected using an enhanced chemiluminescence (ECL) reagent (Bio-Rad, Hercules, CA, USA) using primary antibodies specific to proteins of interest, and proteins were detected using X-ray film and enhanced chemiluminescence reagent. Primary antibodies against the following were used: LAT-1, cleaved PARP (Cell Signaling Technologies, Beverly, MA, USA), GAPDH (Abcam, Cambridge, MA, USA), and β-actin (Sigma Aldrich, St. Louis, MO, USA). Secondary antibodies were purchased from Cell signaling technology.

3.3.4. Cytotoxicity Assay in Pancreatic Cells

Cancer cell viabilities after treatment with gemcitabine or its derivatives were quantified using MTT assay as described previously [25]. Briefly, cells were seeded onto 96-well plates at $3 \times 10^3$ cells per well and incubated at 37 °C. The cells were treated with each compound at indicated concentrations for 48 h and then, 20 μL of MTT labeling mixture was added to each well. After incubation for 4 h, optical densities (OD) were determined using a microplate reader by measuring absorbances at 540 nm.

3.3.5. Terminal Deoxynucleotidyl Transferase–Mediated Nick End Labeling (TUNEL) Assay

TUNEL assay was conducted with the ApopTag® Peroxidase In Situ Apoptosis Detection Kit (Merck Millipore, Burlington, MA, USA). Briefly, BxPC-3 cells were seeded onto 18-mm cover glasses in medium and grown to ~70% confluence over 24 h. Cells were then treated with 1 μM of gemcitabine or Gem-Thr for 24 h, fixed in an ice-cold mixture of acetic acid and ethanol solution, washed with DPBS, TUNEL stained, mounted, and examined under a light microscope for nuclear fragmentation.

*3.4. In Vitro Metabolic Stability of Gem-Thr*

The metabolic stability of Gem-Thr was examined in the presence of PBS (pH 7.4), rat plasma, or liver microsomes as described previously (Maeng et al., 2014). Fresh rat plasma was obtained from sacrificed SD rats (260–280 g) after centrifuging blood (12,000× *g*, 15 min). Rat liver microsomes were obtained from BD Gentest. Gem-Thr (10 μM) was spiked into PBS, blank rat plasma, or liver microsomes (1 mg protein/mL) and incubated in a shaking water bath at 37 °C for 8 h. Aliquots (50-μL) were taken after 0, 15, 30, 60, 120, 240, 260, or 480 min of incubation, immediately pretreated with ice-cold methanolic solution including the internal standard, vortexed for 3 min and centrifuged (12,000× *g*, 10 min). Supernatants were stored at −80 °C until required for analysis.

*3.5. Systemic Pharmacokinetics Study of Gem-Thr in Rats*

This study was performed in male Sprague-Dawley (SD) rats (Orient Bio, Sungnam, Korea), as described previously [26]. Rats were provided water and food *ad libitum* and maintained under a 12:12-h light/dark cycle. On the experimental day, a femoral vein and contralateral artery were cannulated using an Intramedic™ polyethylene tube (PE-50; Becton Dickinson Diagnostics, Sparks, MD, USA) under Zoletil induced anesthesia (50 mg/kg intramuscularly; Virbac, Carros, France). For the intravenous pharmacokinetic study, Gem-Thr or gemcitabine solution were injected into a femoral vein at 4 mg/kg. Blood (~0.22 mL) was taken from the femoral artery at 0 (blank), 1, 5, 15, 30, 60, 120, 240, 480, 720, or 1440 min. To prevent blood loss, the same volume of normal saline was injected intravenously at each time point. Plasma was obtained immediately centrifugation and then stored at −80 °C prior to LC-MS/MS.

The pharmacokinetic parameters of Gem-Thr, gemcitabine, Gem-Thr plus gemcitabine, and free gemcitabine were calculated by non-compartmental analysis using WinNonlin (Version 3.1; Pharsight, Mountain View, CA, USA), as described previously (Park et al., 2016).

*3.6. Analysis of Gem-Thr and Gemcitabine by LC-MS/MS*

To determine concentrations of Gem-Thr and gemcitabine in rat plasma, plasma samples were deproteinized by adding two volumes of acetonitrile containing internal standard (phenacetin). After mixing, mixtures were centrifuged at 14,000× *g* for 10 min.

Aliquots of supernatants (2-μL) were injected into the LC-MS/MS system, which consisted of an Agilent HPLC and an Agilent 6490 QQQ mass spectrometer equipped with an ESI+ Agilent Jet Stream ion source (Agilent Technologies, Santa Clara, CA, USA). The separation of each drug and IS from endogenous plasma substances was achieved on a Synergi Polar-RP 80A column (150 × 2.0 mm, 4 μm; Phenomenex, Torrance, CA, USA). The mobile phase consisted of 0.1% formic acid and acetonitrile (20:80, *v*/*v*) at a flow rate of 0.2 mL/min. The column and autosampler tray were maintained at 25 and

4 °C, respectively. Gemcitabine, Gem-Thr, and the IS were quantified by multiple-reaction monitoring (MRM) in positive electrospray ionization mode. Respective precursor-to-product ion transitions were as follows: Gemcitabine, 264.1→112.1, Gem-Thr, 387.1→343.2, and IS (phenacetin), 180.2→162.2. Data acquisition was performed using Mass Hunter software (ver. A.02.00; Agilent Technologies).

### *3.7. Statistical Analysis*

Results are expressed as means ± standard deviations (SDs). Differences between group means were analyzed using the two-tailed Student's *t*-test. Statistical significance was accepted for $p$ values < 0.05.

## 4. Conclusions

Various amino acid derivatives of gemcitabine were successfully synthesized by forming amide bonds. Of the six derivatives of gemcitabine synthesized, Gem-Thr most effectively killed pancreatic cancer cells, in which LAT-1 was overexpressed. Our in vitro metabolic stability showed Gem-Thr is stable in PBS, plasma and a liver microsomal fraction, which demonstrated Gem-Thr is stable in cancer cells overexpressing amino acid transporter. Interestingly, our systemic pharmacokinetic results suggested that the amide prodrug approach improves the metabolic stability of gemcitabine in our in vivo model due to reduced decreased metabolic clearance. To the best of our knowledge, this is the first report on the cytotoxic effects of an amino acid transporter-targeting gemcitabine prodrug, produced by the introduction of threonine, on pancreatic cancer cells. Although the in vitro anti-cancer effect of Gem-Thr was only slightly superior to that of free gemcitabine, this improved pharmacokinetic property of Gem-Thr may have substantial anti-cancer effects in pancreatic cancer.

**Author Contributions:** S.H., S.-S.H. and H.-J.M. conceived and designed the experiments; H.-Y.J., Z.F. and J.-H.Y. performed the experiments; S.H., S.-S.H., Z.F. and H.-J.M. analyzed the data; S.H. and S.-S.H. contributed reagents/materials/analysis tools; S.H., S.-S.H. and H.-J.M. wrote the paper.

**Acknowledgments:** The authors would like to thank Dr. Kyung Hee Jung for technical assistance on the cytotoxicity assay.

## References

1. Vitellius, C.; Fizanne, L.; Menager-Tabourel, E.; Nader, J.; Baize, N.; Laly, M.; Lermite, E.; Bertrais, S.; Caroli-Bosc, F.X. The combination of everolimus and zoledronic acid increase the efficacy of gemcitabine in a mouse model of pancreatic adenocarcinoma. *Oncotarget* **2018**, *9*, 28069–28082. [CrossRef] [PubMed]
2. Moore, M.J.; Goldstein, D.; Hamm, J.; Figer, A.; Hecht, J.R.; Gallinger, S.; Au, H.J.; Murawa, P.; Walde, D.; Wolff, R.A.; et al. Erlotinib plus gemcitabine compared with gemcitabine alone in patients with advanced pancreatic cancer: A phase III trial of the National Cancer Institute of Canada Clinical Trials Group. *J. Clin. Oncol.* **2007**, *25*, 1960–1966. [CrossRef] [PubMed]
3. Von Hoff, D.D.; Ervin, T.; Arena, F.P.; Chiorean, E.G.; Infante, J.; Moore, M.; Seay, T.; Tjulandin, S.A.; Ma, W.W.; Saleh, M.N.; et al. Increased survival in pancreatic cancer with nab-paclitaxel plus gemcitabine. *N. Engl. J. Med.* **2013**, *369*, 1691–1703. [CrossRef] [PubMed]
4. Ogawa, Y.; Suzuki, E.; Mikata, R.; Yasui, S.; Abe, M.; Iino, Y.; Ohyama, H.; Chiba, T.; Tsuyuguchi, T.; Kato, N. Five Cases of Interstitial Pneumonitis Due to Gemcitabine and Nab-Paclitaxel Combination Treatment in Pancreatic Cancer Patients. *Pancreas* **2018**, *47*, e42–e43. [CrossRef] [PubMed]
5. Huang, J.; Robertson, J.M.; Ye, H.; Margolis, J.; Nadeau, L.; Yan, D. Dose-volume analysis of predictors for gastrointestinal toxicity after concurrent full-dose gemcitabine and radiotherapy for locally advanced pancreatic adenocarcinoma. *Int. J. Radiat. Oncol. Biol. Phys.* **2012**, *83*, 1120–1125. [CrossRef] [PubMed]
6. Bender, D.M.; Bao, J.; Dantzig, A.H.; Diseroad, W.D.; Law, K.L.; Magnus, N.A.; Peterson, J.A.; Perkins, E.J.; Pu, Y.J.; Reutzel-Edens, S.M.; et al. Synthesis, crystallization, and biological evaluation of an orally active prodrug of gemcitabine. *J. Med. Chem.* **2009**, *52*, 6958–6961. [CrossRef] [PubMed]

7. Beumer, J.H.; Eiseman, J.L.; Parise, R.A.; Joseph, E.; Covey, J.M.; Egorin, M.J. Modulation of gemcitabine (2′,2′-difluoro-2′-deoxycytidine) pharmacokinetics, metabolism, and bioavailability in mice by 3,4,5,6-tetrahydrouridine. *Clin. Cancer Res.* **2008**, *14*, 3529–3535. [CrossRef] [PubMed]
8. Wickremsinhe, E.; Bao, J.; Smith, R.; Burton, R.; Dow, S.; Perkins, E. Preclinical absorption, distribution, metabolism, and excretion of an oral amide prodrug of gemcitabine designed to deliver prolonged systemic exposure. *Pharmaceutics* **2013**, *5*, 261–276. [CrossRef] [PubMed]
9. Zhang, Y.; Gao, Y.; Wen, X.; Ma, H. Current prodrug strategies, for improving oral absorption of nucleoside analogues. *Asian J. Pharm. Sci.* **2014**, *9*, 65–74. [CrossRef]
10. Tsume, Y.; Incecayir, T.; Song, X.; Hilfinger, J.M.; Amidon, G.L. The development of orally administrable gemcitabine prodrugs with D-enantiomer amino acids: Enhanced membrane permeability and enzymatic stability. *Eur. J. Pharm. Biopharm.* **2014**, *86*, 514–523. [CrossRef] [PubMed]
11. Song, X.; Lorenzi, P.L.; Landowski, C.P.; Vig, B.S.; Hilfinger, J.M.; Amidon, G.L. Amino acid ester prodrugs of the anticancer agent gemcitabine: Synthesis, bioconversion, metabolic bioevasion, and hPEPT1-mediated transport. *Mol. Pharm.* **2005**, *2*, 157–167. [CrossRef] [PubMed]
12. Zhang, D.; Bender, D.M.; Victor, F.; Peterson, J.A.; Boyer, R.D.; Stephenson, G.A.; Azman, A.; McCarthy, J.R. Facile rearrangement of N4-(α-aminoacyl)cytidines to *N*-(4-cytidinyl)amino acid amides. *Tetrahedron Lett.* **2008**, *49*, 2052–2055. [CrossRef]
13. Wang, G.; Chen, H.; Zhao, D.; Ding, D.; Sun, M.; Kou, L.; Luo, C.; Zhang, D.; Yi, X.; Dong, J.; et al. Combination of l-carnitine with lipophilic linkage-donating gemcitabine derivatives as intestinal novel organic cation transporter 2-targeting oral prodrugs. *J. Med. Chem.* **2017**, *60*, 2552–2561. [CrossRef] [PubMed]
14. Jin, S.E.; Jin, H.E.; Hong, S.S. Targeting L-type amino acid transporter 1 for anticancer therapy: Clinical impact from diagnostics to therapeutics. *Expert Opin. Ther. Targets* **2015**, *19*, 1319–1337. [CrossRef] [PubMed]
15. Kaira, K.; Sunose, Y.; Arakawa, K.; Ogawa, T.; Sunaga, N.; Shimizu, K.; Tominaga, H.; Oriuchi, N.; Itoh, H.; Nagamori, S.; et al. Prognostic significance of L-type amino acid transporter 1 expression in surgically resected pancreatic cancer. *Br. J. Cancer* **2012**, *107*, 632–638. [CrossRef] [PubMed]
16. Yanagisawa, N.; Ichinoe, M.; Mikami, T.; Nakada, N.; Hana, K.; Koizumi, W.; Endou, H.; Okayasu, I. High expression of L-type amino acid transporter 1 (LAT1) predicts poor prognosis in pancreatic ductal adenocarcinomas. *J. Clin. Pathol.* **2012**, *65*, 1019–1023. [CrossRef] [PubMed]
17. Kwak, E.Y.; Shim, W.S.; Chang, J.E.; Chong, S.; Kim, D.D.; Chung, S.J.; Shim, C.K. Enhanced intracellular accumulation of a non-nucleoside anti-cancer agent via increased uptake of its valine ester prodrug through amino acid transporters. *Xenobiotica* **2012**, *42*, 603–613. [CrossRef] [PubMed]
18. Maeng, H.J.; Kim, E.S.; Chough, C.; Joung, M.; Lim, J.W.; Shim, C.K.; Shim, W.S. Addition of amino acid moieties to lapatinib increases the anticancer effect via amino acid transporters. *Biopharm. Drug Dispos.* **2014**, *35*, 60–69. [CrossRef] [PubMed]
19. Park, Y.; Park, J.H.; Park, S.; Lee, S.Y.; Cho, K.H.; Kim, D.D.; Shim, W.S.; Yoon, I.S.; Cho, H.J.; Maeng, H.J. Enhanced Cellular Uptake and Pharmacokinetic Characteristics of Doxorubicin-Valine Amide Prodrug. *Molecules* **2016**, *21*, 1272. [CrossRef] [PubMed]
20. Zhao, C.; Li, Y.; Qin, Y.; Wang, R.; Li, G.; Sun, C.; Qu, X.; Li, W. Pharmacokinetics and metabolism of SL-01, a prodrug of gemcitabine, in rats. *Cancer Chemother. Pharmacol.* **2013**, *71*, 1541–1550. [CrossRef] [PubMed]
21. Son, M.K.; Jung, K.H.; Lee, H.S.; Lee, H.; Kim, S.J.; Yan, H.H.; Ryu, Y.L.; Hong, S.S. SB365, Pulsatilla saponin D suppresses proliferation and induces apoptosis of pancreatic cancer cells. *Oncol. Rep.* **2013**, *30*, 801–808. [CrossRef] [PubMed]
22. Pyo, J.S.; Roh, S.H.; Kim, D.K.; Lee, J.G.; Lee, Y.Y.; Hong, S.S.; Kwon, S.W.; Park, J.H. Anti-cancer effect of Betulin on a human lung cancer cell line: A pharmacoproteomic approach using 2 D SDS PAGE coupled with nano-HPLC tandem Mass Spectrometry. *Planta Med.* **2009**, *75*, 127–131. [CrossRef] [PubMed]
23. Gu, W.W.; Lin, J.; Hong, X.Y. Cyclin A2 regulates homologous recombination DNA repair and sensitivity to DNA damaging agents and poly (ADP-ribose) polymerase (PARP) inhibitors in human breast cancer cells. *Oncotarget* **2017**, *24*, 90842–90851. [CrossRef] [PubMed]
24. Hatiboglu, M.A.; Kocyigit, A.; Guler, E.M.; Akdur, K.; Nalli, A.; Karatas, E.; Tuzgen, S. Thymoquinone Induces Apoptosis in B16-F10 Melanoma Cell Through Inhibition of p-STAT3 and Inhibits Tumor Growth in a Murine Intracerebral Melanoma Model. *World Neurosurg.* **2018**, *114*, e182–e190. [CrossRef] [PubMed]

25. Yun, S.M.; Jung, K.H.; Lee, H.; Son, M.K.; Seo, J.H.; Yan, H.H.; Park, B.H.; Hong, S.; Hong, S.S. Synergistic anticancer activity of HS-173, a novel PI3K inhibitor in combination with Sorafenib against pancreatic cancer cells. *Cancer Lett.* **2013**, *331*, 250–261. [CrossRef] [PubMed]
26. Kim, Y.C.; Kim, I.B.; Noh, C.K.; Quach, H.P.; Yoon, I.S.; Chow, E.C.Y.; Kim, M.; Jin, H.E.; Cho, K.H.; Chung, S.J.; et al. Effects of 1$\alpha$,25-dihydroxyvitamin D3, the natural vitamin D receptor ligand, on the pharmacokinetics of cefdinir and cefadroxil, organic anion transporter substrates, in rat. *J. Pharm. Sci.* **2014**, *103*, 3793–3805. [CrossRef] [PubMed]

# 4

# Unexpected Resistance to Base-Catalyzed Hydrolysis of Nitrogen Pyramidal Amides based on the 7-Azabicyclic[2.2.1]heptane Scaffold

**Diego Antonio Ocampo Gutiérrez de Velasco [1], Aoze Su [1], Luhan Zhai [1], Satowa Kinoshita [1,2], Yuko Otani [1] and Tomohiko Ohwada [1,*]**

1 Laboratory of Organic and Medicinal Chemistry, Graduate School of Pharmaceutical Sciences, University of Tokyo, 7-3-1 Hongo, Bunkyo-ku, Tokyo 113-0033, Japan; daogv1@hotmail.com (D.A.O.G.d.V.); aaronsusu@gmail.com (A.S.); zhailuhanzlh@gmail.com (L.Z.); s.kinoshita13@gmail.com (S.K.); otani@mol.f.u-tokyo.ac.jp (Y.O.)

2 Department of Chemistry, St John's College, University of Cambridge, St John's Street, Cambridge CB2 1TP, UK

* Correspondence: ohwada@mol.f.u-tokyo.ac.jp

Academic Editor: Michal Szostak

**Abstract:** Non-planar amides are usually transitional structures, that are involved in amide bond rotation and inversion of the nitrogen atom, but some ground-minimum non-planar amides have been reported. Non-planar amides are generally sensitive to water or other nucleophiles, so that the amide bond is readily cleaved. In this article, we examine the reactivity profile of the base-catalyzed hydrolysis of 7-azabicyclo[2.2.1]heptane amides, which show pyramidalization of the amide nitrogen atom, and we compare the kinetics of the base-catalyzed hydrolysis of the benzamides of 7-azabicyclo[2.2.1]heptane and related monocyclic compounds. Unexpectedly, non-planar amides based on the 7-azabicyclo[2.2.1]heptane scaffold were found to be resistant to base-catalyzed hydrolysis. The calculated Gibbs free energies were consistent with this experimental finding. The contribution of thermal corrections (entropy term, $-T\Delta S^{\ddagger}$) was large; the entropy term ($\Delta S^{\ddagger}$) took a large negative value, indicating significant order in the transition structure, which includes solvating water molecules.

**Keywords:** non planar amide; base-catalyed hydrolysis; water solvation; entropy

---

## 1. Introduction

In non-planar amides, distortion of the amide bond can arise from both twisting about the C-N bond and pyramidalization at the nitrogen atom (Scheme 1) [1,2]. These transformations of the amide bond are essentially mutually correlated, and the transition states of the amide rotation involved bond twisting and nitrogen pyramidalization at the same time [3] (see also Reference [4]). The partial double-bond character of planar amides limits rotation about the C-N bond, and this feature also contributes to stabilization, due to electron-delocalization. Decrease in $sp^2$ nitrogen character, with increase of $sp^3$ character, tends to weaken the C-N bond and increase the electrophilicity of the carbonyl carbon atom [1].

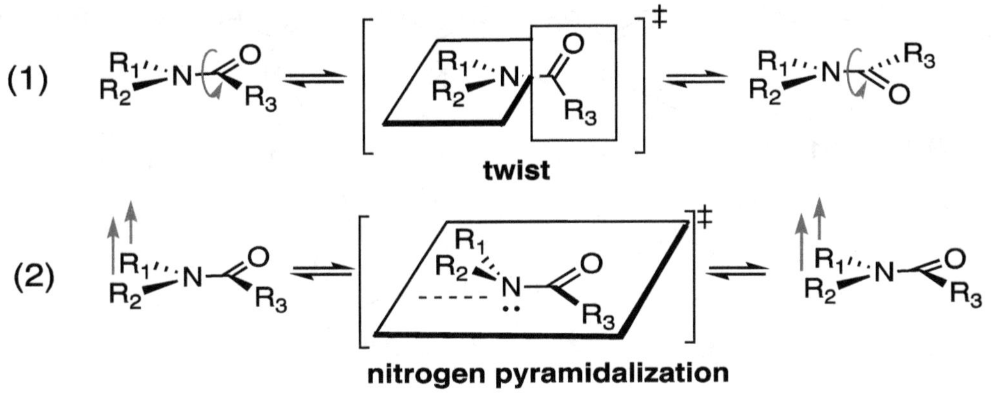

Scheme 1. Amide transformation processes causing non-planarity: (1) N-C bond twisting (rotation) and (2) nitrogen pyramidalization. These transformations are interconnected.

Non-planar amides are usually transitional structures that are involved in amide bond rotation (Scheme 1 (**1**)) and inversion of the nitrogen atom (Scheme 1 (**2**)). However, even in ground-minimum structures, amide distortion can be caused by several different factors, as illustrated in Figure 1A lactam ring strain of the nitrogen atom at a bridgehead position [5–8]; Figure 1B steric repulsion between substituents at the carbonyl and nitrogen positions [9]; Figure 1C angle strain at the nitrogen position [10]; Figure 1D bulkiness of substituents at the nitrogen position [11]; Figure 1E anomeric effect [12]; Figure 1F 1,3-allylic strain with respect to the pseudo C-N double bond [4,13]. These compounds are examples of ground-minimum non-planar amides. One of the most significant consequences of losing planarity of amides is an increase in lability: Reduction of the amide resonance exposes the carbonyl functionality to nucleophilic attack and acyl transfer reaction. In particular, hydrolysis by water under both acidic and basic conditions, and even under neutral conditions, is greatly accelerated when planarity is disrupted [5–9].

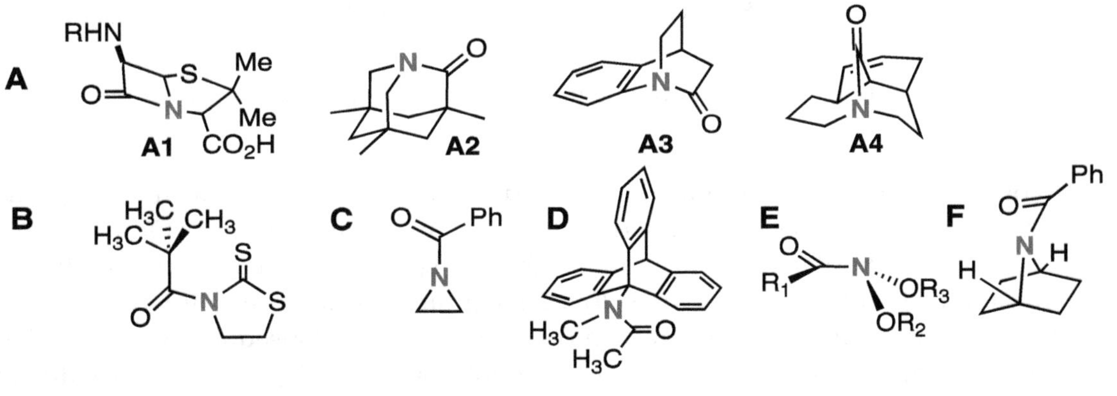

Figure 1. Some examples of non-planar amides.

The torsional angle ($\tau$) (the mean twisting angle around the C-N bond, see Figure 2) for completely planar amides is 0.0°. It is clear that the $\tau$ angle of ground-minimum non-planar amides can adopt values different from zero. Stable ground-state *N*,*N*-disubstituted tertiary amides, such as benzamide derivatives (**1a–1j**) can also take non-zero $\tau$ values (Figure 2), as their calculated structures show distortions from planarity [14,15]. Some of them (**1c**, **1l** and **1j**) are activated for facile cleavage of the amide C-N bond in the presence of various catalysts [16–19].

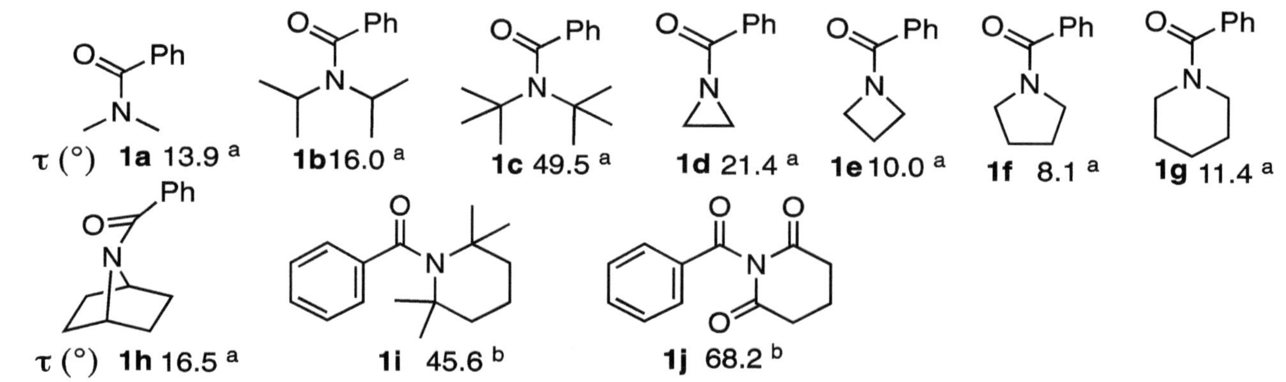

**Figure 2.** Distortion angles (τ) of ground-state-stable benzoyl amides. [a] B3LYP/6-31 + G(d) level of theory (Reference [14]). [b] X-ray data (Reference [15]). Calculated dihedral angle τ. $\tau = (\omega_1 + \omega_2)/2$ ($\omega_1$ = ∠R-N-C-O and $\omega_2$ = ∠R-N-C-R′) [14].

Brown proposed a close relationship between nitrogen pyramidalization, C-N bond length and kinetic reactivity to hydroxy anion attack [7], based on a comparison of the hydrolysis kinetics of analogous planar (molecule **1k** in Figure 3) and non-planar amides (**A3** in Figure 3). The base-catalyzed hydrolysis reaction at 25 °C showed a striking activation by 7 orders of magnitude in passing from the planar to the non-planar structure (see values in Figure 3).

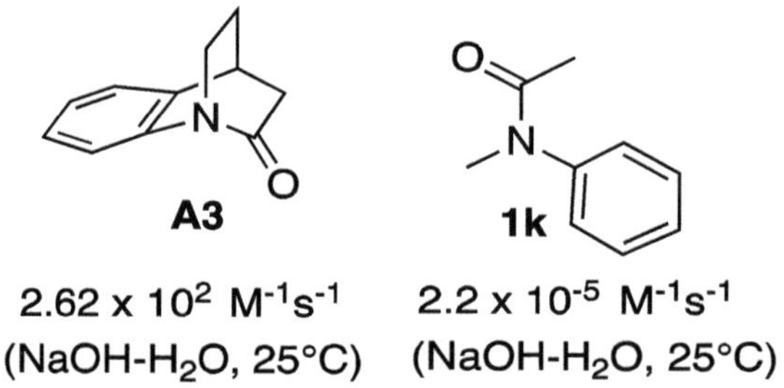

**Figure 3.** Model molecules used by Brown in his comparative kinetics study.

Because of the lability of most non-planar amides in the presence of water [5–9], there have been few applications of these scaffolds (Figure 1) in molecular design. On the other hand, while 7-azabicyclo[2.2.1]heptane amides are highly suspicious of chemical stability, due to nitrogen-pyramidalization [12,13], 7-azabicyclo[2.2.1]heptane amides are of interest, beccause they can be regarded as conformationally constrained β-proline mimics. Consequently, several derivatives have been synthesized, and helical structures of homooligomers of β-proline mimics derived from azabicyclo[2.2.1]heptane amide have been reported [20,21]. The helical structures were stable even in the absence of intramolecular hydrogen bonds [22,23]. By introducing suitable bridgehead substituents, either all-*cis* amide or all-*trans* amide conformations were obtained. Conformational control favoring the *cis*-isomer was achieved by introducing substituents at the C-4 bridgehead position (Figure 4(a)). The *cis*-amide structure is heat-stable and the helical structure remains intact in a variety of solvents (water, alcohol, halogenated solvents and cyclohexane) [24]. On the other hand, conformational control favoring the *trans*-isomer was achieved by introducing substituents at the C-1 bridgehead position (Figure 4(b)). The *trans*-amide structure also proved to be heat-stable and the helical structure remained intact in both hydrophilic and hydrophobic solvents [25].

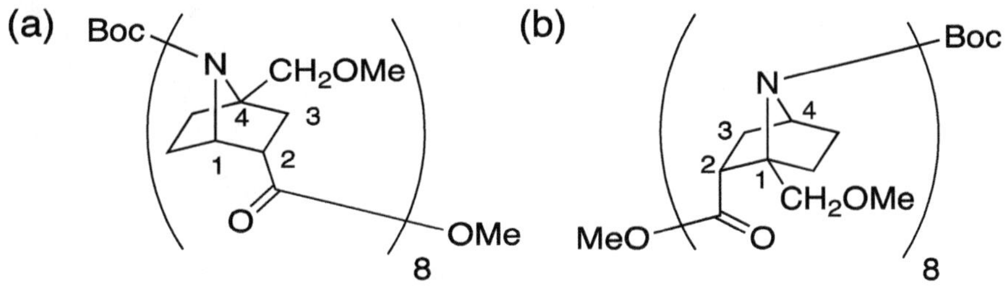

**Figure 4.** (**a**) Schematic representation of 7-azabicyclo[2.2.1]heptane amide *cis*-(*S*)-8mer. (**b**) Schematic representation of the *trans*-(*R*)-8mer.

In the synthesis of oligomers of 7-azabicyclo[2.2.1]heptane amides (Figure 4) [24,25], acid-catalyzed deprotection of the Boc group was compatible with the amide linkage (Figure 5). Therefore, we thought that the bicyclic amide linkage might be stable under acidic conditions and convectional mild reaction conditions. However, to our knowledge, neither qualitative nor quantitative data about the base-catalyzed hydrolytic reactivity of this system have been reported.

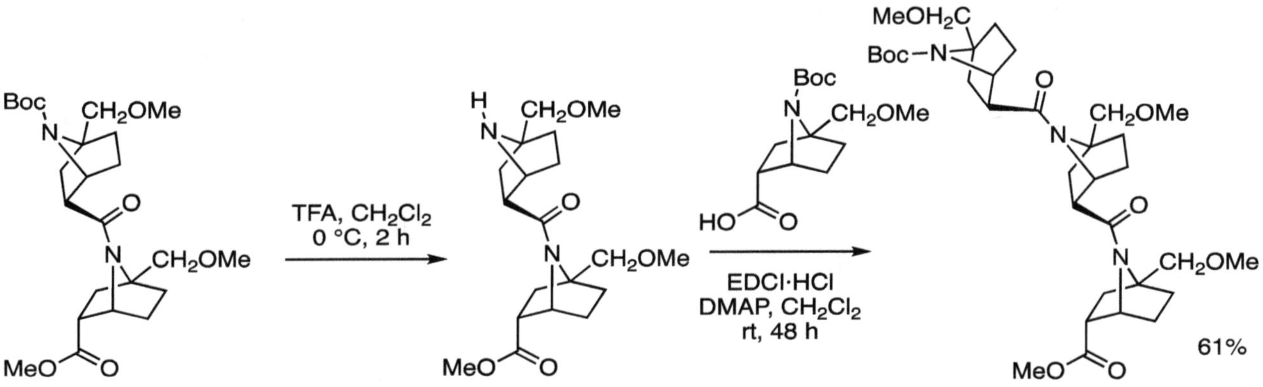

**Figure 5.** Acidic deprotection procedure in the synthesis of the homooligomers.

Therefore, the aim of the present study is to establish the reactivity profile in the base-catalyzed hydrolysis of this 7-azabicyclo[2.2.1]heptane amide system, which might serve as a model for the enzymatic cleavage of peptide bonds. To this end, kinetic studies of the base-catalyzed hydrolysis of the amide of 7-azabicyclo[2.2.1]heptane benzamides were conducted and the results were compared with reported data for related monocyclic amide compounds. Theoretical calculations were also carried out to aid in understanding the unexpectedly low reactivity.

## 2. Results and Discussion

In order to estimate the strength of amide bonding, we compared the reactivities of planar and non-planar amides, specifically pyrrolidine amides (**3a–e**) and 7-azabicyclo[2.2.1]heptane amides (**4a–e**) (Figure 6). We also evaluated the effect of introducing substituents on the bridgehead position of the bicycle (**5a–e**). We utilized azetidine amides (**2a–e**) (Figure 6) as reference compounds for non-planar cyclic amides. In general, we found that the base-catalyzed hydrolysis of the bicyclic amides (**4** and **5**) was rather slow. Among aromatic substituents, we focused on **a** (H), **b** (Cl) and **c** ($NO_2$) (Figure 6), for which the reaction proceeds at acceptable speed, since the reactions in the cases of substituents **d** (Me) and **e** (MeO) are too slow to obtain kinetic data by means of NMR (see below).

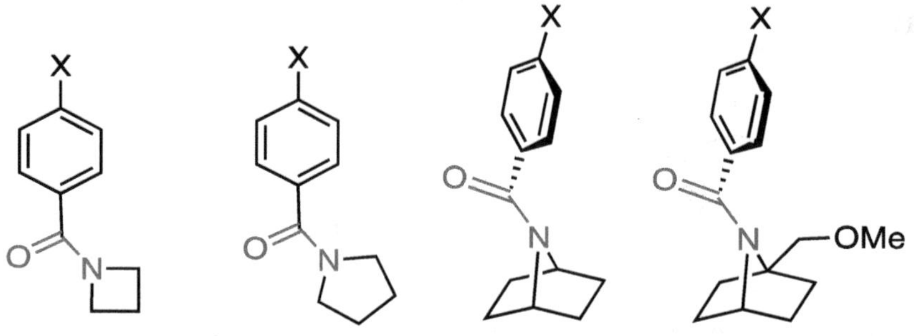

**2a-e** **3a-e** **4a-e** **5a-c**

**a: X=H; b: X=Cl; c: X=$NO_2$; d X=Me, e: X=MeO**

**Figure 6.** Model molecules used in this study. *N*-Benzoylazetidines **(2a–e)** and *N*-benzoylpyrrolidines **(3a–e)** were examined as monocyclic amides. *N*-Benzoyl-7-azabicyclo[2.2.1]heptanes (**4a–e**) and *N*-benzoyl-1-(methoxymethyl)-7-azabicyclo- [2.2.1]heptanes (**5a–c**) were examined as bicyclic amides.

## *2.1. Synthesis*

The monocyclic *N*-benzoylazetidines **2a–e** and *N*-benzoylpyrrolidines **3a–e** were synthesized in a straightforward manner by coupling the corresponding amines with different benzoyl chlorides (Scheme 2). For the synthesis of the azetidine compounds, the chloride salt of the amine was used as the starting material, with 3 equivalents of DIPEA (diispropylethylamine). For the synthesis of pyrrolidine compounds, 1.2 equivalents of DIPEA sufficed. Compounds were obtained in good yields.

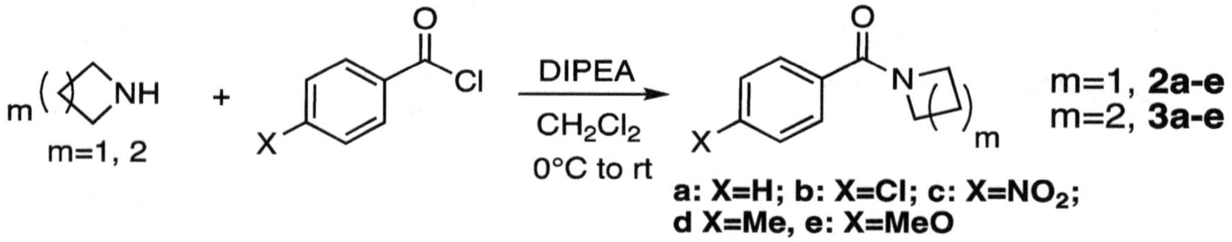

**Scheme 2.** Synthesis of monocyclic amides **2** and **3**.

In addition to the monocyclic amides, we synthesized unsubstituted (**4a–e**) and substituted (**5a–e**) bicyclic amides. *N*-Benzoyl-7-azabicyclo[2.2.1]heptanes were synthesized starting from *trans*-4-aminocyclohexanol (Scheme 3) [4]. The primary amine was substituted by benzyloxycarbonyl chloride (ZCl), and the hydroxy group was changed to toluenesulfonate in order to facilitate bicycle formation. Coupling of benzoyl chloride or *para*-substituted benzoyl chlorides gave the bicyclic amides (**4a–e**).

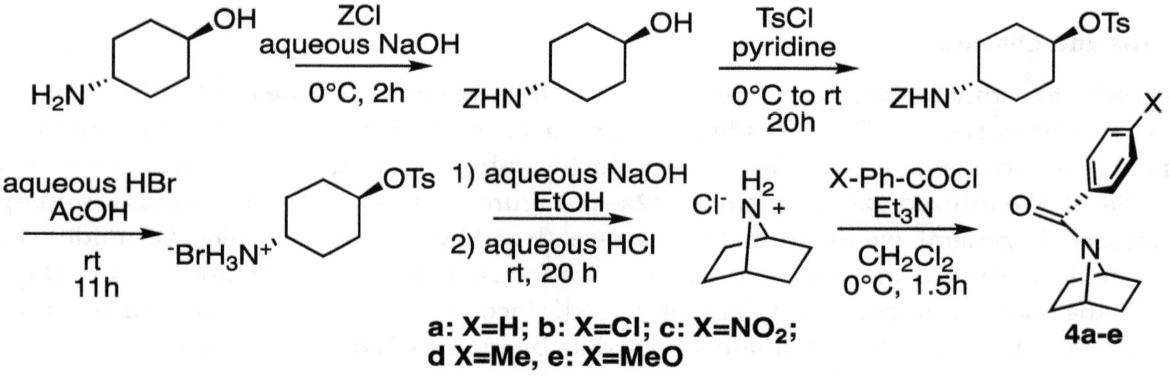

**Scheme 3.** Synthesis of bridgehead-unsubstituted bicyclic amides **4**.

A different strategy was followed for the bridgehead-substituted bicyclic amides **5a–c** (Scheme 4). The hydroxy group was removed from the previously synthesized monomer scaffold by Barton–McCombie deoxygenation using AIBN and tris(trimethylsilyl)silane (TTMSS) [26]. After that, the bridgehead ester functionality was first reduced to alcohol and then changed to ether. After Boc-deprotection, the compounds were coupled with various *para*-substituted benzoyl chlorides to afford the bridgehead-substituted bicyclic amides.

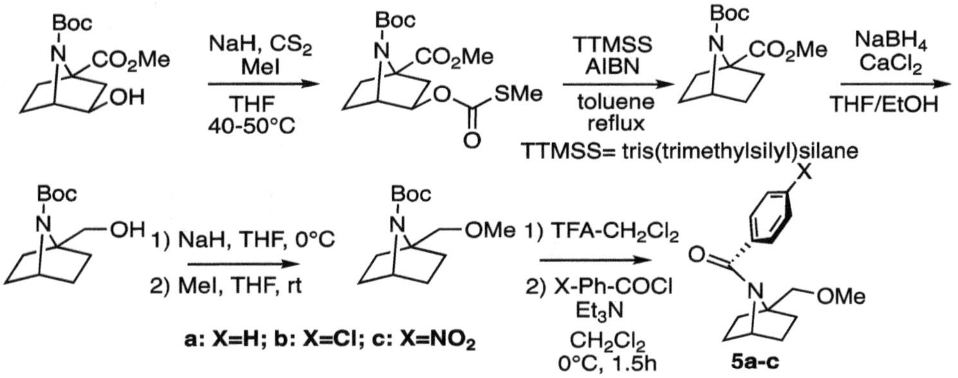

**Scheme 4.** Synthesis of bridgehead-substituted bicyclic amides **5**.

## 2.2. *Alkaline Hydrolysis of Planar Amide 3a*

In order to assess the chemical reactivities of the non-planar 7-azabicyclo[2.2.1]heptane amides **4** and **5** in alkaline conditions, we first examined the kinetics of the planar amide N-benzoyl pyrrolidine **3a** (X=H) in order to optimize the reaction conditions, because the hydrolysis of **3a** is expected to be the slowest among these compounds (**2**, **3**, **4** and **5** in Figure 5).

### 2.2.1. Optimization of Reaction Conditions

Given that hydrolysis involves working with water as a solvent, it was necessary to confirm the solubility of the reactants. In order to carry out the reactions, water-miscible co-solvents had to be chosen. In addition, since alkaline conditions entail high concentrations of hydroxide, some solvents (such as ketones or acetonitrile) are unsuitable. The list of possible co-solvents was narrowed down to 1,4-dioxane, methanol, THF and DMSO.

The first attempts at hydrolysis were conducted with 0.15 mmol (30 mg) of N-benzoylpyrrolidine **3a**, 100 μL of deuterated methanol (as a co-solvent) and 400 μL of a solution of NaOD in $D_2O$ (40 *w/w* %). The procedure was also done using 100 μL of deuterated 1,4-dioxane(1,4-dioxane-$d_8$). The samples were heated in a water bath at 37 °C and subjected to TLC. NMR spectra were recorded after 24 and 48 h. However, no hydrolysis product was detected by $^1$H-NMR, and no new product appeared on TLC. Thus, the hydrolysis reaction did not proceed at 37 °C. Furthermore, the NaOD solution and the 1,4-dioxane solution separated into two phases.

Next, the concentration of the base was reduced to 0.4 M and that of the reactant to 0.1 M (i.e., a 4-fold excess of base over starting material). The total reaction volume was 1 mL (100 μL of co-solvent and 900 μL of $D_2O$).

Heating at 50 or 70 °C was applied, and the hydrolysis of **3a** was monitored by TLC analysis. The reaction proceeded at 70 °C. The starting material was no longer detectable after 48 h, and a single spot corresponding to the hydrolysis product (benzoic acid) appeared. However, at 50 °C the starting amide **3a** was still detectable on the TLC plate after 48 h. In order to assess the working range for the other compounds, a similar test was done at 70 °C for *p*-nitro derivative **3c** (no spot of the starting material was detected after 5 h) and *p*-methoxy derivative **3e** (the starting material was still detected after 65 h). Moreover, at the higher temperature, the 1,4-dioxane solution remained monophasic. Therefore, it was decided to work at 70 °C.

In order to follow the progression of the reaction quantitatively, we recorded NMR spectra of hydrolysis reaction mixtures of **3a–e** every two min at 70 °C. Rate constants were calculated from the decrease of the integrals of the reactant (Figure 7).

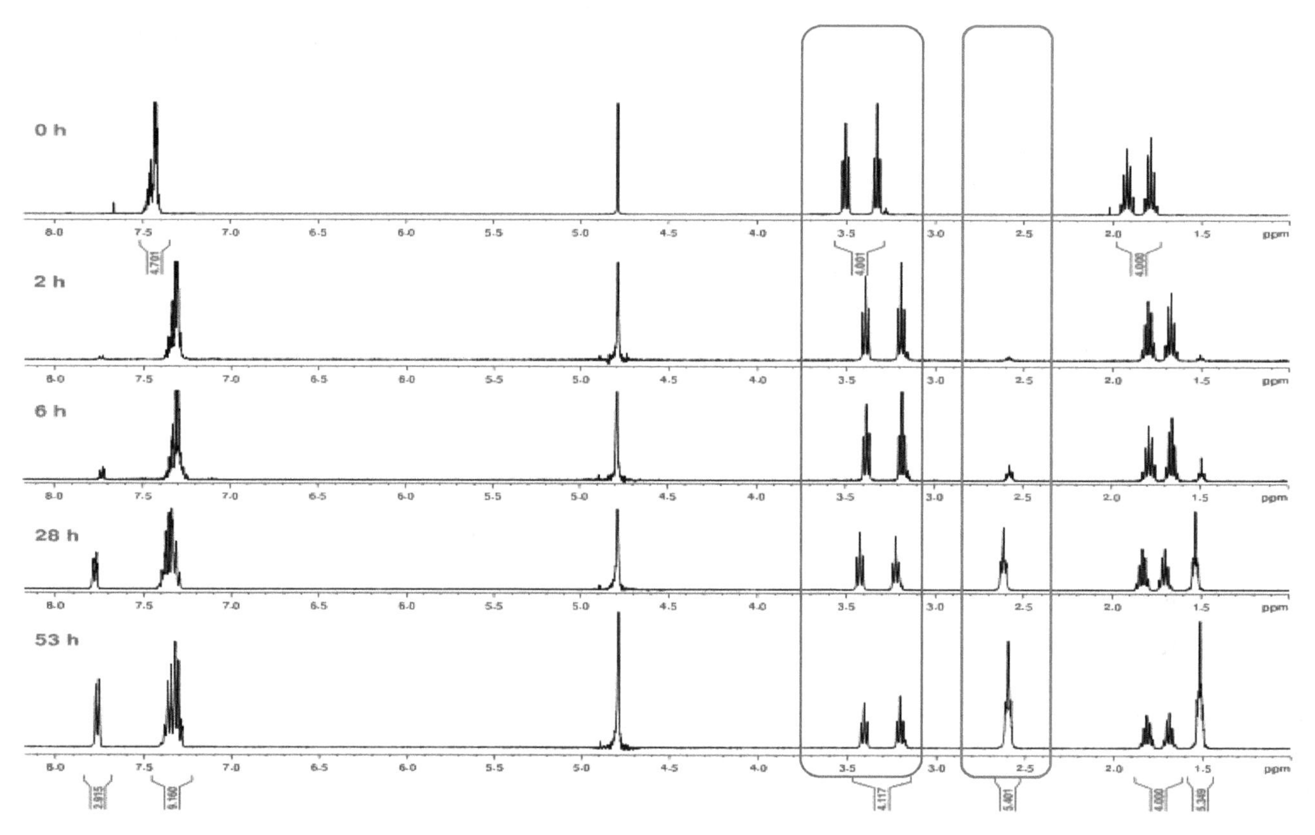


**Figure 7.** $^1$H-NMR monitoring of the hydrolysis of **3a**, with NaOH in $D_2O$ and 1,4-dioxane-$d_8$ at 70 °C. The intensity of amide peaks decreased (red box) over time, and product signals appeared (blue box).

From the $^1$H-NMR integration information we were able to determine the loss of amide over time. Since working with excess deuteroxide anion guarantees pseudo-first-order kinetics, rate constants were calculated using the following first-order equation:

$$\frac{d[\text{Amide}]}{dt} = -k[Amide]. \quad (1)$$

The integration of this rate equation gives the following equation:

$$\ln[\text{Amide}] = -kt + \ln[\text{Amide}]_0 \quad (2)$$

where [Amide] represents the molar concentration (M) of amide, $k$ represents the reaction constant ($s^{-1}$), and $\boldsymbol{t}$ represents time (**s**). A least-squares plot of the natural logarithm of amide concentration *versus* time gave a straight line whose slope equals $-k$. The initial amide concentration corresponds to the value of the y-intercept.

When either the aromatic protons or the pyrrolidine protons were used as a reference for signal integration, all five pyrrolidine amides **3a–e** showed good correlations between concentration and time, and first-order reaction rate constants could be determined (Figure S1). The regression coefficients $R^2$ were high for all five compounds **3a–e** (Figure S1). The reaction showed Hammett-like behaviour, that is the hydrolysis proceeded faster when an electron-withdrawing group was present at the *para* position of the phenyl ring, and slower when an electron-donating group was present.

Based on these results, we next examined, the hydrolysis of the bicyclic compounds under the same conditions. In order to hydrolyze compounds **4a** and **4b** it was necessary to increase the proportion of 1,4-dioxane from 10% to 20%. The reaction proceeded smoothly, and the disappearance of the reactants was successfully monitored by NMR. Unfortunately, the bicyclic compounds were not sufficiently soluble under these conditions. Hence, we decided to increase the proportion of co-solvent. The reaction volume was also scaled down from 1 mL to 500 μL. Solvent conditions were modified on the basis of an examination of the hydrolysis of the *p*-$NO_2$-substituted pyrrolidine benzamide **3c** (Table 1).

**Table 1.** Effect of variations of co-solvent proportions on the hydrolysis rate of **3c**.

| 1,4-Dioxane-$d_8$/$D_2O$ ($v/v$%) | 3c | NaOD | Temperature | $-k_{obs}$ ($s^{-1}$) |
|---|---|---|---|---|
| 25/75 | 0.05 mmol | 0.5 mmol | 70 °C | $3.0 \times 10^{-4}$ |
| 50/50 | 0.05 mmol | 0.5 mmol | 70 °C | $9.6 \times 10^{-5}$ |
| 75/25 | 0.05 mmol | 0.5 mmol | 70 °C | $2.3 \times 10^{-6}$ |

The sample containing 75% 1,4-dioxane did not form a homogenous solution even after heating at 70 °C, possibly due to the high NaOD concentration in the water phase. On the other hand, the use of 25% 1,4-dioxane resulted in a low $R^2$ value (0.943). Despite these setbacks, it was seen that the reaction proceeds faster in more polar solvent systems. This trend was also seen with other co-solvents (Figure 8). The reaction time was shorter in DMSO-$d_6$ than in methanol-$d_4$, which in turn was shorter than in 1,4-dioxane. The reaction was also carried out in THF-$d_8$, but the compound was not sufficiently soluble even at high temperature. The effect of solvent polarity on the hydrolysis rate can be explained by the fact that the amide bond has a polar nature, and charges develop as the bond is broken. Therefore, more polar solvent systems are better at stabilizing the developing charges in the transition states and the products. Although DMSO is a good solvent, the presence of hydroxide anion can produce the basic dimsyl anion ($Na^{+}\,{}^{-}CH_2$-SO-$CH_3$) from DMSO. Therefore, we focused on 1,4-dioxane and methanol, rather than DMSO.

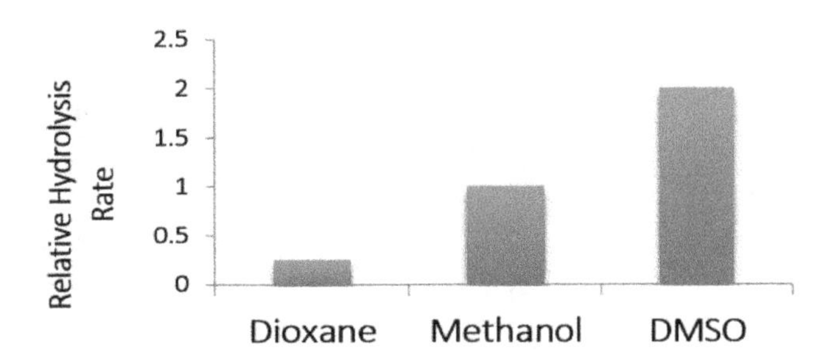


**Figure 8.** Effect of polarity of the solvent system on the rate of alkaline hydrolysis. Higher polarity of the solvent system accelerates the reaction. The rate of hydrolysis in methanol as a solvent was arbitrarily set at unity (1.0).

### 2.2.2. Alkaline Hydrolysis of Amides in Two Solvents

Finally, the set of conditions, shown in Table 2, was selected for all compounds: 250 μL of co-solvent (1,4-dioxane-$d_8$ or methanol-$d_4$), and 250 μL of water (total volume (500 μL)), and 10 equivalents of base with respect to the reactant amide. The reaction was carried out at 70 °C. These conditions provided first-order kinetics with respect to amide concentration.

**Table 2.** Final conditions for alkaline hydrolysis of amide compounds.

| Starting Amide | NaOD 40% wt. | $D_2O$ | Co-solvent | Temperature |
|---|---|---|---|---|
| 0.05 mmol | 0.5 mmol | 250 μL | 250 μL | 70 °C |

### 2.2.3. Alkaline Hydrolysis in Dioxane

Several compounds were subjected to NaOD-catalyzed hydrolysis in 1,4-dioxane-$d_8$-$D_2O$ (1:1) under the conditions, shown in Table 3. The co-solvent was 1,4-dioxane-$d_8$. The plots in Figure S2 are based on the raw data of selected hydrolysis experiments (Figure 9). Hydrolysis was repeated three times for some of the compounds in order to assess the repeatability of the method. It was found that the error when 1,4-dioxane-$d_8$ was used as the co-solvent was ±13.8%. Products (carboxylate and amine) were identified by mass spectrometry. NMR monitoring revealed signals corresponding to the reactant and the hydrolysis products in all cases. The ring-opening product of azetidine amide was not detected. As a general trend, base-catalyzed hydrolysis of azetidine amides (**2**) proceeded more rapidly than that of pyrrolidine amides (**3**), which in turn were hydrolyzed faster than unsubstituted bicyclic amides (**4**), while bridgehead-substituted bicycles (**5**) were least reactive (for example, reaction rate: **2a** > **3a** > **4a** > **5a**; **2b** > **3b** > **4b** > **5b**; **2c** > **3c** > **4c** > **5c**). The expected Hammett-like trend was observed: The electron-withdrawing substituent $NO_2$ (**c**) on the phenyl moiety accelerated the reaction.

**Table 3.** Base-catalyzed hydrolysis rates ($-k_{obs}$ in $M^{-1}s^{-1}$) of amides **2x**–**5x** (**x** = **a**–**e**) in two solvent systems.[a] Average values are shown where possible (see footnotes). Relative reaction rates (referenced to **4**) are shown in parentheses.

| | 1,4-Dioxane-$D_2O$ (1:1), NaOD, 70 °C [a] | | | | Methanol-$D_2O$ (1:1), NaOD, 70 °C [b] | | | |
|---|---|---|---|---|---|---|---|---|
| x= | 2x | 3x | 4x | 5x | 2x | 3x | 4x | 5x |
| a (H) | $2.1 \times 10^{-5}$ (5.8) | $7.8 \times 10^{-6}$ (2.2) | $3.6 \times 10^{-6}$ (1) | $6.5 \times 10^{-7}$ (0.2) | $1.1 \times 10^{-4}$ | $2.9 \times 10^{-5}$ | ND | ND |
| b (Cl) | $2.3 \times 10^{-4}$ (82.1) | $1.8 \times 10^{-5}$ (6.4) | $2.8 \times 10^{-6}$ (1) | $1.2 \times 10^{-6}$ (0.4) | $9.7 \times 10^{-4}$ (88.2) | $7.0 \times 10^{-5}$ (6.4) | $1.1 \times 10^{-5}$(1) | $3.0 \times 10^{-6}$ (0.3) |
| c ($NO_2$) | $9.9 \times 10^{-4}$ (33.0) | $1.0 \times 10^{-4}$ (3.3) | $3.0 \times 10^{-5}$ (1) | $3.7 \times 10^{-5}$ (1.2) | $6.5 \times 10^{-3}$ (92.9) | $4.2 \times 10^{-4}$ (6.0) | $7.0 \times 10^{-5}$(1) | $4.3 \times 10^{-5}$ (0.6) |
| d (Me) | ND | $4.1 \times 10^{-6}$ | NA | NE | ND | $1.7 \times 10^{-5}$ (4.2) | $4.1 \times 10^{-6}$(1) | NE |
| e (MeO) | ND | $4.8 \times 10^{-6}$ | NE | NE | ND | $1.9 \times 10^{-5}$ | NE | NE |

[a] Error estimation: ±13.8% (in the 1,4-dioxane system). (**b**) Error estimation: ±17.4% (in the methanol system). ND = not determined; NA = not available (due to the solubility problem); NE = not executable (due to very slow reaction).

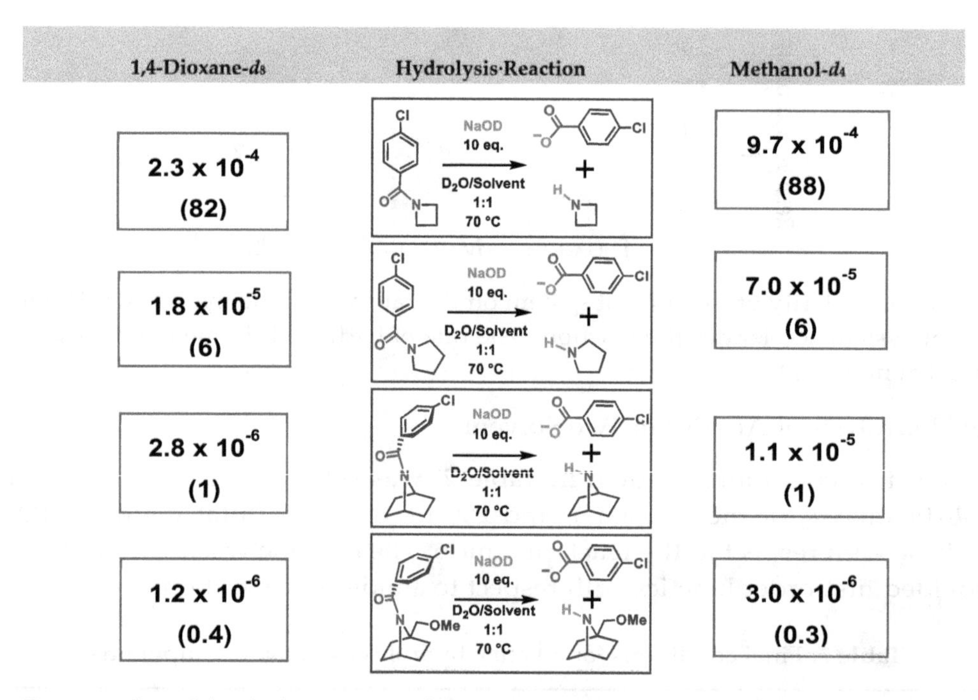


**Figure 9.** Base-catalyzed hydrolysis rates of **2c**, **3c**, **4c** and **5c** in two solvent systems (1,4-dioxane and methanol). Values of $-k_{obs}$ ($s^{-1}$) are shown. Relative reaction rates are shown in parentheses (referenced to **4c**).

For example, *N*-(*p*-nitrobenzoyl-7-azabicyclo[2.2.1]heptane (**4c**) was hydrolyzed in 1,4-dioxane and $D_2O$ at a slower rate than the analogous planar monocycle **2c** (6 times faster than **3c**) or the non-planar monocycle **1c** (82–88 times faster than **3c**). Moreover, a bridgehead substituent (**4c**) further slowed the hydrolysis rate (2.3–3.6 times slower than **3c**) (Figure 9).

### 2.2.4. Alkaline Hydrolysis in Methanol

Several compounds were subjected to hydrolysis in methanol under the final conditions, shown in Table 3 (methanol-$d_4$-$D_2O$ 1:1). The co-solvent used in this case was methanol-$d_4$. The following plot shows the $^1$H-NMR spectral change corresponding to the slow consumption of **5b** in methanol (Figure 10). We could not detect intermediate formation of the methyl ester, which may be formed by the attack of methoxide anion on the amide. Hydrolysis was repeated three times for some of the compounds in order to assess the repeatability of the method. It was found that the error of the method when using methanol-$d_4$ as a co-solvent was $\pm$17.4%. Products were identified by mass spectrometry. Signals corresponding to reactants and hydrolysis products were identified in all cases. The ring-opening product of azetidine amide was not detected. Reactivity followed the same trend as in 1,4-dioxane (reaction rate: **2b** > **3b** > **4b**> **5b**). The reactivity was higher in methanol than in 1,4-dioxane.

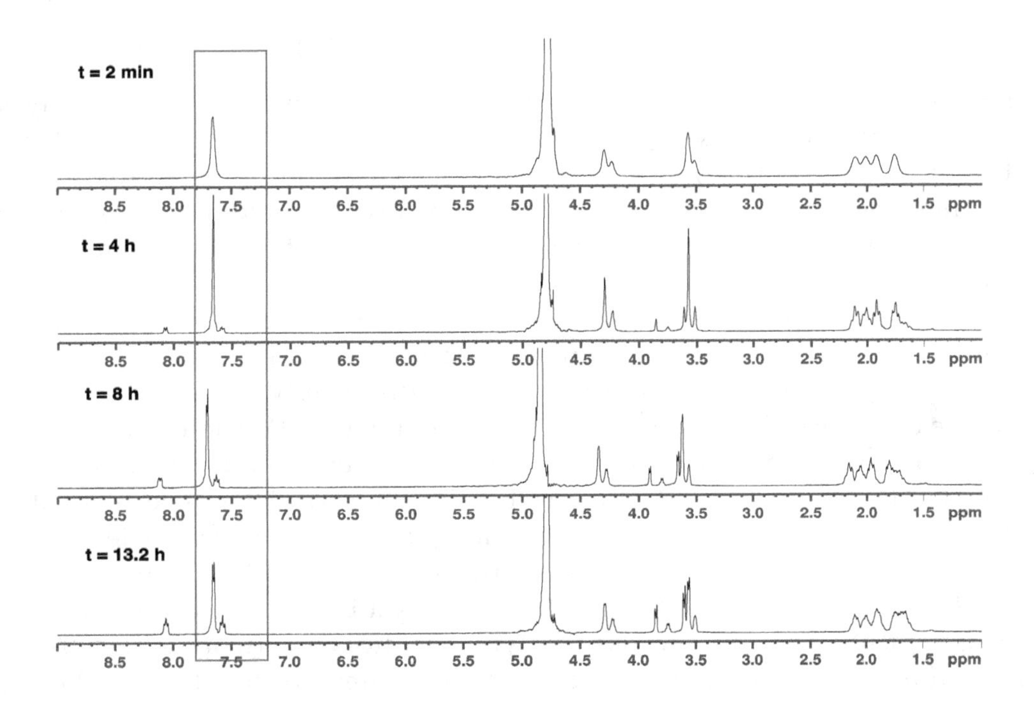


**Figure 10.** Progress of the hydrolysis of **5b** in methanol-$d_4$/$D_2O$ (1:1) at 70 °C, followed by $^1$H-NMR; the intensity of amide peaks decreased, and new product signals appeared as the reaction progressed (red box).

### 2.2.5. Comparison of Kinetic Data

As shown in Table 3, the same general trend was observed irrespective of the solvent system employed. Azetidine amides (**2**) were the most reactive, followed by pyrrolidine amides (**3**), then unsubstituted bicyclic amides (**4**), and finally bridgehead-substituted bicycles (**5**) (reaction rate: **2b** > **3b** > **4b**> **5b**). Phenyl substitution had the expected effects (according to the inductive effects) in all series of amides.

## *2.3. Computational Studies*

### 2.3.1. Reaction Model

It was unexpected to find that non-planar amides based on the 7-azabicyclo[2.2.1]heptane scaffold (**4**) showed such poor susceptibility to base-catalyzed hydrolysis, even upon heating, as compared with the corresponding monocyclic amides (**3**). Bridgehead substitution of the 7-azabicyclo[2.2.1]heptane amides (**5**) also decelerated base-catalyzed hydrolysis of the amide as compared with the unsubstituted bicyclic derivative (**4**). It has been established in previous studies on heavy atom isotope effects that formation of the tetrahedral intermediate is rate-determining in the base-catalyzed hydrolysis of formamide ($HCONH_2$) (Scheme 5) [27].

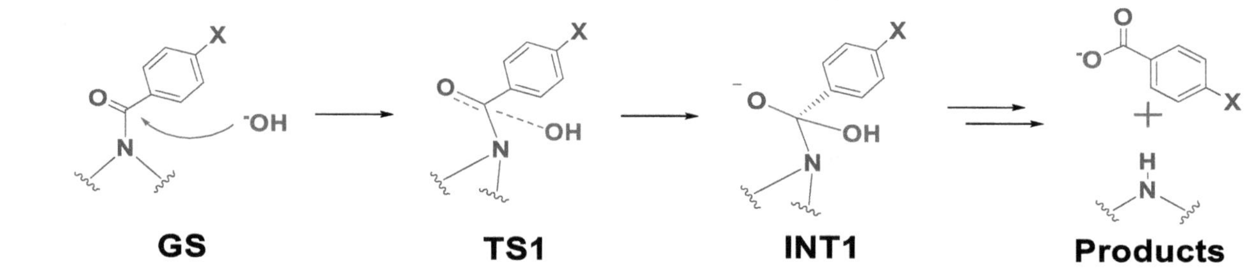

**Scheme 5.** General reaction path for the base-catalyzed hydrolysis of amides.

While there have been several ab initio and DFT calculation studies of the hydrolysis of planar amides and non-planar amides [28–31], base-catalyzed hydrolysis of amides has been relatively little studied until recently, Further, most studies have focused on rather simple amides, such as formamide, N-methylacetamide, DMF, and DMA (dimethylacetamide) [32]. Here, we aimed to rationalize the observed reactivity trends by computational studies of our more complex reactants and transition structures.

It is known that that explicit water solvation is crucial for the calculation of amide hydrolysis. Xiong and Zhan [32] showed that incorporation of five implicit water molecules is required, and there are two kinds of hydrogen-bonding networks of water in the vicinity of the hydroxy ion ($^{-}OH$) and amide group in the presence of five explicit water molecules (Figure 11). These two patterns commonly involve activation of the carbonyl group by hydrogen-bonding of two water molecules to the oxygen atom (increasing its electrophilicity) and hydrogen bonding of three water molecules with the oxygen atom of the hydroxy anion (decreasing its nucleophilicity and at the same time decreasing electronic repulsions). There is a difference in the topology of the hydrogen-bonding networks. Type a is more stable than Type b by approximately 1-2 kcal/mol, but finding a TS of Type a in our experimental amide systems **2–5** was difficult, probably because the Type a hydrogen network is sensitive to the steric interactions encountered in more complicated amide structures (see the detail in the Experimental Section). Therefore, in the present work, we focused on the hydrogen network Type b, which seems relevant to the present compounds.

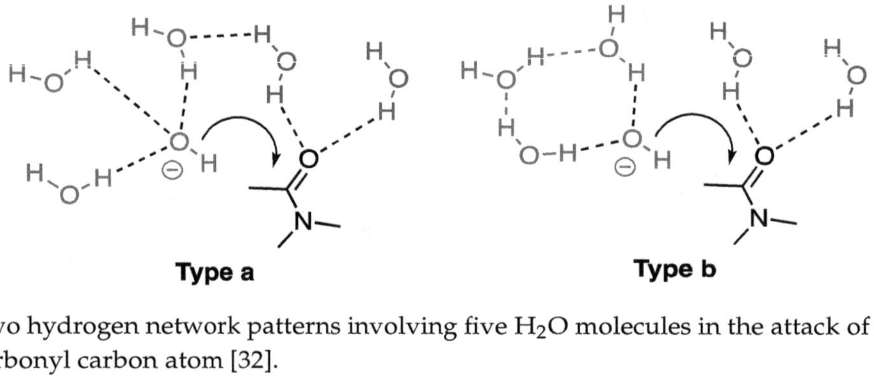

**Figure 11.** Two hydrogen network patterns involving five $H_2O$ molecules in the attack of OH anion on the amide carbonyl carbon atom [32].

Geometry optimizations for the ground states of hydrated reactants (amide and hydroxide anion) and the transition state for the nucleophilic addition of hydroxide anion to the amide carbonyl group were performed in the presence of explicit water molecules at the B3LYP and M06-2X levels of theory with a combination of two basis sets, 6-31+G(d) and 6-311++G(d, p). Bulk solvation effects (self-consistent reaction field, SCRF) were also incorporated by means of IEFPCM (Polarizable Continuum Model, PCM, using the integral equation formalism variant) and SMD methods in water. Vibrational frequency calculations were performed at the same level of theory. The energies were corrected for the zero-point energies and Gibbs free energy at 25 °C (298.15 K), obtained from frequency calculations. Hereafter, we will focus on the calculation values based on M06-2X level with the SMD solvent model. The calculations at the B3LYP level (see Tables S1–S4) were consistent with the trends obtained in the M06-2X calculations.

2.3.2. Model with Five Explicit Water Molecules

Transition structures (TSs) for the hydroxide anion-catalyzed hydrolysis of the amides, including five explicit water molecules were identified with M06-2X//6-31+G(d) (Figure 12). As previously described by Xiong and Zhan [32], we calculated the activation Gibbs energies from the energy difference between the TS structures and the summation of the energies of the two reagents, the amide with two hydrogen-bonded water molecules (amide$(H_2O)_2$) and the hydroxide anion clustered with three water molecules ($^-OH(H_2O)_3$) (Scheme 6). We also estimated Gibbs free energies by single-point calculation with M06-2X//6-311++G(d,p) on the basis of the M06-2X//6-31+G(d) optimized structures (Table 4).

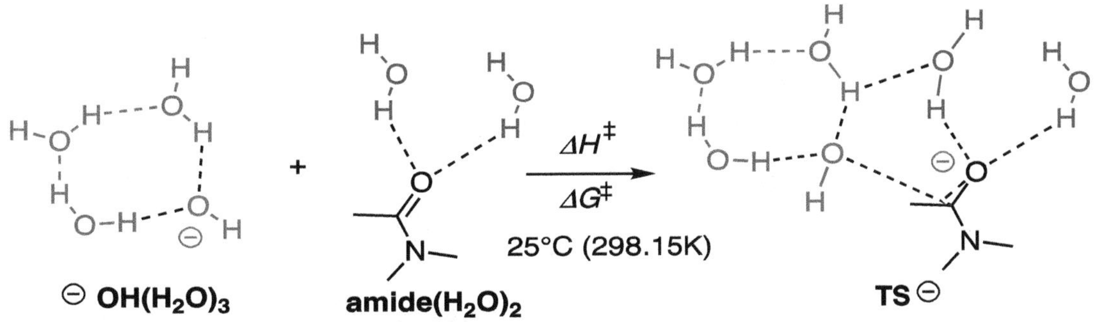

**Scheme 6.** Model of base-catalyzed hydrolysis reaction of amide.

The contribution of thermal corrections (entropy term, $-T\Delta S^{\ddagger}$) was significant. The entropy term ($\Delta S^{\ddagger}$) took a large negative value (Table 2), indicating the presence of significant order in the transition structure. Associated water molecules need to rearrange on the surface of the amide, and thus would contribute to this large negative entropy term. A larger free-energy activation barrier was seen for *N*-benzoyl-7-azabicyclo[2.2.1]heptane (**4a**) than for the monocyclic *N*-benzoylpyrrolidine (**3a**). The pyramidal amide, the azetidine derivative **2a** has the smallest activation energy. The order of the magnitude of the Gibbs activation energies (**4a > 3a, 5a > 2a**, Table 4) is essentially consistent with the experimental reactivity (Figure 8 and Table 3), with the exception of the bridgehead-substituted bicyclic amide **5a**, which is expected to have higher activation energy than **4a**. However, we need to make allowance for the simple harmonic oscillator approximation in the thermal energy correction, and also we need to consider that this thermal energy correction is an approximation of the real entropy change in the solvation process.

The enthalpy terms ($\Delta H^{\ddagger}$) were underestimated (Table 4), but the order of their magnitudes is also essentially consistent with the experimentally observed hydrolysis rates: **2a** > **3a, 4a, 5a**. This trend is consistent with the trajectory in the TS structures (Figure 12): The shorter the distance between the amide carbonyl carbon atom and hydroxide oxygen atom (i.e., later the TS), the larger the enthalpy term ($\Delta H^{\ddagger}$).

**Table 4.** Calculated free energy barrier for the formation of the transition state at 25 °C (298.15 K) at the M06-2X/6-31+G(d) and M06-2X/6-311++G(d, p) levels, considering five water molecules. Solvent effect is SMD (solvent = water) [a].

| Compound | $\Delta H^{\ddagger}$ kcal/mol | $-T\Delta S^{\ddagger}$ kcal/mol | $\Delta S^{\ddagger}$ cal/(mol·K) | $\Delta G^{\ddagger}_{25\,°C}$ kcal/mol |
|---|---|---|---|---|
| | M06-2X/6-31+G(d) SMD=water [b] | | | |
| **2a** | 2.51 | +17.38 | −58.29 | 19.89 |
| **3a** | 5.03 | +17.03 | −57.12 | 22.06 |
| **4a** | 6.80 | +16.57 | −55.58 | 23.37 |
| **5a** | 4.66 | +17.95 | −60.20 | 22.61 |
| | M06-2X/6-311++G(d,p) SMD=water [c] | | | |
| **2a** | 4.26 | +16.16 | −54.20 | 20.42 |
| **3a** | 7.13 | +16.03 | −53.76 | 23.16 |
| **4a** | 8.80 | +15.76 | −52.86 | 24.56 |
| **5a** | 7.50 | +15.44 | −51.79 | 22.94 |

[a] Calculation and experimental Gibbs free energy of hydroxide anion-catalyzed hydrolysis of formamide was 21–22 kcal/mol [32]. [b] Full optimizations. [c] Single-point calculations.

Other reaction models, a model with four explicit water molecules and a model with implicit water molecules, were also examined and the order of the energy demand is consistent with that of the present five-water model (the results are described in Supporting Information).

Non-planar amides based on the 7-azabicyclo[2.2.1]heptane scaffold were found to be rather inert to base-catalyzed hydrolysis. The calculated Gibbs free energies are also consistent with the experimental results.

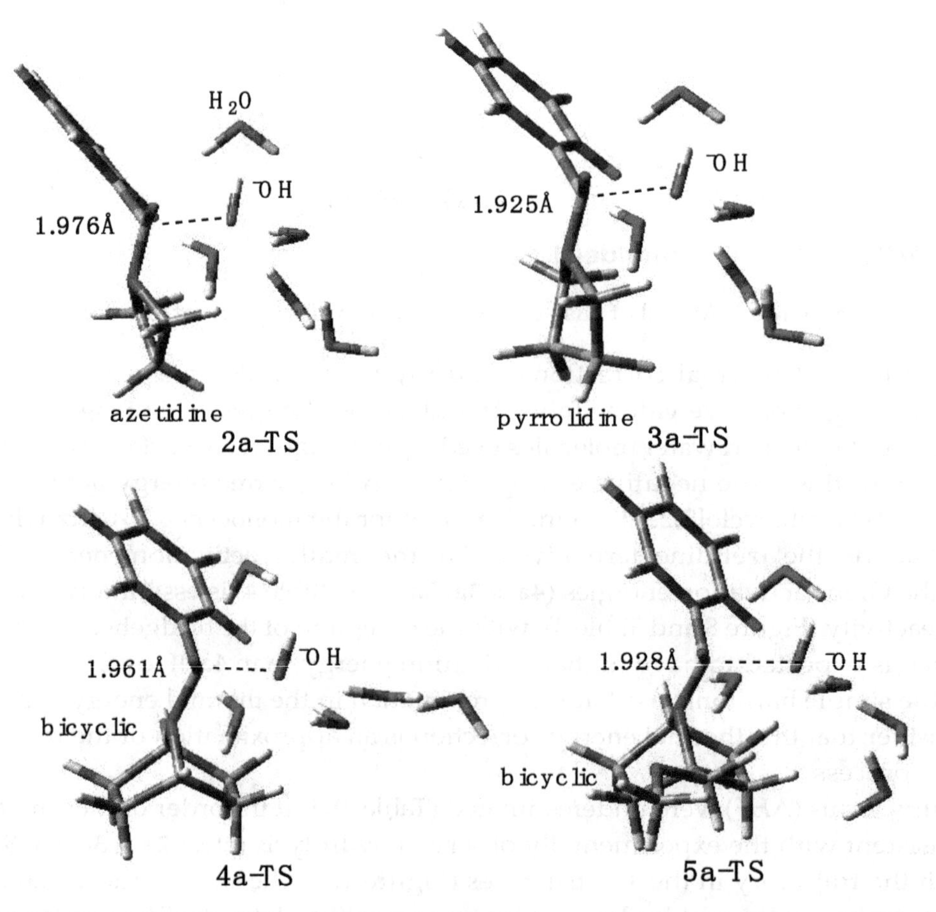


**Figure 12.** M06-2X/6-31+G(d)-optimized transition structures of hydroxide anion addition to the amide. Distance between the amide carbonyl carbon atom and hydroxide oxygen atom was shown.

A close scrutiny of the trajectory in the TS (Figure 12) revealed the optimal trajectory for the hydroxide anion attack on the carbonyl group accompanied with hydrating water molecules. The positions of the hydroxy anion were similar in the respective TS structures. In the bicyclic amides (Figure 13, left), the water molecules in the bridgehead-substituted **5a** were placed a little further from the amide substrate as compared with those in bridgehead-unsubstituted **4a**. A comparison of azetidine **2a** and bridgehead-unsubstituted **4a** also indicated that some of the hydrated waters were placed far from the substrate in the case of **4a** (Figure 13, right). Therefore, these water molecules cannot get close to the substrate, due to greater steric congestion in **4a**, as compared with the azetidine (**2a**). Hydrogen bonding of water molecules stabilized the developing negative charge during the attack of hydroxide anion on the amide carbonyl carbon atom, while the entropy cost compensates for the stabilization. Inefficient hydration is one of the possible reasons that could explain the increase of the activation energy of the bicyclic amides **4** and **5**.

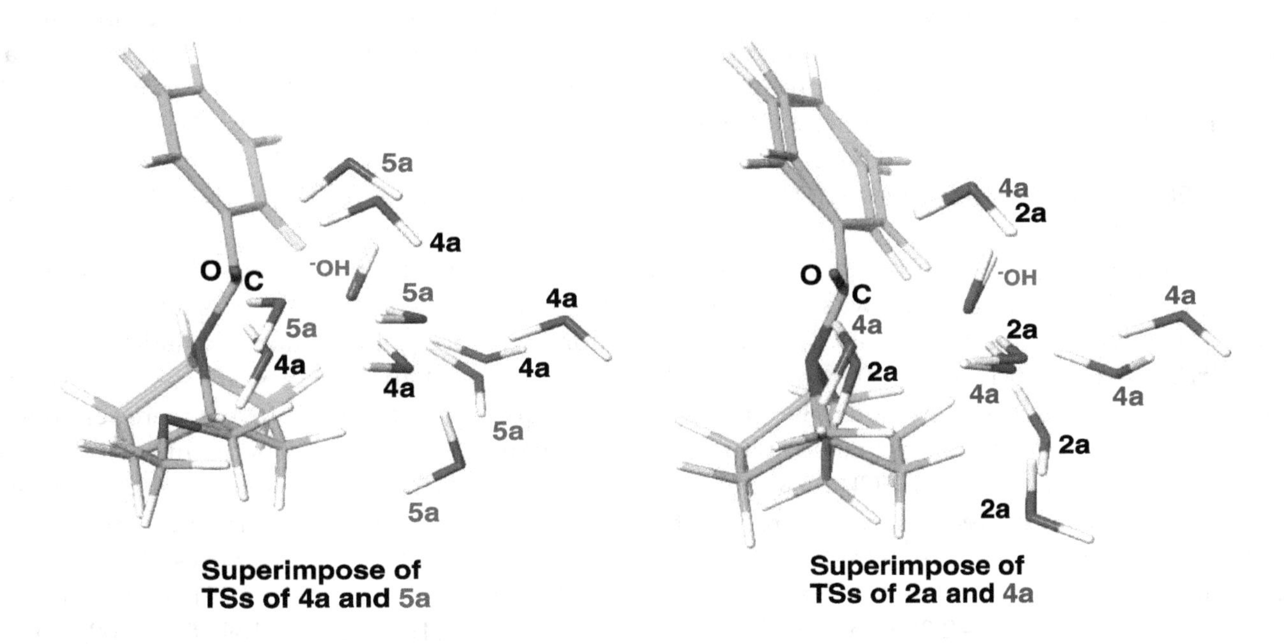


**Figure 13.** Superimposed TS structures of **4a** and **5a** (left) and **2a and 4a** (right), showing the disrupted water network in bulky **5a** (left, magenta) and **4a** (right, blue).

## 3. Materials and Methods

### 3.1. General Procedures

All analyzed compounds were synthesized from commercially available reagents. All compounds were purified before use by column chromatography on silica gel (spherical, neutral silica gel 60 N (100–210 μm), Kanto). Characterization was done by multiple techniques. $^{1}$H- (400 MHz) and $^{13}$C- (100 MHz) NMR spectra were recorded in a 400 MHz Bruker Avance 400 NMR spectrometer at 25 °C. Chemical shifts (δ) are shown in ppm, and coupling constants are given in hertz. Spectral data was obtained using NMR data processing software Brucker TOP-Spin. The NMR probe temperature was calibrated by the temperature-dependent chemical shift difference in ppm between OH proton and $CH_2$ proton of ehtyleneglycol [33].

ESI-TOF mass spectra were recorded in a Bruker Daltonics, micrO-TOF-05. Elemental analyses were done by an independent group in this department and were given within a ± 0.4% error range. Melting points were measured with a Yanaco Micro Melting Point Apparatus and are uncorrected.

*3.2. Synthesis of Amides*

All the amide compounds except **5a–5c** have been synthesized previously [12] and stock samples were used for the present work. Some compounds among **2a–2e**, **3a–3e**, and **4a–4e** were resynthesized, as described below, including the new compounds **5a–5c**.

**Synthesis of *N*-Benzoylazetidines**

*N-Benzoylazetidine* (**2a**). Azetidine chloride (100 mg, 1.1 mmol) was dissolved in dry $CH_2Cl_2$ (5 mL) and the solution was cooled to 0 °C. DIPEA (0.7 mL, 2.9 mmol) was added, and the mixture was stirred for 10 min. Then, benzoyl chloride (186 μl, 1.6 mmol) was added slowly to the solution, and stirring was continued for 30 min. The ice bath was removed and the reaction mixture was allowed to cool to r.t., then quenched by pouring it into water. The aqueous and organic layers were separated, and the aqueous phase was extracted with dichloromethane (3 × 10 mL). The combined organic phase was washed with 0.5 M HCl, 0.5 M aq. $NaHCO_3$ and brine, dried over sodium sulfate, filtered, and evaporated under reduced pressure to afford a yellowish oil. The crude product was purified by open column chromatography (ethyl acetate/DCM, 1:1) to afford **2a** as a transparent oil (130.8 mg, 0.81 mmol, 76%). $^1$H-NMR (400 MHz, $CDCl_3$), δ (ppm): 7.616–7.592 (m, 2H), 7.444–7.353 (m, 3H), 4.296–4.191 (m, 4H), 2.352–2.274 (m, 2H). $^{13}$C-NMR (100 MHz, $CDCl_3$), δ (ppm): 170.39, 133.36, 130.93, 128.40, 127.91, 53.46, 49.01. HRMS (ESI-TOF): $[M + Na]^+$: Calcd. for $C_{10}H_{11}NNaO^+$: 185.0766. Found: 185.0782.

*N-(p-Chlorobenzoyl)azetidine* (**2b**). Azetidine chloride (100 mg, 1.1 mmol) was dissolved in dry $CH_2Cl_2$ (5 mL) and the solution was cooled to 0 °C. DIPEA (0.7 mL, 2.9 mmol) was added, and the mixture was stirred for 10 min. 4-Chlorobenzoyl chloride (205 μL, 1.6 mmol) was added slowly, and stirring was continued for 30 min. The ice bath was removed and the mixture was allowed to warm to r.t., then quenched by pouring it into water. The aqueous and organic layers were separated, and the aqueous phase was extracted with dichloromethane (3 × 10 mL). The combined organic phase was washed with 0.5 M HCl, 0.5 M aq. $NaHCO_3$ and brine, dried over sodium sulfate, filtered, and evaporated under reduced pressure to afford a pale oil. The crude product was purified by open column chromatography (ethyl acetate/DCM, 1:2) to provide **2b** as colorless crystals (171.9 mg, 0.88 mmol, 82%). $^1$H-NMR (400 MHz, $CDCl_3$), δ (ppm): 7.587–7.554 (m, 2H), 7.391–7.358 (m, 2H), 4.309–4.199 (m, 4H), 2.386–2.309 (m, 2H). $^{13}$C-NMR (100 MHz, $CDCl_3$), δ (ppm): 169.18, 137.08, 131.76, 129.39, 128.70, 53.51, 49.09. HRMS (ESI-TOF): $[M + Na]^+$ Calcd. for $C_{10}H_{10}ClNNaO^+$: 218.0343. Found: 218.0360. Anal. Calcd. for $C_{10}H_{10}ClNO$: C, 61.39; H, 5.15; N, 7.16. Found: C, 61.02; H, 5.47; N, 6.96.

*N-(p-Nitrobenzoyl)azetidine* (**2c**). Azetidine chloride (100 mg, 1.1 mmol) was dissolved in dry $CH_2Cl_2$ (5 mL) and the solution was cooled to 0 °C. DIPEA (0.7 mL, 2.9 mmol) was added, and stirring was continued for 10 min. Then, 4-nitrobenzoyl chloride (186 μL, 1.6 mmol) was added slowly, and stirring was continued for 30 min. The ice bath was removed and the mixture was allowed to warm to r.t., then quenched by pouring it into water. The aqueous and organic layers were separated, and the aqueous phase was extracted with dichloromethane (3 × 10 mL). The combined organic phase was washed with 0.5 M HCl, 0.5 M aq. $NaHCO_3$ and brine, dried over sodium sulfate, filtered, and evaporated under reduced pressure to give a yellow oil. The crude product was purified by open column chromatography (ethyl acetate/DCM, 1:2) to afford **2c** as a yellow solid (172.6 mg, 0.84 mmol, 78%). $^1$H-NMR (400 MHz, $CDCl_3$), δ (ppm): 8.268–8.235 (m, 2H), 7.793–7.760 (m, 2H), 4.311–4.231 (m, 4H), 2.425–2.347 (m, 2H). $^{13}$C-NMR (100 MHz, $CDCl_3$), δ (ppm): 168.00, 139.22, 131.19, 129.01, 123.76, 53.41, 49.32. HRMS (ESI-TOF): $[M + Na]^+$ Calcd. for $C_{10}H_{10}N_2NaO_3{}^+$: 229.0585. Found: 229.0585.

**Synthesis of *N*-Benzoylpyrrolidines**

*N-Benzoylpyrrolidine* (**3a**). Pyrrolidine (1 mL, 12.2 mmol) was dissolved in dry $CH_2Cl_2$ (5 mL) and the solution was cooled to 0 °C. DIPEA (3.4 mL, 14.5 mmol) was added, and the mixture was stirred for 10 min. Then, benzoyl chloride (1.7 mL, 14.75 mmol) was added slowly, and stirring was continued for 30 min. The ice bath was removed, and the mixture was allowed to warm to r.t., then quenched by pouring it into water. The aqueous and organic layers were separated, and the aqueous phase was extracted with dichloromethane (3 × 10 mL). The combined organic phase was washed with 0.5 M HCl, 0.5 M aqeouse solution of $NaHCO_3$ and brine, dried over sodium sulfate, filtered, and evaporated under reduced pressure to give a yellow liquid. The crude product was purified by open column chromatography (ethyl acetate) to afford **3a** as a transparent yellow liquid (1.5530, 8.86 mmol, 73%). $^1$H-NMR (400 MHz, $CDCl_3$), δ (ppm): 7.504–7.356 (m, 5H), 3.615 (t, $J$ = 6.8 Hz, 2H), 3.381 (t, $J$ = 6.6 Hz, 2H), 1.929–1.804 (m, 4H). $^{13}$C-NMR (100 MHz, $CDCl_3$), δ (ppm): 169.35, 137.10, 129.51, 128.00, 126.86, 49.33, 45.93, 26.17, 24.21. HRMS (ESI-TOF) $m/z$: [M + Na]$^+$ Calcd. for $C_{11}H_{13}NNaO^+$: 198.0890. Found: 198.0872. Anal. Calcd. for $C_{11}H_{13}NO$: C, 75.40; H, 7.48; N, 7.99. Found: C, 75.21; H, 7.63; N, 7.97.

*N-(p-Chlorobenzoyl)pyrrolidine* (**3b**). Pyrrolidine (0.5 mL, 6.1 mmol) was dissolved in dry $CH_2Cl_2$ (3 mL) and the solution was cooled to 0 °C. DIPEA (1.7 mL, 7.7 mmol) was added, and the mixture was stirred for 10 min. Then, 4-chlorobenzoyl chloride (1.6164 mL, 7.4 mmol) dissolved in $CH_2Cl_2$ (2 mL) was added slowly, and stirring was continued for 1 h. The ice bath was removed, and the mixture was allowed to warm to r.t., then quenched by pouring it into water. The aqueous and organic layers were separated, and the aqueous phase was extracted with dichloromethane (3 × 10 mL). The combined organic phase was washed with 0.5 M HCl, 0.5 M aq. $NaHCO_3$ and brine, dried over sodium sulfate, filtered, and evaporated under reduced pressure to afford a transparent oil. The crude product was purified by open column chromatography (hexane/ethyl acetate) to provide **3b** as a white amorphous solid (444 mg, 2.1 mmol, 35%). m.p.: 62–64 °C. $^1$H-NMR (400 MHz, $CDCl_3$), δ (ppm): 7.410 (dd; $J$ = 4.3 Hz, 1.8 Hz; 2H), 7. 307 (dd; $J$ = 8.8 Hz, 2 Hz; 2H), 3.569 (t, $J$ = 6.6 Hz, 2H), 3.349 (t, $J$ = 6.6 Hz, 2H), 1.930–1.786 (m, 4H). $^{13}$C-NMR (100 MHz, $CDCl_3$), δ (ppm): 168.48, 135.73, 135.50, 128.63, 128.44, 49.55, 46.25, 26.36, 24.36. HRMS (ESI-TOF): [M + Na]$^+$ Calcd. for $C_{11}H_{12}ClNNaO^+$: 232.0500. Found 232.0517. Anal. Calcd. for $C_{11}H_{12}ClNO$: C, 63.01; H, 5.77; N, 6.68. Found: C, 62.95; H, 5.78; N, 6.51.

*N-(p-Nitrobenzoyl)pyrrolidine* (**3c**). Pyrrolidine (0.5 mL, 6.1 mmol) was dissolved in dry $CH_2Cl_2$ (5 mL) at 0 °C. DIPEA (1.7 mL, 7.7 mmol) was added to the solution, and the mixture was stirred for 10 min. Then, 4-nitrobenzoyl chloride (1.1301 g, 7.4 mmol) dissolved in $CH_2Cl_2$ (10 mL) was added slowly, and stirring was continued for 1 h at 0 °C. The ice bath was removed, and the mixture was allowed to warm to r.t., then quenched by pouring it into water. The aqueous and organic layers were separated, and the aqueous phase was extracted with dichloromethane (3 × 30 mL). The combined organic phase was washed with brine, dried over sodium sulfate, filtered, and evaporated under reduced pressure to give a a yellow liquid. The crude product was purified by open column chromatography (ethyl acetate/acetone 1:1) to afford **3c** as a slightly yellow solid (821.6 mg, 3.7 mmol, 61.3%). m.p.: 78–80 °C. $^1$H-NMR (400 MHz, $CDCl_3$), δ (ppm): 8.293–8.207 (m, 2H), 7.698–7.665 (m, 2H), 3.676 (t, $J$ = 7.0 Hz, 2H), 3.382 (t, $J$ = 6.6 Hz, 2H), 2.039–1.887 (m, 4H). $^{13}$C-NMR (100 MHz, $CDCl_3$), δ (ppm): 167.61, 143.19, 128.29, 123.84, 49.61, 46.56, 26.52, 24.50. HRMS (ESI-TOF): [M + Na]$^+$ Calcd. for $C_{11}H_{12}N_2NaO_3{}^+$: 244.0774. Found: 244.0742. Anal. Calcd. for $C_{11}H_{12}N_2O_3$: C, 59.99; H, 5.49; N, 12.72. Found: C, 59.82; H, 5.45; N, 12.65.

*N-(p-Toluoyl)pyrrolidine* (**3d**). Pyrrolidine (0.5 mL, 6.1 mmol) was dissolved in dry $CH_2Cl_2$ (5 mL) and the solution was cooled to 0 °C. DIPEA (1.7 mL, 7.7 mmol) was added, and the mixture was stirred for 10 min. Then, a solution of $p$-tolyl chloride (975 μL, 7.4 mmol) in $CH_2Cl_2$ (10 mL) was slowly added, and the mixture was stirred for 1 h at 0 °C. The ice bath was removed, and the mixture was allowed to warm to r.t., then quenched by pouring it into water. The aqueous and organic layers were separated, and the aqueous phase was extracted with dichloromethane (3 × 30 mL). The combined organic phase was washed with brine, dried over sodium sulfate, filtered, and evaporated under reduced pressure to provide a yellow liquid. The crude product was purified by open column chromatography (hexane/ethyl acetate 4:1) to afford **3d** as a white solid (654.4 mg, 3.5 mmol, 56.8%). M.p.: 75–76 °C. $^1$H-NMR (400 MHz, $CDCl_3$), δ (ppm): 7.438–7.408 (m, 2H), 7.202–7.179 (m, 2H), 3.639 (t, $J$ = 7.0 Hz, 2H), 3.439 (t, $J$ = 6.6 Hz, 2H), 2.372 (s, 3H), 1.986–1.829 (m, 4H). $^{13}$C-NMR (100 MHz, $CDCl_3$), δ (ppm):

169.93, 139.98, 134.52, 128.93, 127.35, 49.75, 46.30, 26.55, 24.59, 21.50 HRMS (ESI-TOF) *m/z*: [M + Na]$^+$ Calcd. for $C_{12}H_{15}NNaO^+$: 212.1046. Found 212.1036. Anal. Calcd. for $C_{12}H_{15}NO$: C, 76.16; H, 7.99; N, 7.40. Found: C, 75.80; H, 8.09; N, 7.33.

*N-(p-Anisoyl)pyrrolidine* (**3e**). Pyrrolidine (0.5 mL, 6.1 mmol) was dissolved in dry $CH_2Cl_2$ (5 mL) and the solution was cooled to 0 °C. DIPEA (1.7 mL, 7.7 mmol) was added, and the mixture was stirred for 10 min. Then, a solution of 4-nitrobenzoyl chloride (1.2573 g, 7.4 mmol) in $CH_2Cl_2$ (10 mL) was slowly added, and stirring was continued for 1 h at 0 °C. The ice bath was removed, and the mixture was allowed to warm to r.t., then quenched by pouring it into water. The aqueous and organic layers were separated, and the aqueous phase was extracted with dichloromethane (3 × 30 mL). The combined organic phase was washed with brine, dried over sodium sulfate, filtered, and evaporated under reduced pressure to provide a yellow liquid. The crude product was purified by open column chromatography (*n*-hexane/ethyl acetate 1:1) to afford **3e** as a pale-brown solid (808.3 mg, 3.9 mmol, 65%). m.p.: 73.5–75.5 °C. $^1$H-NMR (400 MHz, $CDCl_3$), δ (ppm): 7.515 (ddd, *J* = 9.2, 2.4, 2 Hz, 4H), 3.832 (s, 3H), 3.634 (t, *J* = 6.4 Hz, 2H), 3.476 (t, *J* = 6.2 Hz, 2H), 1.966–1.851 (m, 4H). $^{13}$C-NMR (100 MHz, $CDCl_3$), δ (ppm): 169.56, 160.93, 129.63, 129.29, 113.55, 55.45, 49.90, 46.44, 26.64, 24.59. HRMS (ESI-TOF) *m/z*: [M + Na]$^+$ Calcd. for $C_{12}H_{15}NNaO_2{}^+$: 228.0995. Found 228.0993. Anal. Calcd. for $C_{12}H_{15}NO_2$: C, 70.22; H, 7.39; N, 6.82. Found: C, 70.01; H, 7.48; N, 6.76.

**Synthesis of *N*-Benzoylazetidines**

*N-Benzoyl-7-azabicyclo[2.2.1]heptane* (**4a**). *Trans*-4-aminocyclohexyl *p*-toluensulfonate hydrobromide (1 g, 7.5 mmol) was dissolved in ethanol (110 mL) and water (30 mL), and then NaOH 1 M (25 mL) was added. The mixture was stirred at room temperature for 20 h, and then quenched by adding HCl 4 M in 1,4-dioxane (5 mL). Stirring was continued for 20 min at r.t., and the mixture was evaporated under reduced pressure. NaOH 10% (10 mL) was added to the residue, and free amide was extracted with ether (3 × 30 mL). This solution was evaporated, and the residue was redissolved in dry $CH_2Cl_2$ (30 mL). DIPEA (2 mL, 9.3 mmol) was added to the resulting solution, and the mixture was stirred for 10 min at 0 °C. Then, benzoyl chloride (1.1 mL, 9.1 mmol) was slowly added. Stirring was continued for 30 min. The ice bath was removed, and the mixture was allowed to warm to r.t. Stirring was continued for 4 h, and then the mixture was quenched by pouring it into water. The aqueous and organic layers were separated, and the aqueous phase was extracted with dichloromethane (3 × 50 mL). The combined organic phase was washed with brine, dried over sodium sulfate, filtered, and evaporated under reduced pressure to give a yellow oil. The crude product was purified by open column chromatography ($CH_2Cl_2$/ethyl acetate 9:1) to afford **4a** as a white solid (52.0 mg, 35%). $^1$H-NMR (400 MHz, $CDCl_3$), δ (ppm): 7.813–7.696 (m, 2H), 7.506–7.288 (m, 3H) 4.613–4.460 (m, 1H), 4.121–3.900 (m, 1H), 2.239–1.240 (m, 8H). ESI-HRMS (*m/z*): Calculated for $C_{13}H_{15}NNaO^+$ [M + Na]$^+$: 225.1079. Found: 225.1081.

*N-(p-Toluoyl)-7-azabicyclo[2.2.1]heptane* (**4d**). *Trans*-4-aminocyclohexyl *p*-toluensulfonate hydrobromide (1 g, 7.5 mmol) was dissolved in ethanol (110 mL) and water (30 mL), and then NaOH 1 M (25 mL) was added. The solution was stirred at room temperature for 20 h, and then quenched by adding concentrated HCl 4 M (5 mL). Stirring was continued for 20 min at r.t., and then the solution was evaporated under reduced pressure. NaOH 10% (10 mL) was added to the residue. Free amide was extracted with ether (3 × 30 mL), then HCl in 1,4-dioxane (1 mL) was added, and the mixture was evaporated. The residue was redissolved in dry $CH_2Cl_2$ (30 mL). DIPEA (2 mL, 9.3 mmol) was added to the resulting solution, and the mixture was stirred for 10 min at 0 °C. Then, *p*-tolyl benzoyl chloride (1.6 mL, 9.1 mmol) was slowly added. Stirring was continued for 30 min. The ice bath was removed, and the mixture was allowed to warm to r.t., and further stirred for 4 h, then quenched by pouring it into water. The aqueous and organic layers were separated, and the aqueous phase was extracted with dichloromethane (3 × 30 mL). The combined organic phase was washed with brine, dried over sodium sulfate, filtered, and evaporated under reduced pressure to give a yellow oil. The crude product was purified twice by open column chromatography (*n*-hexane/ethyl acetate

(1:1) and dichloromethane/ethyl acetate (9:1)) to afford **4d** as a yellow solid. Recrystallization afforded transparent crystals (238.6 mg, 11%). M.p.: 105–107 °C. $^1$H-NMR (400 MHz, $CDCl_3$), δ (ppm): 7.449 (d, *J* = 8 Hz, 2H), 7.188 (d, *J* = 8Hz, 2H), 4.720 (br. s, 1H), 4.148 (br. s, 1H), 2.377 (s, 3H), 1.899–1.810 (s, 4H), 1.508–1.466 (m, 4H). $^{13}$C-NMR (100 MHz, $CDCl_3$), δ (ppm): 169.03, 140.72, 133.50, 128.96, 127.97, 77.36, 30.60, 28.87, 21.56. ESI-HRMS: Calculated for $C_{14}H_{17}NNaO^+$ $[M + Na]^+$: 238.1202. Found: 238.1217. Anal. Calcd. for $C_{14}H_{17}NO_2$: C, 78.10; H, 7.96; N, 6.51. Found: C, 77.76; H, 7.90; N, 6.58.

**Synthesis of *N*-Benzoyl-1-(Methoxymethyl)-7-azabicyclo[2.2.1]heptanes**

*N-Benzoyl-1-(Methoxymethyl)-7-azabicyclo[2.2.1]heptane* (**5a**). Boc-protected 1-(methoxymethyl)-7-azabicyclo[2.2.1]heptane (114.6 mg, 0.47 mmol) was dissolved in a 1:1 mixture of dry $CH_2Cl_2$ (1 mL) and TFA (1 mL), and the solution was stirred at room temperature for one h. The solvent was removed in vacuo, and the residue was dissolved in $CH_2Cl_2$. The resulting solution was evaporated again (a total of three times). Finally, the residue was dissolved in dry $CH_2Cl_2$ (5 mL) and DIPEA (130 µl, 0.61 mmol) was added to the solution. The mixture was stirred for 10 min at 0 °C, then benzoyl anhydride (170 µL, 0.9 mmol) was slowly added. Stirring was continued for 30 min. The ice bath was removed, and the mixture was allowed to warm to r.t., then quenched by pouring it into water. The aqueous and organic layers were separated, and the aqueous phase was extracted with dichloromethane (3 × 10 mL). The combined organic phases were washed with brine, dried over sodium sulfate, filtered, and evaporated under reduced pressure to give a yellow liquid. The crude product was purified by open column chromatography ($CH_2Cl_2$/ethyl acetate 9:1) to afford **5a** as a white solid (520.0 mg, 35%). M.p.: 53–55°C. $^1$H-NMR (400 MHz, $CDCl_3$), δ (ppm): 7.524–7.512 (m, 2H), 7.437–7.342 (m, 3H), 4.225 (s, 2H), 4.106 (t, *J* = 5 Hz, 1H), 3.425 (s, 3H), 2.013–1.932 (m, 2H), 1.867–1.771 (m, 2H), 1.728–1.667 (m, 2H), 1.487–1.425 (m, 2H). HRMS (ESI-TOF): $[M + Na]^+$ Calcd. for $C_{15}H_{19}NNaO_2{}^+$ $[M + Na]^+$: 268.1308. Found: 268.1312.

*N-(p-Chlorobenzoyl)-(1-(methoxymethyl)-7-azabicyclo[2.2.1]heptanes* (**5b**). To a solution of the corresponding Boc-protected 1-(methoxymethyl)-7- azabicyclo[2.2.1]heptane (41.2 mg) in dry $CH_2Cl_2$ (2 mL) was added 3 mL of TFA. The solution was stirred at room temperature for 2 h The organic solvent was evaporated, and the residue was dissolved in 10 mL of $CH_2Cl_2$. Triethylamine (93.6 µL) was added at 0 °C, followed by *p*-chlorobenzoyl chloride (58.8 mg). The reaction mixture was stirred for 2 h at 0 °C, and then poured into saturated aqueous $NaHCO_3$. The whole was extracted with $CH_2Cl_2$ (30 mL × 3). The combined organic layer was dried over $Na_2SO_4$, and the solvent was evaporated. Column chromatography (*n*-hexane:ethyl acetate = 4:1) of the residue gave **5b** (47.8 mg, 77%). $^1$H-NMR (400 MHz, $CDCl_3$), δ (ppm): 7.489 (2H, d, *J* = 8.4 Hz), 7.356 (2H, d *J* = 8.0 Hz), 4.198 (2H, s), 4.087 (1H, m), 3.423 (3H, s), 2.076–1.571 (6H, m), 1.568–1.369 (2H, m). $^{13}$C-NMR (100 MHz, $CDCl_3$), δ (ppm): 169.11, 136.37, 135.45, 129.20, 128.43, 73.91, 68.03, 61.37, 59.34, 32.65, 29.59. HRMS (ESI-TOF): $[M + Na]^+$ Calcd. for $C_{15}H_{18}ClNNaO_2{}^+$ $[M + Na]^+$: 302.0918. Found: 302.0924. Anal. Calcd. for $C_{15}H_{18}ClNO_2$: C, 64.40; H, 6.49; N, 5.01. Found: C, 64.27; H, 6.52; N, 4.92.

*N-(p-Nitrobenzoyl)-(1-(Methoxymethyl)-7-azabicyclo[2.2.1]heptanes* (**5c**). Boc-protected 1-(methoxymethyl)-7-azabicyclo[2.2.1]heptane (119.3 mg, 0.49 mmol) was dissolved in a 1:1 mixture of dry $CH_2Cl_2$ (1 mL) and TFA (1 mL), and the solution was stirred at room temperature for one h. The solvent was removed in vacuo, then the residue was dissolved in $CH_2Cl_2$ and this solution was evaporated again (a total of three times). Finally, the residue was dissolved in dry $CH_2Cl_2$ (5 mL), DIPEA (132 µL, 0.63 mmol) was added to it, and the mixture was stirred for 10 min at 0 °C. Then, *p*-nitrobenzoyl chloride (91.70 mg, 0.6 mmol) was slowly added, and stirring was continued for 30 min. The ice bath was removed, and the mixture was allowed to warm to r.t. Stirring was continued for 5.5 h, and then the mixture was quenched by pouring it into water. The aqueous and organic layers were separated, and the aqueous phase was extracted with dichloromethane (3 × 10 mL). The combined organic phase was washed with brine, dried over sodium sulfate, filtered, and evaporated under reduced pressure to give a a yellow liquid. The crude residue was purified by open column chromatography (*n*-hexane/ethyl acetate 1:1) to afford **5c** as a white solid (63.9 mg, 45%). m.p.: 140–142 °C. $^1$H-NMR (400 MHz, $CDCl_3$), δ (ppm):

8.256 (dd, $J$ = 2 Hz, 6.8 Hz, 2H), 7.685 (dd, $J$ = 2 Hz, 6.8 Hz, 2H), 4.178 (s, 2H), 4.010 (t, $J$ = 4.4 Hz, 1H), 3.419 (s, 3H), 2.003–1.961 (m, 2H), 1.873–1.783 (m, 2H), 1.750–1.689 (m, 2H), 1.549–1.487 (m, 2H). $^{13}$C-NMR (100 MHz, $CDCl_3$), δ (ppm): 167.52, 148.90, 143.12, 128.68, 123.74, 73.64, 68.66, 61.24, 59.46, 32.79, 29.71. HRMS (ESI-TOF): $[M + Na]^+$ Calcd. for $C_{15}H_{18}N_2NaO_4{}^+$ $[M + Na]^+$: 313.1159. Found: 313.1169. Anal. Calcd. for $C_{15}H_{18}N_2O_4$: C, 62.06; H, 6.25; N, 9.65. Found: C, 61.92; H, 6.28; N, 9.63.

### 3.3. Kinetic Studies

Kinetic data were collected by recording the $^1$H-NMR (400 MHz) spectra at 2- min intervals. Data were obtained at 70.0 °C under a $N_2$ flow of 400 L/h. Deuterated solvents were used for all measurements: Deuterium oxide, 99.9% D from Wako; methanol-$d_4$, 99.8% D from Kanto Chemical Co. Inc.; 1,4-dioxane-$d_8$, 99% D from Cambridge Isotope Laboratories, Inc.; and sodium deuteroxide, 40% wt. % in $D_2O$, 99.5% D from Sigma-Aldrich, Co. Parameters were set using the reaction solution at room temperature. Reactions were initiated by raising the temperature of the NMR machine. Spectral data were then recorded for at least one half-life. Hydrolysis experiments were repeated three times. Hydrolysis products were identified by comparison of the $^1$H-NMR data with those of authentic samples, or by MS-ESI. Pseudo-first-order rate constants ($k_{obs}$) were evaluated by a linear least-squares fitting of a plot of the logarithm of reactant concentration *versus* time.

### 3.4. Computational Studies

Computational studies were carried out using the Gaussian 09 software package [34]. Geometry optimizations for the model molecules were performed at the M06-2X/6-31+G(d) level with the bulk solvent model SMD (solvent = water) together with explicit water molecules. Vibrational frequency calculations were performed at the same level of theory. Optimized geometries were verified by frequency calculations as minima (zero imaginary frequencies) or transition structures (TS, a single imaginary frequency) (the coordinates, frequency and thermodynamic values are shown in Supporting Information). Intrinsic reaction coordinate (IRC) computations of the transition structures verified the reactants and products in the case of simple amide (formamide and acetamide (data not shown)). In the cases of more realistic substrates (**2a**–**5a**), particularly in combination of the solvent model, the IRC calculations were unsuccessful. In all the transition states the validity of transition state structures was confirmed by inspecting the direction of vibration of the negative frequency, which matched the trajectory of the nucleophilic attack of the hydroxy anion onto the carbonyl carbon atom. Geometry minimization of the transition structures lead to the hydroxide-addition adducts, which also supported the validity of the transition state structures, which represented the trajectory of the nucleophilic attack of the hydroxy anion. We also estimated Gibbs free energies by single-point calculations (frequency calculations) with M06-2X//6-311++G(d,p) for the M06-2X//6-31+G(d)-optimized structures. In these single-point frequency calculations, a single negative frequency was found respectively, which also corresponded to the trajectory of the nucleophilic attack of the hydroxy anion onto the carbonyl carbon atom. The energies were corrected for zero-point energies and Gibbs free energy at 25 °C (298.15 K), obtained from frequency calculations. Calculations at the B3LYP level showed similar trends to the M06-2X calculations.

Identification of the transition states structures of the attack of the hydroxide anion to the carbonyl carbon atom of the amide group was carried out in the following multiple procedures: The transition state structures of the attack of the hydroxide anion to the carbonyl carbon atom of the amide were first identified in the absence of the explicit water molecules (data not shown). Addition of explicit water molecules in the arrangement similar to Types a and b (Figure 11) [32], followed by geometry optimization (OPT = TS). Or we reproduced the TSs of the hydrolysis of formamide and acetamide in the presence of four and five water molecules, respectively, which were consistent with the two hydrogen network models (Types a and b, Figure 11), reported in the previous literature [32]. Then the amide functionality was morphed into the realistic amide substrates (**2a**–**5a**). The initial minimum conformation of the neutral amide substrates (**2a**–**5a**) were obtained by conformational search in

Marcomodel software (Schrödinger, LLC, New York, NY, USA), followed by optimization with the DFT methods. The Type a arrangement of waters (Figure 11) did not converge to the optimization or transformed to the Type b arrangement of water molecules. Therefore, we discussed the activation energies on the basis of the Type b hydrogen network model, while the Type **a** hydrogen network model might be the energy minimum (see the main text).

The change of Gibbs free energy for the reaction was calculated on the basis of the reaction model, shown in Scheme 6. The results of the present calculations (Table 4) were reasonable and reliable, because the present calculations gave the activation energies of similar magnitude to the cases of simple amide: The calculated activation energies of the present amides **2a–5a** were in similar magnitude to the calculated and experimental Gibbs free energy change of hydroxide anion-catalyzed hydrolysis of formamide, 21–22 kcal/mol [32], (see also Table 4, footnote a).

## 4. Conclusions

Herein we measured and compared the kinetics of base-catalyzed hydrolysis of non-planar *N*-benzoyl-7-azabicyclo[2.2.1] heptane amides (**4**) and related compounds (**2**, **3** and **5**) under pseudo-first-order conditions at 70 °C. Excess sodium deuteroxide was used as the base in two different solvent systems (methanol or 1,4-dioxane in a 1:1 $D_2O$ solution). Reaction progress was monitored by $^1H$ NMR spectroscopy. Unexpectedly we found that 7-azabicyclo[2.2.1]heptane amides (**4**) were resistant to base-catalyzed hydrolysis. As a general trend, independently of whether methanol or 1,4-dioxane was used as a co-solvent, it was found that the reactivity of nitrogen-pyramidal azetidine amides (**2**) was greater than that of the planar pyrrolidine amides (**3**), followed by unsubstituted bicyclic amides (**4**), while bridgehead-substituted bicycles (**5**) were the least reactive (reaction rate: **2a** > **3a** > **4a** > **5a**, etc.). Phenyl substituents showed the expected electronic trends. We also executed DFT calculations of the rate-determining process, addition of the hydroxide anion to the amide carbonyl group, and found that the experimental kinetic data is consistent with the magnitude of the calculated Gibbs free energies of activation. Our results confirm the stability of the 7-azabicyclo[2.2.1]heptane amides at least at 37 °C. This is important, because it implies that the 7-azabicyclo[2.2.1]heptane amide scaffold is available for practical molecular design, e.g., as an amino acid surrogate.

**Author Contributions:** D.A.O.G.d.V.: synthesis of compounds and kinetics measurement, data analysis; A.S.: synthesis of compounds; L.Z.: synthesis of compounds; S.K.: synthesis of compounds; Y.O.: plotting, synthesis, data analysis, funding, writing manuscript; T.O.: plotting, data analysis, calculations, funding, and writing manuscript.

**Acknowledgments:** The computations were performed at the Research Center for Computational Science, Okazaki, Japan. D.A.O.G.d.V. and A.S. were supported by Japanese Government (MEXT) scholarship.

## References

1. Greenberg, A.; Liebman, J.F. *The Amide Linkage Structural Significance in Chemistry, Biochemistry and Materials Science*; Wiley: New York, NY, USA, 2000.
2. Fischer, G. Chemical Aspects of Peptide Bond Isomerisation. *Chem. Soc. Rev.* **2000**, *29*, 119–127. [CrossRef]
3. Thakkar, B.S.; Svendsen, J.-S.M.; Engh, R.A. Cis/Trans Isomerization in Secondary Amides: Reaction Paths, Nitrogen Inversion, and Relevance to Peptidic Systems. *J. Phys. Chem. A.* **2017**, *121*, 6830–6837. [CrossRef] [PubMed]
4. Otani, Y.; Nagae, O.; Naruse, Y.; Inagaki, S.; Ohno, M.; Yamaguchi, K.; Yamamoto, G.; Uchiyama, M.; Ohwada, T. An Evaluation of Amide Group Planarity in 7-Azabicyclo[2.2.1]heptane Amides. Low Amide Bond Rotation Barrier in Solution. *J. Am. Chem. Soc.* **2003**, *125*, 15191–15199. [CrossRef] [PubMed]
5. Bose, A.K.; Manhas, M.S.; Bank, B.K.; Srirajan, V. B-Lactams: Cyclic amides of distinction, chapter 7. In *The Amide Linkage Structural Significance in Chemistry, Biochemistry and Materials Science*; Greenberg, A., Liebman, J.F., Eds.; Wiley: New York, NY, USA, 2000; pp. 157–214.

6. Komarov, I.V.; Yanik, S.; Ishchenko, A.Y.; Davies, J.E.; Goodman, J.M.; Kirby, A.J. The Most Reactive Amide As a Transition-State Mimic For *cis-trans* Interconversion. *J. Am. Chem. Soc.* **2015**, *137*, 926–930. [CrossRef] [PubMed]
7. Wang, Q.P.; Bennet, A.J.; Brown, R.S.; Santarsiero, B.D. Distorted Amides as Models for Activated Peptide N-C(O) Units. 3. Synthesis, Hydrolytic Profile, and Molecular Structure of 2,3,4,5-Tetrahydro-2-oxo-1,5-propanobenzazepine. *J. Am. Chem. Soc.* **1991**, *113*, 5757–5765. [CrossRef]
8. Aubé, J. A New Twist on Amide Solvolysis. *Angew. Chem. Int. Ed.* **2012**, *51*, 3063–3065. [CrossRef] [PubMed]
9. Yamada, S. Structure and Reactivity of a Highly Twisted Amide. *Angew. Chem. Int. Edit. Engl.* **1993**, *32*, 1083–1085. [CrossRef]
10. Alkorta, I.; Cativela, C.; Elguero, J.; Gil, A.M.; Jiménez, A.I. A Theoretical Study of the Influence of Nitrogen Angular Constraints on the Properties of Amides: Rotation/Inversion Barriers and Hydrogen Bond Accepting Abilities of N-Formylaziridine and—Azirine. *New J. Chem.* **2005**, *29*, 1450–1453. [CrossRef]
11. Yamamoto, G.; Nakajo, F.; Tsubai, N.; Murakami, H.; Mazaki, Y. Structures and Stereodynamics of *N*-9-Triptycylacetamide and Its *N*-Alkyl Derivatives. *Bull. Chem. Soc. Jpn.* **1999**, *72*, 2315–2326. [CrossRef]
12. Glover, S.A.; White, J.M.; Rosser, A.A.; Digianantonio, K.M. Structures of *N,N*-Dialkoxyamides: Pyramidal Anomeric Amides with Low Amidicity. *J. Org. Chem.* **2011**, *76*, 9757–9763. [CrossRef] [PubMed]
13. Ohwada, T.; Achiwa, T.; Okamoto, I.; Shudo, K. On the Planarity of Amide Nitrogen. Intrinsic Pyramidal Nitrogen of N-Acyl-7Azabicyclo[2.2.1]heptanes. *Tetrahedron Lett.* **1998**, *39*, 865–868. [CrossRef]
14. Mujika, J.I.; Matxain, J.M.; Eriksson, L.A.; Lopez, X. Resonance Structures of the Amide Bond: The Advantages of Planarity. *Chem. Eur. J.* **2006**, *12*, 7215–7224. [CrossRef] [PubMed]
15. Bisz, E.; Piontek, A.; Dziuk, B.; Szostak, R.; Szostak, M. Barriers to Rotation in ortho-Substituted Tertiary Aromatic Amides: Effect of Chloro-Substitution on Resonance and Distortion. *J. Org. Chem.* **2018**, *83*, 3159–3163. [CrossRef] [PubMed]
16. Clayden, J.; Foricher, Y.J.Y.; Lam, H.K. Intermolecular Dearomatising Addition of Organolithium Compounds to *N*-Benzoylamides of 2,2,6,6-Tetramethylpiperidine. *Eur. J. Chem.* **2002**, 3558–3565. [CrossRef]
17. Hutchby, M.; Houlden, C.E.; Haddow, M.F.; Tyler, S.N.G.; Llloyd-Jones, G.C.; Booker-Milburn, K.I. Switching Pathways: Room-Temperature Neutral Solvolysis and Substitution of Amides. *Angew. Chem. Int. Ed.* **2012**, *51*, 548–551. [CrossRef] [PubMed]
18. Shi, S.; Meng, G.; Szostak, M. Synthesis of Biaryls through Nickel-Catalyzed Suzuki–Miyaura Coupling of Amides by Carbon–Nitrogen Bond Cleavage. *Angew. Chem. Int. Ed.* **2016**, *55*, 6959–6963. [CrossRef] [PubMed]
19. Liu, C.; Szostak, M. Twisted Amides; From Obscurity to Broadly Useful Transition-Metal-Catalyzed Reactions by N-C Amide Bond Activation. *Chem. Eur. J.* **2017**, *23*, 7157–7173. [CrossRef] [PubMed]
20. Otani, Y.; Futaki, S.; Kiwada, T.; Sugiura, Y.; Muranaka, A.; Kobayashi, N.; Uchiyama, M.; Yamaguchi, K.; Ohwada, T. Oligomers of β-Amino Acid Bearing Non-Planar Amides Form Ordered Structures. *Tetrahedron* **2006**, *62*, 11635–11644. [CrossRef]
21. Hori, T.; Otani, Y.; Kawahata, M.; Yamaguchi, K.; Ohwada, T. Non-planar Structures of Thiomides Derived from 7-Azabicyclo[2.2.1]heptane. Electronically Tunable Planarity of Thioamides. *J. Org. Chem.* **2008**, *73*, 9102–9108. [CrossRef] [PubMed]
22. Otani, Y.; Watanabe, S.; Ohwada, T.; Kitao, A. Molecular Dynamics Study of Nitrogen-Pyramidalized Bicyclic β-Proline Oligomers: Length-Dependent Convergence to Organized Structure. *J. Phys. Chem. B* **2017**, *121*, 100–109. [CrossRef] [PubMed]
23. Alemán, C.; Jiménez, A.I.; Cativela, C.; Pérez, J.J.; Casanovas, J. Unusually High Pyramidal Geometry of the Bicyclic Amide Nitrogen in a Complex 7-Azabicyclo[2.2.1]heptane Derivative: Theoretical Analysis Using a Bottom-up Strategy. *J. Phys. Chem. B* **2005**, *109*, 11836–11841. [CrossRef] [PubMed]
24. Hosoya, M.; Otani, Y.; Kawahata, M.; Yamaguchi, K.; Ohwada, T. Water-Stable Helical Structure of Tertiary Amides of Bicyclic β-Amino Acid Bearing 7-Azabicyclo[2.2.1]heptane. Full Control of Amide Cis-Trans Equilibrium by Bridgehead Substitution. *J. Am. Chem. Soc.* **2010**, *132*, 14780–14789. [CrossRef] [PubMed]
25. Wang, S.; Otani, Y.; Liu, X.; Masatoshi, K.; Yamaguchi, K.; Ohwada, T. Robust *trans*-Amide Helical Structure of Oligomers of Bicyclic Mimics of β-Proline: Impact of Positional Switching of Bridgehead Substituent on Amide *cis-trans* Equilibrium. *J. Org. Chem.* **2014**, *79*, 5287–5300. [CrossRef] [PubMed]

26. Zhai, L.; Wang, S.; Nara, M.; Takeuchi, K.; Shimada, I.; Otani, Y.; Ohwada, T. Application of C-terminal 7-azabicyclo[2.2.1]heptane to stabilize β-strand-like extended conformation of a neighboring α-amino acid. *J. Org. Chem.*. Accepted for publication.
27. Marlier, J.F.; Dopke, N.C.; Johnstone, K.R.; Wirdzig, T.J. A Heavy-Atom Isotope Effect Study of the Hydrolysis of Formamide. *J. Am. Chem. Soc.* **1999**, *121*, 4356–4363. [CrossRef]
28. Mujika, J.I.; Mercero, J.M.; Lopez, X. Water-Promoted Hydrolysis of a Highly Twisted Amides: Rate Acceleration Caused by the Twist of the Amide Bond. *J. Am. Chem. Soc.* **2005**, *127*, 4445–4453. [CrossRef] [PubMed]
29. Gorb, L.; Asensio, A.; Tuñón, I.; Ruiz-López, M.F. The Mechanism of Formamide Hydrolysis in Water from Ab Initio Calculations and Simulations. *Chem. Eur. J.* **2005**, *11*, 6743–6753. [CrossRef] [PubMed]
30. Wang, B.; Cao, Z. Mechanism of Acid-Catalyzed Hydrolysis of Formamide from Cluster-Continuum Model Calculations: Concerted versus Stepwise Pathway. *J. Phys. Chem. A* **2010**, *114*, 12918–12927. [CrossRef] [PubMed]
31. Matsubara, T.; Ueta, C. Computational Study of the Effects of Steric Hindrance on Amide Bond Cleavage. *J. Phys. Chem. A* **2014**, *118*, 8664–8675. [CrossRef] [PubMed]
32. Xiong, Y.; Zhan, C.-G. Theoretical Studies of Transition–State Structures and Free Energy Barriers for Base-Catalyzed Hydrolysis of amides. *J. Phys. Chem. A* **2006**, *110*, 12644–12652. [CrossRef] [PubMed]
33. Ammann, C.; Meier, P.; Merbach, A.E. A simple multinuclear NMR thermometer. *J. Mag. Reson.* **1982**, *46*, 319–321.
34. Frisch, M.J.; Trucks, G.W.; Schlegel, H.B.; Scuseria, G.E.; Robb, M.A.; Cheeseman, J.R.; Scalmani, G.; Barone, V.; Mennucci, B.; Petersson, G.A.; et al. *Gaussian 09*, revision D.01; Gaussian, Inc.: Wallingford, CT, USA, 2013.

5

# Well-Defined Pre-Catalysts in Amide and Ester Bond Activation

**Sandeep R. Vemula, Michael R. Chhoun and Gregory R. Cook ***

Department of Chemistry and Biochemistry, North Dakota State University, Fargo, ND 58108–6050, USA; sandeepreddy.vemula@ndsu.edu (S.R.V.); michael.chhoun@ndsu.edu (M.R.C.)
* Correspondence: gregory.cook@ndsu.edu

Academic Editor: Michal Szostak

**Abstract:** Over the past few decades, transition metal catalysis has witnessed a rapid and extensive development. The discovery and development of cross-coupling reactions is considered to be one of the most important advancements in the field of organic synthesis. The design and synthesis of well-defined and bench-stable transition metal pre-catalysts provide a significant improvement over the current catalytic systems in cross-coupling reactions, avoiding excess use of expensive ligands and harsh conditions for the synthesis of pharmaceuticals, agrochemicals and materials. Among various well-defined pre-catalysts, the use of Pd(II)-NHC, particularly, provided new avenues to expand the scope of cross-coupling reactions incorporating unreactive electrophiles, such as amides and esters. The strong σ-donation and tunable steric bulk of NHC ligands in Pd-NHC complexes facilitate oxidative addition and reductive elimination steps enabling the cross-coupling of broad range of amides and esters using facile conditions contrary to the arduous conditions employed under traditional catalytic conditions. Owing to the favorable catalytic activity of Pd-NHC catalysts, a tremendous progress was made in their utilization for cross-coupling reactions via selective acyl C–X (X=N, O) bond cleavage. This review highlights the recent advances made in the utilization of well-defined pre-catalysts for C–C and C–N bond forming reactions via selective amide and ester bond cleavage.

**Keywords:** pre-catalysts; palladium catalysis; amide bond activation; ester bond activation; cross-coupling

---

## 1. Introduction

Transition metal-catalyzed cross-coupling reactions to form C–C and C–N bonds are a mainstay of organic synthesis for a wide range of academic and industrial applications [1–6]. Due to their wide applicability, these reactions have become a critical arsenal for synthetic chemists and have clearly changed retrosynthetic analysis of complex targets. Since their discovery in the late 1960s, palladium catalyzed cross-coupling reactions has been considerable and continues to be a focus of organometallic research [7–14]. The most active Pd catalysts for cross-coupling reactions involve the use of strong donor ligands to reach a high degree of efficiency. In fact, one of the major advancement in cross-coupling reactions is the synthesis and utilization of specialized electron-rich phosphines and *N*-heterocyclic carbenes (NHC) for the development of active catalytic systems expanding the substrate scope with lower catalyst loadings and milder conditions [15,16]. However, the monetary costs of these specialized ligands are often comparable to the Pd precursor. Therefore, the traditional route of addition of excess ligand for generating the active Pd(0) becomes unattractive [15,17]. Furthermore, in many cross-coupling reactions, the optimal Pd to ligand ratio is 1:1, with the active species proposed to be a monoligated Pd(0). Therefore, the use of well-defined Pd(II) pre-catalysts to facilitate cross-coupling reactions is highly desirable, as they can generate mono ligated active Pd(0) catalysts in solution. Since Herrmann reported that Pd-NHC complexes efficiently catalyzed Heck reaction [18], these complexes found a widespread use for various cross-coupling reactions incorporating previously unreactive coupling partners (Figure 1) [19].

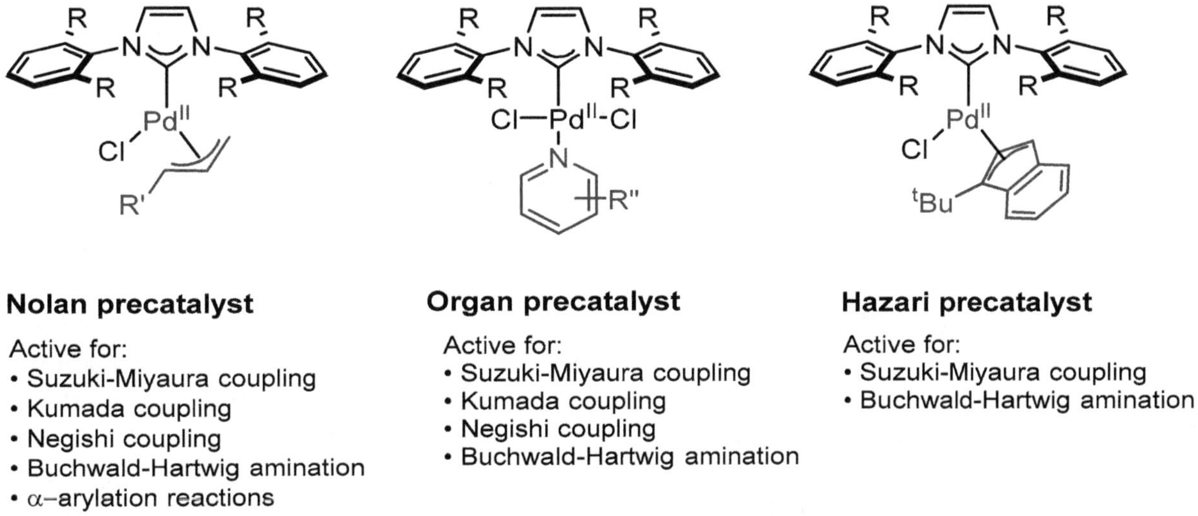

**Nolan precatalyst**

Active for:
- Suzuki-Miyaura coupling
- Kumada coupling
- Negishi coupling
- Buchwald-Hartwig amination
- $\alpha$–arylation reactions

**Organ precatalyst**

Active for:
- Suzuki-Miyaura coupling
- Kumada coupling
- Negishi coupling
- Buchwald-Hartwig amination

**Hazari precatalyst**

Active for:
- Suzuki-Miyaura coupling
- Buchwald-Hartwig amination

**Figure 1.** Overview of cross-coupling reactions catalyzed by Pd(II)-NHC pre-catalysts.

Although various electrophiles are employed in cross-coupling reactions for the construction of C–C and C–N bonds, there is immense interest in increasing the substrate scope to include a wide range of cross-coupling partners [1,4,20]. In recent years, tremendous progress was made to incorporate stable, unreactive, carboxylic acid derivatives, such as amides and esters, in cross-coupling reactions to form ketones or amides [20,21]. The resonance, due to $n_{N\ (or)\ O} \rightarrow \pi^*_{C=O}$ conjugation, makes them a complicated reacting partner in cross-coupling reactions, requiring a high activation energy for the N or O–C(O) bond scission. Destabilization strategies increasing the steric bulk on the amide nitrogen independently developed by the laboratories of Garg [22], Szostak [23], and Zou [24] provide a basis for the development of acyl cross-coupling of amides. Various Pd and Ni catalyst systems with phosphine and NHC ancillary ligands are shown to be effective in utilizing amides and esters as coupling partners in Suzuki-Miyaura coupling and Buchwald-Hartwig amination reactions. However, in this review, we will focus only on describing the recent advances made in the cross-coupling of amides and esters using well-defined pre-catalysts. The individual sections in this review are organized according to the pre-catalysts employed [Pd(allyl)(NHC)Cl, and Pd-PEPPSI] for cross-coupling of amides and esters.

## 2. Pd(allyl)(NHC)Cl Pre-Catalysts in Suzuki-Miyaura Cross-Coupling of Amides and Esters

With a wide functional group tolerance and less sensitivity toward air and moisture, palladium is by far the most commonly used metal for catalysis of cross-coupling reactions. Among the various pre-catalysts developed in the past decade, there is considerable interest and improvement in pre-catalysts based on $\eta^3$-allyl ligands. Aside from the inherent advantages associated with NHC ligands (i.e., strong $\sigma$-donation, tunability, sterics), Nolan demonstrated that $\eta^3$-allyl and $\eta^3$-cinnamyl pre-catalysts with bulky NHC ligands also exhibit high stability toward air and moisture allowing for easy storage and handling [15,25,26]. Furthermore, the commercial availability and ease of synthesis made them an important class of catalysts for reaction screening. The strong $\sigma$-donating nature of NHC facilitates the activation of stable and unreactive bonds. Specifically, the [Pd(cinnamyl)(L)Cl] scaffold (L = IPr* or IPr*OMe) was incorporated into several highly active catalysts for difficult cross-coupling reactions, such as the synthesis of tetra-*ortho*-substituted biaryls using Suzuki-Miyaura coupling [27], Buchwald-Hartwig reactions with secondary amines [28], as well as the use of amides and esters as electrophiles in Suzuki-Miyaura coupling and Buchwald-Hartwi aminations. Figure 2 lists selected and active Pd(allyl)NHC pre-catalysts that were employed for the cross-coupling of amides and esters [19].

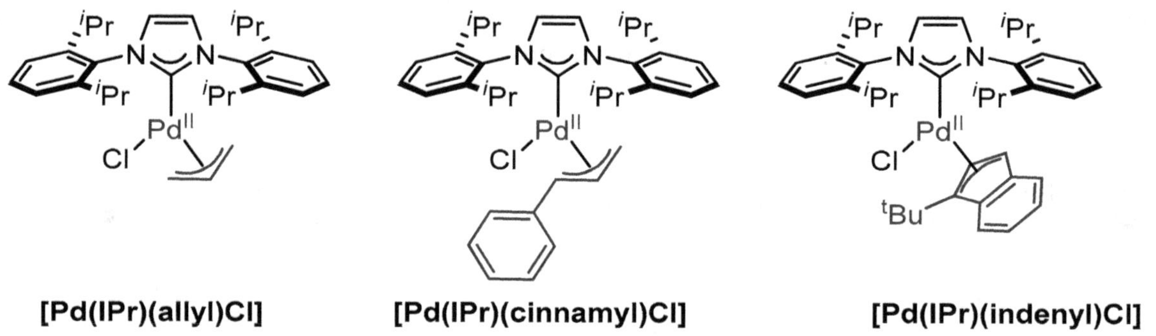

[Pd(IPr)(allyl)Cl] [Pd(IPr)(cinnamyl)Cl] [Pd(IPr)(indenyl)Cl]

**Figure 2.** Pd(allyl)NHC pre-catalysts employed in the cross-coupling of amides.

### *2.1. Pd(allyl)(NHC)Cl Pre-Catalysts in Suzuki-Miyaura Cross-Coupling of Amides*

Given the key role of the amide bond in nature, development of new methods to functionalize amides via C–N bond activation became an active area of research. Mechanistically, activation of the C–N amide bond proceeds through ground-state destabilization (twisting) of the amide bond by steric and/or electronic factors, allowing a facile insertion of a metal into the C–N bond, furnishing the acyl–metal intermediate facilitating coupling reactions [29–31]. Scheme 1 lists various substituents on amide nitrogen employed for the destabilization.

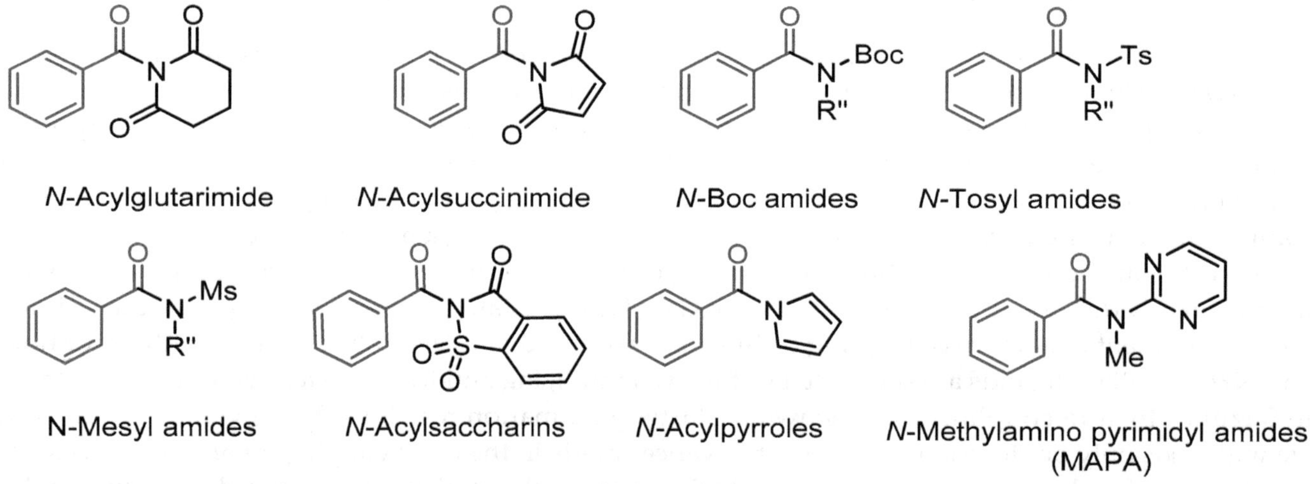

*N*-Acylglutarimide *N*-Acylsuccinimide *N*-Boc amides *N*-Tosyl amides

N-Mesyl amides *N*-Acylsaccharins *N*-Acylpyrroles *N*-Methylamino pyrimidyl amides (MAPA)

**Scheme 1.** Selected *N*-substituents employed for amide C–N bond destabilization.

In 2015, the utilization of amide bond for the Suzuki-Miyaura cross-coupling was reported by Zou and coworkers employing *N*-phenyl-*N*-tosyl substituted amides [24]. The report showcased the effect of substituent electronics on the destabilization of amide resonance facilitating the metal insertion in to the amide C–N bond. Although the methodology showed good functional group tolerance, it suffered from high catalyst loadings and stringent reaction conditions (Scheme 2).

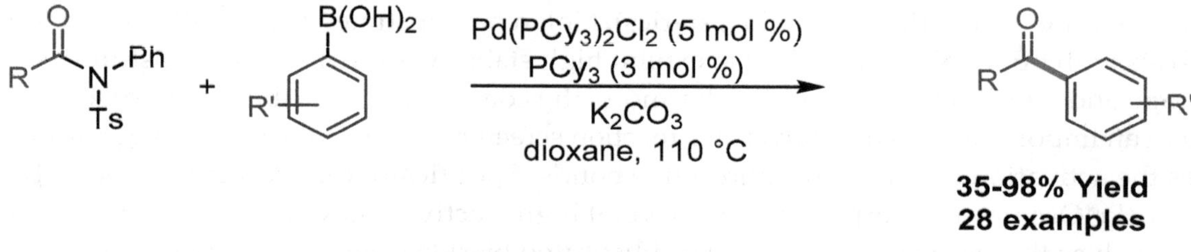

$Pd(PCy_3)_2Cl_2$ (5 mol %), $PCy_3$ (3 mol %), $K_2CO_3$, dioxane, 110 °C

**35-98% Yield**
**28 examples**

**Scheme 2.** Suzuki-Miyaura cross-coupling of *N*-phenyl-*N*-tosyl substituted amides.

In the same year, Szostak and coworkers employed a rather different approach of sterically controlled amide bond destabilization using a variety of *N*-substituted amides and found the best results to be *N*-glutarimide [23]. The methodology provided inherent steric advantages for amide bond distortion, thereby providing a milder reaction conditions and high functional group tolerance (Scheme 3). But the utilization of large excess of expensive ligand (1:4 ratio of $Pd(OAc)_2$ to $PCy_3HBF_4$), and applicability to only highly twisted *N*-glutarimide amides restricted the practical applications of the methodology.

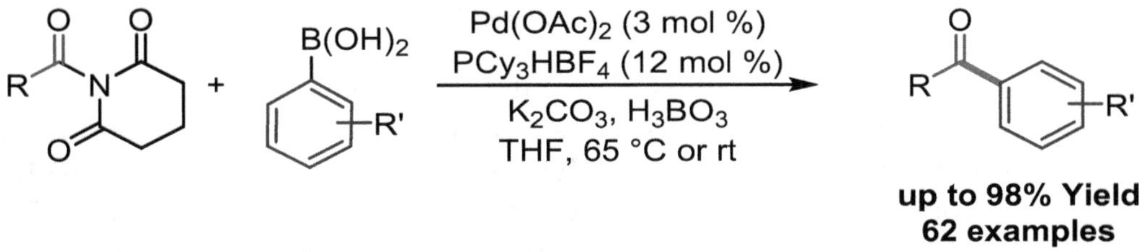

$Pd(OAc)_2$ (3 mol %)
$PCy_3HBF_4$ (12 mol %)
$K_2CO_3$, $H_3BO_3$
THF, 65 °C or rt

**up to 98% Yield**
**62 examples**

**Scheme 3.** Suzuki-Miyaura cross-coupling of sterically substituted twisted amides.

It was not until recently that a more general method incorporating all classes of amides under milder conditions was developed. In 2017, Szostak and coworkers reported the versatility of [Pd(IPr)(cinnamyl)Cl], a well-defined pre-catalyst in the Suzuki-Miyaura cross-coupling of amides under operationally simple conditions (Scheme 4) [23]. The methodology is highly significant, as it promoted the C–N amide bond activation of various amides, including *N*-glutarimide, *N*-Boc-carbamate, and *N*-toluenesulfonamide, under the same reaction conditions with similar yields of the ketone product. The method also showed high functional group tolerance for both reacting partners. The use of easily prepare and common protecting groups (*N*-Boc and *N*-Ts amides) for cross-coupling is appealing, as it brings the amide bond cross-coupling closer to general practical use. As stated previously, it was proposed that strong σ donation of the NHC facilitates oxidative addition, while flexible steric bulk around Pd promotes reductive elimination, triggering high reactivity even with less activated amide precursors.

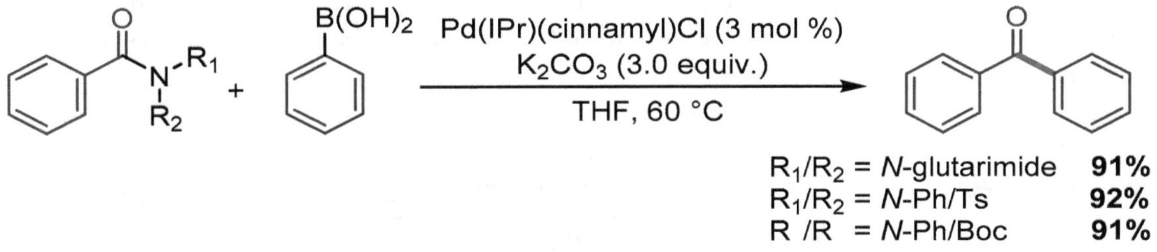

Pd(IPr)(cinnamyl)Cl (3 mol %)
$K_2CO_3$ (3.0 equiv.)
THF, 60 °C

$R_1/R_2$ = *N*-glutarimide **91%**
$R_1/R_2$ = *N*-Ph/Ts **92%**
R /R = *N*-Ph/Boc **91%**

**Scheme 4.** Application of Pd(cinnamyl)(IPr)Cl in amide bond cross-coupling.

While nonplanar amides were found to be very effective and well-documented for cross-coupling reactions by steric distortion, the resonance destabilization of corresponding planar amides was not well developed. To incorporate planar amides for cross-coupling reactions, Szostak and coworkers utilized *N*-acylpyrroles and *N*-acylpyrazoles in Suzuki-Miyaura cross-coupling reactions (Scheme 5) [32]. They found delocalization of $N_{lp}$ into the π-electron system of the pyrrole/pyrazole ring to be sufficient to selectively insert palladium into the amide C–N bond in the absence of steric distortion. The extensive optimization emphasized the importance of well-defined [Pd(cinnamyl)(IPr)Cl] producing ketone in good yields compared to traces of product formation with the traditional catalyst/ancillary ligand system.

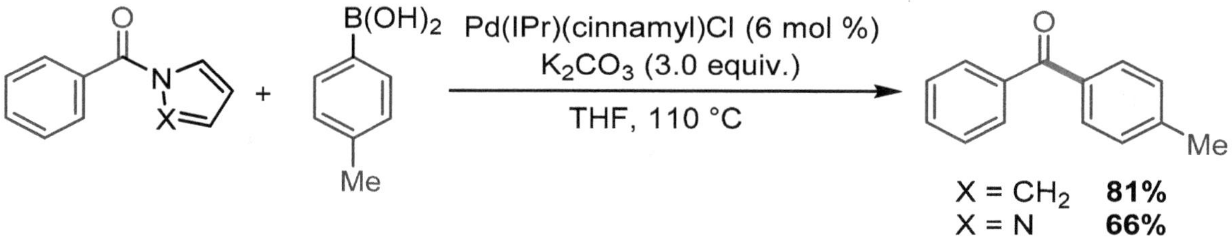

**Scheme 5.** Cross-coupling of planar amides catalyzed by well-defined Pd(cinnamyl)(IPr)Cl.

Compared to both traditional Pd/phosphine catalyst systems or Pd/ancillary NHC catalytic systems, well-defined [Pd(cinnamyl)(IPr)Cl] offers both practical advantages and high activity in cross-coupling reactions of amides. In fact, the high catalytic activity of [Pd(cinnamyl)(IPr)Cl] can be evident by its ability to convert *N*-alkyl-*N*-aryl amides to corresponding ketones. In 2017, Szostak and coworkers introduced *N*-methylaminopyrimidyl amides (MAPA) as highly reactive, electronically activated amides for C–N bond cleavage (Scheme 6) [33,34].

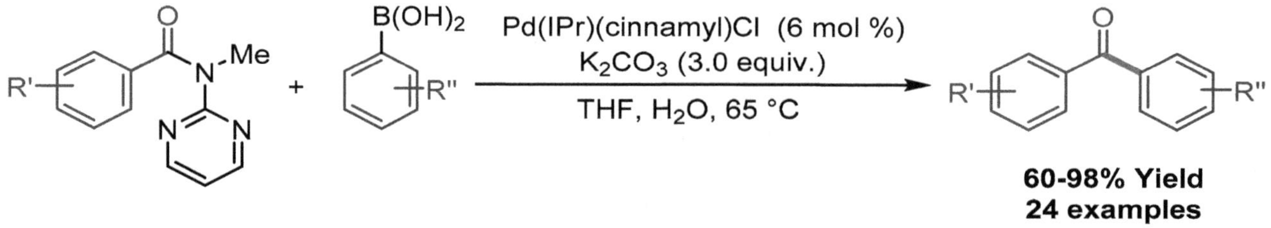

**Scheme 6.** Pd(cinnamyl)(IPr)Cl catalyzed Suzuki-Miyaura cross-coupling of *N*-alkyl-*N*-aryl amides.

Interestingly, Suzuki-Miyaura cross-coupling of both the sterically distorted and/or electronically distorted amides were found to be reactive under similar reaction conditions, further demonstrating how the well-defined nature of a catalyst can enhance the overall catalytic activity and reactivity compared to traditional usage of excess ancillary ligand. This is highly beneficial, as it allows chemists to screen minimal conditions to optimize methodology for new classes of amides with similar distortion angles or resonance energies. In fact, Zeng and coworkers, on their quest to incorporate commercially available and inexpensive *N*-acylhydantoins as amide electrophiles, have also used similar conditions for the formation of ketones indicating the versatility of Pd(allyl)(NHC)Cl pre-catalysts with different substituted amides (Scheme 7) [35].

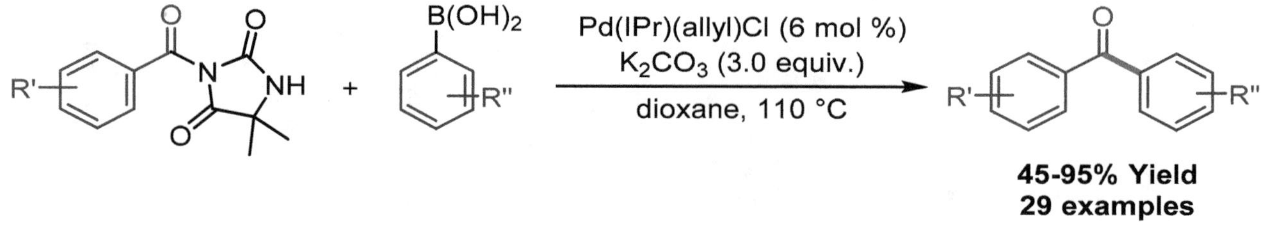

**Scheme 7.** Pd(allyl)(IPr)Cl catalyzed Suzuki-Miyaura cross-coupling of *N*-acylhydantoins.

Although the reported methods to incorporate amide bonds in cross-coupling reactions is quite promising, the synthetic modifications to activate amide bond can impede overall synthetic utility. On the other hand, use of *N*,*N*-di-Boc-activated amides would be highly beneficial, allowing direct engagement of most ubiquitous primary amides. Szostak and coworkers developed a new catalytic system based on [Pd(IPr)(cinnamyl)Cl]/KF for the selective insertion of metal into acyl amide bond (cf. *N*-carbamate bond) [36]. The addition of 5.0 equiv of water was found to be critical for the cross-coupling of di-boc amides, as is evident from decreased yields or no reaction in the absence of water (Scheme 8).

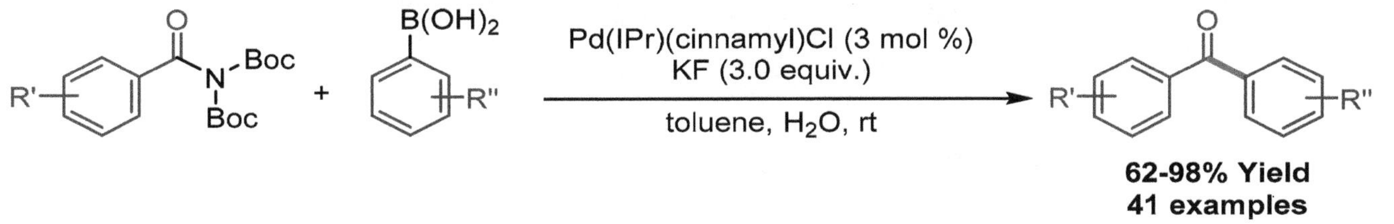

Pd(IPr)(cinnamyl)Cl (3 mol %)
KF (3.0 equiv.)
toluene, $H_2O$, rt

**62-98% Yield**
**41 examples**

**Scheme 8.** Direct cross-coupling of *N,N*-di-Boc-amides under Pd(IPr)(cinnamyl)Cl catalysis.

The high catalytic activity of Pd(IPr)(cinnamyl)Cl allowed for the cross-coupling to be conducted at ambient temperature with even lower catalyst loadings. The robust nature of Pd(IPr)(cinnamyl)Cl also allowed them for the gram scale conversion of primary amides to ketones with just 0.5 mol % of catalyst with a TON of 1220, further increasing the practicality of amide cross-coupling (Scheme 9) [36]. In fact, the high catalytic activity of Pd-IPr complexes also allowed Szostak and coworkers to use easily accessible *N*-acyl amides for the cross-coupling reaction with the highest reported TON (3500), compared to the corresponding traditional Pd-phosphine catalytic systems [37].

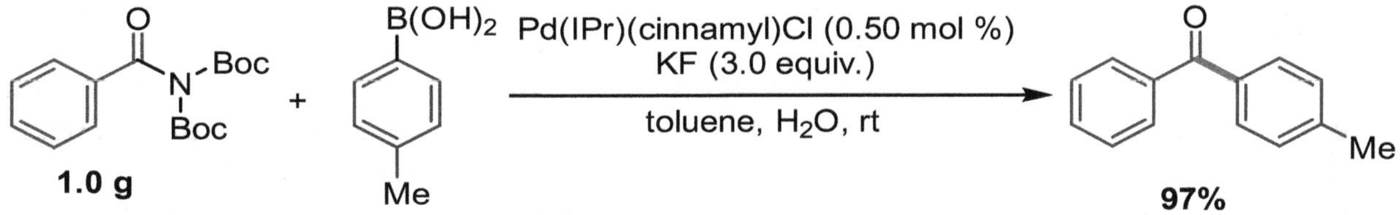

Pd(IPr)(cinnamyl)Cl (0.50 mol %)
KF (3.0 equiv.)
toluene, $H_2O$, rt

**1.0 g** **97%**

**Scheme 9.** Scale-up study in direct cross-coupling of *N,N*-di-Boc-amides under Pd(IPr)(cinnamyl)Cl catalysis.

Recently, employing standard reaction conditions of amide Suzuki-Miyaura cross-coupling under well-defined Pd-NHC catalysis (i.e., Pd(IPr)(cinnamyl)Cl/$K_2CO_3$/THF), Szostak and coworkers also realized challenging C($sp^2$)-C($sp^3$) couplings using *N,N*-di-Boc-amides and B-$sp^3$ alkyl reagents (Scheme 10) [38]. Various substituents on the amide bonds, such as *N*-glutarimide, *N*-Ph-*N*-Boc, and *N*-Ph-*N*-Ms were also found to be reactive under identical conditions albeit at a higher temperature.

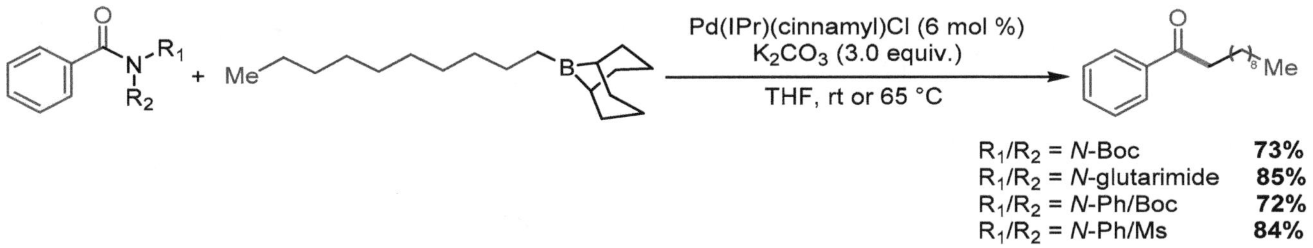

Pd(IPr)(cinnamyl)Cl (6 mol %)
$K_2CO_3$ (3.0 equiv.)
THF, rt or 65 °C

$R_1/R_2$ = *N*-Boc **73%**
$R_1/R_2$ = *N*-glutarimide **85%**
$R_1/R_2$ = *N*-Ph/Boc **72%**
$R_1/R_2$ = *N*-Ph/Ms **84%**

**Scheme 10.** Pd(IPr)(cinnamyl)Cl catalyzed alkylation of amides by C–N bond cleavage.

### *2.2. Pd($\eta^3$-1-t-Bu-indenyl) (NHC)Cl Pre-Catalysts in Suzuki-Miyaura Cross-Coupling of Amides*

The high reactivity of [Pd(allyl)(NHC)Cl] pre-catalysts stems from maintaining the optimal 1:1 Pd to ligand ratio and the fast reduction of Pd(II) to active Pd(0) [15]. It is now widely accepted that base mediated activation of [Pd(allyl)(NHC)Cl] pre-catalysts forms the active ligated Pd(0) catalyst in solution [34]. However, elegant studies by Hazari and coworkers observed a deleterious comproportionation pathway under catalytic conditions to form Pd(I) μ-allyl dimers (Scheme 11) [39,40]. Based on extensive mechanistic investigation, they proposed two strategies to develop even more active catalysts: (1) Increase the barrier to comproportionation between Pd (0) and Pd (II), and (2) develop systems that undergo faster activation so that all of Pd(II) is converted to Pd(0) before comproportionation.

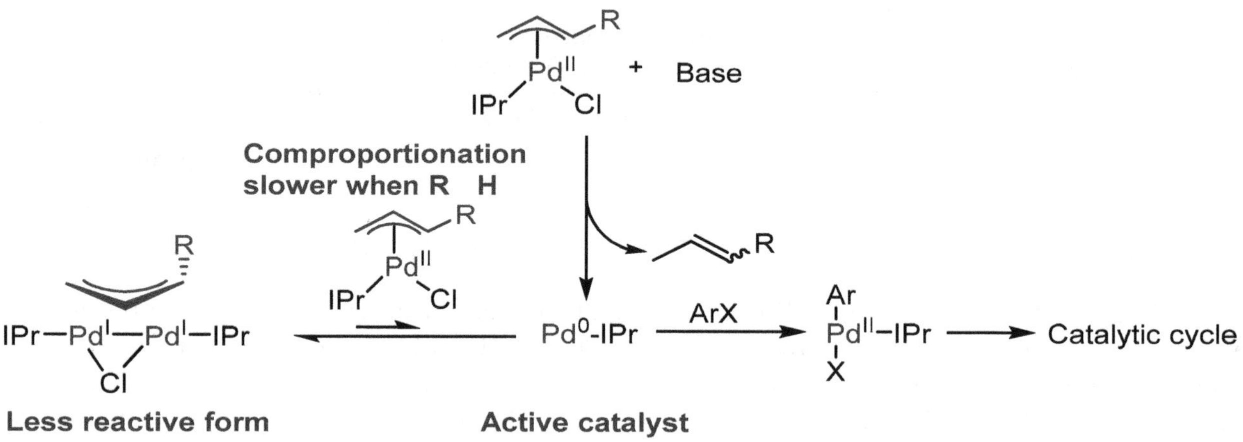

**Scheme 11.** Activation/deactivation pathway during Pd(allyl)(NHC)Cl pre-catalyst activation.

The mechanistic insights on the deactivation pathway led the Hazari group to discover Pd($\eta^3$-1-*t*-Bu-indenyl)(NHC)Cl (Hazari catalyst), a highly efficient pre-catalyst for cross-coupling reactions [41]. As a major advance in Pd-NHC precatalysts, the inability of ($\eta^3$-1-*t*-Bu-indenyl)(NHC)Cl to generate unreactive Pd(I) dimers significantly improved its activity for cross-coupling reactions. In 2017, Szostak and coworkers utilized [Pd(indenyl)(IPr)Cl] for the development of Suzuki-Miyaura cross-coupling of *N*-Ph-*N*-Boc amides under otherwise standard conditions of amide cross-coupling reaction ($K_2CO_3$/THF, Scheme 12) [42]. The unprecedented reactivity of Hazari catalyst by suppressing the deactivation pathway led the amide cross-coupling reaction to occur at room temperature under glovebox-free conditions, increasing the operational simplicity and practicality of the reaction. This reaction is notable as the first example of Suzuki-Miyaura cross-coupling of an amide at room temperature with excellent yield of ketone product. The robust nature of the pre-catalyst allowed the Szostak group to develop the first direct one-pot activation/cross-coupling of secondary amides.

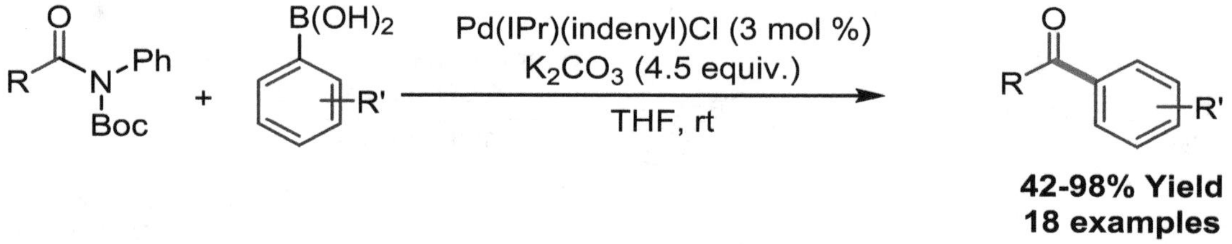

**Scheme 12.** Pd(IPr)(indenyl)Cl catalyzed Suzuki-Miyaura coupling of amides at room temperature.

## 2.3. *Pd(allyl)(NHC)Cl Pre-Catalysts in Suzuki-Miyaura Cross-Coupling of Esters*

The ubiquitous nature of ester bonds attracted the interest of chemists for their synthetic manipulation into useful products. In recent years, the direct use of aromatic esters in cross-coupling reactions to form biaryls or ketones has been demonstrated as a promising area of research [20]. The selective cleavage of ester bonds offers significant advantages in multistep synthesis as they are robust to a range of reaction conditions and can only be activated under specific conditions. Although activation of aryl C–O bonds to form biaryls via decarbonylation is well-documented [43,44] the corresponding synthesis of ketones is still in its infancy. In 2017, Newman, Houk and coworkers reported the first-time utilization of aryl esters as acyl equivalents for the formation of ketones (Scheme 13) [45]. The exceptional bulkiness of NHC ligand on the catalyst hindered the decarbonylation step facilitating ketone formation over the well-known biaryl formation.

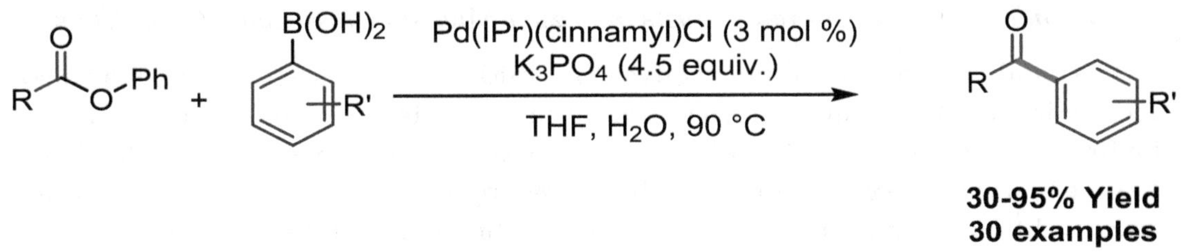

**Scheme 13.** Pd(IPr)(cinnamyl)Cl catalyzed Suzuki-Miyaura cross-coupling of aryl esters.

Optimization studies demonstrated the importance of well-defined pre-catalysts for successful ketone formation (Table 1). When Pd catalysts in combination with excess ancillary ligand were used, very low yield of the product was obtained (Table 1, entry 1–4). On the contrary, when the metal to ligand ratio was decreased to the optimal 1:1 ratio, a substantial increase in the product formation was observed (Table 1, entry 5). Further improvement was reported when a preformed well-defined Pd(IPr)(cinnamyl)Cl was used as the catalyst, forming the cross-coupled product in 95% yield, highlighting the importance of well-defined pre-catalysts as compared to traditional catalyst/ligand system (Table 1, entry 6).

**Table 1.** Optimization study for the Suzuki-Miyaura cross-coupling of aryl esters.

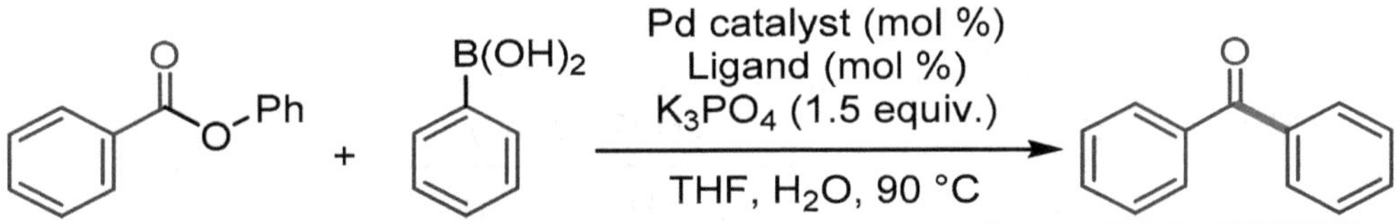

| Entry | Pd Source (mol %) | Ligand (mol %) | Yield (%) |
|---|---|---|---|
| 1 | $Pd(OAc)_2$ (5) | IPr·HCl (10) | 11 |
| 2 | $Pd(dba)_3$ (5) | IPr·HCl (10) | 16 |
| 3 | $[Pd(allyl)Cl]_2$ (5) | IPr·HCl (10) | 19 |
| 4 | $[Pd(cinnamyl)Cl]_2$ (5) | IPr·HCl (10) | 21 |
| 5 | $[Pd(cinnamyl)Cl]_2$ (5) | IPr·HCl (5) | 59 |
| 6 | Pd(IPr)(cinnamyl)Cl (5) | none | 95 |

Very recently, as an improvement from their work on C(sp$^2$)-C(sp$^2$) cross-coupling of esters, the Newman group reported a C(sp$^2$)-C(sp$^3$) cross-coupling of aryl esters with alkyl-BBN to form aryl-alkyl ketones [46]. This reaction is particularly challenging, as alkyl boron reagents, in general, are reluctant to undergo transmetallation relative to their aryl counterparts and the intermediacy of alkylmetal species are prone to $\beta$-hydride elimination. However, the strong $\sigma$-donation and bulky nature of NHC in precatalysts, enabled the reaction to proceed to form ketones. On the contrary, when they employed Pd-ancillary phosphine catalyst conditions, biaryls were formed via decarbonylation and reductive elimination (Scheme 14).

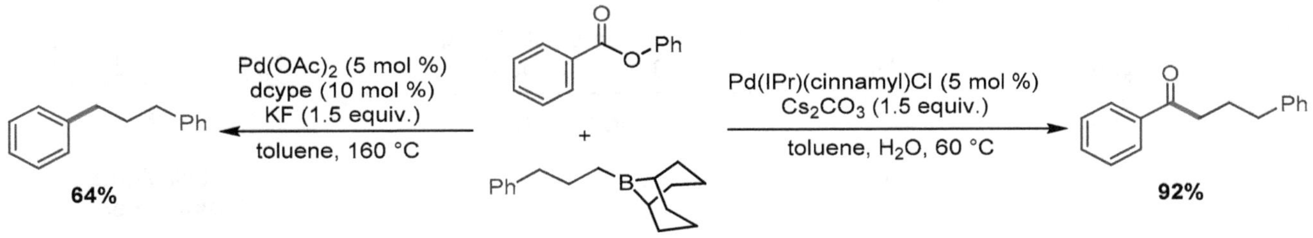

**Scheme 14.** Suzuki-Miyaura coupling of aryl esters with alkyl-BBN.

### 2.4. *Pd($\eta^3$-1-t-Bu-indenyl) (NHC)Cl Pre-Catalysts in Suzuki-Miyaura Cross-Coupling of Esters*

The Hazari catalyst, a highly active catalyst for amide bond cross-coupling, was also found to be very effective in Suzuki-Miyaura cross-coupling of aryl esters. In 2017, Szostak and coworkers reported the first room temperature Suzuki-Miyaura cross-coupling of esters by selective C–O bond activation to give biaryl ketones (Scheme 15) [42]. Interestingly, only a small variation in the amount of base from the conditions employed for amide cross-coupling enabled the cross-coupling of esters.

Pd(IPr)(indenyl)Cl (3 mol %)
$K_2CO_3$ (7.2 equiv.)
THF, rt
60-98% Yield
12 examples

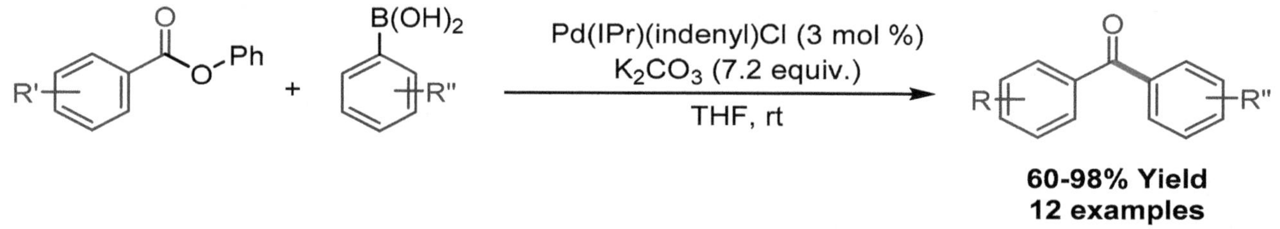

**Scheme 15.** Suzuki-Miyaura cross-coupling of aryl esters catalyzed by Pd(indenyl)(IPr)Cl.

An advancement in the cross-coupling of aryl esters catalyzed by Pd(indenyl)(IPr)Cl was reported by Hazari and coworkers (Scheme 16) [47]. The use of potassium hydroxide base allowed them to develop even milder reaction conditions with lower catalyst loadings (1.0 mol %), and shorter reaction times (6 h to 16 h).

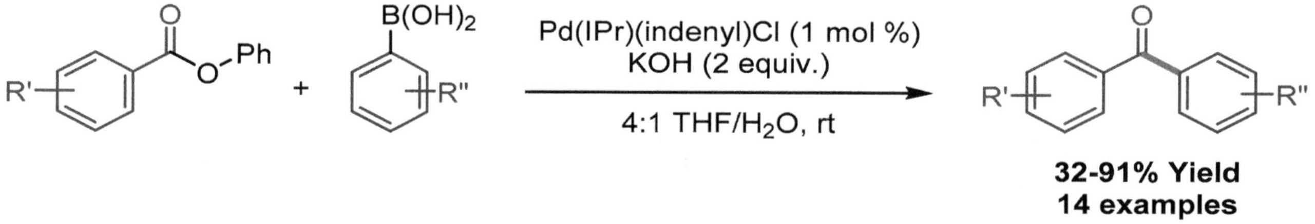

Pd(IPr)(indenyl)Cl (1 mol %)
KOH (2 equiv.)
4:1 THF/$H_2O$, rt
32-91% Yield
14 examples

**Scheme 16.** Hazari protocol for the Suzuki-Miyaura cross-coupling of aryl esters catalyzed by Pd(indenyl)(IPr)Cl using hydroxide bases.

## 3. Pd(allyl)(NHC)Cl Pre-Catalysts in Buchwald-Hartwig Amination of Amides and Esters

### 3.1. *Pd(cinnamyl)(IPr)Cl Pre-Catalyst for the Transamidation of Amides*

The excellent reactivity of Pd(cinnamyl)(IPr)Cl pre-catalysts in amide C–N activation to form acyl-metal species is highly significant, as it can potentially undergo cross-coupling reactions with any nucleophile. In 2017, Szostak and coworkers realized the acyl-metal reaction with other nucleophiles when they reported the first general method for Buchwald-Hartwig amination of acyl-metal species with amines under Pd-NHC catalysis (Scheme 17) [48]. The protocol offered advantages in efficient cross-coupling of both alkyl, aryl, and sterically hindered anilines. The method also showed a broad scope with respect to both the amine and amide components.

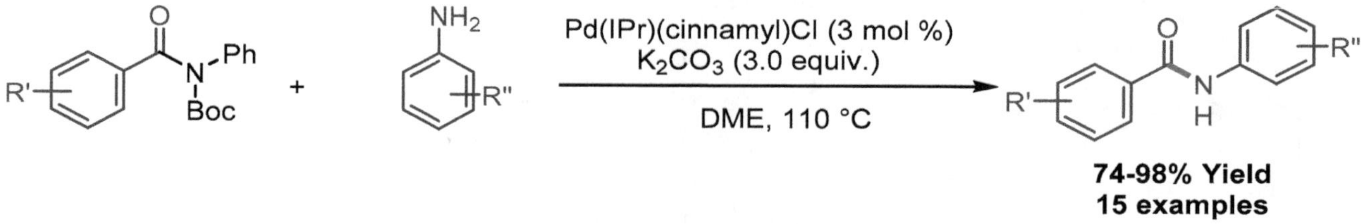

Pd(IPr)(cinnamyl)Cl (3 mol %)
$K_2CO_3$ (3.0 equiv.)
DME, 110 °C
74-98% Yield
15 examples

**Scheme 17.** Pd(IPr)(cinnamyl)Cl catalyzed transamidation of amides.

### *3.2. Pd(allyl)(IPr)Cl Pre-Catalyst for the Transamidation of Esters*

Newman and coworkers reported the first example of Pd-NHC catalyzed amide bond formation directly from aryl esters and anilines. Similar to their observation in Suzuki coupling of aryl esters, a well-defined preassembled NHC-ligated catalyst was essential, as separated components resulted in reduced yields [45]. They found the use of a mild carbonate base and the presence of water was essential for higher conversion, forming the transamidated product in good yields (Scheme 18). The mildness of the protocol was showcased by the reaction of proline ester with anilines and the aminated product was formed with minimal loss of enantiopurity.

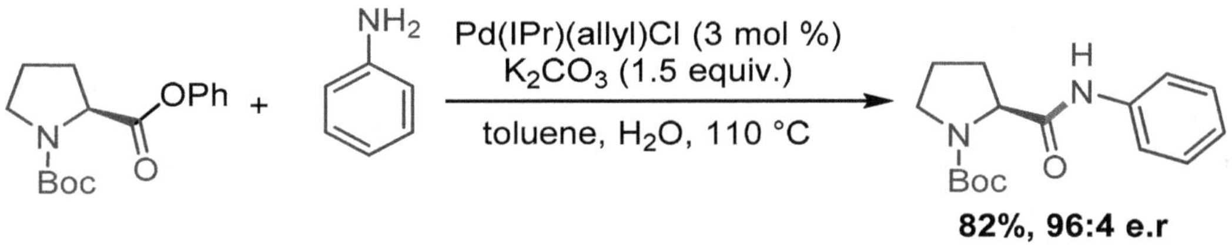

Pd(IPr)(allyl)Cl (3 mol %)
$K_2CO_3$ (1.5 equiv.)
toluene, $H_2O$, 110 °C
82%, 96:4 e.r

**Scheme 18.** Pd(allyl)(IPr)Cl catalyzed amidation of proline esters.

### *3.3. Pd(indenyl)(SIPr)Cl Pre-Catalyst for the Transamidation of Esters*

Given the exceptional catalytic activity of Pd(indenyl)(IPr)(Cl) for Suzuki-Miyaura reactions involving aryl esters, Hazari and coworkers also explored Buchwald-Hartwig amination of aryl esters. Using 1 mol % of Pd(indenyl)(SIPr)(Cl) (SIPr = 1,3-bis(2,6-diisopropylphenyl)imidazolidin-2-ylidene) as the pre-catalyst, they were able to couple phenyl benzoate and aniline in essentially quantitative yield at 40 °C, using a 4:1 $H_2O$/THF solvent mixture (Scheme 19) [47]. They also found that the coupling was highly sensitive to the ligand used, with SIPr affording the best yields, whereas other NHC or phosphine ligands were ineffective.

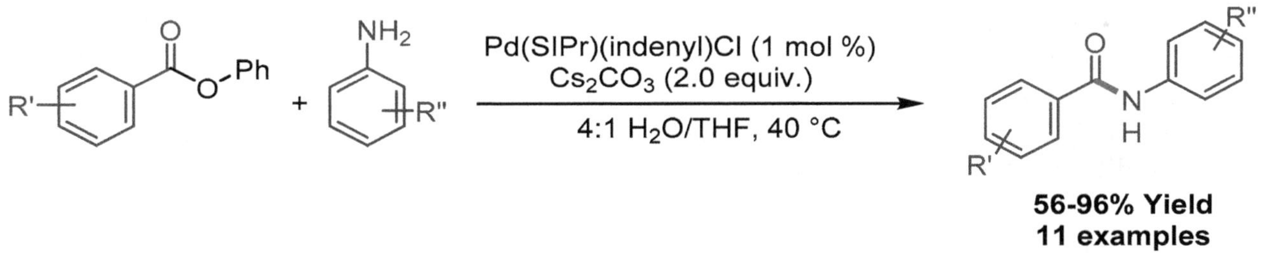

Pd(SIPr)(indenyl)Cl (1 mol %)
$Cs_2CO_3$ (2.0 equiv.)
4:1 $H_2O$/THF, 40 °C
56-96% Yield
11 examples

**Scheme 19.** Pd(indenyl)(SIPr)Cl pre-catalyst for the Buchwald-Hartwig amination of aryl esters.

## 4. Pd-PEPPSI Pre-Catalysts in the Suzuki-Miyaura Cross-Coupling of Amides and Esters

The last two decades saw tremendous growth in the development of highly active, generally applicable, and functional group-tolerant catalytic systems employing NHC ligands. One highly active well-defined pre-catalyst system is Pd-PEPPSI (pyridine-enhanced pre-catalyst preparation, stabilization and initiation), developed by the Organ group [49]. As Figure 1 illustrates, Pd-PEPPSI catalysts found their use in various cross-coupling reactions, such as Suzuki-Miyaura, Negishi, Kumada-Tamao-Corriu, and Buchwald-Hartwig aminations [50]. As with all the Pd-NHC precatalysts, Pd-PEPPSI has two major components: The bulky NHC ligand, which contains strong $\sigma$-donor properties which help promote the oxidative addition, and the sterics aid reductive elimination. The second component, the pyridine, often referred as a "throw away ligand," aids not only in the synthesis of pre-catalyst, but also increases the stability of the isolated complexes (Figure 3). The easy synthesis, high stability toward air and moisture and the exceptional reactivity of Pd-PEPPSI complexes led synthetic chemists to utilize these pre-catalysts in the cross-coupling of difficult and unreactive electrophiles, such as amides and esters (*vide infra*).

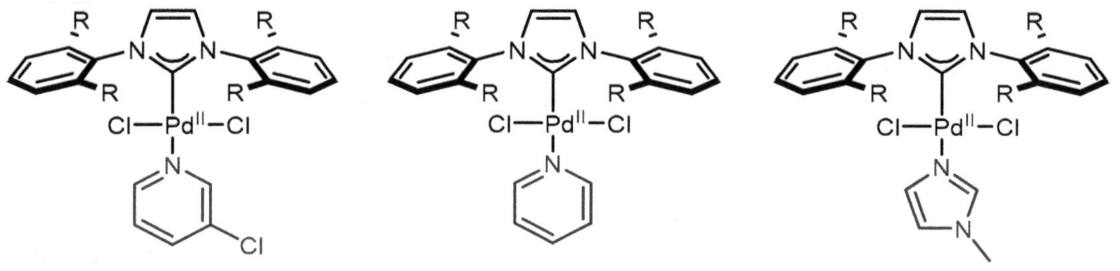

**Figure 3.** Well-defined Pd-PEPPSI pre-catalysts employed in the cross-coupling reactions.

### 4.1. *Pd-PEPPSI Pre-Catalysts in the Suzuki-Miyaura Cross-Coupling of Amides*

Szostak and coworkers were the first to report the use of Pd-PEPPSI pre-catalysts in Suzuki-Miyaura cross-coupling of amides [51]. In agreement with high catalytic activity of in situ generated Pd-NHC complexes, identical reaction conditions could be employed for the cross-coupling of various destabilized amides (Scheme 20).

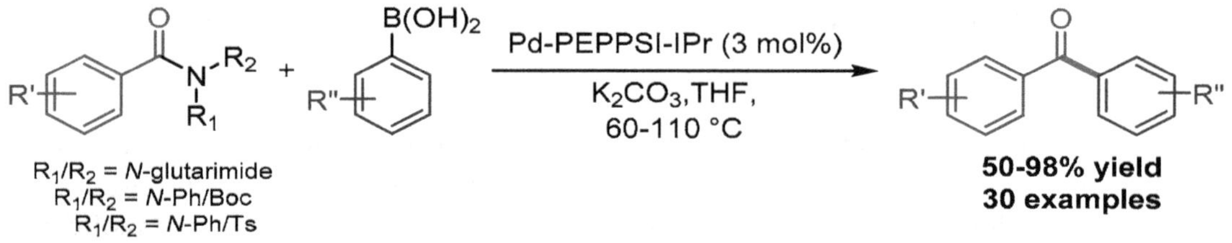

Pd-PEPPSI-IPr (3 mol%)
$K_2CO_3$, THF, 60-110 °C

$R_1/R_2$ = *N*-glutarimide
$R_1/R_2$ = *N*-Ph/Boc
$R_1/R_2$ = *N*-Ph/Ts

**50-98% yield**
**30 examples**

**Scheme 20.** Pd-PEPPSI catalyzed Suzuki-Miyaura cross-coupling of amides.

In accordance with the amide C–N bond destabilization, kinetic studies with *N*-glutarimide, *N*-Ph-*N*-Ts, and *N*-Ph-*N*-Boc amides showed similar reactivity to their cross-coupling reactivity scale with *N*-glutarimide being the fastest and *N*-Ph-*N*-Ts being the slowest [52]. The kinetic data between Pd-PEPPSI and Pd(IPr)(cinnamyl)Cl precatalysts under identical conditions suggested tha while the reaction rates of *N*-glutarimide and *N*-Ph-*N*-Boc amides were higher with Pd-PEPPSI, *N*-Ph-*N*-Ts amides are shown to have faster reaction rates under Pd(IPr)(cinnamyl)Cl catalysis (Figure 4) [19]. The data is highly beneficial, as it provides a general idea for chemists to choose from both the catalyst systems for different type of amides employed in cross-coupling reactions.

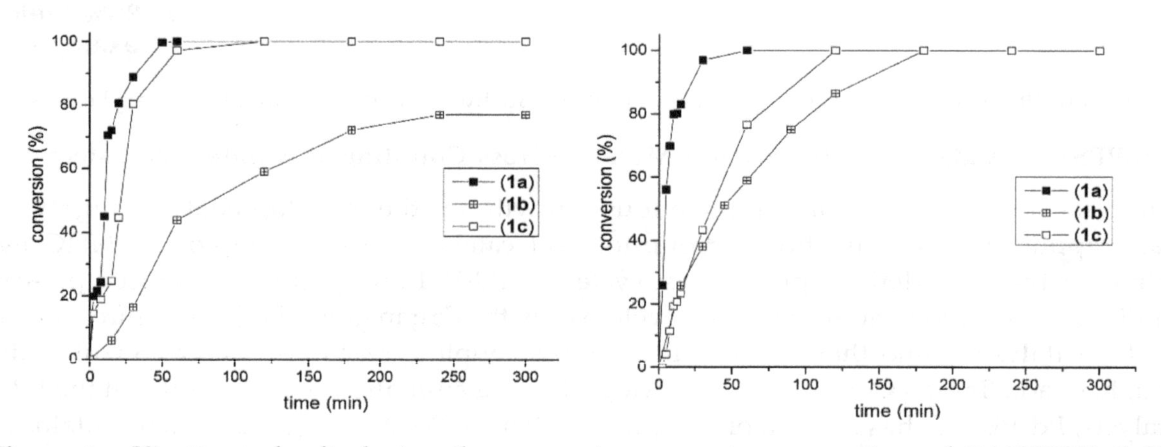


**Figure 4.** Kinetic study displaying the conversion percentage over time of Pd-PEPPSI (**left**), and Pd(IPr)(Cinnamyl)Cl (**right**) pre-catalysts. Activated amides used were *N*-Glutarimide (**1a**), *N*-Ph-*N*-Ts (**1b**), and *N*-Ph-*N*-Boc (**1c**). Reprinted with permission from *J. Org. Chem.*

Very recently, Zou and coworkers reported the Suzuki-Miyaura cross-coupling of less reactive *N*-alkyl-*N*-Ts amides with diarylborinic acids under Pd-PEPPSI catalysis (Scheme 21) [53]. The use of stable diarylborinic acids was advantageous, as it extended nucleophile counterparts in cross-coupling reactions. Under otherwise standard reaction conditions of amide cross-coupling reactions, they were successful in cross-coupling less reactive *N*-Me-*N*-Ts amides to form ketone products. Notably, no reaction ensued under Pd/phosphine catalysis and only low Pd-PEPPSI catalyst loadings (1.0 mol %) were needed, highlighting the high activity of Pd-NHC pre-catalysts. The protocol also offered a broad functional group tolerance, with both electron donating and electron withdrawing groups on both coupling partners.

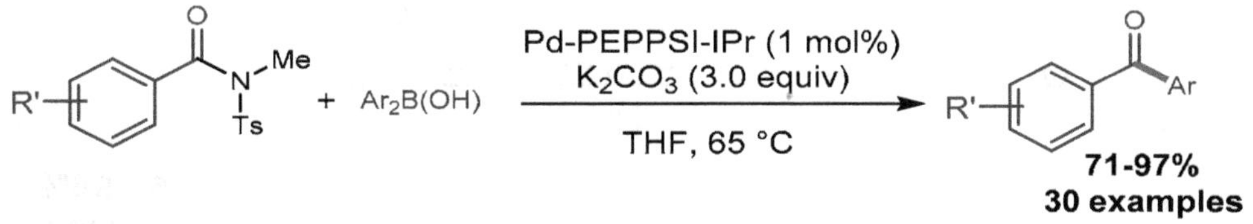

Pd-PEPPSI-IPr (1 mol%)
$K_2CO_3$ (3.0 equiv)
THF, 65 °C
$Ar_2B(OH)$
71-97%
30 examples

**Scheme 21.** Pd-PEPPSI catalyzed Suzuki-Miyaura cross-coupling of amides with diarylborinic acids.

Zou and coworkers further advanced the Suzuki-Miyaura coupling of amides to form alkyl ketones using trialkylboranes as coupling partners under Pd-PEPPSI catalysis [54]. Although Pd/phosphine catalysis proved ineffective, 5 mol % of Pd-PEPPSI was very effective in transforming *N*-Me-*N*-Ts amides to form alkyl ketones (Scheme 22). Unlike the high-order arylboron compounds, in which all the aryl groups react effectively, only one of the three primary alkyl groups in alkylboranes could be used as alkyl source for the acyl alkylation.

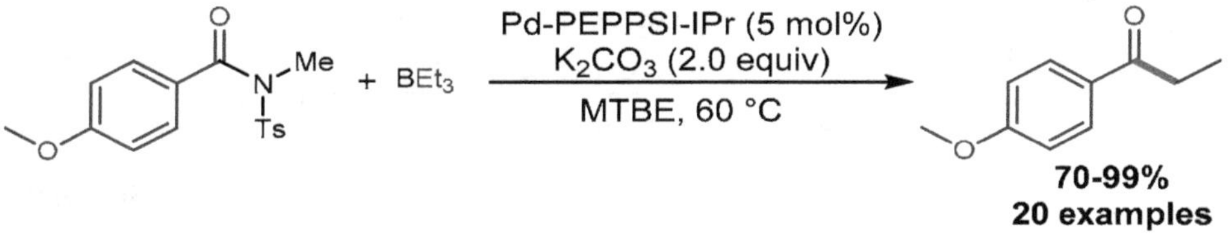

Pd-PEPPSI-IPr (5 mol%)
$K_2CO_3$ (2.0 equiv)
MTBE, 60 °C
$BEt_3$
70-99%
20 examples

**Scheme 22.** Pd-PEPPSI catalyzed alkylation of *N*-Me-*N*-Ts amide.

Wang, Liu, and coworkers recently reported a new series of Pd-NHC pre-catalysts using various *N*-heterocycles [55] as "throwaway ligands" with benzothiazole being the most effective for the Suzuki-Miyaura cross coupling of *N*-succinimide amides (Scheme 23) [56]. Change in the "throw-away ligand" was found to have a profound effect on the overall yield of the reaction, with benzothiazole ligand outperforming traditional 3-chloro pyridine ligand.

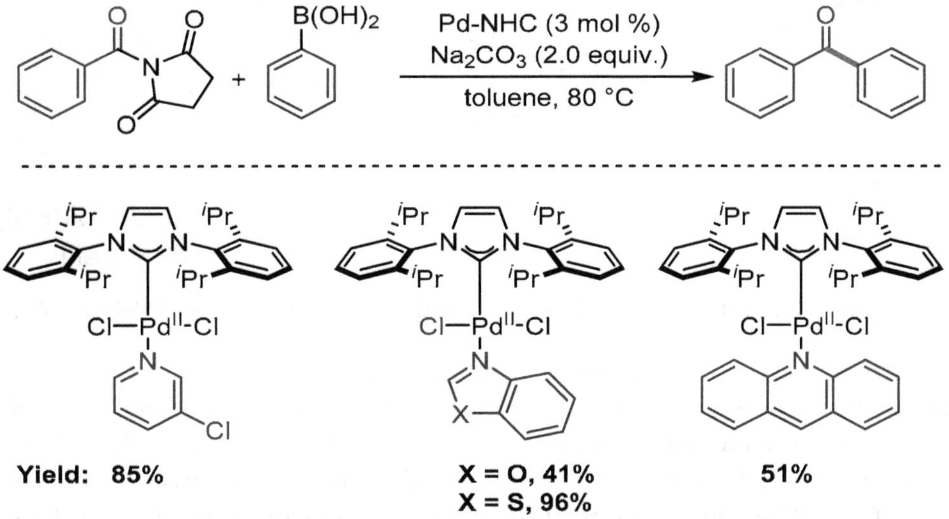

$B(OH)_2$
Pd-NHC (3 mol %)
$Na_2CO_3$ (2.0 equiv.)
toluene, 80 °C

Yield: 85% | X = O, 41% X = S, 96% | 51%

**Scheme 23.** NHC-Pd(II) catalyzed acylative Suzuki-Miyaura cross-coupling of amides.

*4.2. Pd-PEPPSI Pre-Catalysts in the Suzuki-Miyaura Cross-Coupling of Esters*

Pd-PEPPSI pre-catalysts were also found to be effective in catalyzing the cross-coupling of esters. In most cases, the reactivity was comparable to that of [Pd(IPr)(cinnamyl)Cl]. Szostak and coworkers reported a direct Suzuki-Miyaura cross-coupling of aryl esters under Pd-PEPPSI catalysis and found that the rates were similar to those reported for the cross-coupling of *N*-Ph/Boc, and *N*-Ph/Ts amides under identical reaction conditions (Scheme 24) [57].

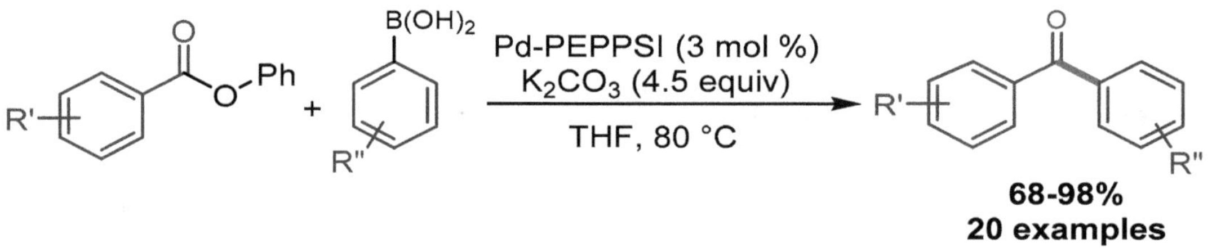

Pd-PEPPSI (3 mol %)
$K_2CO_3$ (4.5 equiv)
THF, 80 °C
**68-98%**
**20 examples**

**Scheme 24.** Pd-PEPPSI catalyzed Suzuki-Miyaura cross-coupling of aryl esters.

To further advance cross-coupling of esters, and to find a more general method, Szostak and coworkers reported a water assisted Suzuki-Miyaura cross-coupling of aryl esters at room temperature (Scheme 25) [58]. They demonstrated that the addition of water (5.0 equiv) was able to facilitate the reaction under very mild reaction conditions with only 1 mol % catalyst loading and at room temperature.

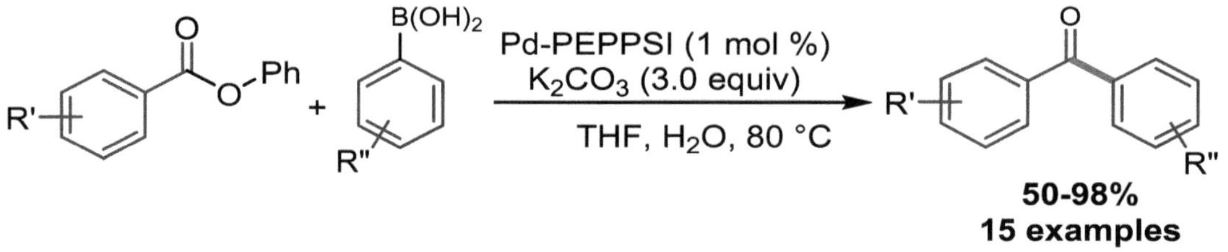

Pd-PEPPSI (1 mol %)
$K_2CO_3$ (3.0 equiv)
THF, $H_2O$, 80 °C
**50-98%**
**15 examples**

**Scheme 25.** Water assisted, Pd-PEPPSI catalyzed Suzuki-Miyaura coupling of esters.

To elucidate the role of water additive, they subjected the Pd-PEPPSI pre-catalyst under optimized reaction conditions without the coupling partners to observe the formation of hydroxide dimer, $[Pd(\mu\text{-}OH)Cl(IPr)]_2$ in appreciable 32% yield (Scheme 26).

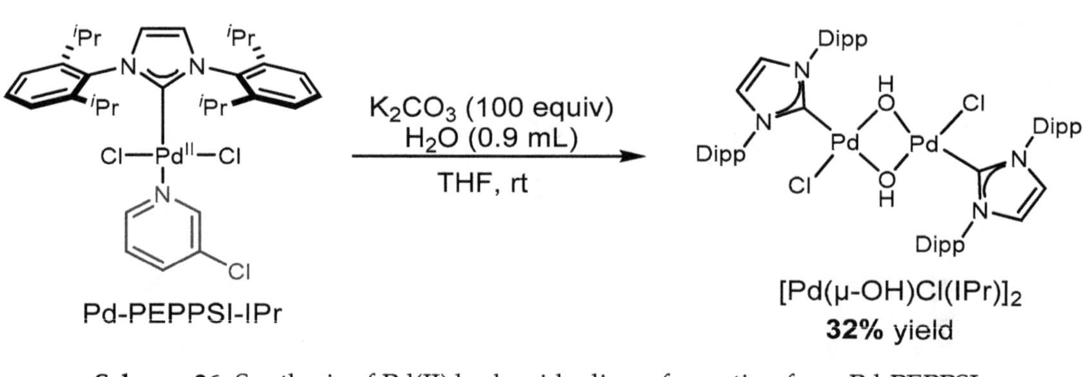

Pd-PEPPSI-IPr
$K_2CO_3$ (100 equiv)
$H_2O$ (0.9 mL)
THF, rt
$[Pd(\mu\text{-}OH)Cl(IPr)]_2$
**32%** yield

**Scheme 26.** Synthesis of Pd(II) hydroxide dimer formation from Pd-PEPPSI.

They also performed kinetic studies to probe the catalytic activity of preformed dimer and observed faster reaction rates compared to the Pd-PEPPSI conditions. Notably, negligible product formation in the absence of water additive indicated the importance of Pd-hydroxide dimer formation prior to the reduction of Pd(II) to active Pd(0) species (Figure 5). This protocol has also allowed the first achievement of a TON > 1000 in the Suzuki-Miyaura cross-coupling of aryl esters.

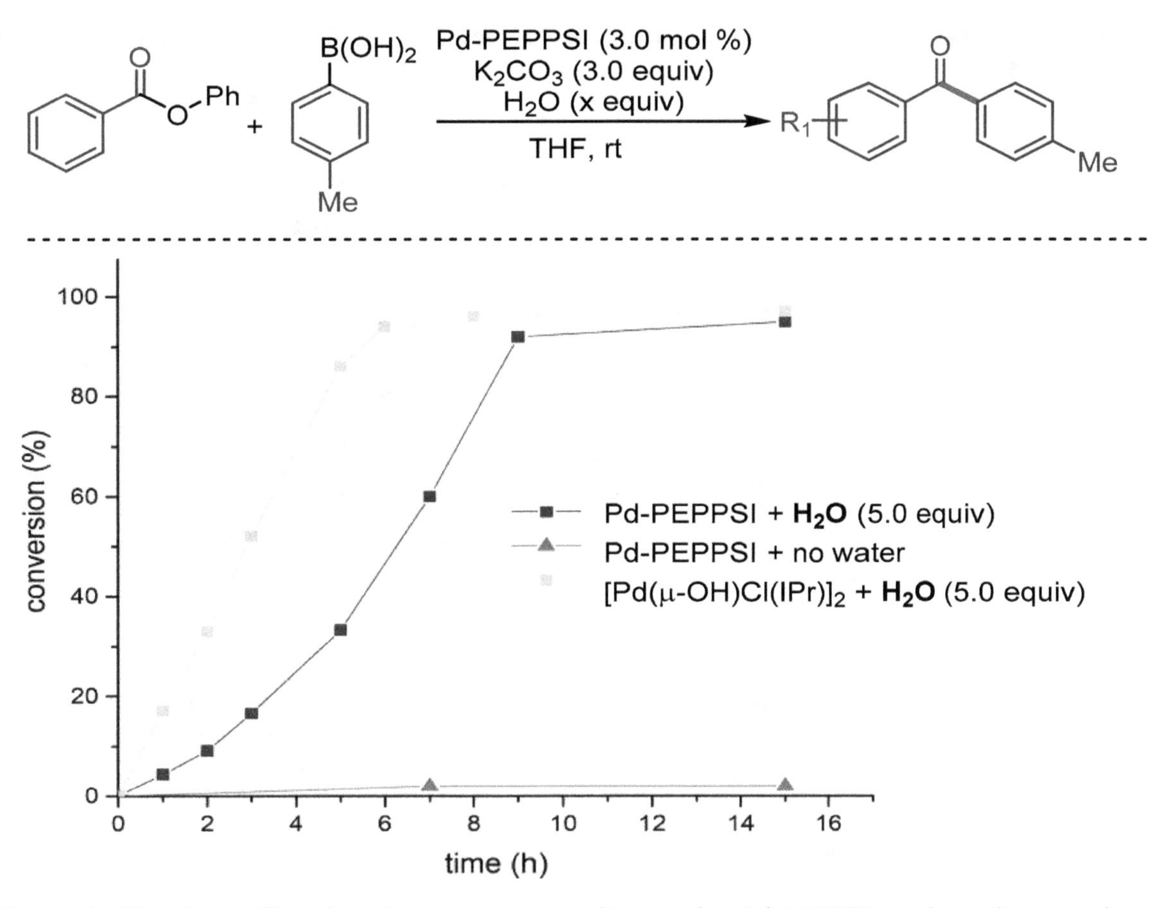


**Figure 5.** Kinetic profile of aryl ester cross-coupling under Pd-PEPPSI catalysis Reprinted with permission from *Adv. Synth. Catal.* **2018**, *360*, 1538–1543.

Very recently, while preparing this manuscript, Szostak and coworkers reported Suzuki-Miyaura cross-coupling of esters using pentafluorophenyl substituents. Although the reported method was primarily focused on the utilization of traditional Pd-ancillary phosphine catalysis, the high catalytic activity of Pd-PEPPSI catalysts allowed them to use even milder conditions with better yields [59].

## 5. Pd–PEPPSI Pre-Catalysts in Buchwald–Hartwig Amidation of Esters and Amides

Szostak and Shi also reported the first Pd-PEPPSI catalyzed Buchwald-Hartwig amination of phenyl esters and activated amides. In their work, they reported the chemo-selective acyl C–O/C–N amination with anilines using the Pd-PEPPSI pre-catalyst [60]. Phenyl esters, *N*-Ph-*N*-Boc, and *N*-Ph-*N*-Ts amides were successfully converted to amides with excellent yield. However, conditions reported were harsher than conditions reported for ketone synthesis (Scheme 27).

Pd-PEPPSI (3 mol%)
$K_2CO_3$ (3.0 equiv)
DME, 110 °C
**50 to 97% yield**
**25 examples**

**Scheme 27.** Pd-PEPPSI catalyzed Buchwald-Hartwig amination of amides and esters.

## 6. Palladium(II)/N-Heterocyclic Carbene-Catalyzed Direct C–H Acylation of Heteroarenes with N-Acylsaccharins

Following up on the work from the Szostak group, Gandhi and coworkers reported the use of pyrene based Pd-PEPPSI catalysts (Figure 6) with wingtip substituents, such as Ar/alkyl and alkyl groups for the cross-coupling of *N*-acylsaccharin amides with azoles and azole derivatives [61].

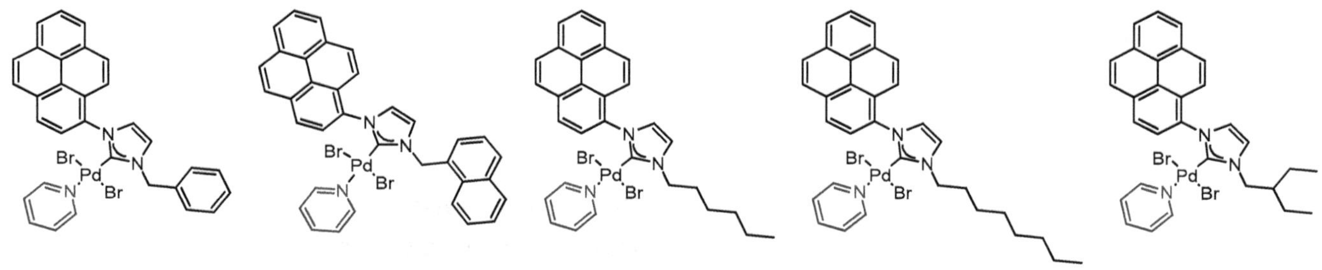

**Figure 6.** Pyrene based Pd-PEPPSI catalysts employed in the cross-coupling reactions.

*N*-acyl saccharin, a commonly used twisted amide, was used to accomplish the difficult cross coupling reactions incorporating amide bond cleavage and C-H activation reactions. The work focused on the cross-coupling of *N*-functionalized saccharins with benzoxazole and its derivatives through a C-H activation process (Scheme 28). Not surprisingly, other activated amides such as *N*-aryl-*N*-Ts were found ineffective, resulting in no product.

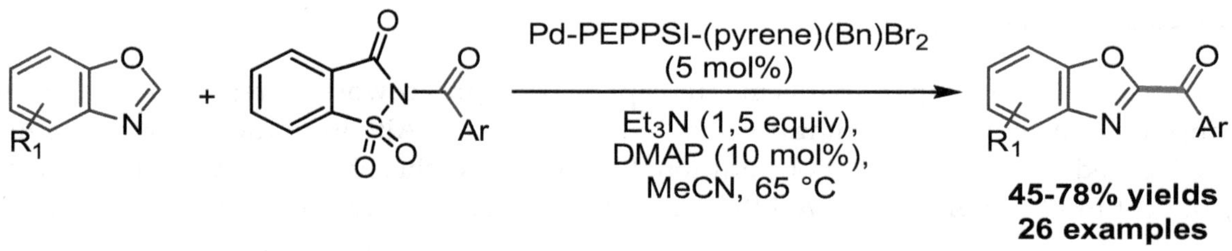

Pd-PEPPSI-(pyrene)(Bn)$Br_2$ (5 mol%)
$Et_3N$ (1,5 equiv), DMAP (10 mol%), MeCN, 65 °C
**45-78% yields**
**26 examples**

**Scheme 28.** Pd-PEPPSI-pyrene pre-catalysts in the C-H acylation of benzoxazole.

## 7. Conclusions

In summary, the use of well-defined Pd(II)-NHC pre-catalysts offers significant advantages over the traditional approach of adding extra ligand to standard Pd catalysts. In the past few years, these catalysts were found to be exceptional in enabling the cross-coupling reactions of amides and esters via C–N(O) bond activation to form ketones and amides. The easy preparation, commercial availability, and stability toward air and moisture of the pre-catalysts facilitate the development of operationally simple and practical cross-coupling methods. The strong σ-bond donating nature of NHC ligands increases the reactivity of Pd center toward oxidative addition of difficult electrophiles, such as amides and esters, increasing the scope and generality of the cross-coupling reactions. Further exciting developments are expected in this relatively new area of employing amides and esters as acyl electrophiles, with recent breakthroughs enabled by Pd(II)-NHC catalysis.

**Author Contributions:** S.R.V., M.R.C. collected the literature references. S.R.V., M.R.C., and G.R.C. contributed to the manuscript writing.

## References


1. Alberico, D.; Scott, M.E.; Lautens, M. Aryl−Aryl Bond Formation by Transition-Metal-Catalyzed Direct Arylation. *Chem. Rev.* **2007**, *107*, 174–238. [CrossRef] [PubMed]
2. Stuart, D.R.; Fagnou, K. The Catalytic Cross-Coupling of Unactivated Arenes. *Science* **2007**, *316*, 1172–1175. [CrossRef]
3. Kumar, D.; Vemula, S.R.; Cook, G.R. Merging C–H Bond Functionalization with Amide Alcoholysis: En Route to 2-Aminopyridines. *ACS Catal.* **2016**, *6*, 3531–3536. [CrossRef]
4. Park, Y.; Kim, Y.; Chang, S. Transition Metal-Catalyzed C–H Amination: Scope, Mechanism, and Applications. *Chem. Rev.* **2017**, *117*, 9247–9301. [CrossRef] [PubMed]
5. Kumar, D.; Vemula, S.R.; Balasubramanian, N.; Cook, G.R. Indium-Mediated Stereoselective Allylation. *Acc. Chem. Res.* **2016**, *49*, 2169–2178. [CrossRef] [PubMed]
6. Kumar, D.; Vemula, S.R.; Cook, G.R. Recent Advances in the Catalytic Synthesis of α-Ketoamides. *ACS Catal.* **2016**, *6*, 4920–4945. [CrossRef]
7. Kumar, D.; Vemula, S.R.; Cook, G.R. Highly Chemo- and Regioselective Allylic Substitution with Tautomerizable Heteroarenes. *Green Chem.* **2015**, *17*, 4300–4306. [CrossRef]
8. Vemula, S.R.; Kumar, D.; Cook, G.R. Palladium-Catalyzed Allylic Amidation with N-Heterocycles via Sp 3 C–H Oxidation. *ACS Catal.* **2016**, *6*, 5295–5301. [CrossRef]
9. Satoh, T.; Miura, M. Catalytic Direct Arylation of Heteroaromatic Compounds. *Chem. Lett.* **2007**, *36*, 200–205. [CrossRef]
10. Tamao, K.; Sumitani, K.; Kumada, M. Selective Carbon-Carbon Bond Formation by Cross-Coupling of Grignard Reagents with Organic Halides. Catalysis by Nickel-Phosphine Complexes. *J. Am. Chem. Soc.* **1972**, *94*, 4374–4376. [CrossRef]
11. Corriu, R.J.P.; Masse, J.P. Activation of Grignard Reagents by Transition-Metal Complexes. A New and Simple Synthesis of Trans-Stilbenes and Polyphenyls. *J. Chem. Soc. Chem. Commun.* **1972**, 144a. [CrossRef]
12. Milstein, D.; Stille, J.K. A General, Selective, and Facile Method for Ketone Synthesis from Acid Chlorides and Organotin Compounds Catalyzed by Palladium. *J. Am. Chem. Soc.* **1978**, *100*, 3636–3638. [CrossRef]
13. Miyaura, N.; Yamada, K.; Suzuki, A. A New Stereospecific Cross-Coupling by the Palladium-Catalyzed Reaction of 1-Alkenylboranes with 1-Alkenyl or 1-Alkynyl Halides. *Tetrahedron Lett.* **1979**, *20*, 3437–3440. [CrossRef]
14. Trost, B.M.; Fullerton, T.J. New Synthetic Reactions. Allylic Alkylation. *J. Am. Chem. Soc.* **1973**, *95*, 292–294. [CrossRef]
15. Marion, N.; Nolan, S.P. Well-Defined N-Heterocyclic Carbenes−Palladium(II) Precatalysts for Cross-Coupling Reactions. *Acc. Chem. Res.* **2008**, *41*, 1440–1449. [CrossRef] [PubMed]
16. Valdés, H.; Canseco-González, D.; Germán-Acacio, J.M.; Morales-Morales, D. Xanthine Based N-Heterocyclic Carbene (NHC) Complexes. *J. Organomet. Chem.* **2018**, *867*, 51–54. [CrossRef]
17. Hazari, N.; Melvin, P.R.; Beromi, M.M. Well-Defined Nickel and Palladium Precatalysts for Cross-Coupling. *Nat. Rev. Chem.* **2017**, *1*, 0025. [CrossRef]
18. Herrmann, W.A.; Elison, M.; Fischer, J.; Köcher, C.; Artus, G.R.J. Metal Complexes of N-Heterocyclic Carbenes—A New Structural Principle for Catalysts in Homogeneous Catalysis. *Angew. Chem. Int. Ed. English* **1995**, *34*, 2371–2374. [CrossRef]
19. Shi, S.; Nolan, S.P.; Szostak, M. Well-Defined Palladium(II)–NHC Precatalysts for Cross-Coupling Reactions of Amides and Esters by Selective N–C/O–C Cleavage. *Acc. Chem. Res.* **2018**, *51*, 2589–2599. [CrossRef]
20. Takise, R.; Muto, K.; Yamaguchi, J. Cross-Coupling of Aromatic Esters and Amides. *Chem. Soc. Rev.* **2017**, *46*, 5864–5888. [CrossRef]
21. Vemula, S.R.; Kumar, D.; Cook, G.R. N-Boc-Glycine-Assisted Indium-Mediated Allylation Reaction: A Sustainable Approach. *Tetrahedron Lett.* **2015**, *56*, 3322–3325. [CrossRef]
22. Hie, L.; Fine Nathel, N.F.; Shah, T.K.; Baker, E.L.; Hong, X.; Yang, Y.-F.; Liu, P.; Houk, K.N.; Garg, N.K. Conversion of Amides to Esters by the Nickel-Catalysed Activation of Amide C–N Bonds. *Nature* **2015**, *524*, 79–83. [CrossRef] [PubMed]
23. Meng, G.; Szostak, M. Sterically Controlled Pd-Catalyzed Chemoselective Ketone Synthesis via N–C Cleavage in Twisted Amides. *Org. Lett.* **2015**, *17*, 4364–4367. [CrossRef] [PubMed]

24. Li, X.; Zou, G. Acylative Suzuki Coupling of Amides: Acyl-Nitrogen Activation via Synergy of Independently Modifiable Activating Groups. *Chem. Commun.* **2015**, *51*, 5089–5092. [CrossRef] [PubMed]
25. Marion, N.; Navarro, O.; Mei, J.; Stevens, E.D.; Scott, N.M.; Nolan, S.P. Modified (NHC)Pd(Allyl)Cl (NHC = N -Heterocyclic Carbene) Complexes for Room-Temperature Suzuki−Miyaura and Buchwald−Hartwig Reactions. *J. Am. Chem. Soc.* **2006**, *128*, 4101–4111. [CrossRef] [PubMed]
26. Navarro, O.; Marion, N.; Mei, J.; Nolan, S.P. Rapid Room Temperature Buchwald–Hartwig and Suzuki–Miyaura Couplings of Heteroaromatic Compounds Employing Low Catalyst Loadings. *Chem. A Eur. J.* **2006**, *12*, 5142–5148. [CrossRef]
27. Chartoire, A.; Lesieur, M.; Falivene, L.; Slawin, A.M.Z.; Cavallo, L.; Cazin, C.S.J.; Nolan, S.P. [Pd(IPr*)(Cinnamyl)Cl]: An Efficient Pre-Catalyst for the Preparation of Tetra-Ortho-Substituted Biaryls by Suzuki-Miyaura Cross-Coupling. *Chem. A Eur. J.* **2012**, *18*, 4517–4521. [CrossRef]
28. Chartoire, A.; Frogneux, X.; Nolan, S.P. An Efficient Palladium-NHC (NHC=N-Heterocyclic Carbene) and Aryl Amination Pre-Catalyst: [Pd(IPr*)(Cinnamyl)Cl]. *Adv. Synth. Catal.* **2012**, *354*, 1897–1901. [CrossRef]
29. Szostak, M.; Aubé, J. Chemistry of Bridged Lactams and Related Heterocycles. *Chem. Rev.* **2013**, *113*, 5701–5765. [CrossRef]
30. Liu, C.; Szostak, M. Twisted Amides: From Obscurity to Broadly Useful Transition-Metal-Catalyzed Reactions by N−C Amide Bond Activation. *Chem. A Eur. J.* **2017**, *23*, 7157–7173. [CrossRef]
31. Meng, G.; Shi, S.; Lalancette, R.; Szostak, R.; Szostak, M. Reversible Twisting of Primary Amides via Ground State N–C(O) Destabilization: Highly Twisted Rotationally Inverted Acyclic Amides. *J. Am. Chem. Soc.* **2018**, *140*, 727–734. [CrossRef] [PubMed]
32. Meng, G.; Szostak, R.; Szostak, M. Suzuki-Miyaura Cross-Coupling of N-Acylpyrroles and Pyrazoles: Planar, Electronically Activated Amides in Catalytic N-C Cleavage. *Org. Lett.* **2017**, *19*, 3596–3599. [CrossRef] [PubMed]
33. Meng, G.; Lalancette, R.; Szostak, R.; Szostak, M. N-Methylamino Pyrimidyl Amides (MAPA): Highly Reactive, Electronically-Activated Amides in Catalytic N-C(O) Cleavage. *Org. Lett.* **2017**, *19*, 4656–4659. [CrossRef] [PubMed]
34. Li, G.; Lei, P.; Szostak, M.; Casals-Cruañas, E.; Poater, A.; Cavallo, L.; Nolan, S.P. Mechanistic Study of Suzuki–Miyaura Cross-Coupling Reactions of Amides Mediated by [Pd(NHC)(Allyl)Cl] Precatalysts. *ChemCatChem* **2018**, *10*, 3096–3106. [CrossRef]
35. Luo, Z.; Liu, T.; Guo, W.; Wang, Z.; Huang, J.; Zhu, Y.; Zeng, Z. N -Acyl-5,5-Dimethylhydantoin, a New Mild Acyl-Transfer Reagent in Pd Catalysis: Highly Efficient Synthesis of Functionalized Ketones. *Org. Proc. Res. Dev.* **2018**, *22*, 1188–1199. [CrossRef]
36. Lei, P.; Meng, G.; Ling, Y.; An, J.; Nolan, S.P.; Szostak, M. General Method for the Suzuki-Miyaura Cross-Coupling of Primary Amide-Derived Electrophiles Enabled by [Pd(NHC)(Cin)Cl] at Room Temperature. *Org. Lett.* **2017**, *19*, 6510–6513. [CrossRef] [PubMed]
37. Liu, C.; Li, G.; Shi, S.; Meng, G.; Lalancette, R.; Szostak, R.; Szostak, M. Acyl and Decarbonylative Suzuki Coupling of N -Acetyl Amides: Electronic Tuning of Twisted, Acyclic Amides in Catalytic Carbon–Nitrogen Bond Cleavage. *ACS Catal.* **2018**, *8*, 9131–9139. [CrossRef]
38. Meng, G.; Szostak, M. Palladium/NHC (NHC = N -Heterocyclic Carbene)-Catalyzed B-Alkyl Suzuki Cross-Coupling of Amides by Selective N–C Bond Cleavage. *Org. Lett.* **2018**, *20*, 6789–6793. [CrossRef]
39. Hruszkewycz, D.P.; Balcells, D.; Guard, L.M.; Hazari, N.; Tilset, M. Insight into the Efficiency of Cinnamyl-Supported Precatalysts for the Suzuki–Miyaura Reaction: Observation of Pd(I) Dimers with Bridging Allyl Ligands During Catalysis. *J. Am. Chem. Soc.* **2014**, *136*, 7300–7316. [CrossRef]
40. Hruszkewycz, D.P.; Guard, L.M.; Balcells, D.; Feldman, N.; Hazari, N.; Tilset, M. Effect of 2-Substituents on Allyl-Supported Precatalysts for the Suzuki–Miyaura Reaction: Relating Catalytic Efficiency to the Stability of Palladium(I) Bridging Allyl Dimers. *Organometallics* **2015**, *34*, 381–394. [CrossRef]
41. Melvin, P.R.; Nova, A.; Balcells, D.; Dai, W.; Hazari, N.; Hruszkewycz, D.P.; Shah, H.P.; Tudge, M.T. Design of a Versatile and Improved Precatalyst Scaffold for Palladium-Catalyzed Cross-Coupling: (η 3 -1- t Bu-Indenyl) 2 (μ-Cl) 2 Pd 2. *ACS Catal.* **2015**, *5*, 3680–3688. [CrossRef]
42. Lei, P.; Meng, G.; Shi, S.; Ling, Y.; An, J.; Szostak, R.; Szostak, M. Suzuki–Miyaura Cross-Coupling of Amides and Esters at Room Temperature: Correlation with Barriers to Rotation around C–N and C–O Bonds. *Chem. Sci.* **2017**, *8*, 6525–6530. [CrossRef] [PubMed]

43. Ishizu, J.; Yamamoto, T.; Yamamoto, A. Selective Cleavage of C–O Bonds In Esters Through Oxidative Addition To Nickel(0) Complexes. *Chem. Lett.* **1976**, *5*, 1091–1094. [CrossRef]
44. Yamaguchi, J.; Muto, K.; Itami, K. Recent Progress in Nickel-Catalyzed Biaryl Coupling. *Eur. J. Org. Chem.* **2013**, *2013*, 19–30. [CrossRef]
45. Ben Halima, T.; Vandavasi, J.K.; Shkoor, M.; Newman, S.G. A Cross-Coupling Approach to Amide Bond Formation from Esters. *ACS Catal.* **2017**, *7*, 2176–2180. [CrossRef]
46. Masson-Makdissi, J.; Vandavasi, J.K.; Newman, S.G. Switchable Selectivity in the Pd-Catalyzed Alkylative Cross-Coupling of Esters. *Org. Lett.* **2018**, *20*, 4094–4098. [CrossRef]
47. Dardir, A.H.; Melvin, P.R.; Davis, R.M.; Hazari, N.; Mohadjer Beromi, M. Rapidly Activating Pd-Precatalyst for Suzuki-Miyaura and Buchwald-Hartwig Couplings of Aryl Esters. *J. Org. Chem.* **2018**, *83*, 469–477. [CrossRef]
48. Meng, G.; Lei, P.; Szostak, M. A General Method for Two-Step Transamidation of Secondary Amides Using Commercially Available, Air- and Moisture-Stable Palladium/NHC (N-Heterocyclic Carbene) Complexes. *Org. Lett.* **2017**, *19*, 2158–2161. [CrossRef]
49. Valente, C.; Çalimsiz, S.; Hoi, K.H.; Mallik, D.; Sayah, M.; Organ, M.G. The Development of Bulky Palladium NHC Complexes for the Most-Challenging Cross-Coupling Reactions. *Angew. Chem. Int. Ed.* **2012**, *51*, 3314–3332. [CrossRef]
50. Froese, R.D.J.; Lombardi, C.; Pompeo, M.; Rucker, R.P.; Organ, M.G. Designing Pd–N-Heterocyclic Carbene Complexes for High Reactivity and Selectivity for Cross-Coupling Applications. *Acc. Chem. Res.* **2017**, *50*, 2244–2253. [CrossRef]
51. Lei, P.; Meng, G.; Ling, Y.; An, J.; Szostak, M. Pd-PEPPSI: Pd-NHC Precatalyst for Suzuki–Miyaura Cross-Coupling Reactions of Amides. *J. Org. Chem.* **2017**, *82*, 6638–6646. [CrossRef] [PubMed]
52. Pace, V.; Holzer, W.; Meng, G.; Shi, S.; Lalancette, R.; Szostak, R.; Szostak, M. Corrigendum to: Structures of Highly Twisted Amides Relevant to Amide N–C Cross-Coupling: Evidence for Ground-State Amide Destabilization. *Chem. A Eur. J.* **2017**, *23*, 3496. [CrossRef] [PubMed]
53. Wang, C.; Huang, L.; Wang, F.; Zou, G. Highly Efficient Synthesis of Aryl Ketones by PEPPSI-Palladium Catalyzed Acylative Suzuki Coupling of Amides with Diarylborinic Acids. *Tetrahedron Lett.* **2018**, *59*, 2299–2301. [CrossRef]
54. Shi, W.; Zou, G. Palladium-Catalyzed Room Temperature Acylative Cross-Coupling of Activated Amides with Trialkylboranes. *Molecules* **2018**, *23*, 2412. [CrossRef] [PubMed]
55. Vemula, S.R.; Kumar, D.; Cook, G.R. Bismuth-Catalyzed Synthesis of 2-Substituted Quinazolinones. *Tetrahedron Lett.* **2018**, *59*, 3801–3805. [CrossRef]
56. Wang, T.; Guo, J.; Wang, H.; Guo, H.; Jia, D.; Zhang, W.; Liu, L. N-Heterocyclic Carbene Palladium(II)-Catalyzed Suzuki-Miyaura Cross Coupling of N-Acylsuccinimides by C–N Cleavage. *J. Organomet. Chem.* **2018**, *877*, 80–84. [CrossRef]
57. Shi, S.; Lei, P.; Szostak, M. Pd-PEPPSI: A General Pd-NHC Precatalyst for Suzuki–Miyaura Cross-Coupling of Esters by C–O Cleavage. *Organometallics* **2017**, *36*, 3784–3789. [CrossRef]
58. Li, G.; Shi, S.; Lei, P.; Szostak, M. Pd-PEPPSI: Water-Assisted Suzuki–Miyaura Cross-Coupling of Aryl Esters at Room Temperature Using a Practical Palladium-NHC (NHC=N-Heterocyclic Carbene) Precatalyst. *Adv. Synth. Catal.* **2018**, *360*, 1538–1543. [CrossRef]
59. Buchspies, J.; Pyle, D.J.; He, H.; Szostak, M. Pd-Catalyzed Suzuki-Miyaura Cross-Coupling of Pentafluorophenyl Esters. *Molecules* **2018**, *23*, 3134. [CrossRef]
60. Shi, S.; Szostak, M. Pd–PEPPSI: A General Pd–NHC Precatalyst for Buchwald–Hartwig Cross-Coupling of Esters and Amides (Transamidation) under the Same Reaction Conditions. *Chem. Commun.* **2017**, *53*, 10584–10587. [CrossRef]
61. Karthik, S.; Gandhi, T. Palladium(II)/ N-Heterocyclic Carbene-Catalyzed Direct C–H Acylation of Heteroarenes with N -Acylsaccharins. *Org. Lett.* **2017**, *19*, 5486–5489. [CrossRef] [PubMed]

6

# Recent uses of *N,N*-Dimethylformamide and *N,N*-Dimethylacetamide as Reagents

**Jean Le Bras and Jacques Muzart** *

Institut de Chimie Moléculaire de Reims, CNRS—Université de Reims Champagne-Ardenne, B.P. 1039, 51687 Reims CEDEX 2, France; jean.lebras@univ-reims.fr
* Correspondence: jacques.muzart@univ-reims.fr

Academic Editor: Michal Szostak

**Abstract:** *N,N*-Dimethylformamide and *N,N*-dimethylacetamide are multipurpose reagents which deliver their own H, C, N and O atoms for the synthesis of a variety of compounds under a number of different experimental conditions. The review mainly highlights the corresponding literature published over the last years.

**Keywords:** *N,N*-dimethylformamide; DMF; *N,N*-dimethylacetamide; DMAc; amination; amidation; thioamidation; formylation; carbonylation; cyanation; insertion; cyclization

## 1. Introduction

The organic, organometallic and bioorganic transformations are extensively carried out in *N,N*-dimethylformamide (DMF) or *N,N*-dimethylacetamide (DMAc). These two polar solvents are not only use for their dissolution properties, but also as multipurpose reagents. They participate in a number of processes and serve as a source of various building blocks giving one or more of their own atoms (Scheme 1).

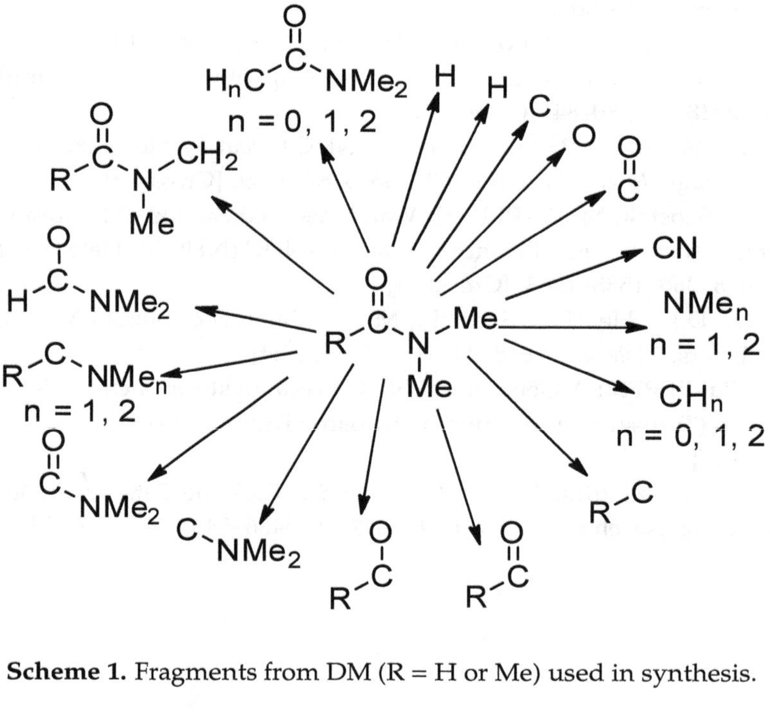

**Scheme 1.** Fragments from DM (R = H or Me) used in synthesis.

In 2009, one of us reviewed the different roles of DMF, highlighting that DMF is much more than a solvent [1]. Subsequently, this topic has been documented by the teams of Jiao [2] and Sing [3]. For of a book devoted to solvents as reagents in organic synthesis, we wrote a chapter summarizing the reactions consuming DMF and DMAc as carbon, hydrogen, nitrogen and/or oxygen sources [4]. This book chapter tentatively covered the literature up to middle 2015. The present mini-review focuses on recent reactions which involve DM (DM = DMF or DMAc) as a reagent although some key older papers are also included for context. Processes which necessitate the prerequisite synthesis of DM derivatives such as the Vilsmeier-Haack reagents [5] and DMF dimethyl acetal [6] are not surveyed, but a few reactions of the present review involve the in-situ formation of a Vilsmeier-type intermediate (Vilsmeier-type reagents have been extensively used. Search on 26 June 2018 for "Vilsmeier" with SciFinder led to 4379 entries). Color equations, based on literature proposals, are used to easily visualize the DM atom origin. When uncertainty is expressed by the authors or suspected by us, the atom is typed in italic. Mechanistic schemes are not reported, but Scheme 2 [7–35] summarizes different proposed reactions of DM with the corresponding literature references, where DM acts as either a nucleophilic or electrophilic reagent, or leads to neutral, ionic or radical species. The review is divided in Sections depending on the DM fragment(s) which is (are) incorporated into the reaction product.

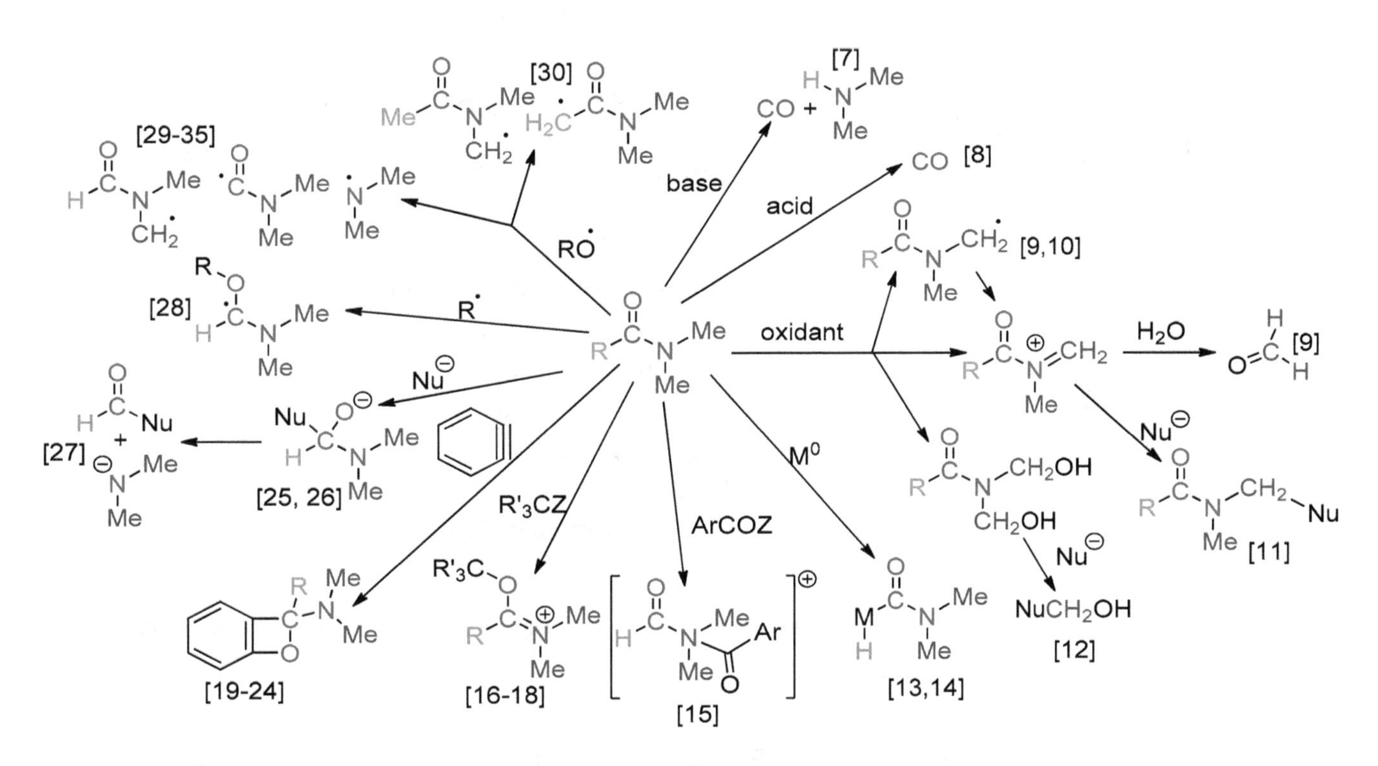


**Scheme 2.** Reactions of DM (R = H or Me).

## 2. C Fragment

Aerobic carbonylation under nickel/copper or palladium/silver synergistic catalysis occurred efficiently using the Me group of DMF as the C source, affording cyclic carbonylated compounds, via the directing group-assisted activation of a C($sp^2$)–H or C($sp^3$)-H bond (Equations (1) and (2) [36], Equations (3) and (4) [37]). Shifting from DMF to DMAc greatly decreased the yields (Equations (1) and (3)).

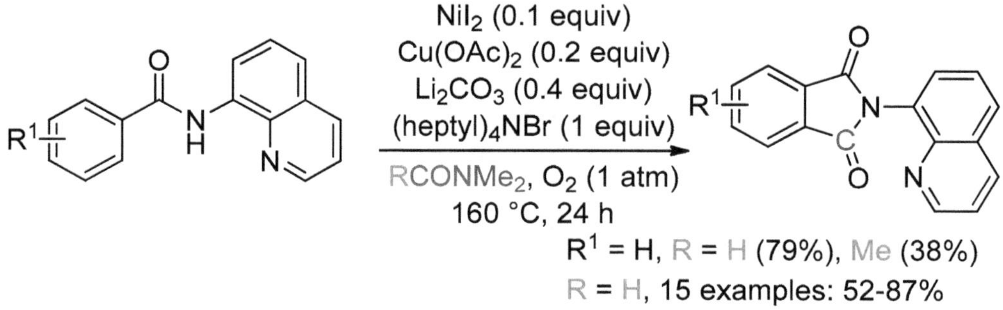

$NiI_2$ (0.1 equiv), $Cu(OAc)_2$ (0.2 equiv), $Li_2CO_3$ (0.4 equiv), $(heptyl)_4NBr$ (1 equiv), $RCONMe_2$, $O_2$ (1 atm), 160 °C, 24 h (1)

$R^1$ = H, R = H (79%), Me (38%)
R = H, 15 examples: 52-87%

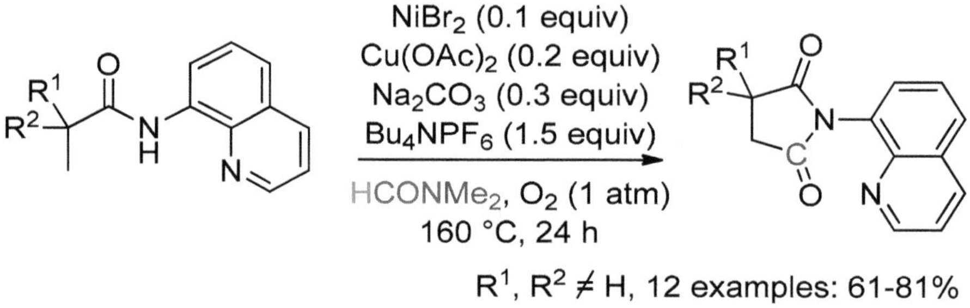

$NiBr_2$ (0.1 equiv), $Cu(OAc)_2$ (0.2 equiv), $Na_2CO_3$ (0.3 equiv), $Bu_4NPF_6$ (1.5 equiv), $HCONMe_2$, $O_2$ (1 atm), 160 °C, 24 h (2)

$R^1$, $R^2 \neq$ H, 12 examples: 61-81%

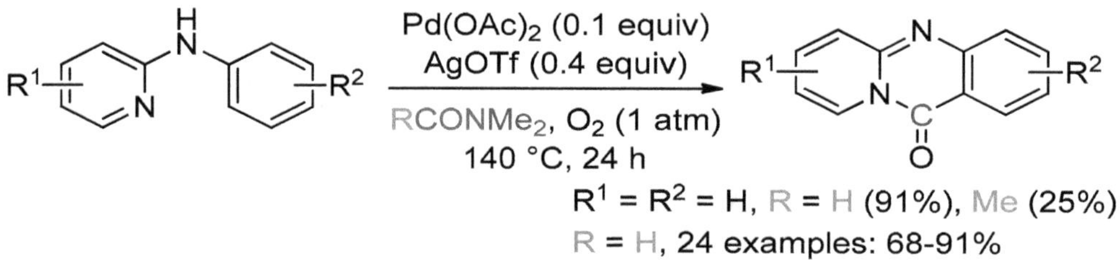

$Pd(OAc)_2$ (0.1 equiv), AgOTf (0.4 equiv), $RCONMe_2$, $O_2$ (1 atm), 140 °C, 24 h (3)

$R^1 = R^2$ = H, R = H (91%), Me (25%)
R = H, 24 examples: 68-91%

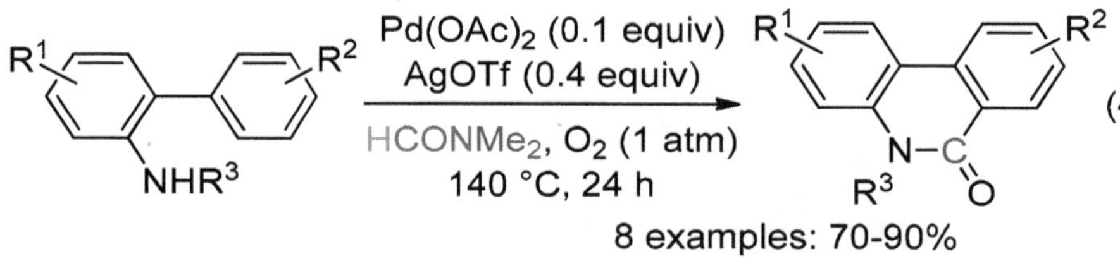

$Pd(OAc)_2$ (0.1 equiv), AgOTf (0.4 equiv), $HCONMe_2$, $O_2$ (1 atm), 140 °C, 24 h (4)

8 examples: 70-90%

The Me group of DMF was also involved in the cyanation of the C($sp^2$)-H bond of arenes catalyzed with an heterogeneous copper catalyst (Equation (5)) [38].

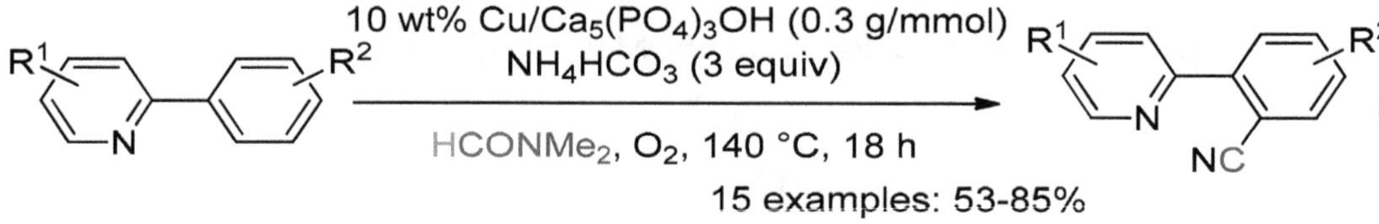

10 wt% $Cu/Ca_5(PO_4)_3OH$ (0.3 g/mmol), $NH_4HCO_3$ (3 equiv), $HCONMe_2$, $O_2$, 140 °C, 18 h (5)

15 examples: 53-85%

## 3. CH Fragment

Treatment of indole at 130 °C with suprastoichiometric amounts of CuI, *t*-BuOOH and AcOH in DMAc under air afforded the corresponding C3-formylation product (Equation (6)) [39]. Such a reaction also occurred with *N*-methylindole using CuI and $CF_3CO_2H$ in DMAc under oxygen [40]. The CH fragment came from the $NMe_2$ moiety [39,40]. In DMF, both procedures led to C3-cyanation (see below, Equation (25)).

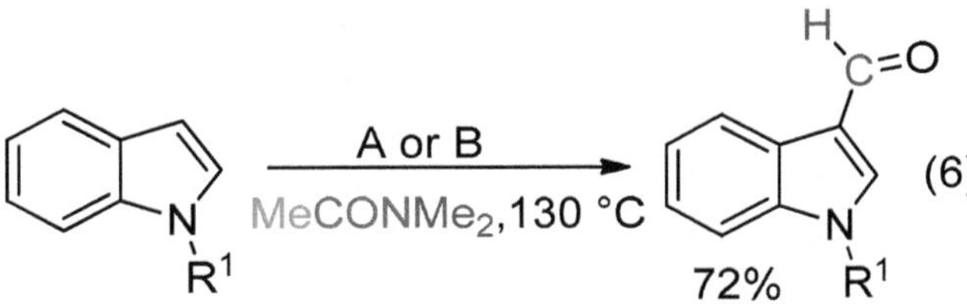

A or B

$MeCONMe_2$, 130 °C (6)

72%

A. $R^1$ = H; CuI (1.2 equiv), *t*-BuOOH (2 equiv), AcOH (8 equiv), air, 48 h: 72%

B. $R^1$ = Me; CuI (1 equiv), $CF_3CO_2H$ (1.2 equiv) $O_2$ (1 atm), 36 h: 42%

Cycloadditions leading to symmetrical tetrasubstituted pyridines using the Me group of DMF as the CH source have been carried out using either ketoxime carboxylates with Ru catalysis (Equation (7)) [41], or arones with both iodine and ammonium persulfate mediation (Equation (8)) [42].

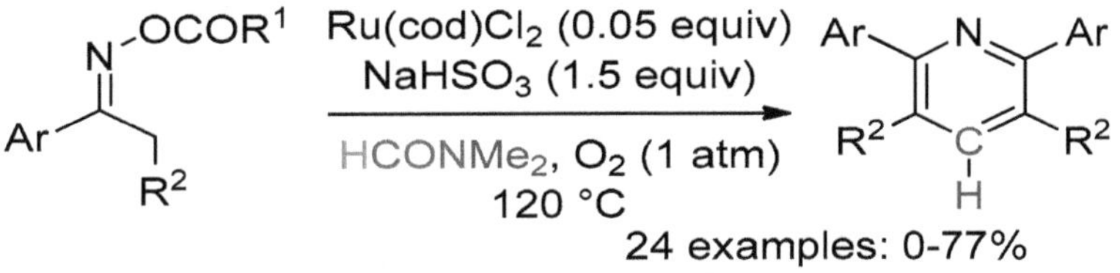

$Ru(cod)Cl_2$ (0.05 equiv), $NaHSO_3$ (1.5 equiv), $HCONMe_2$, $O_2$ (1 atm), 120 °C (7)

24 examples: 0-77%

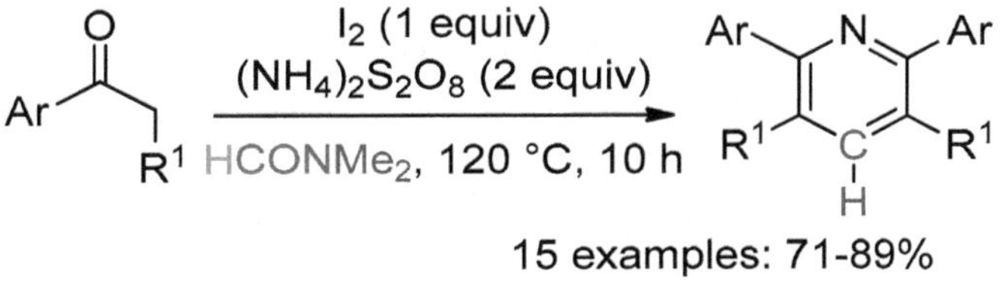

$I_2$ (1 equiv), $(NH_4)_2S_2O_8$ (2 equiv), $HCONMe_2$, 120 °C, 10 h (8)

15 examples: 71-89%

2,4-Diarylpyridines were also synthetized from Ru-catalyzed reaction of acetophenones with ammonium acetate and DMF as source of the N and CH atoms, respectively (Equation (9)) [43].

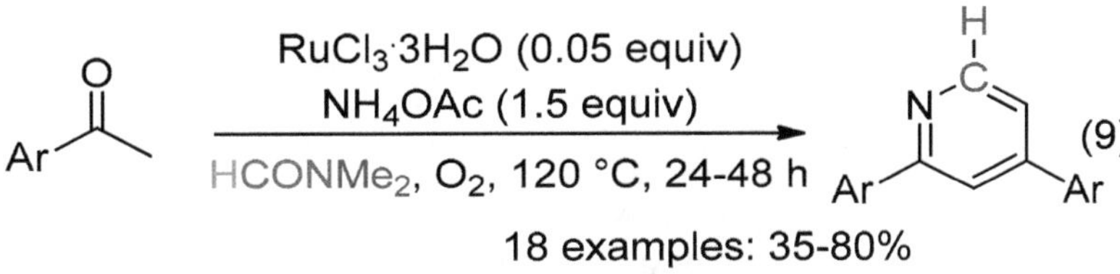

$RuCl_3 \cdot 3H_2O$ (0.05 equiv), $NH_4OAc$ (1.5 equiv), $HCONMe_2$, $O_2$, 120 °C, 24-48 h (9)

18 examples: 35-80%

With sodium azide as the nitrogen source, DMAc was superior to DMF to deliver the CH fragment of the copper-catalyzed domino reactions of aryl halides which led to imidazo [1,2-*c*]quinazolines, quinazolinones or imidazo[4,5-*c*]quinolones (Equations (10)–(12)) [44].

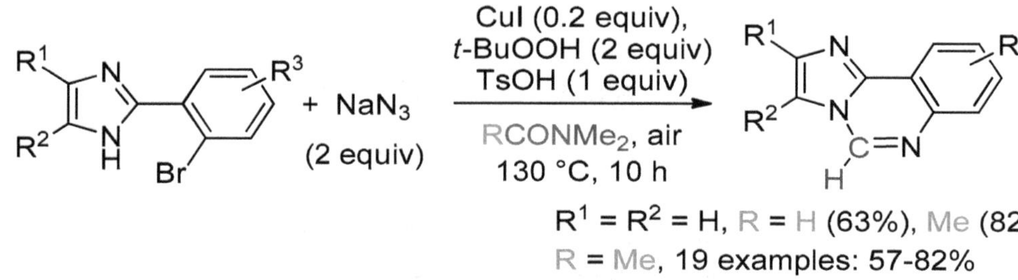

CuI (0.2 equiv), $t$-BuOOH (2 equiv), TsOH (1 equiv), $RCONMe_2$, air, 130 °C, 10 h; + $NaN_3$ (2 equiv) (10)

$R^1 = R^2 = H$, R = H (63%), Me (82%)
R = Me, 19 examples: 57-82%

CuI (0.2 equiv), $t$-BuOOH (2 equiv), TsOH (1 equiv), $MeCONMe_2$, air, 130 °C, 10 h; + $NaN_3$ (2 equiv) (11)

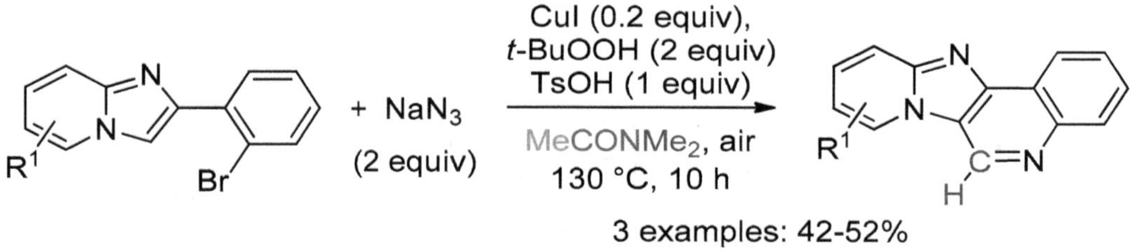

8 examples: 37-71%

CuI (0.2 equiv), $t$-BuOOH (2 equiv), TsOH (1 equiv), $MeCONMe_2$, air, 130 °C, 10 h; + $NaN_3$ (2 equiv) (12)

3 examples: 42-52%

In contrast to the above examples, the Cu-catalyzed cyclization leading to 6*H*-chromeno[4,3-*b*]quinolin-6-ones (Equation (13)) with incorporation of a CH belonging to DMF occurred with low yield when $t$-BuOOH was the oxidant. Shifting to $t$-butyl perbenzoate allowed an effective reaction [45]. For the synthesis of 4-acyl-1,2,3-triazoles from Cu-catalyzed cycloaddition to acetophenones (Equation (14)), $K_2S_2O_8$ was superior to $t$-BuOOH, $(t\text{-BuO})_2$ and $(PhCO_2)_2$ [46]. Yields decreased with DMAc instead of DMF (Equations (13) and (14)).

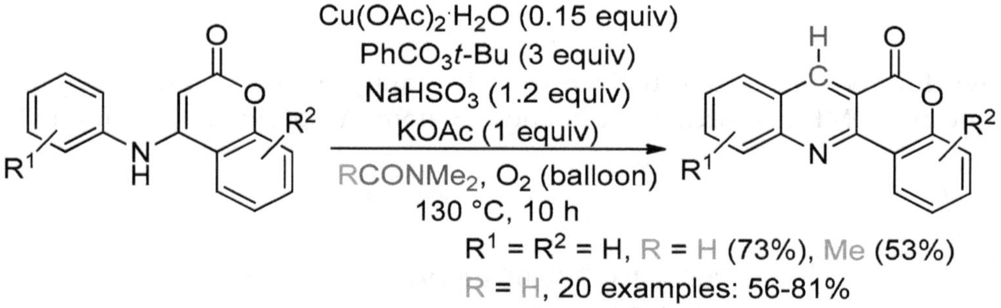

$Cu(OAc)_2 \cdot H_2O$ (0.15 equiv), $PhCO_3t$-Bu (3 equiv), $NaHSO_3$ (1.2 equiv), KOAc (1 equiv), $RCONMe_2$, $O_2$ (balloon), 130 °C, 10 h (13)

$R^1 = R^2 = H$, R = H (73%), Me (53%)
R = H, 20 examples: 56-81%

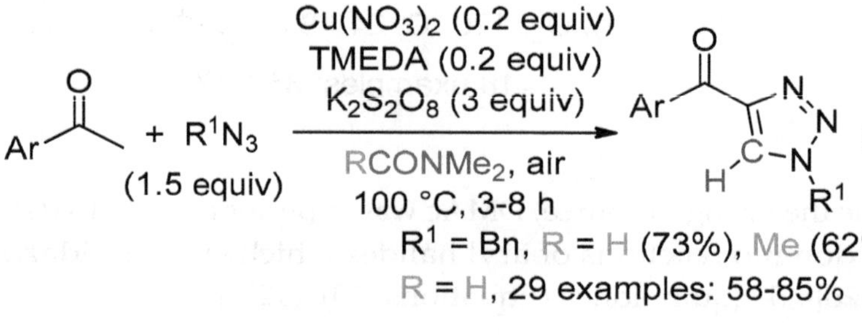

$Cu(NO_3)_2$ (0.2 equiv), TMEDA (0.2 equiv), $K_2S_2O_8$ (3 equiv), $RCONMe_2$, air, 100 °C, 3-8 h; + $R^1N_3$ (1.5 equiv) (14)

$R^1$ = Bn, R = H (73%), Me (62%)
R = H, 29 examples: 58-85%

The insertion of a CH from the $NMe_2$ of DM under metal-free conditions has been reported for the synthesis of cyclic compounds such as:

- pyrimidines from *t*-BuOOH-mediated reaction between acetophenones, amidines and DMF (Equation (15)) [47],

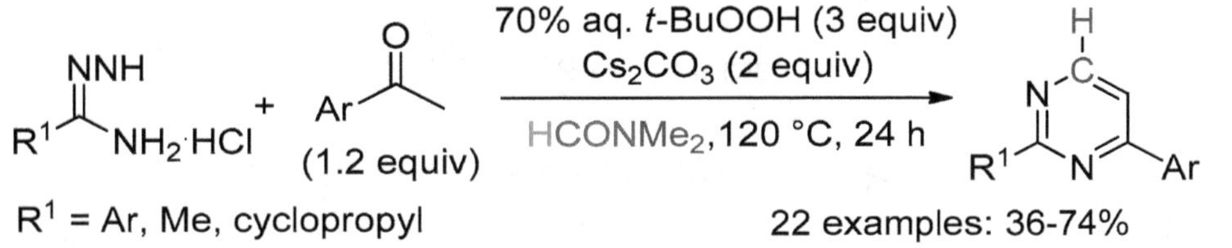

NNH; $R^1$ NH$_2$·HCl + Ar(C=O)Me (1.2 equiv) → 70% aq. *t*-BuOOH (3 equiv), $Cs_2CO_3$ (2 equiv), HCONMe$_2$, 120 °C, 24 h (15)

$R^1$ = Ar, Me, cyclopropyl

22 examples: 36-74%

- substituted phenols also from three components cycloadditions (Equation (16)) [48],

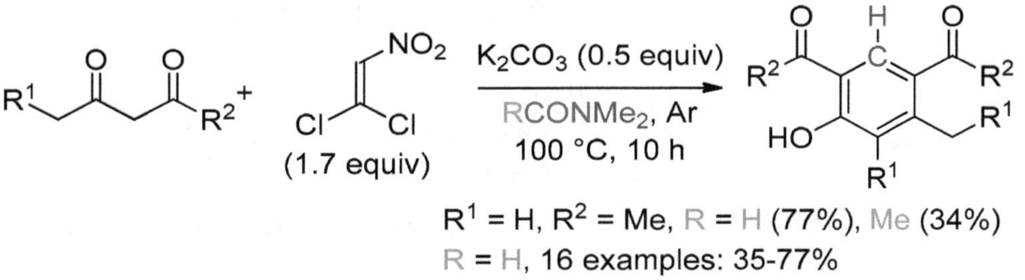

(1.7 equiv) → $K_2CO_3$ (0.5 equiv), RCONMe$_2$, Ar, 100 °C, 10 h (16)

$R^1$ = H, $R^2$ = Me, R = H (77%), Me (34%)

R = H, 16 examples: 35-77%

- 3-acylindoles from 2-alkenylanilines (Equation (17)) [49],

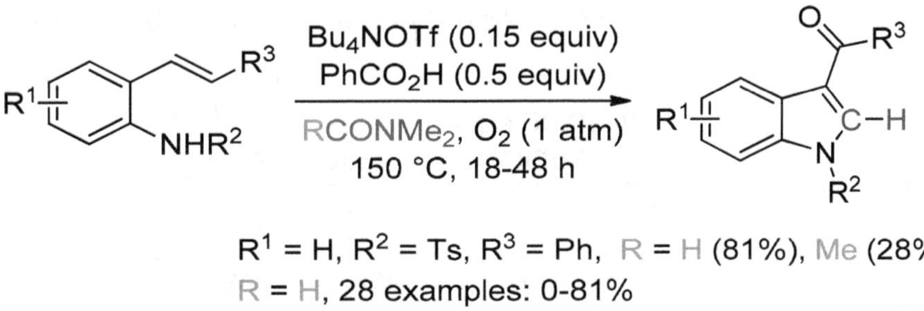

Bu$_4$NOTf (0.15 equiv), PhCO$_2$H (0.5 equiv), RCONMe$_2$, $O_2$ (1 atm), 150 °C, 18-48 h (17)

$R^1$ = H, $R^2$ = Ts, $R^3$ = Ph, R = H (81%), Me (28%)

R = H, 28 examples: 0-81%

- benzimidazoles and benzothiazole from *o*-phenylenediamine or 2-aminobenzenethiol through carbon dioxide-mediated cyclization (Equation (18)) [50].

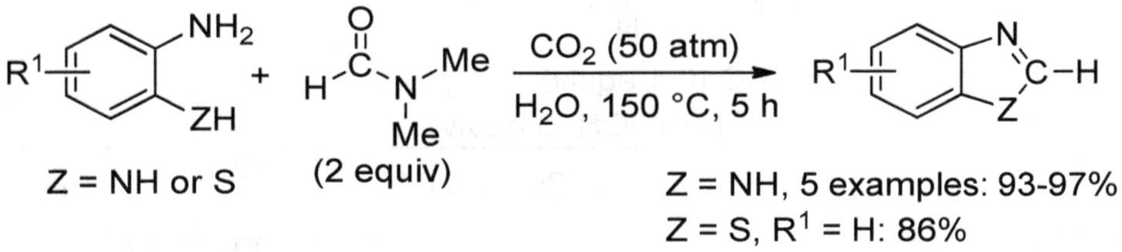

+ HC(O)NMe$_2$ (2 equiv) → $CO_2$ (50 atm), $H_2O$, 150 °C, 5 h (18)

Z = NH or S

Z = NH, 5 examples: 93-97%

Z = S, $R^1$ = H: 86%

**4. CH$_2$ Fragment**

The coupling of indoles or imidazo[1,2-*a*]pyridines to afford heterodiarylmethanes with DMF as the methylenating reagent occurred in fair to high yields with a Cu$^I$ catalyst associated to *t*-BuOOH [51] or $K_2S_2O_8$ [52] (Equations (19)–(21)). Use of DMAc was less efficient.

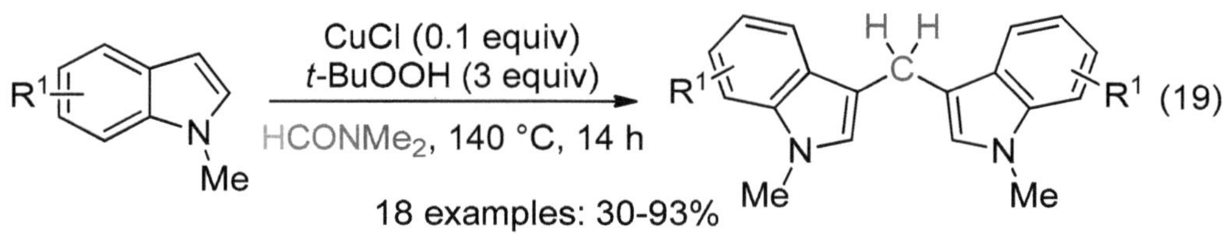

CuCl (0.1 equiv), *t*-BuOOH (3 equiv), $HCONMe_2$, 140 °C, 14 h

18 examples: 30-93% (19)

CuI (0.1 equiv), $K_2S_2O_8$ (1 equiv), $RCONMe_2$, 80 °C, 24 h; indole (1.2 equiv)

$R^1 = R^2 = H$, R = H (68%), Me (20%)

R = H, 18 examples: 55-82% (20)

CuI (0.1 equiv), $K_2S_2O_8$ (1 equiv), $HCONMe_2$, 80 °C, 8 h; $PhNMe_2$ (2.3 equiv)

8 examples: 72-87% (21)

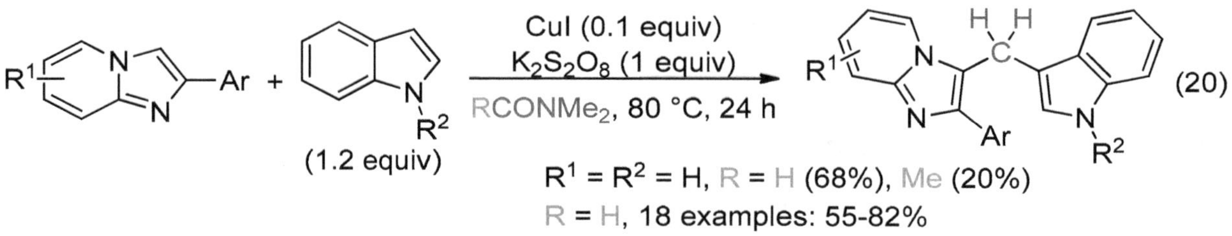

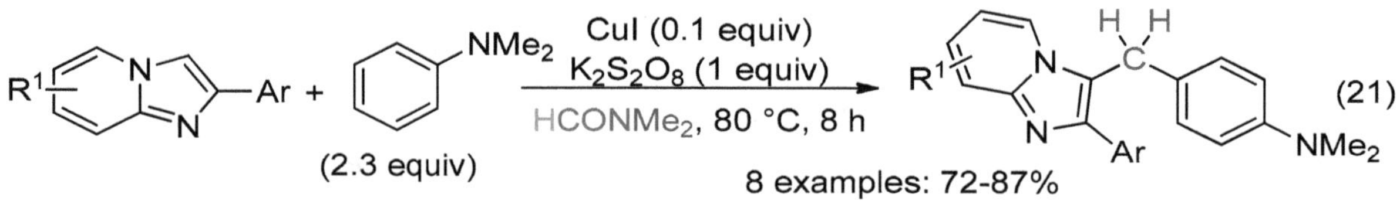

In DMAc, the $I_2$/*t*-BuOOH association catalyzed the formation of methylene-bridged bis-1,3-dicarbonyl compounds from aryl β-ketoesters or β-ketoamides (Equation (22)). Lower yields were obtained with $I_2/K_2S_2O_8$ in DMAc or DMF [9]. Subjection 1,3-diphenylpropane-1,3-dione or ethyl 3-oxobutanoate to the $I_2$/*t*-BuOOH/DMAc did not afford the bridged compounds.

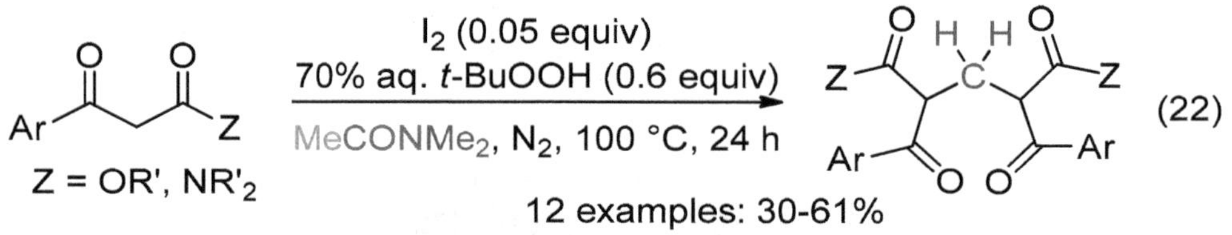

$I_2$ (0.05 equiv), 70% aq. *t*-BuOOH (0.6 equiv), $MeCONMe_2$, $N_2$, 100 °C, 24 h

Z = OR', $NR'_2$

12 examples: 30-61% (22)

A Mannich reaction leading to β-amino ketones with DMF as the formaldehyde source has been reported in the presence of *t*-BuOOH and catalytic amounts of an *N*-heterocyclic carbene, $SnCl_2$ and $NEt_3$ (Equation (23)) [11].

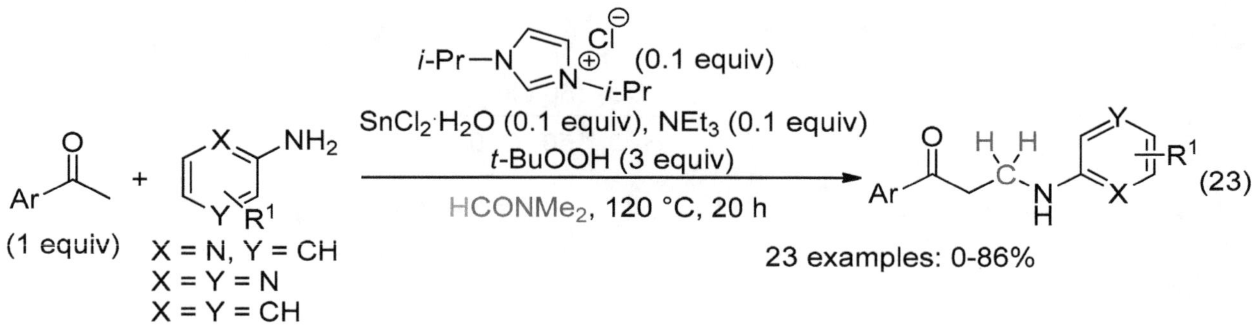

*i*-Pr imidazolium chloride (0.1 equiv), $SnCl_2 \cdot H_2O$ (0.1 equiv), $NEt_3$ (0.1 equiv), *t*-BuOOH (3 equiv), $HCONMe_2$, 120 °C, 20 h

(1 equiv); X = N, Y = CH; X = Y = N; X = Y = CH

23 examples: 0-86% (23)

The study of an unexpected reaction due to the oxidation of DMAc with aqueous *t*-BuOOH (Equation (24)) showed the formation of $MeCONMe(CH_2OH)$ and $MeCON(CH_2OH)_2$, these unusual species delivering the methylene group [12].

70% aq. $t$-BuOOH (10 equiv), $MeCONMe_2$, 75 °C, overnight; 92% (24)

## 5. NC Fragment

While CuI under oxidative and acidic conditions led, in DMAc, to the C3-formylation of indole and $N$-methylindole (Equation (6)), reactions in DMF led to C3-cyanations (Equation (25)) [39,40]. Cyanation of electron-rich arenes and benzaldehydes was also carried out (Equation (26)) [39]. Monitoring the course of the reaction indicated a cyanation arising via the formyl compounds [39,40]. Moreover, 3-iodo indole could also be involved in the formation of the cyano product [40].

A or B, $HCONMe_2$ (25)

A. $R^1$ = H; CuI (1.2 equiv), $t$-BuOOH (2 equiv), AcOH (4 equiv), air, 120-130 °C, 48 h; 15 examples: 30-86%
B. $R^1$ = Me; CuI (1 equiv), $CF_3CO_2H$ (1.2 equiv), $O_2$ (1 atm), 130-150 °C, 36-96 h; 15 examples: 25-90%

A: Z = H or C: Z = CHO; $HCONMe_2$, air, 130-140 °C, 48 h (26)

A. CuI (1.2 equiv), $t$-BuOOH (2 equiv), AcOH (4 equiv)
$R^1 = R^4$ = OMe, $R^2 = R^3$ = H: 80%; $R^1$ = Me, $R^2$ = H, $R^3 = R^4$: OMe: 35%;
$R^1 = R^2 = R^3$ = H, $R^4$ = OMe: 51%
C. $Cu(NO_3)_2 \cdot 3H_2O$ (1 equiv), $t$-BuOOH (2 equiv), AcOH (8 equiv)
$R^1 = R^2 = R^3 = R^4$ = H: 70%; $R^1 = R^4$ = H, $R^2 = R^3$ = OMe: 90%

Laser ablation of silver nitrate in DMF led to silver cyanide (Equation (27)) [53].

$$AgNO_3 \xrightarrow[\text{pulsed laser ablation}]{HCONMe_2} AgCN \quad (27)$$

## 6. $NMe_2$ Fragment

This chapter is divided in sections corresponding to the type of function reacting with DM.

### *6.1. Aryl Halides*

Refluxing chloropyridines in DM or DMAc afforded the corresponding aminopyridines (Equation (28)) [25]. The amination of aryl chlorides and 3-pyridinyl chloride with DMF occurred at room temperature in the presence of potassium $t$-butoxide and a carbenic palladium catalyst (Equation (29)) [27].

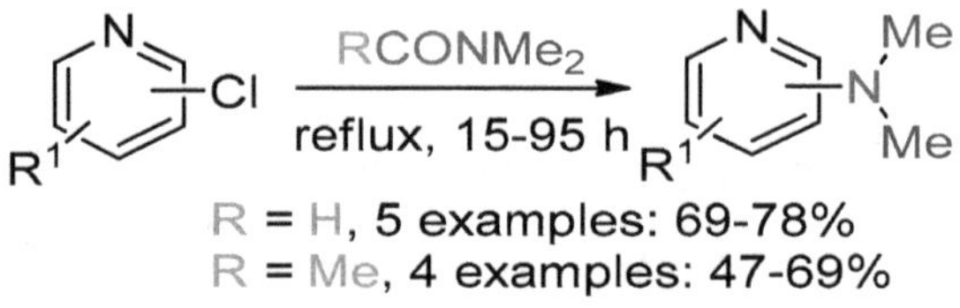

$RCONMe_2$, reflux, 15-95 h (28)

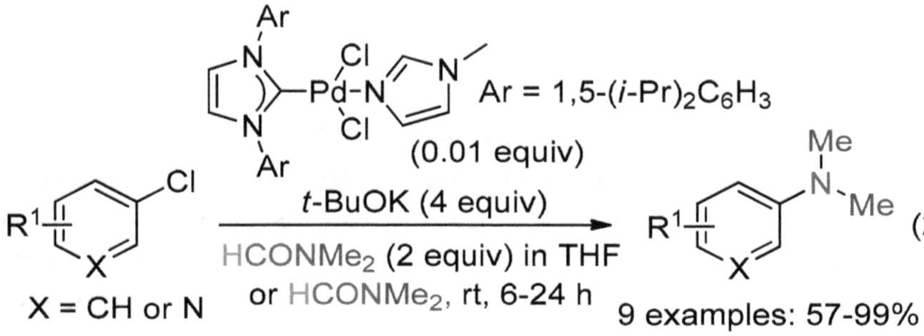

R = H, 5 examples: 69-78%
R = Me, 4 examples: 47-69%

Ar = 1,5-($i$-Pr)$_2C_6H_3$ (0.01 equiv)
$t$-BuOK (4 equiv)
$HCONMe_2$ (2 equiv) in THF or $HCONMe_2$, rt, 6-24 h (29)

X = CH or N

9 examples: 57-99%

*6.2. Alkylarenes*

Oxidation of methylarenes and ethylarenes at 80 °C in DMF using catalytic amounts of both $I_2$ and NaOH [31] or $n$-$Bu_4NI$ [32] associated to aqueous $t$-BuOOH under air led to benzylic oxidation and incorporation of the $NMe_2$ fragment to afford benzamides (Equation (30)) or α-ketoamides (Equation (31)).

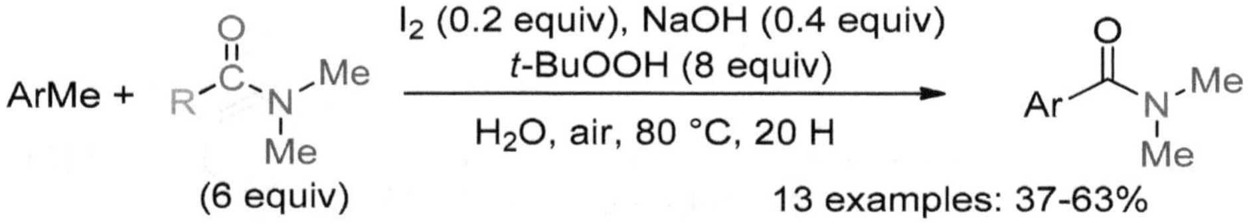

ArMe + RC(O)NMe$_2$ (6 equiv); $I_2$ (0.2 equiv), NaOH (0.4 equiv), $t$-BuOOH (8 equiv), $H_2O$, air, 80 °C, 20 H (30)

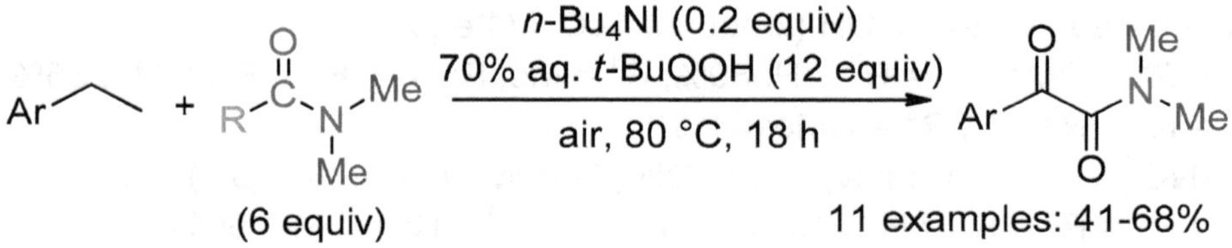

13 examples: 37-63%

ArEt + RC(O)NMe$_2$ (6 equiv); $n$-$Bu_4NI$ (0.2 equiv), 70% aq. $t$-BuOOH (12 equiv), air, 80 °C, 18 h (31)

11 examples: 41-68%

*6.3. Alkenes*

Hydrocarbonylation of terminal alkenes and norbornene followed by acyl metathesis with DM occurred under Pd catalysis, CO pressure and in the presence of ammonium chloride or $N$-methyl-2-pyrrolidone hydrochloride (NMP·HCl) (Equations (32) and (33)) [54]. From alkenes, the selectivity towards linear and branched products depended on the catalytic system (Equation (32)). DMF and DMAc afforded similar results.

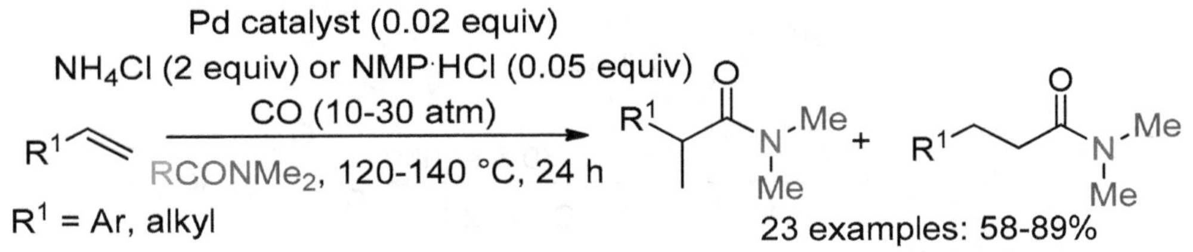

Pd catalyst (0.02 equiv)
$NH_4Cl$ (2 equiv) or NMP·HCl (0.05 equiv)
CO (10-30 atm)
$RCONMe_2$, 120-140 °C, 24 h

$R^1$ = Ar, alkyl

23 examples: 58-89% (32)

$R^1$ = Ph, $Pd(t\text{-}Bu_3P)_2$, $NH_4Cl$, CO (30 atm), 120 °C: R = H: 85%, 98:2; R = Me: 81%, 91:9

$R^1$ = Ph, $PdCl_2(MeCN)_2$ (0.02 equiv), $NH_4Cl$, XantPhos (0.025 equiv), CO (10 atm), 120 °C: R = H: 81%, 10:90; R = Me: 82%, 9:91

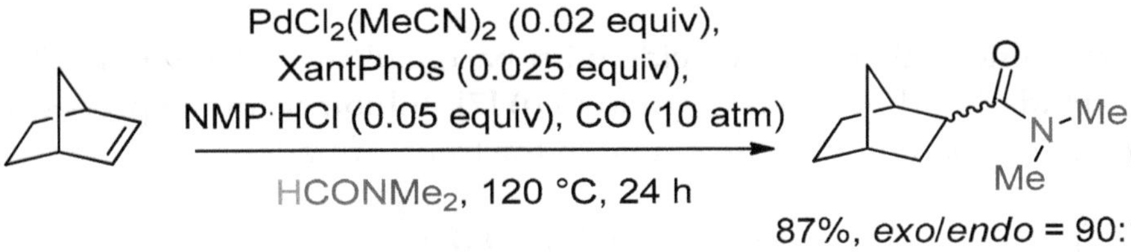

$PdCl_2(MeCN)_2$ (0.02 equiv),
XantPhos (0.025 equiv),
NMP·HCl (0.05 equiv), CO (10 atm)
$HCONMe_2$, 120 °C, 24 h

87%, *exo*/*endo* = 90:10 (33)

*6.4. Acids*

Copper, palladium and ruthenium catalysts associated to oxidants and DMF were used for the amidation of cinnamic acids [29] and carboxylic acids [55] (Equations (34) and (35)). *N,N*-Dimethylbenzamide was one of the products obtained from the $Cu^{II}$-catalyzed oxidation of flavonol [56].

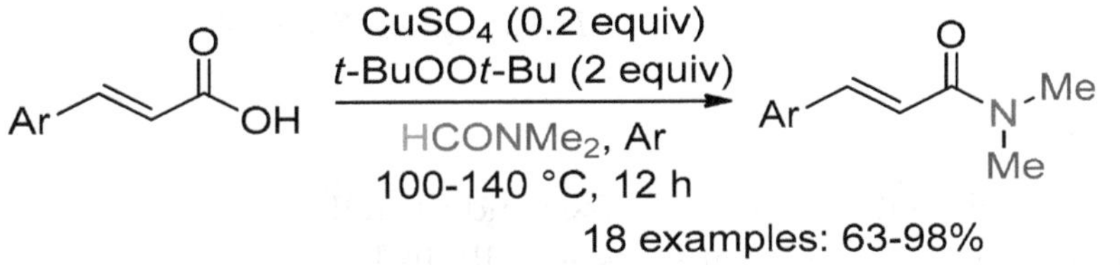

$CuSO_4$ (0.2 equiv)
*t*-BuOO*t*-Bu (2 equiv)
$HCONMe_2$, Ar
100-140 °C, 12 h

18 examples: 63-98% (34)

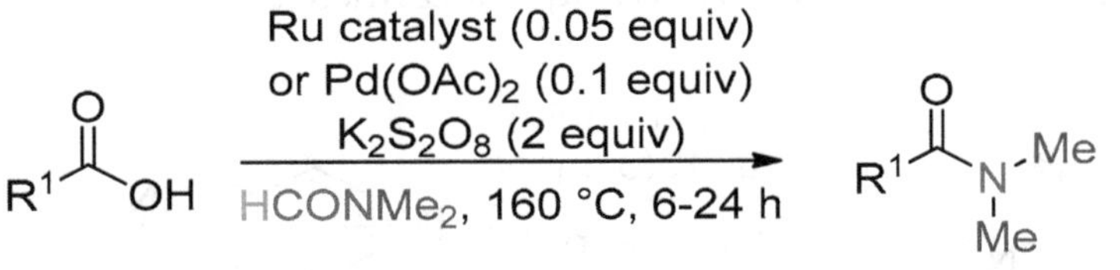

Ru catalyst (0.05 equiv)
or $Pd(OAc)_2$ (0.1 equiv)
$K_2S_2O_8$ (2 equiv)
$HCONMe_2$, 160 °C, 6-24 h (35)

$R^1$ = Ar, alkyl, $RuCl_2$(*p*-cymene), Xantphos (0.1 equiv):
23 examples: 26-93% ($R^1$ = Ph: 88%)
$R^1$ = Ph, $RuCl_3 \cdot 3H_2O$: 64%
$R^1$ = Ph, $Pd(OAc)_2$: 85%

Metal-free conditions and DMF were used for:

- the amidation of acids promoted with propylphosphonic anhydride associated to HCl at 130 °C (Equation (36)) [15],
- the amination of acids employing a hypervalent iodine reagent at room temperature (Equation (37)) [35]. Mesityliodine diacetate was superior to the other hypervalent iodine reagents, while oxidants such as $I_2$, *t*-BuOOH, $NaIO_4$ or $K_2S_2O_8$ did not mediate the amidation reaction [35].

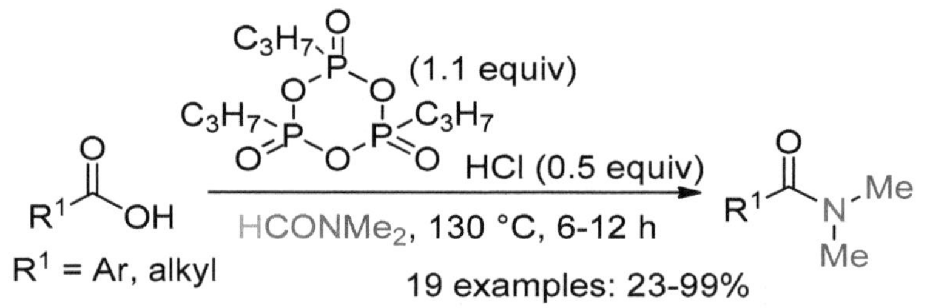

$$R^1COOH \xrightarrow[HCONMe_2,\ 130\ °C,\ 6\text{-}12\ h]{(C_3H_7PO_2)_3\ (1.1\ equiv),\ HCl\ (0.5\ equiv)} R^1CONMe_2 \quad (36)$$

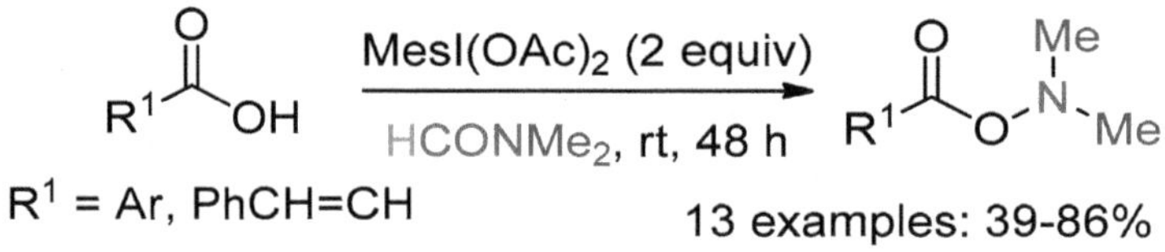

$R^1$ = Ar, alkyl; 19 examples: 23-99%

$$R^1COOH \xrightarrow[HCONMe_2,\ rt,\ 48\ h]{MesI(OAc)_2\ (2\ equiv)} R^1C(O)ONMe_2 \quad (37)$$

$R^1$ = Ar, PhCH=CH; 13 examples: 39-86%

Treatment of arylacetic and cinnamic acids with base, sulfur and DMF at 100–120 °C led to decarboxylative thioamidation (Equations (38) and (39)) [7]. Inhibition of the process in the presence of TEMPO or BHT indicated a radical involvement in the transformations.

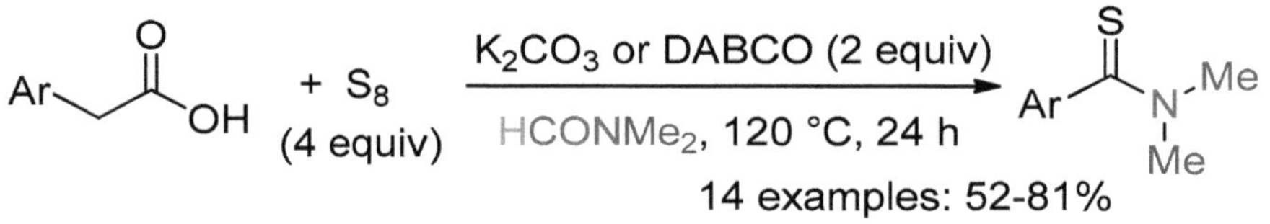

$$ArCH_2COOH + S_8\ (4\ equiv) \xrightarrow[HCONMe_2,\ 120\ °C,\ 24\ h]{K_2CO_3\ or\ DABCO\ (2\ equiv)} ArC(S)NMe_2 \quad (38)$$

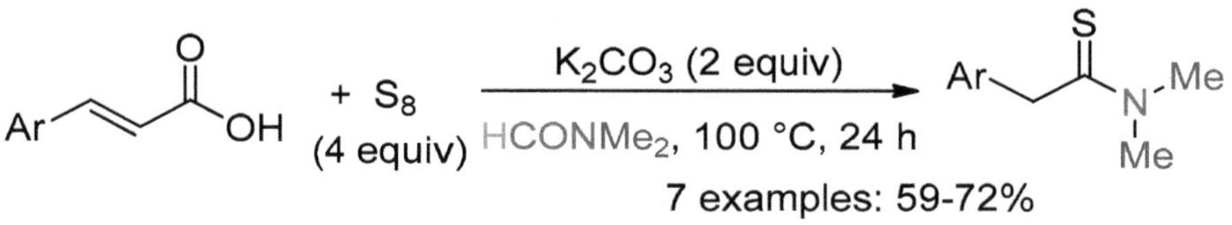

14 examples: 52-81%

$$ArCH{=}CHCOOH + S_8\ (4\ equiv) \xrightarrow[HCONMe_2,\ 100\ °C,\ 24\ h]{K_2CO_3\ (2\ equiv)} ArCH_2C(S)NMe_2 \quad (39)$$

7 examples: 59-72%

*6.5. Carbonylated Compounds*

Reaction of 2-arylquinazolin-4(3*H*)-ones with TsCl and *t*-BuOK in DM provided the corresponding 4-(dimethylamino)quinazolines in good yields, especially in DMF (Equation (40)). These reactions occurred via the formation of the 2-aryl-4-(tosyloxy)quinazolines [57].

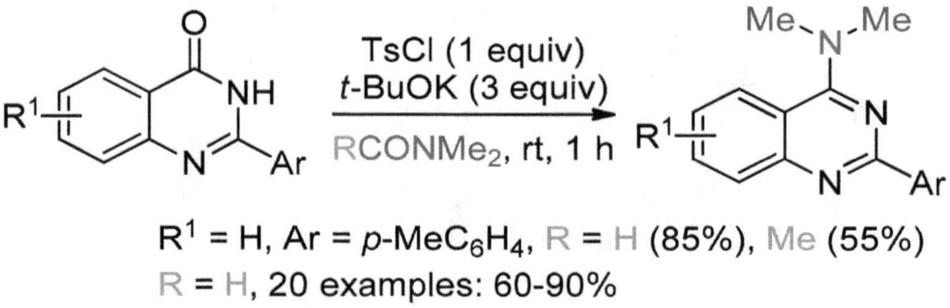

$$\text{2-Ar-}R^1\text{-quinazolin-4(3H)-one} \xrightarrow[RCONMe_2,\ rt,\ 1\ h]{TsCl\ (1\ equiv),\ t\text{-}BuOK\ (3\ equiv)} \text{4-}NMe_2\text{-2-Ar-}R^1\text{-quinazoline} \quad (40)$$

$R^1$ = H, Ar = $p$-$MeC_6H_4$, R = H (85%), Me (55%)
R = H, 20 examples: 60-90%

Various amides have been synthetized from aldehydes and DMF using *t*-BuOOH and a recyclable heterogeneous catalyst—a carbon–nitrogen embedded cobalt nanoparticle denoted as Co@C-N600 (Equation (41)) [33]. The same transformation of benzaldehydes was subsequently reported using Co/Al hydrotalcite-derived catalysts [58].

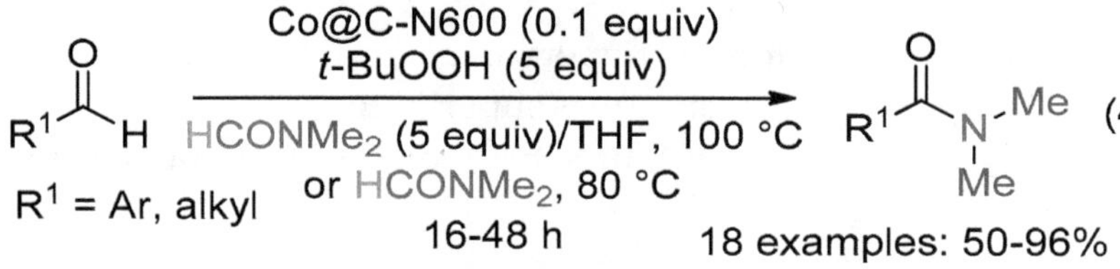

$$R^1CHO \xrightarrow[HCONMe_2\ (5\ equiv)/THF,\ 100\ °C\ or\ HCONMe_2,\ 80\ °C,\ 16\text{-}48\ h]{Co@C\text{-}N600\ (0.1\ equiv),\ t\text{-}BuOOH\ (5\ equiv)} R^1CONMe_2 \quad (41)$$

$R^1$ = Ar, alkyl; 18 examples: 50-96%

Copper oxide and iodine mediated the reaction of acetophenones with sulfur and DMF to afford α-arylketothioamides (Equation (42)) via the formation of α-iodoacetophenones [59].

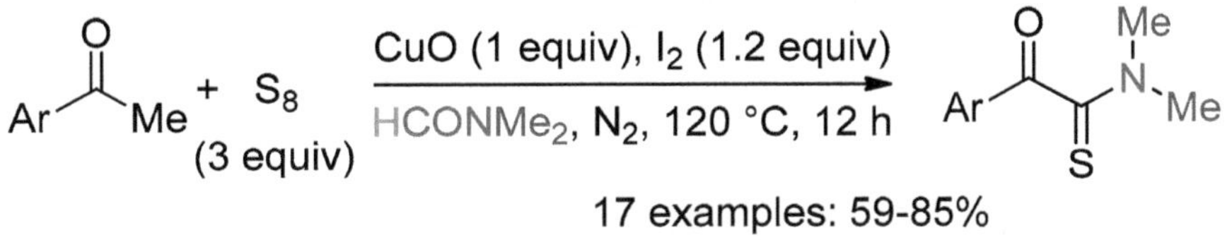

$$\text{ArC(O)Me} + S_8 \ (3 \text{ equiv}) \xrightarrow[\text{HCONMe}_2,\ N_2,\ 120\ °C,\ 12\ h]{\text{CuO (1 equiv)},\ I_2\ (1.2 \text{ equiv})} \text{ArC(O)C(S)NMe}_2 \quad (42)$$

17 examples: 59-85%

Elemental sulfur and the $NMe_2$ moiety of DMF or DMAc was also used for the DBU-promoted synthesis of thioamides from aldehydes (Equation (43)) or arones (Equation (44)) [60], the latter undergoing an efficient Willgerodt-Kindler reaction [61,62].

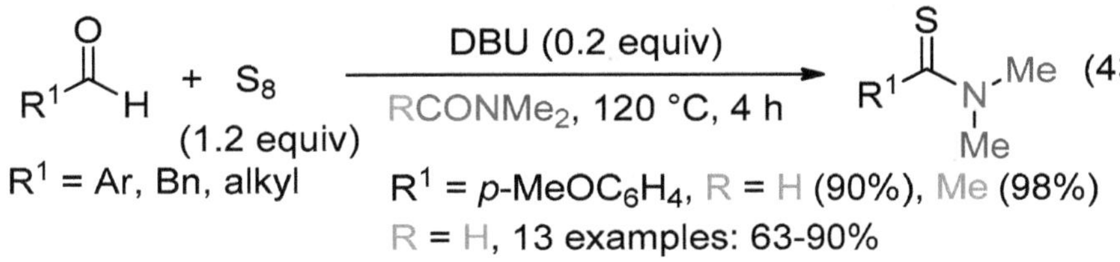

$$R^1\text{CHO} + S_8 \ (1.2 \text{ equiv}) \xrightarrow[\text{RCONMe}_2,\ 120\ °C,\ 4\ h]{\text{DBU (0.2 equiv)}} R^1\text{C(S)NMe}_2 \quad (43)$$

$R^1$ = Ar, Bn, alkyl

$R^1$ = $p$-MeOC$_6$H$_4$, R = H (90%), Me (98%)

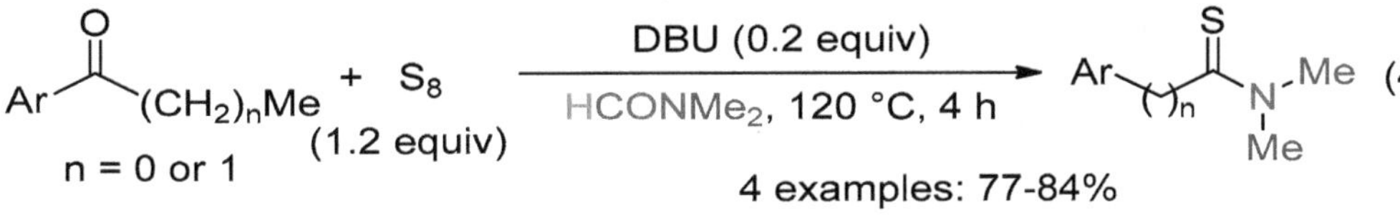

R = H, 13 examples: 63-90%

$$\text{ArC(O)(CH}_2)_n\text{Me} + S_8 \ (1.2 \text{ equiv}) \xrightarrow[\text{HCONMe}_2,\ 120\ °C,\ 4\ h]{\text{DBU (0.2 equiv)}} \text{Ar(CH}_2)_n\text{CH}_2\text{C(S)NMe}_2 \quad (44)$$

n = 0 or 1

4 examples: 77-84%

### 6.6. Benzyl Amines

The recyclable Co/Al catalysts used above in DMF for the amidation of benzaldehydes also led to benzamides from benzylamines and $t$-BuOOH (Equation (45)). These transformations would involve benzaldehydes as intermediates [58].

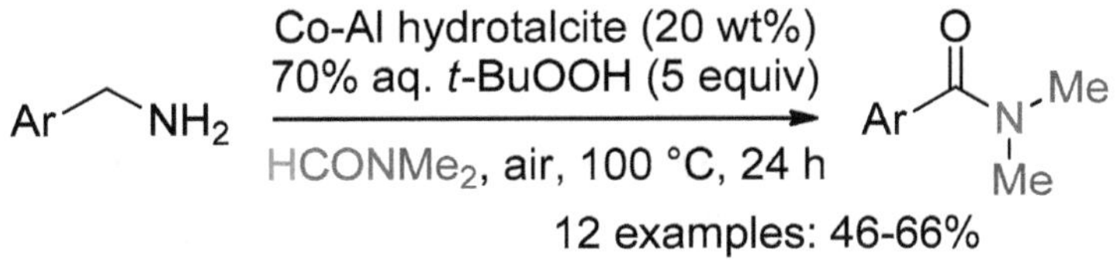

$$\text{ArCH}_2\text{NH}_2 \xrightarrow[\text{HCONMe}_2,\ \text{air},\ 100\ °C,\ 24\ h]{\text{Co-Al hydrotalcite (20 wt\%)},\ 70\%\ \text{aq. } t\text{-BuOOH (5 equiv)}} \text{ArC(O)NMe}_2 \quad (45)$$

12 examples: 46-66%

### 6.7. Nitriles

NaOH mediated, at room temperature, the efficient reaction of the CN group of 4-oxo-2,4-diphenylbutanenitrile with DMF to afford the corresponding γ-ketoamide (Equation (46)) [63]. Such compounds were also obtained from the domino reaction of chalcones with malononitrile and NaOH in DMF [63].

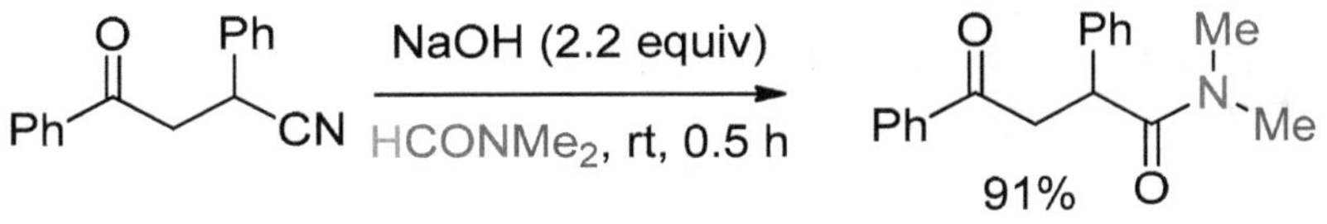

$$\text{PhC(O)CH}_2\text{CH(Ph)CN} \xrightarrow[\text{HCONMe}_2,\ \text{rt},\ 0.5\ h]{\text{NaOH (2.2 equiv)}} \text{PhC(O)CH}_2\text{CH(Ph)C(O)NMe}_2 \quad (46)$$

91%

### 6.8. Sulfur Compounds

Sulfonamides were synthetized:

- from thiophenols, DMF and air via an oxygen-activated radical process mediated by copper salts and cinnamic acid (Equation (47)) [64],

ArSH → [CuCl (1 equiv), $Cu(OAc)_2$ (1 equiv), cinnamic acid (1 equiv); $HCONMe_2$, air, 110 °C, 24 h] → $ArSO_2NMe_2$ (47)
12 examples, 46-90%

or reaction of sodium sulfonates with *N*-iodosuccinimide and DMF pretreated with *t*-BuOK (Equation (48)) via, probably, sulfonyl iodides (Equation (49)) [65].

$HC(O)NMe_2$ → [1) *t*-BuOK (4 equiv), MeCN, 50 °C, 30 mn; 2) $ArSO_2Na$, NIS (2 equiv), 50 °C, 12 h] → $ArSO_2NMe_2$ (48)
18 examples: 27-85%

$PhSO_2I$ + $HC(O)NMe_2$ → [*t*-BuOK (4 equiv); MeCN, 50 °C, 12 h] → $PhSO_2NMe_2$ (49)
50%

**7. O Fragment**

DMF delivered its oxygen atom to 1,2-cyclic sulfamidates via nucleophilic displacement at the quaternary center to afford, after hydrolysis, an aminoalcohol (Equation (50)) [17].

1) $HCONMe_2$, 70 °C, 6 h; 2) aq. $H_2SO_4$, $CH_2Cl_2$ (50)
97%

DMF was also the oxygen source leading to an imidazolinone from the reaction with the Cu-carbene complex and the borate salt depicted in Equation (51) [66].

$[CF_3B(OMe)_3]K$ (6 equiv.); $HCONMe_2$, Ar, 30 °C, 10 h (51)
X = Cl (56%), O*t*-Bu (99%), F (66%)

The $I_2$/CuO association allowed the α-hydroxylation of arones in abstracting, via the α-iodoarone, the oxygen atom of DMF (Equation (52)) [18].

$ArC(O)CH_2R^1$ → [CuO (1 equiv), $I_2$ (1.2 equiv); $HCONMe_2$, $N_2$, 100 °C, 24 h] → $ArC(O)CH(OH)R^1$ (52)
15 examples: 57-86%

## 8. C=O Fragment

With DMF as the CO surrogate, quinazolinones have been prepared at 140–150 °C

- via C($sp^2$)-H bond activation and annulation using Pd/C [67] or $Pd(OAc)_2$ [8], in the presence of $K_2S_2O_8$, $CF_3CO_2H$ and $O_2$ (Equation (53)),

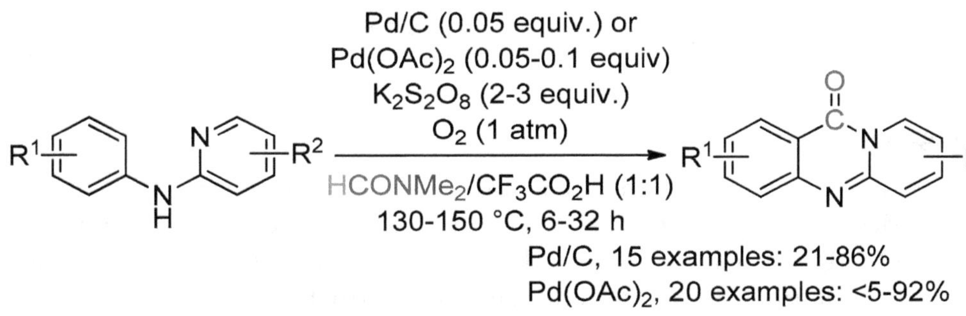

Pd/C (0.05 equiv.) or
$Pd(OAc)_2$ (0.05-0.1 equiv)
$K_2S_2O_8$ (2-3 equiv.)
$O_2$ (1 atm)
$HCONMe_2/CF_3CO_2H$ (1:1)
130-150 °C, 6-32 h
(53)
Pd/C, 15 examples: 21-86%
$Pd(OAc)_2$, 20 examples: <5-92%

or carbon dioxide-mediated cyclization of 2-aminobenzonitrile (Equation (54)) [50]. This latter reaction would involve a Vilsmeier-Haack type intermediate and did not occur with DMAc.

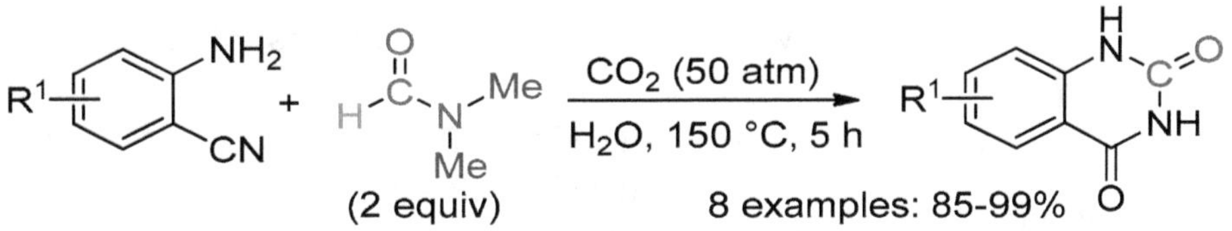

(2 equiv)
$CO_2$ (50 atm)
$H_2O$, 150 °C, 5 h
(54)
8 examples: 85-99%

A carbonylative Suzuki-type reaction leading to diarylketones arose in DMF under Ni catalysis at 100 °C (Equation (55)), IPr = bis(2,6-diisopropylphenyl)imidazol-2-ylidene) [13], or Pd catalysis and UV light assistance at room temperature (Equation (56)) [68].

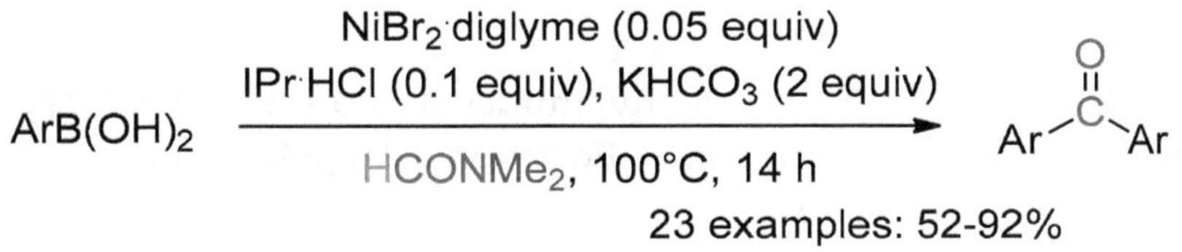

$ArB(OH)_2$ → $ArC(O)Ar$ (55) 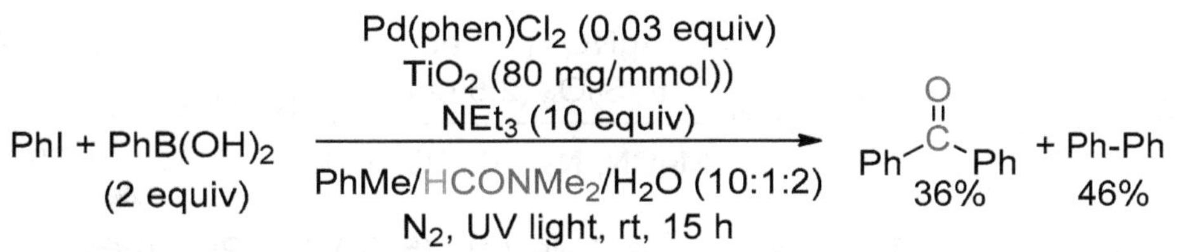
$NiBr_2$·diglyme (0.05 equiv)
IPr·HCl (0.1 equiv), $KHCO_3$ (2 equiv)
$HCONMe_2$, 100°C, 14 h
23 examples: 52-92%

PhI + $PhB(OH)_2$ (2 equiv) → PhC(O)Ph 36% + Ph-Ph 46% (56)
$Pd(phen)Cl_2$ (0.03 equiv)
$TiO_2$ (80 mg/mmol))
$NEt_3$ (10 equiv)
$PhMe/HCONMe_2/H_2O$ (10:1:2)
$N_2$, UV light, rt, 15 h

Catalytic amounts of a ruthenium pincer complex and *t*-BuOK led, at 165 °C, to symmetric and unsymmetric *N,N′*-disubstituted ureas from primary amines and DMF (Equation (57)) [69].

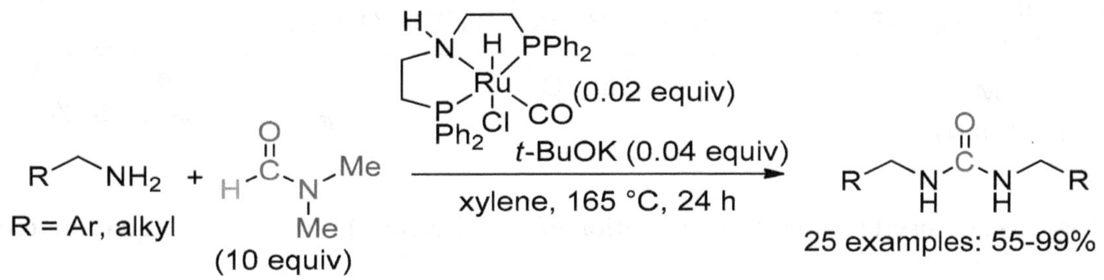

$RCH_2NH_2$ + $HC(O)NMe_2$ (10 equiv) → $RCH_2NHC(O)NHCH_2R$ (57)
R = Ar, alkyl
Ru pincer complex (0.02 equiv)
*t*-BuOK (0.04 equiv)
xylene, 165 °C, 24 h
25 examples: 55-99%

At 120 °C under $CuBr_2$ catalysis, *o*-iodoanilines reacted with potassium sulfide and DMF, leading to benzothiazolones (Equation (58)) [70].

$CuBr_2$ (0.1 equiv), $HCONMe_2$, $N_2$, 120 °C 15 h; $K_2S$ (3 equiv) (58)
26 examples: 14-83%

## 9. C=ONMe$_2$ Fragment

Potassium persulfate-promoted the reaction of pyridines with DMF to provide *N,N*-dimethylpicolinamides (Equation (59)) [71], while the oxidative carbamoylation of isoquinoline *N*-oxides, with also DMF, was catalyzed by Pd in the presence of ytterbium oxide as base and tetrabutylammonium acetate, the latter mediating the N-O reduction (Equation (60)) [72].

$K_2S_2O_8$ (2 equiv), air, 70 °C, 12 h; DMF (20 equiv) (59)
8 examples: 42-57%

$PdCl_2(MeCN)_2$ (0.1 equiv), *n*-$Bu_4NOAc$ (2 equiv), $Yb_2O_3$ (2 equiv), neat or PhMe, air, 120 °C, 24-40 h; DMF (75 equiv) (60)
12 examples: 50-67%

Alkynylation of DMF leading to *N,N*-dimethylamides was produced with peroxides and either an hypervalent alkynyl iodide under Ag catalysis (Equation (61)) [73], or a terminal alkyne under Cu catalysis (Equation (62)) [34].

$AgNO_3$ (0.1 equiv), $K_2S_2O_8$ (2.5 equiv), MeCN, $N_2$, 100 °C, 24 h; DMF (13 equiv) (61)
$R^1$ = $SiMe_2t$-Bu (51%), Si(*i*-Pr)$_3$ (68%)

$CuCl_2.2H_2O$ (0.05 equiv), 2,6-bis(benzimidazol-2'-yl)pyridine (0.06 equiv), *t*-BuOLi (0.6 equiv), 70% aq. *t*-BuOOH (1.5 equiv), Ar, 60 °C, 1 h; DMF (32 equiv) (62)
4 examples: 32-52%

A peroxide was also used for the carbamoylation of 4-arylcoumarin with DMF (Equation (63)) [74].

Ph, $R^1$, + H–C(=O)–N(Me)Me (26 equiv); $PhCO_3t$-Bu (3 equiv), 120 °C, 12 h → Ph, $R^1$, C(=O)–N(Me)Me (63)
6 examples: 47-75%

α-Ketoamides were obtained from the domino reaction of toluenes with DMF using $(t\text{-BuO})_2$, $Cs_2CO_3$ and catalytic amounts of $n$-$Bu_4NI$ (Equation (64)) [75].

ArMe; $n$-$Bu_4NI$ (0.1 equiv), $t$-BuO$Ot$-Bu (8 equiv), $Cs_2CO_3$ (2 equiv), $HCONMe_2$, Ar, 80 °C, 12 h → Ar–C(=O)–C(=O)–N(Me)Me (64)
13 examples: 0-85%

At 100 °C in DMF, Cu catalyst associated to $t$-BuOOH led to unsymmetrical ureas from 2-oxindoles (Equation (65)) [76]. The peroxide would mediate the cleavage reaction, and was the oxygen source of the benzylic carbonyl. That resulted in a ketoamine which undergone the Cu-catalyzed reaction with DMF/$t$-BuOOH, leading to the urea.

$R^1$, $R^2$, C=O, N–H; $Cu(OAc)_2$ (0.2 equiv), $t$-BuOOH (4 equiv), $HCONMe_2$, 100 °C, 1 h → $R^1$, C(=O)$R^2$, N–H, O=C–N(Me)Me (65)
10 examples: 45-77%

## 10. H Fragment

Semihydrogenation of diaryl alkynes occurred under Ru (Equation (66)) [77] and Pd [78] catalysis with DMF and water as hydrogen source.

$Ar^1$–≡–$Ar^2$; $Ru_3(CO)_{12}$ (0.03 equiv), $PPh_3$ (0.05 equiv), AcOH (1.5 equiv), $H_2O$ (2 equiv), $HCONMe_2$, 145 °C, 24 h → $Ar^1$(H)C=C(H)$Ar^2$ + $Ar^1$(H)C=C(H)$Ar^2$ (66)
9 examples: 40-98%, $Z/E$ = 75:25-99:1

Cobalt porphyrins catalyzed the hydrogenation transfer from DMF to the C($sp^3$)-C($sp^3$) bond of [2.2]paracyclophane (Equation (67)) [79]. DMF was also involved in the Ni-catalyzed intramolecular hydroarylations depicted in Equation (68) [14].

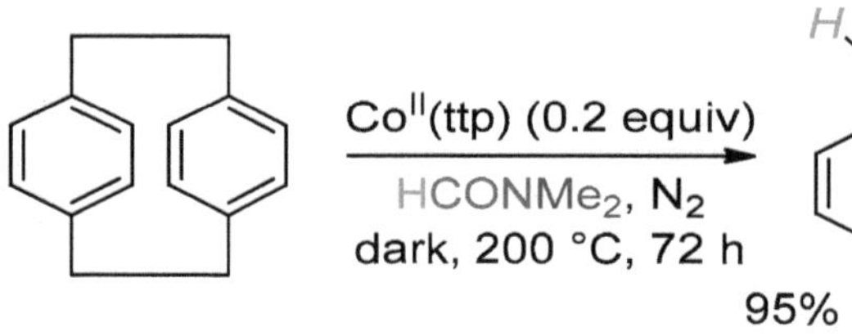

$Co^{II}$(ttp) (0.2 equiv), $HCONMe_2$, $N_2$, dark, 200 °C, 72 h; 95% (67)

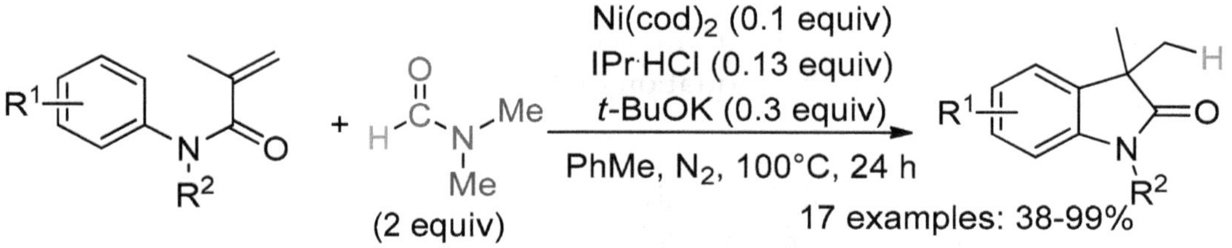

$Ni(cod)_2$ (0.1 equiv), IPr·HCl (0.13 equiv), *t*-BuOK (0.3 equiv), PhMe, $N_2$, 100°C, 24 h; + $HC(O)NMe_2$ (2 equiv); 17 examples: 38-99% (68)

## 11. RC Fragment

New metal-catalyzed and metal-free conditions involving the CH of the formyl group of DMF have been reported for cyclizations leading to heterocycles (Equations (69) [80–82] and (70) [83]).

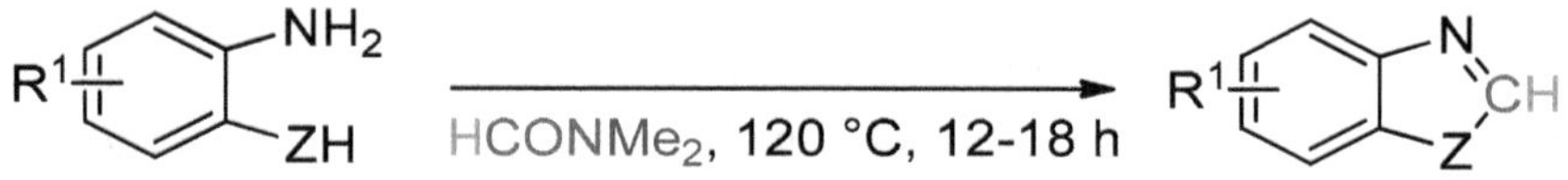

$HCONMe_2$, 120 °C, 12-18 h (69)

$Zn(OAc)_2 \cdot 2H_2O$ (0.05 equiv), poly(methylhydrosiloxane) (5 equiv) [80]: $Z = NR^2$, 12 examples: 0-95%; Z = S, $R^1$ = H: 83%; Z = O, 2 examples: 70-81%

$B(C_6F_5)_3$ (0.05 equiv), $H_2SiEt_2$ or $HSi(OEt)_3$ (5 equiv), $CO_2$ (1 atm) [81]: $Z = NR^2$, 17 examples: 68-99%; Z = S, 10 examples: 68-99%; Z = O, $R^1$ = H: 70%

$PhSiH_3$ (4 equiv) [82]: $Z = NR^2$, 13 examples: 31-95%; Z = S, $R^1$ = H: 83%

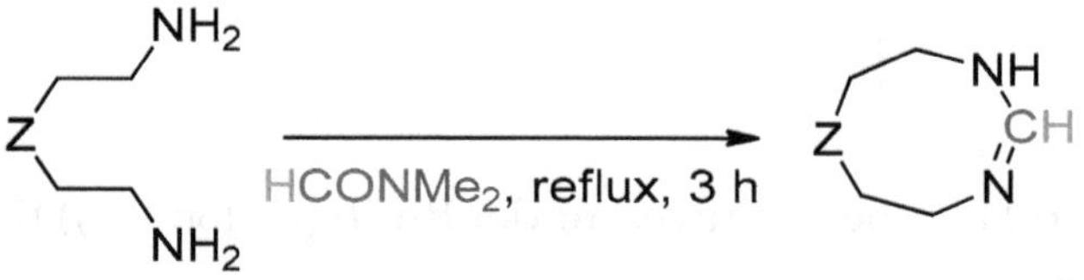

$HCONMe_2$, reflux, 3 h (70)

Z = $CH_2$ (95%), NH (94%), $HNCH_2CH_2NH$ (92%)

2-Methylbenzimidazoles were obtained from $PhSiH_3$-assisted delivery of the CMe of DMAc to benzene-1,2-diamines (Equation (71)) [82].

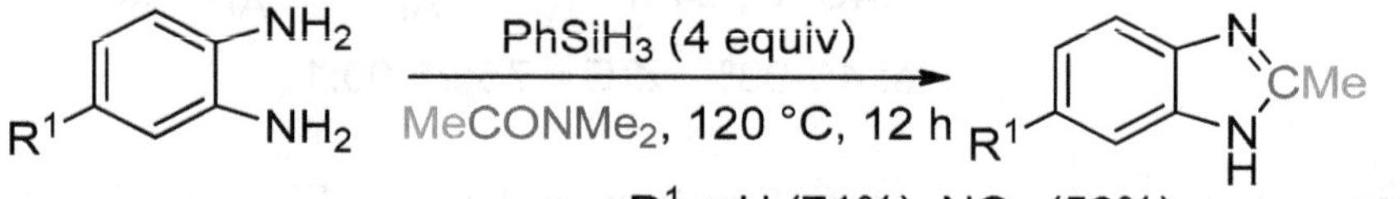

$PhSiH_3$ (4 equiv), $MeCONMe_2$, 120 °C, 12 h (71)

$R^1$ = H (71%), $NO_2$ (58%)

Addition of *p*-tolyllithium to DMF followed by reaction with hydroxylamine hydrochloride afforded 4-methylbenzaldehyde oxime (Equation (72)). The latter underwent cycloaddition with diphenylphosphoryl azide or, in the presence of Oxone®, with diethylacetylene dicarboxylate to provide the corresponding 5-aryltetrazole or 3-arylisoxazole, respectively [84].

1) *n*-BuLi (1.2 equiv), THF, -70 °C, 0.5 h
2) $HCONMe_2$ (1.2 equiv), rt, 1 h
3) $NH_2OH{\cdot}HCl$ (1.5 equiv), $NEt_3$ (1.5 equiv), rt, 2 h
90% (72)

## 12. $RCNMe_2$ Fragment

Dihydropyrrolizino[3,2-*b*]indol-10-ones were isolated in fair to high yields from a $Cs_2CO_3$-promoted domino reaction leading to the formation of three bonds with incorporation of the HCNMe of DMF. Such a reaction-type with incorporation of the MeCNMe also occurred in DMAc but with a low yield (Equation (73)) [26].

$Cs_2CO_3$ (2 equiv), $RCONMe_2$, $N_2$, 120 °C, 24 h (73)
R = H, 12 examples: 32-93%
$R^1 = R^3 = H$, $R^2 = Me$, R = H (87%), Me (7%)

## 13. RC-O Fragment

A 1:1 mixture of CuO and $I_2$ led the $\alpha$-formyloxylation or $\alpha$-acetoxylation of methylketones by DMF or DMAc, respectively (Equation (74)). $\alpha$-Iodoketones would be the intermediates as indicated by the reaction of 2-iodo-1-(4-methoxyphenyl)ethanone (Equation (75)). Traces of water delivered the carbonyl oxygen [16].

CuO (1 equiv.), $I_2$ (1 equiv.), $RCONMe_2$, air, 110 °C, 1-6 h (74)
$R^1$ = Ar, ArCH=CH
R = H, 24 examples: 50-92%
R = Me, 13 examples: 52-80%

CuO (0 or 1 equiv.), $HCONMe_2$, air, 110 °C, 2 h (75)
with CuO: 75%
without CuO: 50%

Stereoinversion of the secondary alcohols of a number of carbocyclic substrates was carried out via their triflylation followed by treatment with aqueous DMF (Equation (76)) and subsequent methanolysis [85]. A one pot stereoinversion process was reported.

$HCONMe_2/H_2O$ (1:1), Ar, 100 °C, 4 h (76)
65%

The formyl group of DMF was involved in the triflic anhydride-mediated domino reaction depicted in Equation (77) [86].

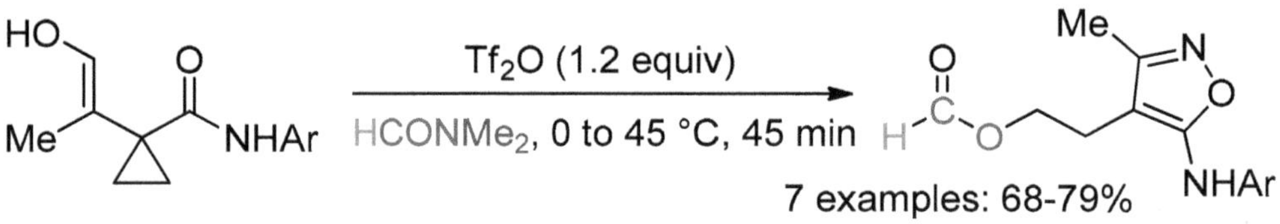

$Tf_2O$ (1.2 equiv), $HCONMe_2$, 0 to 45 °C, 45 min (77)

7 examples: 68-79%

Chloroformyloxylation and chloroacetoxylation of olefinic substrates were performed with $PhICl_2$ and either wet DMF or DMAc (Equation (78)) [87]. Styrenes suffered also difunctionalization using aryl diazonium salts and Ru photocatalysis in wet DM (Equation (79)) [28].

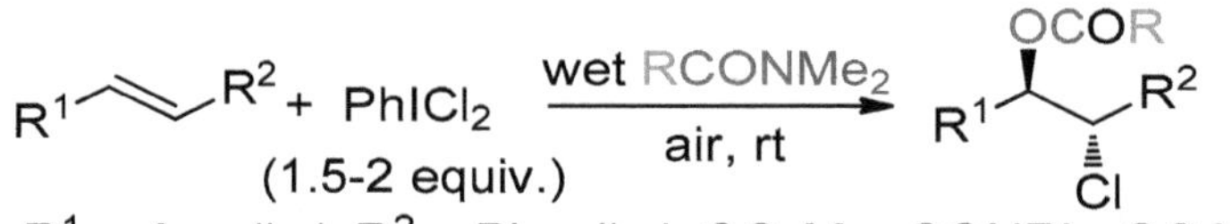

$R^1$–CH=CH–$R^2$ + $PhICl_2$ (1.5-2 equiv.), wet $RCONMe_2$, air, rt (78)

$R^1$ = Ar, alkyl; $R^2$ = Ph, alkyl, $CO_2Me$, $CONR'_2$, COAr

R = H, 12 examples: 60-95%, dr = 65:35-96:4
R = Me, 3 examples: 83-90%, dr = 88:12-95:5

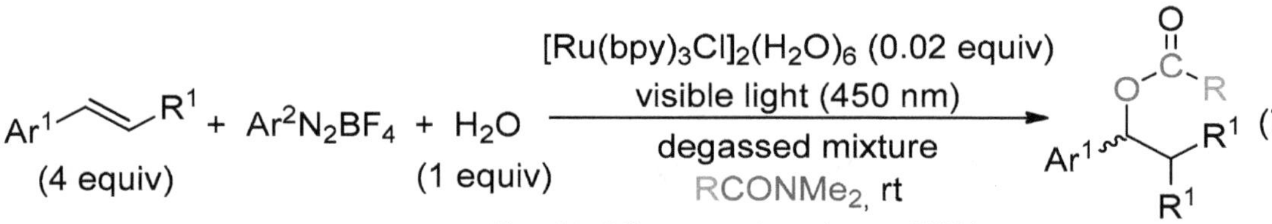

$Ar^1$–CH=CH–$R^1$ (4 equiv) + $Ar^2N_2BF_4$ + $H_2O$ (1 equiv), $[Ru(bpy)_3Cl]_2(H_2O)_6$ (0.02 equiv), visible light (450 nm), degassed mixture, $RCONMe_2$, rt (79)

R = H, 22 examples: trace-76%
$Ar^1$ = Ph, $R^1$ = p-$ClC_6H_4$, R = H (76%), Me (52%)

**14. RC=O Fragment**

Metal-catalyzed or $CO_2$-mediated C–N and N–H bond metathesis reactions between primary or secondary amines and DM provided the transamidation products, that is formamides or acetamides (Equation (80)) [88–91].

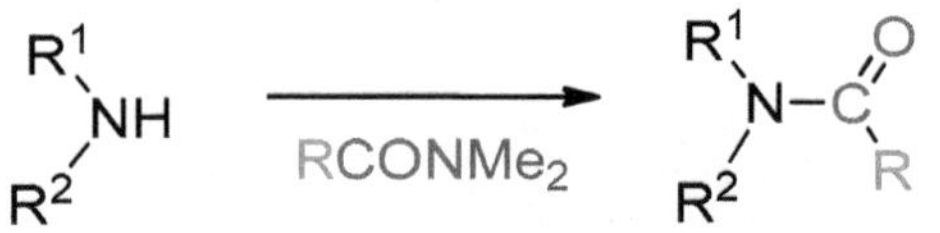

$R^1R^2NH$ → ($RCONMe_2$) $R^1R^2N$–C(=O)R (80)

$R^1$ = Ar, alkyl, Bn; $R^2$ = H, alkyl, Bn

$CO_2$ (1 bar), 100 °C, 24-48 h [88]: R = H, 23 examples: 68-99%
$CeO_2$ (cat), 150 °C, 24 h [89]: R = H, 17 examples: 49-99%
$Nd_2Na_8(OCH_2CF_3)_{14}(THF)_6$ (0.04 equiv), Ar, 120 °C, 5-10 h [90]: R = H, $R^1$ = H, $R^2$ = Bn (78%), $(CH_2)_{10}Me$ (76%); R = Me, $R^1$ = H, $R^2$ = Bn (86%), $(CH_2)_{10}Me$ (83%) 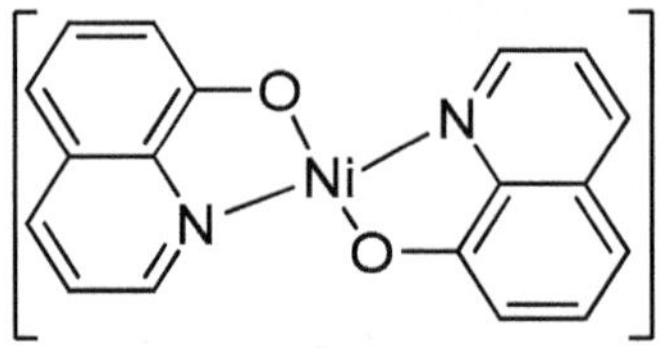
[Ni(8-quinolinolate)$_2$]·$2H_2O$ (0.05 equiv), imidazole (3 equiv), 150 °C, 2-6 h [91]: R = H, 21 examples: 50-99%; R = Me, 14 examples: 53-98%

Formylation of aromatic substrates resulted from their treatment with LDA [92] or n-BuLi [93] and, subsequently, DMF (Equation (81)).

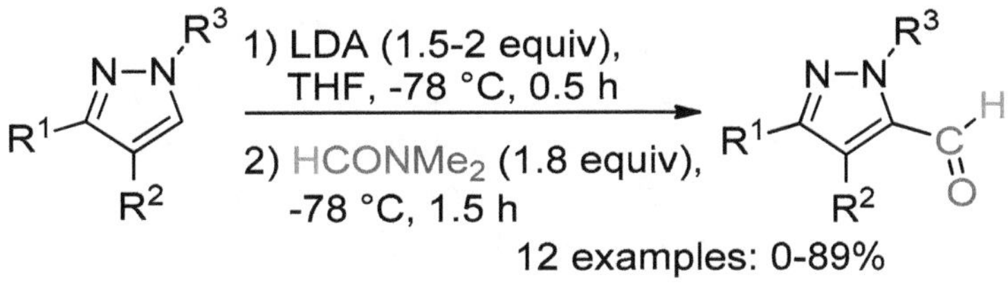

1) LDA (1.5-2 equiv), THF, -78 °C, 0.5 h
2) $HCONMe_2$ (1.8 equiv), -78 °C, 1.5 h
12 examples: 0-89% (81)

Graphene oxide reacted with DBU and DM to afford, in presence of trace of water, *N*-(3-(2-oxoazepan-1-yl)propyl)formamide or the corresponding acetamide (Equation (82)) [94].

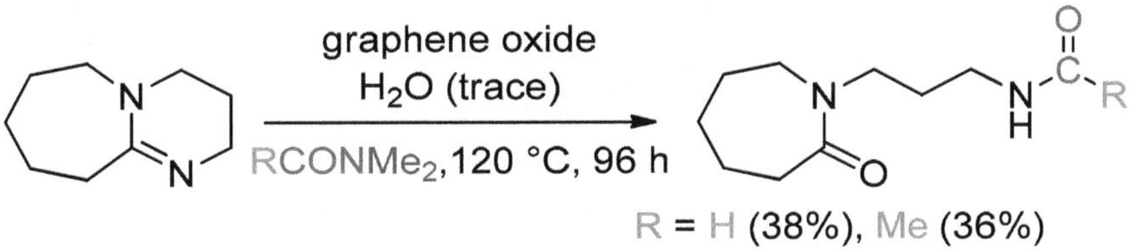

graphene oxide
$H_2O$ (trace)
$RCONMe_2$, 120 °C, 96 h
R = H (38%), Me (36%) (82)

## 15. RC=ON(CH₂)Me Fragment

*N*-Amidoalkylation of imidazoles and 1,2,3-triazoles with DM effectively arose under various experimental conditions (Equations (83) [95] and (84) [10]).

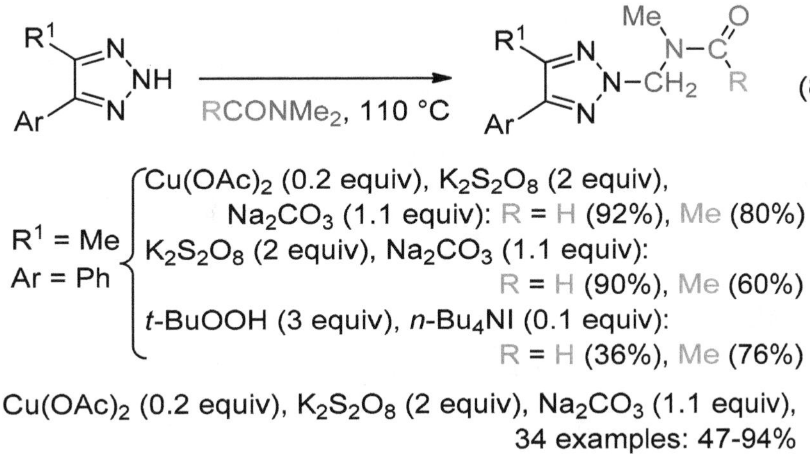

$RCONMe_2$, 110 °C (83)

$R^1$ = Me, Ar = Ph:
- $Cu(OAc)_2$ (0.2 equiv), $K_2S_2O_8$ (2 equiv), $Na_2CO_3$ (1.1 equiv): R = H (92%), Me (80%)
- $K_2S_2O_8$ (2 equiv), $Na_2CO_3$ (1.1 equiv): R = H (90%), Me (60%)
- *t*-BuOOH (3 equiv), *n*-$Bu_4NI$ (0.1 equiv): R = H (36%), Me (76%)

$Cu(OAc)_2$ (0.2 equiv), $K_2S_2O_8$ (2 equiv), $Na_2CO_3$ (1.1 equiv), 34 examples: 47-94%

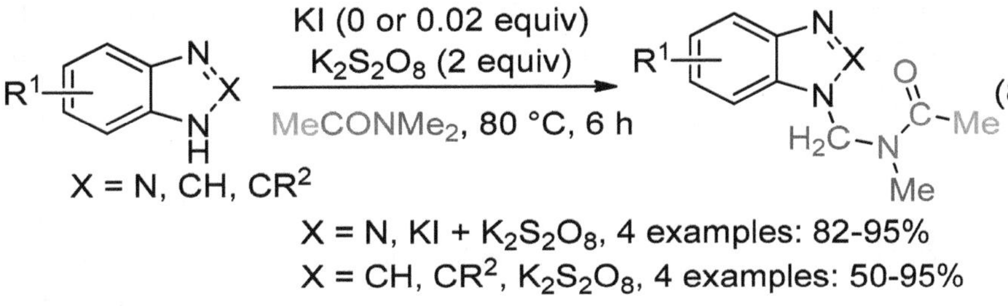

KI (0 or 0.02 equiv)
$K_2S_2O_8$ (2 equiv)
$MeCONMe_2$, 80 °C, 6 h (84)

X = N, CH, $CR^2$

X = N, KI + $K_2S_2O_8$, 4 examples: 82-95%
X = CH, $CR^2$, $K_2S_2O_8$, 4 examples: 50-95%

## 16. HC-ONMe₂ Fragment

Catalysis with $Mo(CO)_6$ of the reduction of DMF with triethylsilane afforded a siloxymethylamine (Equation (85)), which was used as a Mannich reagent [96].

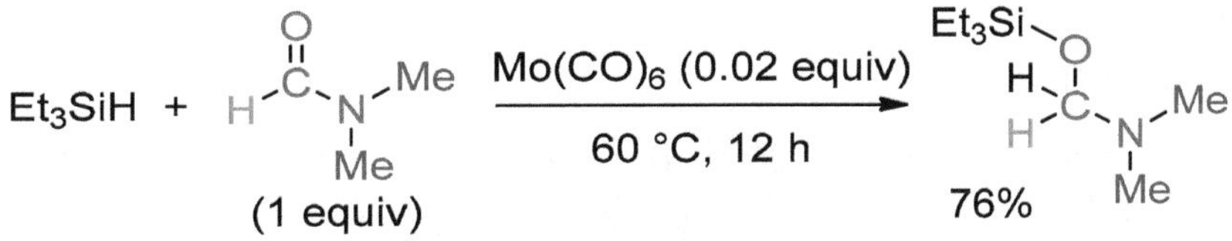

$Et_3SiH$ + $HC(O)NMe_2$ (1 equiv) → [$Mo(CO)_6$ (0.02 equiv), 60 °C, 12 h] → $Et_3SiOCH_2NMe_2$ 76% (85)

## 17. RC and O Fragment

Arynes, which are easily obtained from, for example 2-(trimethylsilyl)phenyl trifluoromethanesulfonate, undergone a [2 + 2] cyclization with DM giving a benzoxetene and its isomer, the *ortho*-quinone methide (Scheme 3). Trapping of these intermediates provides various products, which contain the formyl or acetyl CH part and the O atom of DM (Equations (86) [19], (87) [20], (88) [21] and (89) [97]), or the $HCNMe_2$ and O fragments of DMF (see Section 18).

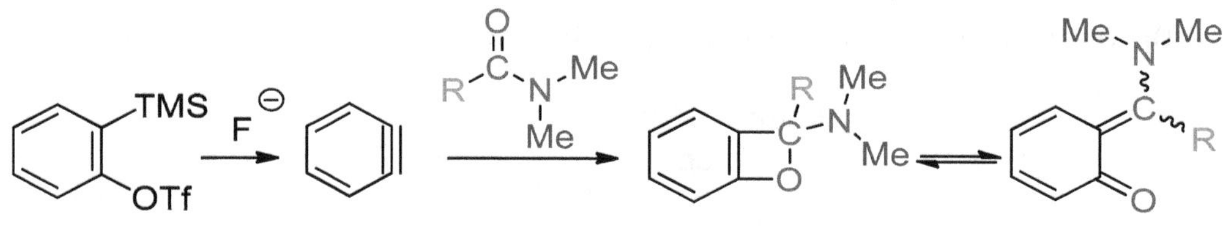

TMS OTf $F^{\ominus}$ R C(O) N Me Me R Me C N Me O Me N Me C R O

**Scheme 3.** Aryne formation and reaction with DM.

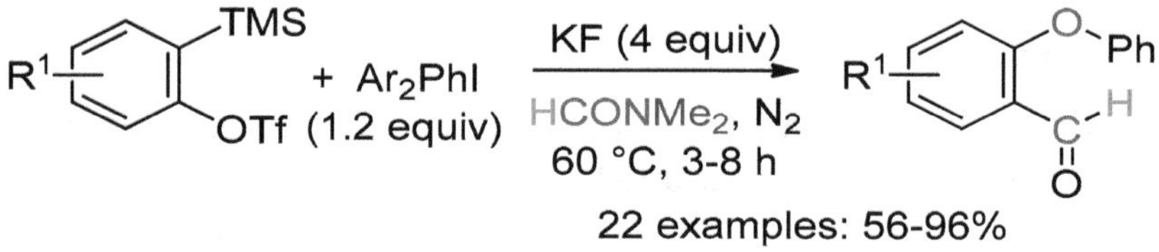

$R^1$ TMS OTf + $Ar_2PhI$ (1.2 equiv) KF (4 equiv) $HCONMe_2$, $N_2$ 60 °C, 3-8 h $R^1$ O Ph C H O (86)

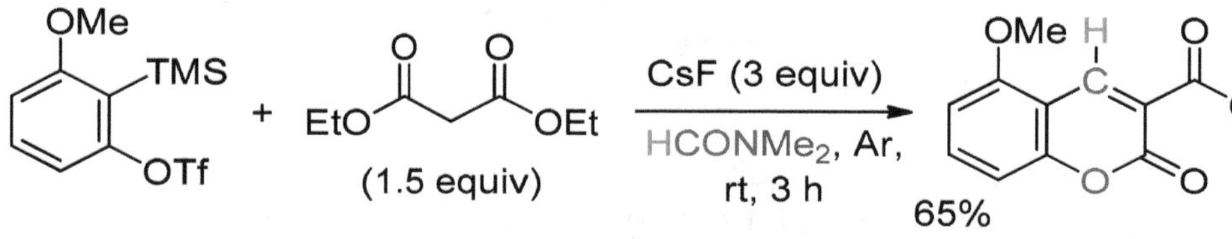

22 examples: 56-96%

OMe TMS OTf + EtO O O OEt (1.5 equiv) CsF (3 equiv) $HCONMe_2$, Ar, rt, 3 h OMe H C O OEt O O (87) 65%

OMe TMS OTf + O O (1 equiv) $n$-$Bu_4NF$ (3 equiv) $HCONMe_2$, Ar, rt, 3 h OMe H C O O OH 53% (88)

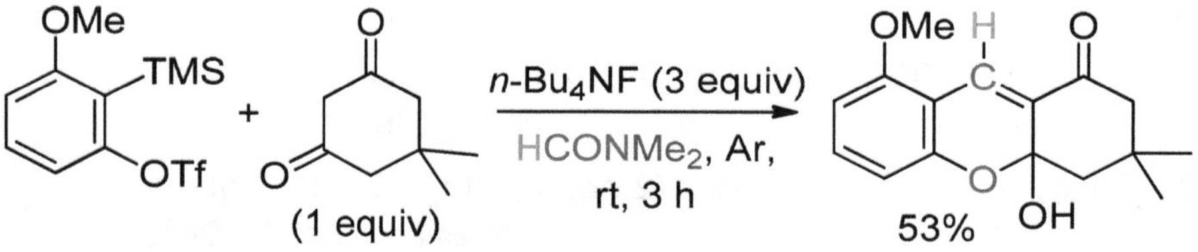

At 115 °C in wet toluene, the hexadehydro-Diels–Alder of tetraynes depicted in Equation (89) was in-situ followed by [2 + 2] cycloaddition reaction with DM leading to multifunctionalized salicylaldehydes and salicylketones [97].

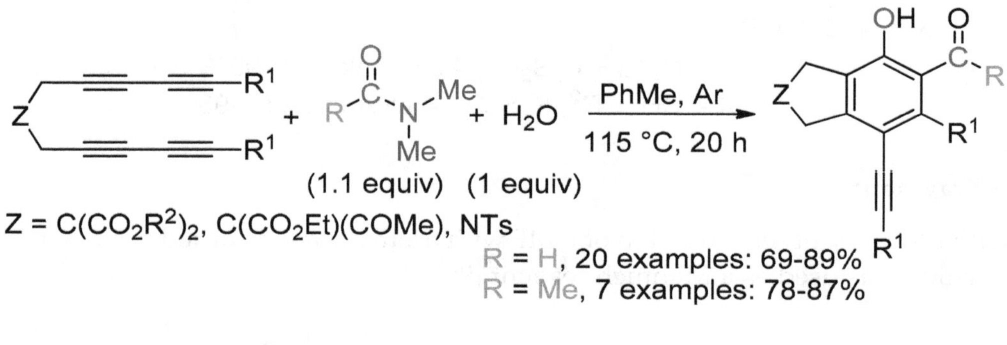

Z $R^1$ $R^1$ + R C(O) N Me Me (1.1 equiv) + $H_2O$ (1 equiv) PhMe, Ar 115 °C, 20 h Z OH O C R $R^1$ $R^1$ (89)

Z = $C(CO_2R^2)_2$, $C(CO_2Et)(COMe)$, NTs

R = H, 20 examples: 69-89%
R = Me, 7 examples: 78-87%

## 18. $RCNMe_2$ and O Fragment

All atoms of DMF were inserted in the polyfunctionalized compounds produced from domino reactions involving formation of arynes, cycloaddition with DMF and subsequent trapping with α-chloro β-diesters (Equation (90)) [22], aroyl cyanides (Equation (91)) [23] or diesters of acetylenedicarboxylic acid (Equation (92)) [24].

(1.2 equiv) + $R^2O_2C$–CH(Cl)–$CO_2R^2$ → CsF (6 equiv), $ZnEt_2$ (1 equiv), $HCONMe_2$, Ar, -40 °C to rt; 5 examples: 39-89% (90)

(2 equiv) + ArCOCN → CsF (4 equiv), MS 4Å, $HCONMe_2$, $N_2$, 60 °C, 3 h; 20 examples: 29-88% (91)

+ $R^1O_2C$–C≡C–$CO_2R^1$ (2 equiv) → CsF (3 equiv), $HCONMe_2$, 25 °C, 1-5 h; $R^1$ = Me (85%), Et (76%), *t*-Bu (71%) (92)

## 19. HC=O and HC Fragment

Lithium thioanisole biscarbanion reacted with two molecules of DMF to afford benzo[*b*]thiophene-2-carbaldehyde (Equation (93)) [98].

1) TMEDA (3 equiv), *n*-BuLi (5 equiv), TBME, rt, 3 h; 2) $HCONMe_2$ (10 equiv), rt, 24 h; 75% (93)

## 20. H and $NMe_2$ Fragment

The reaction of DMF with sodium and subsequent addition of terminal activated alkynes afforded the corresponding hydroamination compounds (Equation (94)) [99].

1) Na, 5 min; 2) ≡–$COR^1$; $R^1$ = Me (98%), OMe (100%), OEt (100%) (94)

## 21. H and $C{=}ONMe_2$ Fragment

Semicarbazides have been synthetized from additions, mediated with $(t\text{-}BuO)_2$ and catalytic amounts of both NaI and PhCOCl, of the H and $CONMe_2$ moieties of DMF to the extremities of the N=N bond of azoarenes (Equation (95)) [100]. The role of NaI and PhCOCl is not clear and, furthermore, exchange of NaI for imidazole led to formylhydrazines (Equation (96)) [100]. The corresponding acetylhydrazine was not formed in DMAc (Equation (96)). The $(t\text{-}BuO)_2$/NaI/PhCOCl/DMF system led to the addition of H and $CONMe_2$ to the N=C bond of *N*-benzylideneaniline but with low yield (Equation (97)) [100].

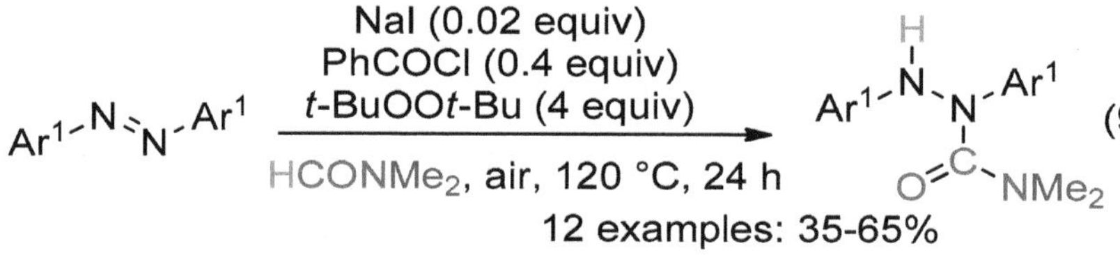

(95)

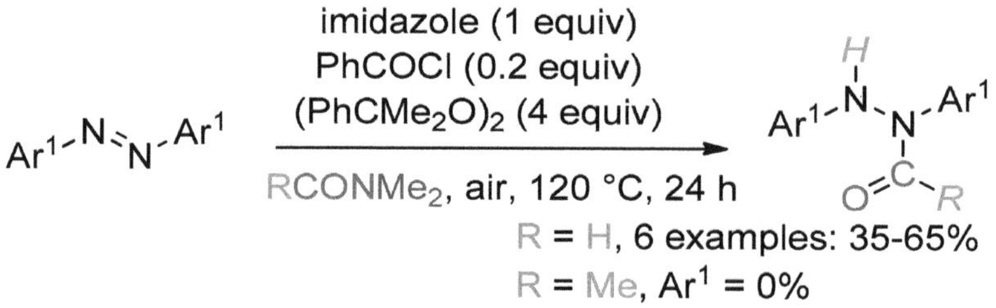

(96)

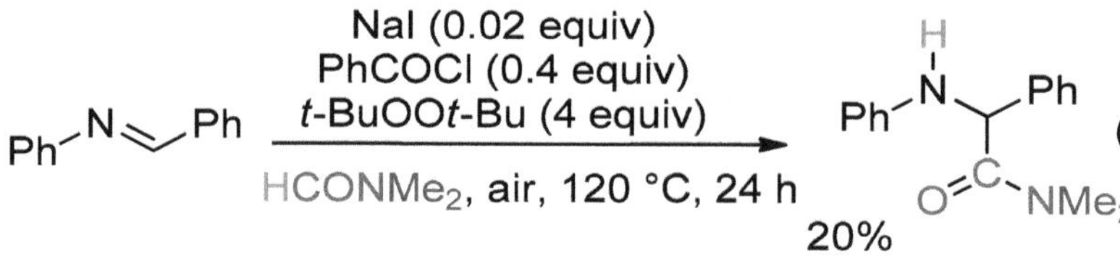

(97)

The Ru-catalyzed hydrocarbamoylative cyclization of 1,6-diynes proceeded in DMF to afford cyclic $\alpha,\beta,\delta,\gamma$-unsaturated amides (Equation (98)) [101–103].

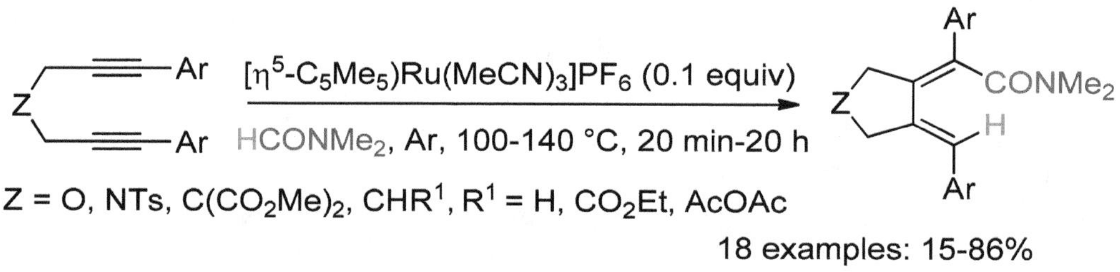

(98)

## 22. H, C=ONMe₂ and NMe₂ Fragment

$Re_2(CO)_8[\mu\text{-}\eta^2\text{-}C(H){=}C(H)Bu](\mu\text{-}H)$ undergone reaction with DMF leading to hexenyl/$CONMe_2$ and CO/$HNMe_2$ exchange of ligands (Equation (99)) [104].

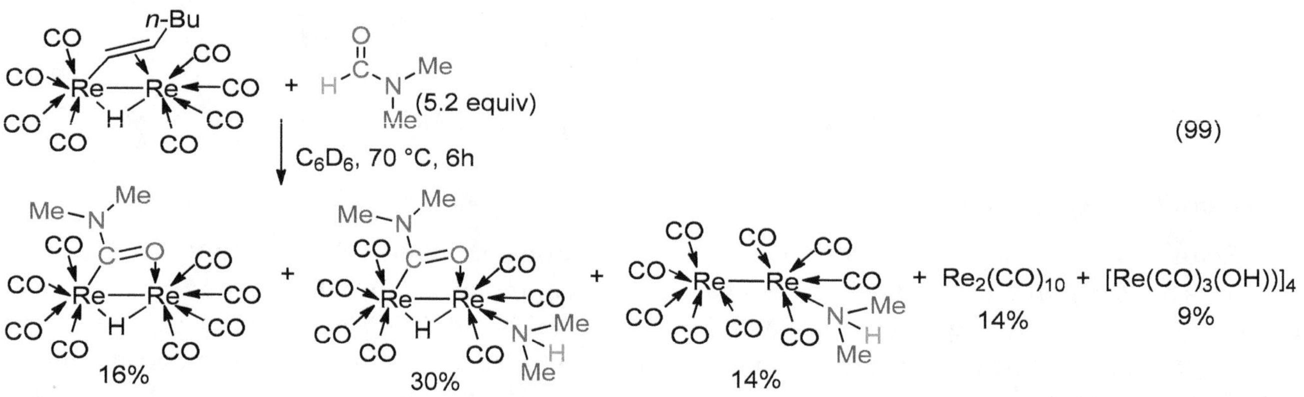

(99)

## 23. C=ONMe$_2$ and CH Fragment

Couplings between amidines, styrenes and fragments of two molecules of DMF in the presence of *t*-BuOOH and a Pd$^{II}$ catalyst provided pyrimidine carboxamides (Equation (100)) [105]. DMAc may also be the CH source as exemplified with the formation of the *N,N*-diethyl-2,4-diphenylpyrimidine-5-carboxamide when *N,N*-diethylformamide was the source of the amide moiety (Equation (101)).

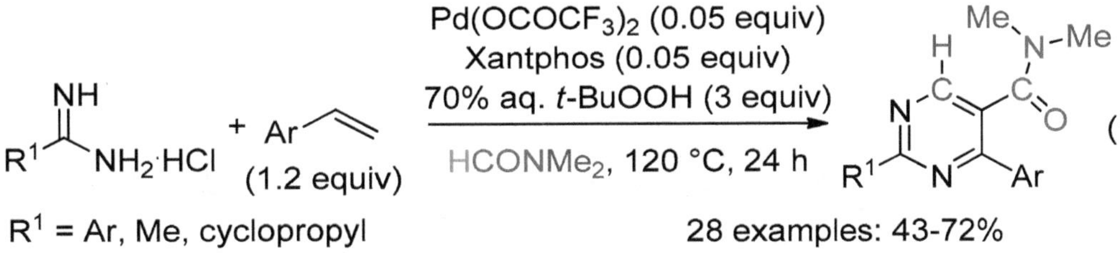

$$R^1C(=NH)NH_2\cdot HCl + ArCH=CH_2 \, (1.2 \text{ equiv}) \xrightarrow[\text{HCONMe}_2,\ 120\ °\text{C},\ 24\ \text{h}]{\text{Pd(OCOCF}_3)_2\ (0.05\ \text{equiv}),\ \text{Xantphos}\ (0.05\ \text{equiv}),\ 70\%\ \text{aq.}\ t\text{-BuOOH}\ (3\ \text{equiv})} \text{2-}R^1\text{-4-Ar-pyrimidine-5-C(=O)NMe}_2 \quad (100)$$

$R^1$ = Ar, Me, cyclopropyl

28 examples: 43-72%

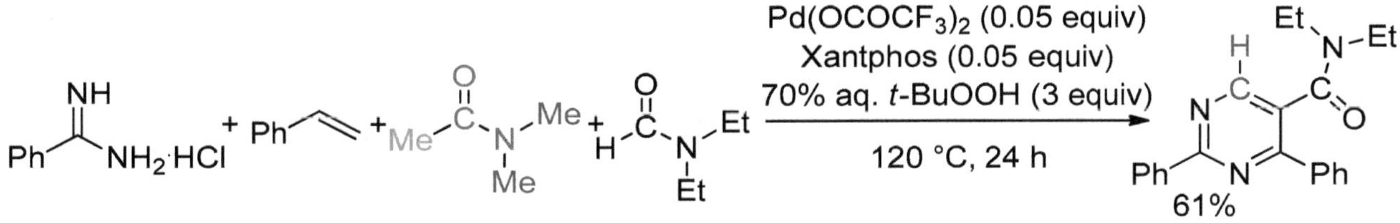

$$PhC(=NH)NH_2\cdot HCl + PhCH=CH_2 + MeC(=O)NMe_2 + HC(=O)NEt_2 \xrightarrow[120\ °\text{C},\ 24\ \text{h}]{\text{Pd(OCOCF}_3)_2\ (0.05\ \text{equiv}),\ \text{Xantphos}\ (0.05\ \text{equiv}),\ 70\%\ \text{aq.}\ t\text{-BuOOH}\ (3\ \text{equiv})} \text{2,4-diphenylpyrimidine-5-C(=O)NEt}_2 \quad (101)$$

61%

## 24. Reducing or Stabilizing Agent

DMF is a powerful reducing agent of metal salts, hence its use for the preparation of metal colloids [106]. In wet DMF, $PdCl_2$ led to carbamic acid and Pd(0) nanoparticles (Equation (102)) [107]. The latter have been associated with the metal-organic framework $Cu_2(BDC)_2(DABCO)$ (BDC = 1,4-benzenedicarboxylate), leading to a catalytic system with high activity and recyclability for the aerobic oxidation of benzyl alcohols to aldehydes [108] and Suzuki-Miyaura cross-coupling reactions [107].

$$PdX_2 + HCONMe_2 + H_2O \longrightarrow Pd(0) + HOCONMe_2 + 2\,HX \quad (102)$$

In addition, DMF can act as stabilizing agent of metal colloids to afford effective and recyclable catalysts, based for examples:

- on iron for the hydrosilylation of alkenes (Equation (103)) [109],

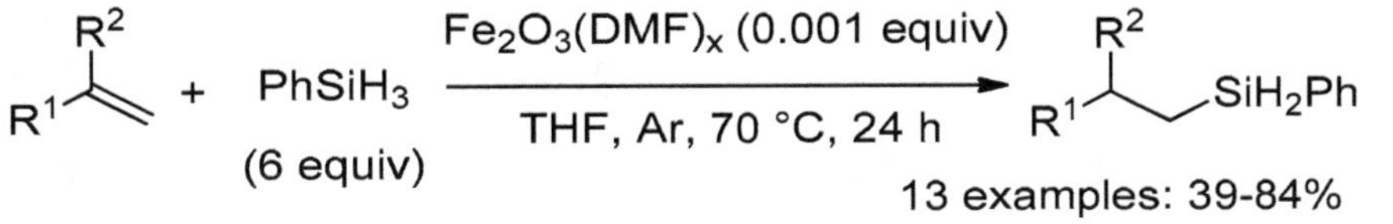

$$R^1C(R^2)=CH_2 + PhSiH_3\ (6\ \text{equiv}) \xrightarrow[\text{THF, Ar, 70 °C, 24 h}]{Fe_2O_3(DMF)_x\ (0.001\ \text{equiv})} R^1CH(R^2)CH_2SiH_2Ph \quad (103)$$

13 examples: 39-84%

- on palladium for the synthesis of 2,3-disubstituted indoles from 2-haloanilines and alkynes (Equation (104)) [110],

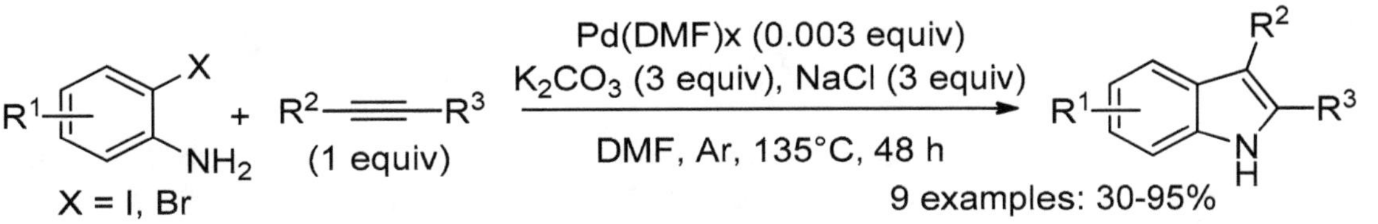

$$R^1\text{-}C_6H_3(X)(NH_2) + R^2C\equiv CR^3\ (1\ \text{equiv}) \xrightarrow[\text{DMF, Ar, 135°C, 48 h}]{\text{Pd(DMF)x (0.003 equiv)},\ K_2CO_3\ (3\ \text{equiv}),\ \text{NaCl (3 equiv)}} R^1\text{-indole-2-}R^3\text{-3-}R^2 \quad (104)$$

X = I, Br

9 examples: 30-95%

- on copper for Sonogashira–Hagihara cross-coupling reactions (Equation (105)) [111],

ArX + H—≡—R → Ar—≡—R (105)
Cu(DMF)x (0.002 equiv), $PPh_3$ (0.1 equiv), $K_2CO_3$ (2 equiv); DMF, Ar, 135°C, 48 h
X = I, Br; (1.5 equiv)
13 examples: 57-97%

- on iridium for methylation of alcohols (Equation (106)) and amines (Equation (107)), using methanol as the C1 source [112].

Ir(DMF)x (0.001 equiv), $Cs_2CO_3$ (1-3 equiv); MeOH, Ar, 150 °C, 24-48 h (106)
5 examples: 54-94%

Ir(DMF)x (0.001 equiv), $Cs_2CO_3$ (1-3 equiv); MeOH, Ar, 150 °C, 24-48 h (107)
R = H (78%), *p*-Me (80%), *p*-Cl (87%)

Thermal decomposition of DMF leads to CO, which reacts with water under $CuFe_2O_4$ catalysis to produce hydrogen [113]. In the presence of 2-nitroanilines, this water gas shift reaction was part of a domino reaction involving the reduction of the nitro group followed by cyclisation into benzimidazoles using a CH from the $NMe_2$ of DMF (Equation (108)) [113]. Such cyclisation is above documented under different experimental conditions (Equation (18)) [50].

$CuFe_2O_4$ (cat); $HCONMe_2/H_2O$ (2:1), 180 °C, 12 h (108)

## 25. Conclusions

This minireview highlights recent uses of DMF and DMAc as sources of building blocks in various reactions of the organic synthesis. We assume that new uses of these multipurpose reagents will be reported.

**Author Contributions:** J. Muzart collected the literature references. J. Muzart and J. Le Bras contributed to the manuscript writing.

## References

1. Muzart, J. *N,N*-Dimethylformamide: Much more than a solvent. *Tetrahedron* **2009**, *65*, 8313–8323. [CrossRef]
2. Ding, S.; Jiao, N. *N,N*-Dimethylformamide: A multipurpose building block. *Angew. Chem. Int. Ed.* **2012**, *51*, 9226–9231. [CrossRef] [PubMed]
3. Batra, A.; Singh, P.; Singh, K.N. Cross dehydrogenative coupling (CDC) reactions of *N,N*-disubstituted formamides, benzaldehydes and cycloalkanes. *Eur. J. Org. Chem.* **2016**, *2016*, 4927–4947. [CrossRef]

4. Le Bras, J.; Muzart, J. *N,N*-Dimethylformamide and *N,N*-dimethylacetamide as carbon, hydrogen, nitrogen and/or oxygen sources. In *Solvents as Reagents in Organic Synthesis, Reactions and Applications*; Wu, X.-F., Ed.; Wiley-VCH: Weinheim, Germany, 2017; pp. 199–314.
5. Kolosov, M.A.; Shvets, E.H.; Manuenkov, D.A.; Vlasenko, S.A.; Omelchenko, I.V.; Shishkina, S.V.; Orlov, V.D. A synthesis of 6-functionalized 4,7-dihydro[1,2,4]triazolo[1,5-*a*]pyrimidines. *Tetrahedron Lett.* **2017**, *58*, 1207–1210. [CrossRef]
6. Borah, A.; Goswami, L.; Neog, K.; Gogoi, P. DMF dimethyl acetal as carbon source for α-methylation of ketones: A hydrogenation-hydrogenolysis strategy of enaminones. *J. Org. Chem.* **2015**, *80*, 4722–4728. [CrossRef] [PubMed]
7. Kumar, S.; Vanjari, R.; Guntreddi, T.; Singh, K.N. Sulfur promoted decarboxylative thioamidation of carboxylic acids using formamides as amine proxy. *Tetrahedron* **2016**, *72*, 2012–2017. [CrossRef]
8. Chen, J.; Feng, J.-B.; Natte, K.; Wu, X.-F. Palladium-catalyzed carbonylative cyclization of arenes by C-H bond activation with DMF as the carbonyl source. *Chem. Eur. J.* **2015**, *21*, 16370–16373. [CrossRef] [PubMed]
9. Guo, S.; Zhu, Z.; Lu, L.; Zhang, W.; Gong, J.; Cai, H. Metal-free $Csp^3$-N bond cleavage of amides using *tert*-butyl hydroperoxide as oxidant. *Synlett* **2015**, *26*, 543–546. [CrossRef]
10. Zhu, Z.; Wang, Y.; Yang, M.; Huang, L.; Gong, J.; Guo, S.; Cai, H. A metal-free cross-dehydrogenative coupling reaction of amides to access *N*-alkylazoles. *Synlett* **2016**, *27*, 2705–2708.
11. Alanthadka, A.; Devi, E.S.; Selvi, A.T.; Nagarajan, S.; Sridharan, V.; Maheswari, C.U. *N*-Heterocyclic carbene-catalyzed Mannich reaction for the synthesis of β-amino ketones: *N,N*-dimethylformamide as carbon source. *Adv. Synth. Catal.* **2017**, *359*, 2369–2374. [CrossRef]
12. Zewge, D.; Bu, X.; Sheng, H.; Liu, Y.; Liu, Z.; Harman, B.; Reibarkh, M.; Gong, X. Mechanistic insight into oxidized *N,N*-dimethylacetamide as a source of formaldehyde related derivatives. *React. Chem. Eng.* **2018**, *3*, 146–150. [CrossRef]
13. Li, Y.; Tu, D.-H.; Wang, B.; Lu, J.-Y.; Wang, Y.-Y.; Liu, Z.-T.; Liu, Z.-W.; Lu, J. Nickel-catalyzed carbonylation of arylboronic acids with DMF as CO source. *Org. Chem. Front.* **2017**, *4*, 569–572. [CrossRef]
14. Lu, K.; Han, X.-W.; Yao, W.-W.; Luan, Y.-X.; Wang, Y.-X.; Chen, H.; Xu, X.-T.; Zhang, K.; Ye, M. DMF-promoted redox-neutral Ni-catalyzed intramolecular hydroarylation of alkene with simple arene. *ACS Catal.* **2018**, *8*, 3913–3917. [CrossRef]
15. Bannwart, L.; Abele, S.; Tortoioli, S. Metal-free amidation of acids with formamides and T3P®. *Synthesis* **2016**, *48*, 2069–2078.
16. Zhu, Y.-P.; Gao, Q.-H.; Lian, M.; Yuan, J.-J.; Liu, M.-C.; Zhao, Q.; Yang, Y.; Wu, A.-X. A sustainable byproduct catalyzed domino strategy: Facile synthesis of α-formyloxy and acetoxy ketones via iodination/nucleophilic substitution/hydrolyzation/oxidation sequences. *Chem. Commun.* **2011**, *47*, 12700–12702. [CrossRef] [PubMed]
17. Mata, L.; Avenoza, A.; Busto, J.H.; Peregrina, J.M. Chemoselectivity control in the reactions of 1,2-cyclic sulfamidates with amines. *Chem. Eur. J.* **2013**, *19*, 6831–6839. [CrossRef] [PubMed]
18. Liu, W.; Chen, C.; Zhou, P. *N,N*-Dimethylformamide (DMF) as a source of oxygen to access α-hydroxy arones via the α-hydroxylation of arones. *J. Org. Chem.* **2017**, *82*, 2219–2222. [CrossRef] [PubMed]
19. Liu, F.; Yang, H.; Hu, X.; Jiang, G. Metal-free synthesis of *ortho*-CHO diaryl ethers by a three-component sequential coupling. *Org. Lett.* **2014**, *16*, 6408–6411. [CrossRef] [PubMed]
20. Yoshioka, E.; Kohtani, S.; Miyabe, H. Three-component coupling reactions of arynes for the synthesis of benzofurans and coumarins. *Molecules* **2014**, *19*, 863–880. [CrossRef] [PubMed]
21. Yoshioka, E.; Kohtani, S.; Miyabe, H. 2,3,4,9-Tetrahydro-9-(3-hydroxy-1,4-dioxo-1*H*-dihydronaphthalen-2-yl)-8-methoxy-3,3-dimethyl-1*H*-xanthen-1-one. *Molbank* **2015**, *2015*, M841. [CrossRef]
22. Yoshioka, E.; Tanaka, H.; Kohtani, S.; Miyabe, H. Straightforward synthesis of dihydrobenzofurans and benzofurans from arynes. *Org. Lett.* **2013**, *15*, 3938–3941. [CrossRef] [PubMed]
23. Zhou, C.; Wang, J.; Jin, J.; Lu, P.; Wang, Y. Three-component synthesis of α-amino-α-aryl carbonitriles from arynes, aroyl cyanides, and *N,N*-dimethylformamide. *Eur. J. Org. Chem.* **2014**, *2014*, 1832–1835. [CrossRef]
24. Yoshioka, E.; Tamenaga, H.; Miyabe, H. [4+2] Cycloaddition of intermediates generated from arynes and DMF. *Tetrahedron Lett.* **2014**, *55*, 1402–1405. [CrossRef]
25. Kodimuthali, A.; Mungara, A.; Prasunamba, P.L.; Pal, M. A simple synthesis of aminopyridines: Use of amides as amine source. *J. Braz. Chem. Soc.* **2010**, *21*, 1439–1445. [CrossRef]

26. Zhang, Q.; Song, C.; Huang, H.; Zhang, K.; Chang, J. Cesium carbonate promoted cascade reaction involving DMF as a reactant for the synthesis of dihydropyrrolizino[3,2-*b*]indol-10-ones. *Org. Chem. Front.* **2018**, *5*, 80–87. [CrossRef]
27. Chen, W.-X.; Shao, L.-X. *N*-Heterocyclic carbene–palladium(II)-1-methylimidazole complex catalyzed amination between aryl chlorides and amides. *J. Org. Chem.* **2012**, *77*, 9236–9239. [CrossRef] [PubMed]
28. Yao, C.-J.; Sun, Q.; Rastogi, N.; Koenig, B. Intermolecular formyloxyarylation of alkenes by photoredox Meerwein reaction. *ACS Catal.* **2015**, *5*, 2935–2938. [CrossRef]
29. Yan, H.; Yang, H.; Lu, L.; Liu, D.; Rong, G.; Mao, J. Copper-catalyzed synthesis of α,β-unsaturated acylamides via direct amidation from cinnamic acids and *N*-substituted formamides. *Tetrahedron* **2013**, *69*, 7258–7263. [CrossRef]
30. Salamone, M.; Milan, M.; DiLabio, G.A.; Bietti, M. Reactions of the cumyloxyl and benzyloxyl radicals with tertiary amides. Hydrogen abstraction selectivity and the role of specific substrate-radical hydrogen bonding. *J. Org. Chem.* **2013**, *78*, 5909–5917. [CrossRef] [PubMed]
31. Du, B.; Sun, P. Syntheses of amides via iodine-catalyzed multiple $sp^3$ C-H bonds oxidation of methylarenes and sequential coupling with *N,N*-dialkylformamides. *Sci. China Chem.* **2014**, *57*, 1176–1182. [CrossRef]
32. Du, B.; Jin, B.; Sun, P. The syntheses of α-ketoamides via $n$-$Bu_4NI$-catalyzed multiple $sp^3$C-H bond oxidation of ethylarenes and sequential coupling with dialkylformamides. *Org. Biomol. Chem.* **2014**, *12*, 4586–4589. [CrossRef] [PubMed]
33. Bai, C.; Yao, X.; Li, Y. Easy access to amides through aldehydic C-H bond functionalization catalyzed by heterogeneous Co-based catalysts. *ACS Catal.* **2015**, *5*, 884–891. [CrossRef]
34. Wu, J.-J.; Li, Y.; Zhou, H.-Y.; Wen, A.-H.; Lun, C.-C.; Yao, S.-Y.; Ke, Z.; Ye, B.-H. Copper-catalyzed carbamoylation of terminal alkynes with formamides via cross-dehydrogenative coupling. *ACS Catal.* **2016**, *6*, 1263–1267. [CrossRef]
35. Zhang, C.; Yue, Q.; Xiao, Z.; Wang, X.; Zhang, Q.; Li, D. Synthesis of *O*-aroyl-*N,N*-dimethylhydroxylamines through hypervalent iodine-mediated amination of carboxylic acids with *N,N*-dimethylformamide. *Synthesis* **2017**, *49*, 4303–4308.
36. Wu, X.; Zhao, Y.; Ge, H. Direct aerobic carbonylation of $C(sp^2)$-H and $C(sp^3)$-H bonds through Ni/Cu synergistic catalysis with DMF as the carbonyl source. *J. Am. Chem. Soc.* **2015**, *137*, 4924–4927. [CrossRef] [PubMed]
37. Rao, D.N.; Rasheed, S.; Das, P. Palladium/silver synergistic catalysis in direct aerobic carbonylation of $C(sp^2)$–H bonds using DMF as a carbon source: Synthesis of pyrido-fused quinazolinones and phenanthridinones. *Org. Lett.* **2016**, *18*, 3142–3145.
38. Boosa, V.; Bilakanti, V.; Gutta, N.; Velisoju, V.K.; Medak, S.; Ramineni, K.; Jorge, B.; Muxina, K.; Akula, V. C–H bond cyanation of arenes using *N,N*-dimethylformamide and $NH_4HCO_3$ as a CN source over a hydroxyapatite supported copper catalyst. *Catal. Sci. Technol.* **2016**, *6*, 8055–8062.
39. Zhang, L.; Lu, P.; Wang, Y. Copper-mediated cyanation of indoles and electron-rich arenes using DMF as a single surrogate. *Org. Biomol. Chem.* **2015**, *13*, 8322–8329, Correction in *Org. Biomol. Chem.* **2016**, *14*, 1840. [CrossRef] [PubMed]
40. Xiao, J.; Li, Q.; Chen, T.; Han, L.-B. Copper-mediated selective aerobic oxidative C3-cyanation of indoles with DMF. *Tetrahedron Lett.* **2015**, *56*, 5937–5940. [CrossRef]
41. Zhao, M.-N.; Hui, R.-R.; Ren, Z.-H.; Wang, Y.-Y.; Guan, Z.-H. Ruthenium-catalyzed cyclization of ketoxime acetates with DMF for synthesis of symmetrical pyridines. *Org. Lett.* **2014**, *16*, 3082–3085. [CrossRef] [PubMed]
42. Liu, W.; Tan, H.; Chen, C.; Pan, Y. A method to access symmetrical tetrasubstituted pyridines via iodine and ammonium persulfate mediated [2+2+1+1]-cycloaddition reaction. *Adv. Synth. Catal.* **2017**, *359*, 1594–1598. [CrossRef]
43. Bai, Y.; Tang, L.; Huang, H.; Deng, G.-J. Synthesis of 2,4-diarylsubstituted-pyridines through a Ru-catalyzed four component reaction. *Org. Biomol. Chem.* **2015**, *13*, 4404–4407. [CrossRef] [PubMed]
44. Nandwana, N.K.; Dhiman, S.; Saini, H.K.; Kumar, I.; Kumar, A. Synthesis of quinazolinones, imidazo[1,2-*c*]quinazolines and imidazo[4,5-*c*]quinolines through tandem reductive amination of aryl halides and oxidative amination of $C(sp^3)$–H bonds. *Eur. J. Org. Chem.* **2017**, *2017*, 514–522. [CrossRef]

45. Weng, Y.; Zhou, H.; Sun, C.; Xie, Y.; Su, W. Copper-catalyzed cyclization for access to 6*H*-chromeno[4,3-*b*]quinolin-6-ones employing DMF as the carbon source. *J. Org. Chem.* **2017**, *82*, 9047–9053. [CrossRef] [PubMed]
46. Liu, Y.; Nie, G.; Zhou, Z.; Jia, L.; Chen, Y. Copper-catalyzed oxidative cross-dehydrogenative coupling/oxidative cycloaddition: Synthesis of 4-acyl-1,2,3-triazoles. *J. Org. Chem.* **2017**, *82*, 9198–9203. [CrossRef] [PubMed]
47. Zheng, L.-Y.; Guo, W.; Fan, X.-L. Metal-free, TBHP-mediated, [3+2+1]-type intermolecular cycloaddition reaction: Synthesis of pyrimidines from amidines, ketones, and DMF through C($sp^3$)–H activation. *Asian J. Org. Chem.* **2017**, *6*, 837–840. [CrossRef]
48. Wang, F.; Wu, W.; Xu, X.; Shao, X.; Li, Z. Synthesis of substituted phenols via 1,1-dichloro-2-nitroethene promoted condensation of carbonyl compounds with DMF. *Tetrahedron Lett.* **2018**, *59*, 2506–2510. [CrossRef]
49. Wang, J.-B.; Li, Y.-L.; Deng, J. Metal-free activation of DMF by dioxygen: A cascade multiple-bond-formation reaction to synthesize 3-acylindoles from 2-alkenylanilines. *Adv. Synth. Catal.* **2017**, *359*, 3460–3467. [CrossRef]
50. Rasal, K.B.; Yadav, G.D. Carbon dioxide mediated novel synthesis of quinazoline-2,4(1*H*,3*H*)-dione in water. *Org. Process Res. Dev.* **2016**, *20*, 2067–2073. [CrossRef]
51. Pu, F.; Li, Y.; Song, Y.-H.; Xiao, J.; Liu, Z.-W.; Wang, C.; Liu, Z.-T.; Chen, J.-G.; Lu, J. Copper-catalyzed coupling of indoles with dimethylformamide as a methylenating reagent. *Adv. Synth. Catal.* **2016**, *358*, 539–542. [CrossRef]
52. Mondal, S.; Samanta, S.; Santra, S.; Bagdi, A.K.; Hajra, A. *N,N*-Dimethylformamide as a methylenating reagent: Synthesis of heterodiarylmethanes *via* copper-catalyzed coupling between imidazo[1,2-*a*]pyridines and indoles/*N,N*-dimethylaniline. *Adv. Synth. Catal.* **2016**, *358*, 3633–3641. [CrossRef]
53. Lee, S.; Jung, H.J.; Choi, H.C.; Hwang, Y.S.; Choi, M.Y. Solvent acting as a precursor: Synthesis of AgCN from $AgNO_3$ in *N,N*-DMF solvent by laser ablation. *Bull. Korean Chem. Soc.* **2017**, *38*, 136–139. [CrossRef]
54. Zhou, X.; Zhang, G.; Gao, B.; Huang, H. Palladium-catalyzed hydrocarbonylative C-N coupling of alkenes with amides. *Org. Lett.* **2018**, *20*, 2208–2212. [CrossRef] [PubMed]
55. Bi, X.; Li, J.; Shi, E.; Wang, H.; Gao, R.; Xiao, J. Ru-catalyzed direct amidation of carboxylic acids with *N*-substituted formamides. *Tetrahedron* **2016**, *72*, 8210–8214. [CrossRef]
56. Sun, Y.-J.; Li, P.; Huang, Q.-Q.; Zhang, J.-J.; Itoh, S. Dioxygenation of flavonol catalyzed by copper(II) complexes supported by carboxylate-containing ligands: Structural and functional models of quercetin 2,4-dioxygenase. *Eur. J. Inorg. Chem.* **2017**, *2017*, 1845–1854. [CrossRef]
57. Chen, X.; Yang, Q.; Zhou, Y.; Deng, Z.; Mao, X.; Peng, Y. Synthesis of 4-(dimethylamino)quinazoline via direct amination of quinazolin-4(3*H*)-one using *N,N*-dimethylformamide as a nitrogen source at room temperature. *Synthesis* **2015**, *47*, 2055–2062. [CrossRef]
58. Gupta, S.S.R.; Nakhate, A.V.; Rasal, K.B.; Deshmukh, G.P.; Mannepalli, L.K. Oxidative amidation of benzaldehydes and benzylamines with *N*-substituted formamides over a Co/Al hydrotalcite-derived catalyst. *New J. Chem.* **2017**, *41*, 15268–15276. [CrossRef]
59. Liu, W.; Chen, C.; Zhou, P. Concise access to α-arylketothioamides by redox reaction between acetophenones, elemental sulfur and DMF. *ChemistrySelect* **2017**, *2*, 5532–5535. [CrossRef]
60. Liu, W.; Chen, C.; Liu, H. Dimethylamine as the key intermediate generated in situ from dimethylformamide (DMF) for the synthesis of thioamides. *Beilstein J. Org. Chem.* **2015**, *11*, 1721–1726. [CrossRef] [PubMed]
61. Amupitan, J.O. An extension of the Willgerodt-Kindler reaction. *Synthesis* **1983**, *1983*, 730. [CrossRef]
62. Nooshabadi, M.; Aghapoor, K.; Darabi, H.R.; Mojtahedi, M.M. The rapid synthesis of thiomorpholides by Willgerodt-Kindler reaction under microwave heating. *Tetrahedron Lett.* **1999**, *40*, 7549–7552. [CrossRef]
63. Wei, E.; Liu, B.; Lin, S.; Liang, F. Multicomponent reaction of chalcones, malononitrile and DMF leading to γ-ketoamides. *Org. Biomol. Chem.* **2014**, *12*, 6389–6392. [CrossRef] [PubMed]
64. Huang, X.; Wang, J.; Ni, Z.; Wang, S.; Pan, Y. Copper-mediated S–N formation via an oxygen-activated radical process: A new synthesis method for sulfonamides. *Chem. Commun.* **2014**, *50*, 4582–4584. [CrossRef] [PubMed]
65. Bao, X.; Rong, X.; Liu, Z.; Gu, Y.; Liang, G.; Xia, Q. Potassium *tert*-butoxide-mediated metal-free synthesis of sulfonamides from sodium sulfinates and *N,N*-disubstituted formamides. *Tetrahedron Lett.* **2018**, *59*, 2853–2858. [CrossRef]

66. Zeng, W.; Wang, E.Y.; Qiu, R.; Sohail, M.; Wu, S.; Chen, F.X. Oxygen-atom insertion of NHC-copper complex: The source of oxygen from *N,N*-dimethylformamide. *J. Organomet. Chem.* **2013**, *743*, 44–48. [CrossRef]
67. Chen, J.; Natte, K.; Wu, X.-F. Pd/C-catalyzed carbonylative C-H activation with DMF as the CO source. *Tetrahedron Lett.* **2015**, *56*, 6413–6416. [CrossRef]
68. Feng, X.; Li, Z. Photocatalytic promoting dimethylformamide (DMF) decomposition to in-situ generation of self-supplied CO for carbonylative Suzuki reaction. *J. Photochem. Photobiol. A Chem.* **2017**, *337*, 19–24. [CrossRef]
69. Krishnakumar, V.; Chatterjee, B.; Gunanathan, C. Ruthenium-catalyzed urea synthesis by N-H activation of amines. *Inorg. Chem.* **2017**, *56*, 7278–7284. [CrossRef] [PubMed]
70. Yang, Y.; Zhang, X.; Zeng, W.; Huang, H.; Liang, Y. Copper catalyzed three-component synthesis of benzothiazolones from *o*-iodoanilines, DMF, and potassium sulfide. *RSC Adv.* **2014**, *4*, 6090–6093. [CrossRef]
71. Mete, T.B.; Singh, A.; Bhat, R.G. Transition-metal-free synthesis of primary to tertiary carboxamides: A quick access to prodrug-pyrazinecarboxamide. *Tetrahedron Lett.* **2017**, *58*, 4709–4712. [CrossRef]
72. Yao, B.; Deng, C.-L.; Liu, Y.; Tang, R.-Y.; Zhang, X.-G.; Li, J.-H. Palladium-catalyzed oxidative carbamoylation of isoquinoline *N*-oxides with formylamides by means of dual C-H oxidative coupling. *Chem. Commun.* **2015**, *51*, 4097–4100. [CrossRef] [PubMed]
73. Wang, H.; Guo, L.-N.; Wang, S.; Duan, X.-H. Decarboxylative alkynylation of α-keto acids and oxamic acids in aqueous media. *Org. Lett.* **2015**, *17*, 3054–3057. [CrossRef] [PubMed]
74. Singh, S.J.; Mir, B.A.; Patel, B.K. A TBPB-mediated C-3 cycloalkylation and formamidation of 4-arylcoumarin. *Eur. J. Org. Chem.* **2018**, *2018*, 1026–1033. [CrossRef]
75. Fan, W.; Shi, D.; Feng, B. TBAI-catalyzed synthesis of α-ketoamides via $sp^3$ C-H radical/radical cross-coupling and domino aerobic oxidation. *Tetrahedron Lett.* **2015**, *56*, 4638–4641. [CrossRef]
76. Chen, W.-T.; Bao, W.-H.; Ying, W.-W.; Zhu, W.M.; Liang, H.; Wei, W.-T. Copper-promoted tandem radical reaction of 2-oxindoles with formamides: Facile synthesis of unsymmetrical urea derivatives. *Asian J. Org. Chem.* **2018**, *7*, 1057–1060. [CrossRef]
77. Li, J.; Hua, R. Stereodivergent ruthenium-catalyzed transfer semihydrogenation of diaryl alkynes. *Chem. Eur. J.* **2011**, *17*, 8462–8465. [CrossRef] [PubMed]
78. Li, J.; Hua, R.; Liu, T. Highly chemo- and stereoselective palladium-catalyzed transfer semihydrogenation of internal alkynes affording *cis*-alkenes. *J. Org. Chem.* **2010**, *75*, 2966–2970. [CrossRef] [PubMed]
79. Tam, C.M.; To, C.T.; Chan, K.S. Carbon-carbon σ-bond transfer hydrogenation with DMF catalyzed by cobalt porphyrins. *Organometallics* **2016**, *35*, 2174–2177. [CrossRef]
80. Nale, D.B.; Bhanage, B.M. *N*-substituted formamides as C1-sources for the synthesis of benzimidazole and benzothiazole derivatives by using zinc catalysts. *Synlett* **2015**, *26*, 2835–2842. [CrossRef]
81. Gao, X.; Yu, B.; Mei, Q.; Yang, Z.; Zhao, Y.; Zhang, H.; Hao, L.; Liu, Z. Atmospheric $CO_2$ promoted synthesis of *N*-containing heterocycles over $B(C_6F_5)_3$ catalyst. *New J. Chem.* **2016**, *40*, 8282–8287. [CrossRef]
82. Zhu, J.; Zhang, Z.; Miao, C.; Liu, W.; Sun, W. Synthesis of benzimidazoles from *o*-phenylenediamines and DMF derivatives in the presence of $PhSiH_3$. *Tetrahedron* **2017**, *73*, 3458–3462. [CrossRef]
83. Simion, C.; Gherase, D.; Sima, S.; Simion, A.M. Serendipitous synthesis of a cyclic formamidine. Application to the synthesis of polyazamacrocycles. *C. R. Chim.* **2015**, *18*, 611–613. [CrossRef]
84. Kobayashi, E.; Togo, H. Facile one-pot preparation of 5-aryltetrazoles and 3-arylisoxazoles from aryl bromides. *Tetrahedron* **2018**, *74*, 4226–4235. [CrossRef]
85. Ochiai, H.; Niwa, T.; Hosoya, T. Stereoinversion of stereocongested carbocyclic alcohols via triflylation and subsequent treatment with aqueous *N,N*-dimethylformamide. *Org. Lett.* **2016**, *18*, 5982–5985. [CrossRef] [PubMed]
86. Yu, M.; Zhang, Q.; Wang, J.; Huang, P.; Yan, P.; Zhang, R.; Dong, D. Triflic anhydride mediated ring-opening/recyclization reaction of α-carbamoyl α-oximyl cyclopropanes with DMF: Synthetic route to 5-aminoisoxazoles. *Synthesis* **2016**, *48*, 1934–1938. [CrossRef]
87. Liu, L.; Zhang-Negrerie, D.; Du, F.; Zhao, K. $PhICl_2$ and wet DMF: An efficient system for regioselective chloroformyloxylation/α-chlorination of alkenes/α,β-unsaturated compounds. *Org. Lett.* **2014**, *16*, 436–439. [CrossRef] [PubMed]
88. Wang, Y.; Zhang, J.; Liu, J.; Zhang, C.; Zhang, Z.; Xu, J.; Xu, S.; Wang, F.; Wang, F. C-N and N-H bond metathesis reactions mediated by carbon dioxide. *ChemSusChem* **2015**, *8*, 2066–2072. [CrossRef] [PubMed]

89. Wang, Y.; Wang, F.; Zhang, C.; Zhang, J.; Li, M.; Xu, J. Transformylating amine with DMF to formamide over $CeO_2$ catalyst. *Chem. Commun.* **2014**, *50*, 2438–2441. [CrossRef] [PubMed]
90. Sheng, H.; Zeng, R.; Wang, W.; Luo, S.; Feng, Y.; Liu, J.; Chen, W.; Zhu, M.; Guo, Q. An efficient heterobimetallic lanthanide alkoxide catalyst for transamidation of amides under solvent-free conditions. *Adv. Synth. Catal.* **2017**, *359*, 302–313. [CrossRef]
91. Sonawane, R.B.; Rasal, N.K.; Jagtap, S.V. Nickel-(II)-catalyzed *N*-formylation and *N*-acylation of amines. *Org. Lett.* **2017**, *19*, 2078–2081. [CrossRef] [PubMed]
92. Lindsay-Scott, P.J.; Charlesworth, N.G.; Grozavu, A. A flexible strategy for the regiocontrolled synthesis of pyrazolo[1,5-*a*]pyrazines. *J. Org. Chem.* **2017**, *82*, 11295–11303. [CrossRef] [PubMed]
93. Nakagawa-Goto, K.; Kobayashi, T.; Kawasaki, T.; Somei, M. Lithiation of 1-alkoxyindole derivatives. *Heterocycles* **2018**. [CrossRef]
94. Ramírez-Jiménez, R.; Franco, M.; Rodrigo, E.; Sainz, R.; Ferritto, R.; Lamsabhi, A.M.; Aceña, J.L.; Cid, M.B. Unexpected reactivity of graphene oxide with DBU and DMF. *J. Mater. Chem. A* **2018**, *6*, 12637–12646. [CrossRef]
95. Deng, X.; Lei, X.; Nie, G.; Jia, L.; Li, Y.; Chen, Y. Copper-catalyzed cross-dehydrogenative $N^2$-coupling of *NH*-1,2,3-triazoles with *N,N*-dialkylamides: *N*-amidoalkylation of *NH*-1,2,3-triazoles. *J. Org. Chem.* **2017**, *82*, 6163–6171. [CrossRef] [PubMed]
96. Gonzalez, P.E.; Sharma, H.K.; Chakrabarty, S.; Metta-Magaña, A.; Pannell, K.H. Triethylsiloxymethyl-*N,N*-dimethylamine, $Et_3SiOCH_2NMe_2$: A dimethylaminomethylation (Mannich) reagent for O–H, S–H, P–H and aromatic C–H systems. *Eur. J. Org. Chem.* **2017**, *2017*, 5610–5616. [CrossRef]
97. Hu, Y.; Hu, Y.; Hu, Q.; Ma, J.; Lv, S.; Liu, B.; Wang, S. Direct access to fused salicylaldehydes and salicylketones from tetraynes. *Chem. Eur. J.* **2017**, *23*, 4065–4072. [CrossRef] [PubMed]
98. Zhu, R.; Liu, Z.; Chen, J.; Xiong, X.; Wang, Y.; Huang, L.; Bai, J.; Dang, Y.; Huang, J. Preparation of thioanisole biscarbanion and C–H lithiation/annulation reactions for the access of five-membered heterocycles. *Org. Lett.* **2018**, *20*, 3161–3165. [CrossRef] [PubMed]
99. Zeng, R.; Sheng, H.; Rao, B.; Feng, Y.; Wang, H.; Sun, Y.; Chen, M.; Zhu, M. An efficient and green approach to synthesizing enamines by intermolecular hydroamination of activated alkynes. *Chem. Res. Chin. Univ.* **2015**, *31*, 212–217. [CrossRef]
100. Liu, D.; Mao, J.; Rong, G.; Yan, H.; Zheng, Y.; Chen, J. Concise synthesis of semicarbazides and formylhydrazines via direct addition reaction between aromatic azoarenes and *N*-substituted formamides. *RSC Adv.* **2015**, *5*, 19301–19305. [CrossRef]
101. Mori, S.; Shibuya, M.; Yamamoto, Y. Ruthenium-catalyzed hydrocarbamoylative cyclization of 1,6-diynes with formamides. *Chem. Lett.* **2017**, *46*, 207–210. [CrossRef]
102. Yamamoto, Y.; Okude, Y.; Mori, S.; Shibuya, M. Combined experimental and computational study on ruthenium(II)-catalyzed reactions of diynes with aldehydes and *N,N*-dimethylformamide. *J. Org. Chem.* **2017**, *82*, 7964–7973. [CrossRef] [PubMed]
103. Yamamoto, Y. Theoretical study on ruthenium-catalyzed hydrocarbamoylative cyclization of 1,6-diyne with dimethylformamide. *Organometallics* **2017**, *36*, 1154–1163. [CrossRef]
104. Adams, R.D.; Dhull, P. Formyl C-H activation in *N,N*-dimethylformamide by a dirhenium carbonyl complex. *J. Organomet. Chem.* **2017**, *849–850*, 228–232. [CrossRef]
105. Guo, W.; Liao, J.; Liu, D.; Li, J.; Ji, F.; Wu, W.; Jiang, H. A four-component reaction strategy for pyrimidine carboxamide synthesis. *Angew. Chem. Int. Ed.* **2017**, *56*, 1289–1293. [CrossRef] [PubMed]
106. Pastoriza-Santos, I.; Liz-Marzán, L.M. *N,N*-Dimethylformamide as a reaction medium for metal nanoparticle synthesis. *Adv. Funct. Mater.* **2009**, *19*, 679–688. [CrossRef]
107. Akbari, S.; Mokhtari, J.; Mirjafari, Z. Solvent-free and melt aerobic oxidation of benzyl alcohols using $Pd/Cu_2(BDC)_2DABCO$–MOF prepared by one-step and through reduction by dimethylformamide. *RSC Adv.* **2017**, *7*, 40881–40886. [CrossRef]
108. Tahmasebi, S.; Mokhtari, J.; Naimi-Jamal, M.R.; Khosravi, A.; Panahi, L. One-step synthesis of $Pd\text{-}NPs@Cu_2(BDC)_2DABCO$ as efficient heterogeneous catalyst for the Suzuki-Miyaura cross-coupling reaction. *J. Organomet. Chem.* **2017**, *853*, 35–41. [CrossRef]
109. Azuma, R.; Nakamichi, S.; Kimura, J.; Yano, H.; Kawasaki, H.; Suzuki, T.; Kondo, R.; Kanda, Y.; Shimizu, K.-I.; Kato, K.; et al. Solution synthesis of *N,N*-dimethylformamide-stabilized iron-oxide nanoparticles as an efficient and recyclable catalyst for alkene hydrosilylation. *ChemCatChem* **2018**, *10*, 2378–2382. [CrossRef]

110. Onishi, K.; Oikawa, K.; Yano, H.; Suzuki, T.; Obora, Y. *N,N*-Dimethylformamide-stabilized palladium nanoclusters as a catalyst for Larock indole synthesis. *RSC Adv.* **2018**, *8*, 11324–11329. [CrossRef]
111. Oka, H.; Kitai, K.; Suzuki, T.; Obora, Y. *N,N*-Dimethylformamide-stabilized copper nanoparticles as a catalyst precursor for Sonogashira–Hagihara cross coupling. *RSC Adv.* **2017**, *7*, 22869–22874. [CrossRef]
112. Oikawa, K.; Itoh, S.; Yano, H.; Kawasaki, H.; Obora, Y. Preparation and use of DMF-stabilized iridium nanoclusters as methylation catalysts using methanol as the C1 source. *Chem. Commun.* **2017**, *53*, 1080–1083. [CrossRef] [PubMed]
113. Rasal, K.B.; Yadav, G.D. One-pot synthesis of benzimidazole using DMF as a multitasking reagent in presence $CuFe_2O_4$ as catalyst. *Catal. Today* **2018**, *309*, 51–60. [CrossRef]

7

# $K_2S_2O_8$-Promoted Aryl Thioamides Synthesis from Aryl Aldehydes using Thiourea as the Sulfur Source

**Yongjun Bian [*,†], Xingyu Qu [†], Yongqiang Chen, Jun Li and Leng Liu**

College of Chemistry and Chemical Engineering, Jinzhong University, Yuci 030619, China; quxy@jzxy.edu.cn (X.Q.); chenyongqiang@jzxy.edu.cn (Y.C.); hxx406@126.com (J.L.); liuleng@jzxy.edu.cn (L.L.)
* Correspondence: yjbian2013@jzxy.edu.cn
† These authors contributed equally to this work

**Abstract:** Thiourea as a sulfur atom transfer reagent was applied for the synthesis of aryl thioamides through a three-component coupling reaction with aryl aldehydes and *N,N*-dimethylformamide (DMF) or *N,N*-dimethylacetamide (DMAC). The reaction could tolerate various functional groups and gave moderate to good yields of desired products under the transition-metal-free condition.

**Keywords:** aryl thioamides; thiourea; C-H/C-N activation; C-S formation; transition-metal-free

---

## 1. Introduction

The synthesis of sulfur-containing organic compounds has received much attention in recent years, due to their wide applications in biology, chemistry, and materials science [1–10]. There are many sulfur reagents for their synthesis, such as $P_2S_5$ [11], Lawesson's reagent [12], disulfides [13–15], thiols [16–19], sulfonyl hydrazides [20–23], sodium sulfonate [24–28], and elemental sulfur [29–32]. Among them, both $P_2S_5$ and Lawesson's reagent are the most widely used reagents, and yet they have an obvious drawback of being sensitive to moisture. Therefore, much better sulfur reagents have been pursued by the organic chemists for the past decades [33]. Thioamides, as an important class of sulfur-containing organic compounds, have been synthesized by applying different sulfur reagents as the sulfur source [1,34–39]. For example, Jiang et al. [35] reported that sodium sulfide as a sulfur source was applied for the synthesis of thioamides using aldehydes and *N*-substituted formamides. More recently, a coupling reaction between quaternary ammoniums, *N*-substituted formamides, and sodium disulfide was accomplished for rapid access to aryl thioamides [36]. Thiourea as an inexpensive and easy-to-handle sulfur atom transfer reagent, was used extensively as well, mainly for the synthesis of inorganic metal sulfides [40–42], organic thioethers [43–47], and thioesters [48,49]. As far as we known, a similar reaction using thiourea and aldehydes to prepare thioamides has not yet been introduced. Hence, we want to report a new three-component coupling reaction between aryl aldehydes, thiourea as an effective sulfur source, and DMF or DMAC, for the synthesis of various aryl thioamides.

## 2. Results and Discussion

Initially, we treated the reaction of 4-chlorobenzaldehyde **1a** in DMF and $H_2O$ at 125 °C in the presence of thiourea using the benzoyl peroxide (BPO) as an oxidant. After 24 h, the desired thioamide product **3a** was isolated in 58% yield (Table 1, Entry 1). Subsequently, various oxidants, which are commonly used in C-H activation, such as *p*-benzoquinone (BQ), di-*t*-butyl peroxide (DTBP), *tert*-butyl hydroperoxide (TBHP), $K_2S_2O_8$, or $(NH_4)_2S_2O_8$, were attempted, to optimize the reaction condition (Table 1, Entries 2–6). Among them, $K_2S_2O_8$ proved to be best to give the desired thioamide product **3a** in 69% yield (Table 1, Entry 5). For this transformation, $H_2O$ played an extremely important role. No desired product **3a** was observed when increasing the concentration of $H_2O$ to 42 M or without addition of $H_2O$ (Table 1, Entries 7 and 9). Slightly enhancing or reducing the loading amount of $K_2S_2O_8$, the yield of **3a** was not obviously changeable (Table 1, Entries 10–11). When the 20% of $Cu(OAc)_2$ were used as a catalyst, only 54% yield of **3a** was afforded (Table 1, Entry 12) [36]. To our delight, the yield of **3a** was further promoted to 80% when 5 equiv. of pyridine (Py) as an additive were added (Table 1, Entry 13) [35].

**Table 1.** Optimization of reaction conditions [a].

**1a** + **2** → **3a** ($CS(NH_2)_2$, Oxidant; $H_2O$, 125°C)

| Entry | Oxidant (Equiv) | Concentration of $H_2O$ (M) | Yield (%) [b] |
|---|---|---|---|
| 1 | BPO (2) | 14 | 58 |
| 2 | BQ (2) | 14 | 0 |
| 3 | DTBP (2) | 14 | <5 |
| 4 | TBHP (2) | 14 | 20 |
| 5 | $K_2S_2O_8$ (2) | 14 | 69 |
| 6 | $(NH_4)_2S_2O_8$ (2) | 14 | 55 |
| 7 | $K_2S_2O_8$ (2) | 42 | 0 [c] |
| 8 | $K_2S_2O_8$ (2) | 8 | 65 [d] |
| 9 | $K_2S_2O_8$ (2) | 0 | 0 [e] |
| 10 | $K_2S_2O_8$ (3) | 14 | 68 |
| 11 | $K_2S_2O_8$ (1.8) | 14 | 61 |
| 12 | $K_2S_2O_8$ (2) | 14 | 54 [f] |
| 13 | $K_2S_2O_8$ (2) | 14 | 80 [g] |

[a] Condition: **1a** (0.25 mmol), **2** (9.6 M), thiourea (0.5 mmol), oxidant, $H_2O$, 125 °C, 24 h. [b] Isolated yield. [c] 3.4 M of DMF were used. [d] 11 M of DMF were used. [e] 13 M of DMF were used. [f] 20 mol% of $Cu(OAc)_2$ were added. [g] 5 equiv. of pyridine (Py) were added.

After establishing the optimized conditions, this procedure was applied to access a variety of aryl thioamide derivatives. Several different aryl aldehydes could undergo this transformation smoothly in a mild condition to give the desired products **3a–r** (Scheme 1). The results indicated that many popular functional groups were well tolerated, such as methyl, methoxyl, chloro, bromo, fluoro, trifluoromethyl, and *tert*-butyl. Furthermore, the substrate bearing a sensitive hydroxy group, which was generally protected in the presence of an oxidant could be also tolerated in this transformation, and afforded the desired product **3o** in 83% yield. A substituted amino was suitable as well, and gave 88% yield of **3p**. The substituents on aromatic aldehydes had a certain influence on this transformation. When the substituents were strong electron-withdrawing groups, either lower yield of desired products, or no desired products were obtained (**3j** and **3r**). The desired product **3m** was not afforded possibly due to the steric hindrance. Moreover, our experiments demonstrated that 2-naphthaldehyde was a suitable substrate for this transformation, and gave the desired product **3q** a good yield.

To further expand the substrate scope, some selected heterocyclic aldehydes and an aliphatic aldehyde were examined under the optimal condition (Scheme 2). Generally, five or six members heterocyclic derivatives were suitable for this transformation, such as furan-2-carbaldehyde, thiophene-2-carbaldehyde, and nicotinaldehyde, giving the corresponding thioamide products **3s**, **3t**, and **3u** 54%, 74%, and 37% yields, respectively. In addition, aliphatic aldehyde did not accomplish this transformation (**3w**).

We attempted to explore the different *N*-substituted formamides for this transformation in additional solvent. No good results were provided when the normal solvents such as

*N*-methyl-2-pyrrolidone (NMP), 1,4-dioxane, 1,2-dichloroethane (DCE), toluene, chlorobenzene (PhCl), dimethyl sulfoxide (DMSO), ethylene glycol, were used (see Supporting Information). Unexpectedly, *N,N*-dimethylacetamide (DMAC), which was seldom used as an amine source by the acyl C-N bond activation [37], could replace DMF to give the same desired product in good yield. Subsequently, the reactions of aryl aldehydes with thiourea in DMAC were examined under the similar reaction condition (Table 2). The results demonstrated that various groups were tolerated well, such as methyl, chloro, and bromo, and the good yields of the desired products were isolated.

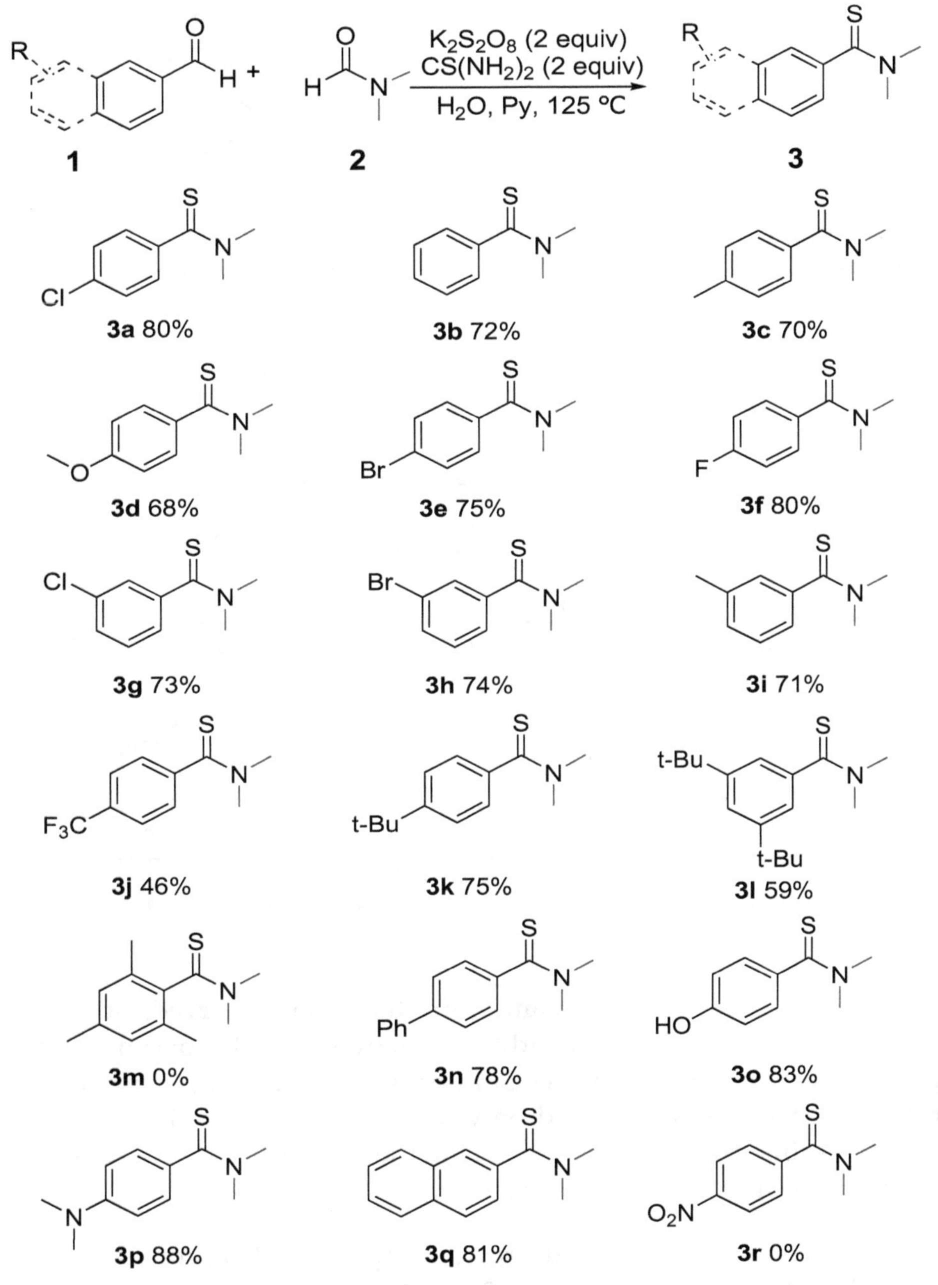

$K_2S_2O_8$ (2 equiv)
$CS(NH_2)_2$ (2 equiv)
$H_2O$, Py, 125 °C

**1** + **2** → **3**

**3a** 80% **3b** 72% **3c** 70%
**3d** 68% **3e** 75% **3f** 80%
**3g** 73% **3h** 74% **3i** 71%
**3j** 46% **3k** 75% **3l** 59%
**3m** 0% **3n** 78% **3o** 83%
**3p** 88% **3q** 81% **3r** 0%

**Scheme 1.** The substrate scope of substituted benzaldehydes. Reaction condition: aryl aldehyde **1** (0.25 mmol), thiourea (0.5 mmol), $H_2O$ (14 M), $K_2S_2O_8$ (0.5 mmol) and Py (5 equiv.) in DMF (1.5 mL) at 125 °C for 24 h in sealed tube. Isolated yields were given.

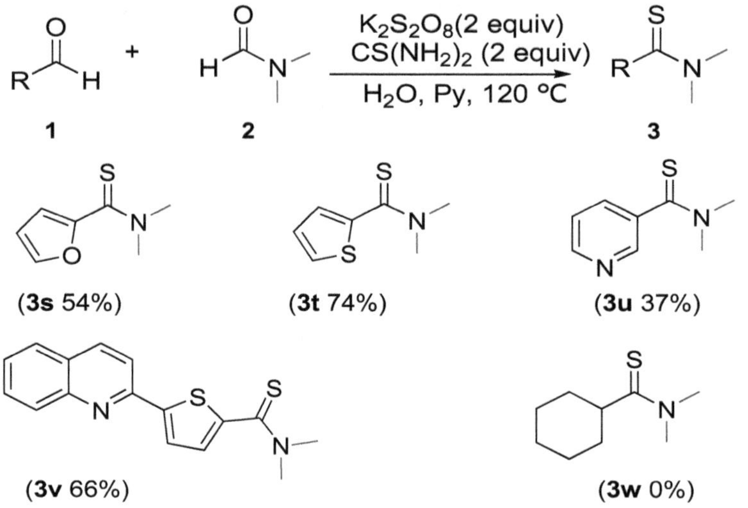

**Scheme 2.** The substrate scope of other aldehydes. Reaction condition: aldehyde **1** (0.25 mmol), thiourea (0.5 mmol), $H_2O$ (14 M), $K_2S_2O_8$ (0.5 mmol), and Py (5 equiv.) in DMF (1.5 mL) at 125 °C for 24 h in sealed tube. Isolated yields were given.

**Table 2.** The synthesis of aryl thioamides by DMAC [a].

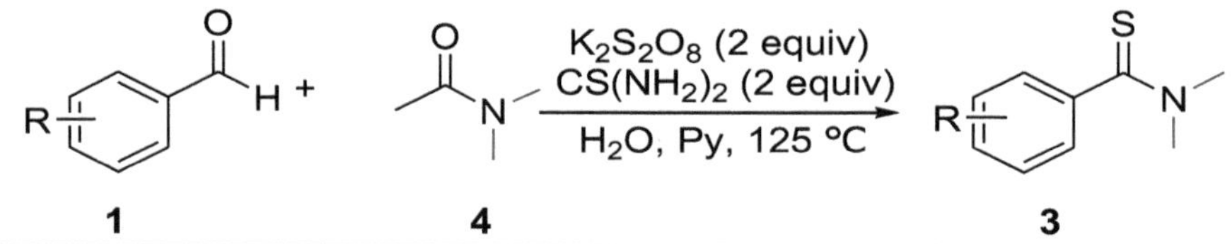

| Entry | R | Yield (%) of 3 [b] |
|---|---|---|
| 1 | 4-Cl | 63 (**3a**) |
| 2 | H | 52 (**3b**) |
| 3 | 4-$CH_3$ | 55 (**3c**) |
| 4 | 3-Br | 48 (**3h**) |
| 5 | 3-$CH_3$ | 51 (**3i**) |
| 6 | 4-Ph | 60 (**3n**) |

[a] Condition: 1 (0.25 mmol), 4 (1.5 mL), thiourea (0.5 mmol), $K_2S_2O_8$ (0.5 mmol), $H_2O$ (0.5 mL), Py (1.25 mmol), 125 °C, 36 h. [b] Isolated yield.

In addition, extremely small amounts of amide products were observed, along with the generation of thioamide products under the optimal condition. So, two control experiments were conducted to explain the tentative reaction mechanism (see Supporting Information). First, no thioamide product was formed in the absence of thiourea, and only trace amounts of amide product were observed. Second, when the *N,N*-dimethylbenzamide replacing the benzaldehyde was manipulated under the standard condition, no desired thioamide was observed.

Based on our experimental results and previous reports [35], a proposed reaction mechanism for this transformation is described in Scheme 3. First, an aryl aldehyde undergoes a nucleophilic attack by a dimethylamine, which is from the hydrolysis of DMF, to generate iminium intermediate **A**. The iminium **A** then is directly attacked by thiourea to form the intermediate **B**, together with the release of urea [43,47,49]. Finally, intermediate **B** is oxidized by $K_2S_2O_8$ to afford the desired thioamide product.

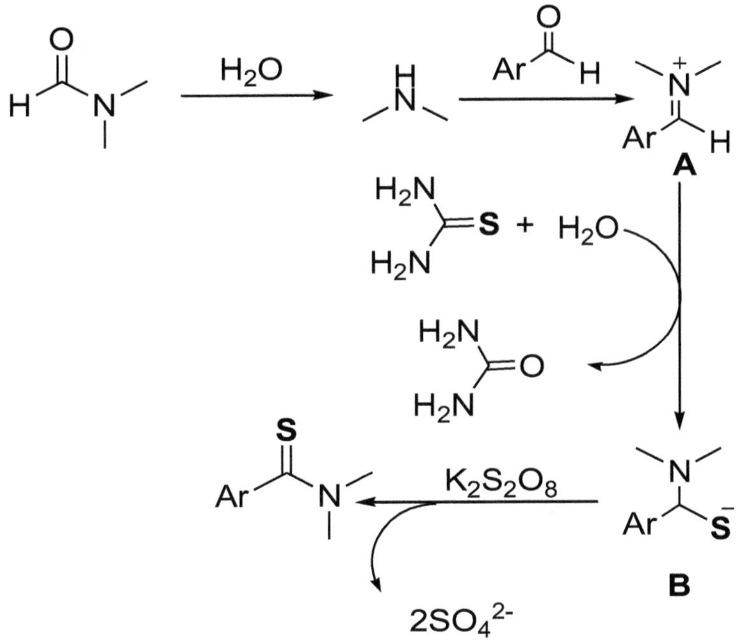

**Scheme 3.** Plausible reaction pathway.

## 3. Materials and Methods

Unless otherwise stated, all reagents and solvents were purchased from commercial suppliers, and were used without further purification. Reactions were monitored by thin layer chromatography (TLC) analysis on silica gel 60 F254, and visualization was accomplished by irradiation with short wave UV light at 254 nm. $^{1}$H-NMR and $^{13}$C-NMR spectra were recorded on a Bruker Avance 400 or a 500 MHz spectrometer (Bruker, Karlsruhe, Germany), with tetramethylsilane (TMS) as the internal standard. The coupling constants $J$ are given in Hz. Mass spectra were measured with the Thermo Scientific LTQ Orbitrap XL MS spectrometer (Thermo Fisher Scientific, Waltham, MA, USA) or GC-MS QP2010 (Comfort Technology Limited, Kowloon, Hong Kong).

A mixture of aldehyde **1** (0.25 mmol), thiourea (0.5 mmol), $K_2S_2O_8$ (0.5 mmol), and Py (1.25 mmol) in 2.0 mL DMF/$H_2O$ ($v/v$ = 3:1) was stirred in a sealed tube under air at 125 °C for 24 h. After the reaction was achieved, the crude mixture was purified by column chromatography (silica gel, EtOAc/petroleum ether) to afford the desired product **3**.

*4-Chloro-N,N-dimethylbenzothioamide* (**3a**) [35]. $^{1}$H-NMR ($CDCl_3$, 400 MHz): δ (ppm): 7.36–7.33 (m, 2H), 7.29–7.26 (m, 2H), 3.61 (s, 3H), 3.19 (s, 3H); $^{13}$C-NMR ($CDCl_3$, 100 MHz): δ (ppm): 199.91, 141.66, 134.60, 128.58, 127.29, 44.18, 43.32; HRMS (ESI) $m/z$ calculated (calcd.) for $C_9H_{11}ClNS^+$ $(M + H)^+$ 200.02952, found 200.02971.

*N,N-Dimethylbenzothioamide* (**3b**) [35]. $^{1}$H-NMR ($CDCl_3$, 400 MHz): δ (ppm): 7.37–7.30 (m, 5H), 3.62 (s, 3H), 3.19 (s, 3H); $^{13}$C-NMR ($CDCl_3$, 100 MHz): δ (ppm): 201.65, 143.69, 128.89, 128.64, 126.04, 44.44, 43.53; GC-MS (EI) $m/z$ (%) 165.10 (70, $M^+$), 164.05 (98), 121.05 (100), 77.00 (46).

*4-Methyl-N,N-dimethylbenzothioamide* **3c** [35]. $^{1}$H-NMR ($CDCl_3$, 400 MHz): δ (ppm): 7.23 (d, $J$ = 8 Hz, 2H), 7.16 (d, $J$ = 8 Hz, 2H), 3.61 (s, 3H), 3.20 (s, 3H), 2.36 (s, 3H); $^{13}$C-NMR ($CDCl_3$, 100 MHz): δ (ppm): 201.55, 140.59, 138.68, 128.89, 125.89, 44.22, 43.35, 21.27; GC-MS (EI) $m/z$ (%) 179.05 (80, $M^+$). 178.05 (100), 145.10 (50), 135.05 (98), 91.05 (45).

*4-Methoxy-N,N-dimethylbenzothioamide* (**3d**) [35]. $^{1}$H-NMR ($CDCl_3$, 400 MHz): δ (ppm): 7.31 (d, $J$ = 12 Hz, 2H), 6.87 (d, $J$ = 12 Hz, 2H), 3.82 (s, 3H), 3.59 (s, 3H), 3.22 (s, 3H); $^{13}$C-NMR ($CDCl_3$, 100 MHz): δ (ppm): 201.31, 160.02, 135.82, 127.90, 113.51, 55.40, 44.37, 43.59; GC-MS (EI) $m/z$ (%) 195.05 ($M^+$, 84), 194.05 (93), 151.05 (100).

*4-Bromo-N,N-dimethylbenzothioamide* (**3e**) [35]. $^1$H-NMR ($CDCl_3$, 400 MHz): δ (ppm): 7.53–7.50 (m, 2H), 7.23–7.19 (m, 2H), 3.60 (s, 3H), 3.19 (s, 3H); $^{13}$C-NMR ($CDCl_3$, 100 MHz): δ (ppm): 199.79, 142.12, 131.53, 127.52, 122.73, 44.20, 44.30; GC-MS (EI) $m/z$ (%) 242.95 (63, $M^+$), 243.90 (100), 242.95 (63), 241.90 (92), 200.85 (47), 198.90 (48), 120.00 (54).

*4-Fluoro-N,N-dimethylbenzothioamide* (**3f**) [35]. $^1$H-NMR ($CDCl_3$, 400 MHz): δ (ppm): 7.32–7.28 (m, 2 H), 7.05–6.99 (m, 2H), 3.58 (s, 3H), 3.16 (s, 3H); $^{13}$C-NMR ($CDCl_3$, 100 MHz): δ (ppm): 200.19, 162.66 (d, $J$ = 248 Hz), 139.44 (d, $J$ = 4 Hz), 128.00 (d, $J$ = 9 Hz), 115.33 (d, $J$ = 22 Hz), 44.24, 43.45.

*3-Chloro-N,N-dimethylbenzothioamide* (**3g**) [35]. $^1$H-NMR ($CDCl_3$, 400 MHz): δ (ppm): 7.33–7.25 (m, 3H), 7.20–7.18 (m, 1H), 3.60 (s, 3H), 3.19 (s, 3H); $^{13}$C-NMR ($CDCl_3$, 100 MHz): δ (ppm): 199.11, 144.78, 134.29, 129.76, 128.63, 125.89, 123.83, 44.18, 43.18; GC-MS (EI) $m/z$ (%) 200.00 (43), 199.0 (69, $M^+$). 198.00 (100), 165.00 (32), 157.00 (26), 155.00 (74), 111.00 (24).

*3-Bromo-N,N-dimethylbenzothioamide* (**3h**) [35]. $^1$H-NMR ($CDCl_3$, 400 MHz): δ (ppm): 7.50–7.46 (m, 2H), 7.27–7.22 (m, 2H), 3.60 (s, 3H), 3.19 (s, 3H); $^{13}$C-NMR ($CDCl_3$, 100 MHz): δ (ppm): 199.01, 144.98, 131.56, 129.99, 128.68, 124.30, 122.38, 44.19, 43.19; GC-MS (EI) $m/z$ (%) 244.95 (64), 243.95 (100), 242.95 (65, $M^+$), 241.95 (95), 200.90 (43), 198.90 (43), 120.00 (51).

*3-Methyl-N,N-dimethylbenzothioamide* (**3i**) [35]. $^1$H-NMR ($CDCl_3$, 400 MHz): δ (ppm): 7.34–7.24 (m, 1H), 7.17 (d, $J$ = 6.0 Hz, 2H), 7.11 (d, $J$ = 7.5 Hz, 1H), 3.64 (s, 3H), 3.21 (s, 3H), 2.40 (s, 3H); $^{13}$C-NMR ($CDCl_3$, 100 MHz): δ (ppm): 201.2, 143.2, 138.0, 129.1, 128.0, 126.1, 122.4, 44.0, 43.0, 21.2.

*4-(Trifluoromethyl)-N,N-dimethylbenzothioamide* (**3j**) [35]. $^1$H-NMR ($CDCl_3$, 400 MHz): δ (ppm): 7.64 (d, $J$ = 8 Hz, 2H), 7.43 (d, $J$ = 12 Hz, 2H), 3.63 (s, 3H), 3.18 (s, 3H); $^{13}$C-NMR ($CDCl_3$, 100 MHz): δ (ppm): 199.30, 146.56,130.46 (q, $J$ = 32.6 Hz),126.02, 125.56 (q, $J$ = 3.8 Hz), 123.73 (q, $J$ = 270.5 Hz), 44.07, 43.09; GC-MS (EI) $m/z$ (%) 233.00 ($M^+$, 71), 232.00 (100), 199.05 (37), 189.00 (64).

*4-tert-Butyl-N,N-dimethylbenzothioamide* (**3k**) [37]. $^1$H-NMR ($CDCl_3$, 400 MHz): δ (ppm): 7.38–7.35 (m, 2H), 7.28–7.24 (m, 2H), 3.61 (s, 3H), 3.21 (s, 3H), 1.32 (s, 9H); $^{13}$C-NMR ($CDCl_3$, 100 MHz): δ (ppm): 201.64, 151.78, 140.49, 125.67, 125.22, 44.27, 43.34, 34.70, 31.23. GC-MS (EI) $m/z$ (%) 221.10 (70, $M^+$), 220.05 (70), 147.05 (24).

*3,5-di-tert-Butyl-N,N-dimethylbenzothioamide* (**3l**). $^1$H-NMR ($CDCl_3$, 400 MHz): δ (ppm): 7.39 (t, $J$ = 4 Hz, 1H), 7.15 (d, $J$ = 4 Hz, 2H), 3.63 (s, 3H), 3.16 (s, 3H), 1.33 (s, 18H); $^{13}$C-NMR ($CDCl_3$, 100 MHz): δ (ppm): 202.78, 150.70, 142.63, 122.75, 120.13, 44.21, 43.32, 34.92, 31.40. GC-MS (EI) $m/z$ (%) 277.15 (86, $M^+$), 276.15 (100), 220.05 (80). HRMS (ESI) $m/z$ calcd. for $C_{17}H_{28}NS^+$ $(M + H)^+$ 278.19370, found 278.19366.

*4-Benzyl-N,N-dimethylbenzothioamide* (**3n**) [38]. $^1$H-NMR ($CDCl_3$, 400 MHz): δ (ppm): 7.59–7.56 (m, 4H), 7.47–7.43 (m, 2H), 7.41–7.34 (m, 3H),3.62 (s, 3H), 3.23 (s, 3H); $^{13}$C-NMR ($CDCl_3$, 100 MHz): δ (ppm): 201.04, 142.17, 141.56, 140.33, 128.89, 127.69, 127.10, 127.09, 126.39, 44.28, 43.32. GC-MS (EI) $m/z$ (%) 241.05 (78, $M^+$), 240.05 (100), 197.00 (58), 181.05 (66), 152.05 (74).

*4-Hydroxy-N,N-dimethylbenzothioamide* (**3o**) [35]. $^1$H-NMR ($CDCl_3$, 400 MHz): δ (ppm): 7.21–7.15 (m, 2H), 6.75–6.70 (m, 2H), 5.89 (br s, 1H), 3.59 (s, 3H), 3.21 (s, 3H); $^{13}$C-NMR ($CDCl_3$, 100 MHz): δ (ppm): 200.97, 156. 03, 136. 00, 127.45, 114.85, 43.97, 43.20; GC-MS (EI) $m/z$ (%) 181.05 ($M^+$, 74), 180.05 (79), 147.10 (31), 137.05 (100).

*4-(Dimethylamino)-N,N-dimethylbenzothioamide* (**3p**) [50]. $^1$H-NMR ($CDCl_3$, 400 MHz): δ (ppm): 7.32 (d, $J$ = 8 Hz, 2H), 7.63 (d, $J$ = 8 Hz, 2H), 3.59 (s, 3H), 3.28 (s, 3H), 2.99 (s, 6H); $^{13}$C-NMR ($CDCl_3$, 100 MHz): δ (ppm): 201.64, 150.54, 130.30, 127.98, 110.56, 44.19, 43.40, 39.86; GC-MS (EI) $m/z$ (%) 208.05 ($M^+$, 74), 207.05 (46), 164.05 (100), 148.10 (57).

*N,N-Dimethylnaphthalene-2-carbothioamide* (**3q**) [35]. $^1$H-NMR ($CDCl_3$, 400 MHz): δ (ppm): 7.82–7.81 (m, 3H), 7.77 (s, 1H), 7.53–7.46 (m, 2H), 7.44–7.41(m, 1H), 3.64 (s, 3H), 3.19 (s, 3H); $^{13}$C-NMR ($CDCl_3$, 100 MHz): δ (ppm): 201.21, 140.60, 128.38, 128.20, 127.75, 126.78, 126.74, 124.73, 123.94, 44.27, 43.31.

*N,N-Dimethylfuran-2-carbothioamide* (**3s**) [51]. $^1$H-NMR ($CDCl_3$, 400 MHz): δ (ppm): 7.47 (d, $J$ = 4 Hz, 1H), 7.09 (d, $J$ = 4 Hz, 1H), 6.46 (d, $J$ = 4 Hz, 1H),3.56 (s, 3H), 3.45 (s, 3H); $^{13}$C-NMR ($CDCl_3$, 100 MHz): δ (ppm): 158.31, 151.98, 142.80, 117.27, 111.49, 44.01, 43.80. GC-MS (EI) $m/z$ (%) 155.05 (100, $M^+$), 111.00 (70), 73.95 (27).

*N,N-Dimethylthiophene-2-carbothioamide* (**3t**) [35]. $^1$H-NMR ($CDCl_3$, 400 MHz): δ (ppm): 7.41 (dd, $J$ = 4.8, 0.8 Hz, 1H), 7.12 (dd, $J$ = 4.0, 1.2 Hz, 1H), 6.99–6.97 (m, 1H), 3.58 (s, 3H), 3.45 (s, 3H); $^{13}$C-NMR ($CDCl_3$, 100 MHz): δ (ppm):191.63, 145.23, 129.30, 126.50, 126.45, 44.65. GC-MS (EI) $m/z$ (%) 171.00 (77, $M^+$), 127.00 (100).

*N,N-Dimethylpyridine-3-carbothioamide* (**3u**) [52]. $^1$H-NMR ($CDCl_3$, 400 MHz): δ (ppm): 8.59–8.57 (m, 2H),7.69 (dt, $J$ = 8.0, 2.0 Hz, 1H), 7.33–7.30 (m, 1H), 3.63 (s, 3H), 3.23 (s, 3H); $^{13}$C-NMR ($CDCl_3$, 100 MHz): δ (ppm):197.59, 149.63, 146.19, 133.70, 123.18, 44.25, 43.37. GC-MS (EI) $m/z$ (%) 166.05 (85, $M^+$), 165.05 (88), 149.10 (35), 122.00 (100), 106.05 (38), 78.00 (62).

*N,N-Dimethyl-5-(quinolin-2-yl) thiophene-2-carbothioamide* (**3v**). $^1$H-NMR ($CDCl_3$, 400 MHz): δ (ppm): 8.15 (d, $J$ = 8 Hz, 1H), 8.08 (d, $J$ = 8 Hz, 1H), 7.78 (t, $J$ = 8 Hz, 2H), 7.74–7.69 (m, 1H), 7.59 (d, $J$ = 4 Hz, 1H), 7.54–7.49 (m, 1H), 7.19 (d, $J$ = 4 Hz, 1H), 3.56 (d, $J$ = 40 Hz, 6H); $^{13}$C-NMR ($CDCl_3$, 100 MHz): δ (ppm): 191.56, 151.54, 148.18, 148.08, 146.86, 136.77, 130.01, 129.37, 127.82, 127.53, 127.40, 126.49, 124.94, 117.42, 44.54. HRMS (ESI) $m/z$ calculated (calcd.) for $C_{16}H_{15}N_2S_2^+$ $(M + H)^+$ 299.06712, found 299.06706.

## 4. Conclusions

In conclusion, we have demonstrated an efficient and transitional-metal-free method for the synthesis of aryl thioamides derived from aryl aldehydes using thiourea as a sulfur source in the presence of potassium persulfate, in DMF or DMAC. This strategy has the advantages of good functional-group tolerance and gives moderate to good yields of desired products. Further studies on synthetic applications are currently under way.

**Author Contributions:** X.Q. conceived and designed the experiments; Y.B. performed the experiments; Y.C. contributed reagents/materials; J.L. and L.L. wrote the paper.

## References

1. Petrov, K.A.; Andreev, L.N. The Chemical Properties of Thioamides. *Russ. Chem. Rev.* **1971**, *40*, 505–524. [CrossRef]
2. Cremlyn, R.J. *An Introduction to Organo-Sulfur Chemistry*; Wiley & Sons: New York, NY, USA, 1996.
3. Jiang, W.; Li, Y.; Wang, Z. Heteroarenes as high performance organic semiconductors. *Chem. Soc. Rev.* **2013**, *42*, 6113–6127. [CrossRef] [PubMed]
4. Jagodziński, T.S. Thioamides as Useful Synthons in the Synthesis of Heterocycles. *Chem. Rev.* **2003**, *103*, 197–228. [CrossRef] [PubMed]
5. Lincke, T.; Behnken, S.; Ishida, K.; Roth, M.; Hertweck, C. Closthioamide: An Unprecedented Polythioamide Antibiotic from the Strictly Anaerobic Bacterium *Clostridium cellulolyticum*. *Angew. Chem. Int. Ed.* **2010**, *49*, 2011–2013. [CrossRef] [PubMed]
6. Shen, C.; Zhang, P.F.; Sun, Q.; Bai, S.Q.; Andy Hor, T.S.; Liu, X.G. Recent advances in C–S bond formation via C–H bond functionalization and decarboxylation. *Chem. Soc. Rev.* **2015**, *44*, 291–314. [CrossRef] [PubMed]
7. Anthony, J.E. Functionalized Acenes and Heteroacenes for Organic Electronics. *Chem. Rev.* **2006**, *106*, 5028–5048. [CrossRef] [PubMed]
8. Ashfaq, M.; Shah, S.S.A.; Najam, T.; Ahmad, M.M.; Tabassum, R.; Rivera, G. Synthetic Thioamide, Benzimidazole, Quinolone and Derivatives with Carboxylic Acid and Ester Moieties: A Strategy in the Design of Antituberculosis Agents. *Curr. Med. Chem.* **2014**, *21*, 911–931. [CrossRef] [PubMed]

9. Zoumpoulakis, P.; Camoutsis, C.; Pairas, G.; Sokovic, M.; Glamoclija, J.; Potamitis, C.; Pitsas, A. Synthesis of novel sulfonamide-1,2,4-triazoles, 1,3,4-thiadiazoles and 1,3,4-oxadiazoles, as potential antibacterial and antifungal agents. Biological evaluation and conformational analysis studies. *Bioorg. Med. Chem.* **2012**, *20*, 1569–1583. [CrossRef] [PubMed]
10. Guo, W.; Fu, Y.Z. A Perspective on Energy Densities of Rechargeable Li-S Batteries and Alternative Sulfur-Based Cathode Materials. *Energy Environ. Mater.* **2018**, *1*, 20–27. [CrossRef]
11. Polshettiwar, V. Phosphorus Pentasulfide ($P_4S_{10}$). *Synlett* **2004**, *12*, 2245–2246. [CrossRef]
12. Ozturk, T.; Ertas, E.; Mert, O. Use of Lawesson's Reagent in Organic Syntheses. *Chem. Rev.* **2007**, *107*, 5210–5278. [CrossRef] [PubMed]
13. Vásquez-Céspedes, S.; Ferry, A.; Candish, L.; Glorius, F. Heterogeneously Catalyzed Direct C–H Thiolation of Heteroarenes. *Angew. Chem., Int. Ed.* **2015**, *54*, 5772–5776. [CrossRef] [PubMed]
14. Jiao, J.; Wei, L.; Ji, X.M.; Hu, M.L.; Tang, R.Y. Direct Introduction of Dithiocarbamates onto Imidazoheterocycles under Mild Conditions. *Adv. Synth. Catal.* **2016**, *358*, 268–275. [CrossRef]
15. Rafique, J.; Saba, S.; Rosário, A.R.; Braga, A.L. Regioselective, Solvent- and Metal-Free Chalcogenation of Imidazo[1,2-*a*] pyridines by Employing $I_2$/DMSO as the Catalytic Oxidation System. *Chem. Eur. J.* **2016**, *22*, 11854–11862. [CrossRef] [PubMed]
16. Ding, Q.P.; Cao, B.P.; Yuan, J.J.; Liu, X.J.; Peng, Y.Y. Synthesis of thioethers via metal-free reductive coupling of tosylhydrazones with thiols. *Org. Biomol. Chem.* **2011**, *9*, 748–751. [CrossRef] [PubMed]
17. Ravi, C.; Mohan, D.C.; Adimurthy, S. *N*-Chlorosuccinimide-Promoted Regioselective Sulfenylation of Imidazoheterocycles at Room Temperature. *Org. Lett.* **2014**, *16*, 2978–2981. [CrossRef] [PubMed]
18. Hiebel, M.A.; Berteina-Raboin, S. Iodine-catalyzed regioselective sulfenylation of imidazoheterocycles in $PEG_{400}$. *Green Chem.* **2015**, *17*, 937–944. [CrossRef]
19. Siddaraju, Y.; Prabhu, K.R. Iodine-Catalyzed Cross Dehydrogenative Coupling Reaction: A Regioselective Sulfenylation of Imidazoheterocycles Using Dimethyl Sulfoxide as an Oxidant. *J. Org. Chem.* **2016**, *81*, 7838–7846. [CrossRef] [PubMed]
20. Yang, F.L.; Tian, S.K. Iodine-Catalyzed Regioselective Sulfenylation of Indoles with Sulfonyl Hydrazides. *Angew. Chem. Int. Ed.* **2013**, *52*, 4929–4932. [CrossRef] [PubMed]
21. Yang, Y.; Zhang, S.; Tang, L.; Hu, Y.B.; Zha, Z.G.; Wang, Z.Y. Catalyst-free thiolation of indoles with sulfonyl hydrazides for the synthesis of 3-sulfenylindoles in water. *Green Chem.* **2016**, *18*, 2609–2613. [CrossRef]
22. Singh, R.; Allam, K.B.; Singh, N.; Kumari, K.; Singh, S.K. A Direct Metal-Free Decarboxylative Sulfono Functionalization (DSF) of Cinnamic Acids to $\alpha$, $\beta$-Unsaturated Phenyl Sulfones. *Org. Lett.* **2015**, *17*, 2656–2659. [CrossRef] [PubMed]
23. Senadi, G.C.; Guo, B.C.; Hu, W.P.; Wang, J.J. Iodine-promoted cyclization of *N*-propynyl amides and *N*-allyl amides via sulfonylation and sulfenylation. *Chem. Commun.* **2016**, *52*, 11410–11413. [CrossRef] [PubMed]
24. Handa, S.; Fennewald, J.C.; Lipshutz, B.H. Aerobic Oxidation in Nanomicelles of Aryl Alkynes, in Water at Room Temperature. *Angew. Chem. Int. Ed.* **2014**, *53*, 3432–3435. [CrossRef] [PubMed]
25. Rao, W.H.; Shi, B.F. Copper(II)-Catalyzed Direct Sulfonylation of C($sp^2$)–H Bonds with Sodium Sulfinates. *Org. Lett.* **2015**, *17*, 2784–2787. [CrossRef] [PubMed]
26. Ding, Y.; Wu, W.; Zhao, W.; Li, Y.; Xie, P.; Huang, Y.; Liu, Y.; Zhou, A. Generation of thioethers via direct C–H functionalization with sodium benzenesulfinate as a sulfur source. *Org. Biomol. Chem.* **2016**, *14*, 1428–1431. [CrossRef] [PubMed]
27. Xiao, F.; Chen, S.; Tian, J.; Huang, H.; Liu, Y.; Deng, G.J. Chemoselective cross-coupling reaction of sodium sulfinates with phenols under aqueous conditions. *Green Chem.* **2016**, *18*, 1538–1546. [CrossRef]
28. Guo, Y.J.; Lu, S.; Tian, L.L.; Huang, E.L.; Hao, X.Q.; Zhu, X.J.; Shao, T.; Song, M.P. Iodine-Mediated Difunctionalization of Imidazopyridines with Sodium Sulfinates: Synthesis of Sulfones and Sulfides. *J. Org. Chem.* **2018**, *83*, 338–349. [CrossRef] [PubMed]
29. Zhou, Z.; Liu, Y.; Chen, J.F.; Yao, E.; Cheng, J. Multicomponent Coupling Reactions of Two *N*-Tosyl Hydrazones and Elemental Sulfur: Selective Denitrogenation Pathway toward Unsymmetric 2,5-Disubstituted 1,3,4-Thiadiazoles. *Org. Lett.* **2016**, *18*, 5268–5271. [CrossRef] [PubMed]
30. Ravi, C.; Reddy, N.N.K.; Pappula, V.; Samanta, S.; Adimurthy, S. Copper-Catalyzed Three-Component System for Arylsulfenylation of Imidazopyridines with Elemental Sulfur. *J. Org. Chem.* **2016**, *81*, 9964–9972. [CrossRef] [PubMed]

31. Zhang, J.R.; Liao, Y.Y.; Deng, J.C.; Feng, K.Y.; Zhang, M.; Ning, Y.Y.; Lin, Z.W.; Tang, R.Y. Oxidative dual C–H thiolation of imidazopyridines with ethers or alkanes using elemental sulphur. *Chem. Commun.* **2017**, *53*, 7784–7787. [CrossRef] [PubMed]
32. Zhu, X.M.; Yang, Y.Z.; Xiao, G.H.; Song, J.X.; Liang, Y.; Deng, G.B. Double C–S bond formation via C–H bond functionalization: Synthesis of benzothiazoles and naphtho[2,1-*d*]thiazoles from *N*-substituted arylamines and elemental sulfur. *Chem. Commun.* **2017**, *53*, 11917–11920. [CrossRef] [PubMed]
33. Bergman, J. Comparison of Two Reagents for Thionations. *Synthesis* **2018**, *50*, 2323–2328. [CrossRef]
34. Hurd, R.N.; Delamater, G. The Preparation and Chemical Properties of Thionamides. *Chem. Rev.* **1961**, *61*, 45–86. [CrossRef]
35. Wei, J.P.; Li, Y.M.; Jiang, X.F. Aqueous Compatible Protocol to Both Alkyl and Aryl Thioamide Synthesis. *Org. Lett.* **2016**, *18*, 340–343. [CrossRef] [PubMed]
36. Zhou, Z.; Yu, J.T.; Zhou, Y.N.; Jiang, Y.; Cheng, J. Aqueous MCRs of quaternary ammoniums, *N*-substituted formamides and sodium disulfide towards aryl thioamides. *Org. Chem. Front.* **2017**, *4*, 413–416. [CrossRef]
37. Xu, K.; Li, Z.Y.; Cheng, F.Y.; Zuo, Z.Z.; Wang, T.; Wang, M.C.; Liu, L.T. Transition-Metal-Free Cleavage of C–C Triple Bonds in Aromatic Alkynes with $S_8$ and Amides Leading to Aryl Thioamides. *Org. Lett.* **2018**, *20*, 2228–2231. [CrossRef] [PubMed]
38. Kumar, S.; Vanjari, R.; Guntreddi, T.; Singh, K.N. Sulfur promoted decarboxylative thioamidation of carboxylic acids using formamides as amine proxy. *Tetrahedron* **2016**, *72*, 2012–2017. [CrossRef]
39. Nguyen, T.B.; Tran, M.Q.; Ermolenko, L.; Al-Mourabit, A. Three-Component Reaction between Alkynes, Elemental Sulfur, and Aliphatic Amines: A General, Straightforward, and Atom Economical Approach to Thioamides. *Org. Lett.* **2014**, *16*, 310–313. [CrossRef] [PubMed]
40. Rao, M.M.; Jayalakshmi, M.; Reddy, R.S. Time-selective Hydrothermal Synthesis of SnS Nanorods and Nanoparticles by Thiourea Hydrolysis. *Chem. Lett.* **2004**, *33*, 1044–1045. [CrossRef]
41. Jayalakshmi, M.; Rao, M.M. Synthesis of zinc sulphide nanoparticles by thiourea hydrolysis and their characterization for electrochemical capacitor applications. *J. Power Sources* **2006**, *157*, 624–629. [CrossRef]
42. Zhang, K.; Han, Q.; Wang, X.; Zhu, J. One-Step Synthesis of $Bi_2S_3$/BiOX and $Bi_2S_3/(BiO)_2CO_3$ Heterojunction Photocatalysts by Using Aqueous Thiourea Solution as Both Solvent and Sulfur Source. *ChemistrySelect* **2016**, *1*, 6136–6145. [CrossRef]
43. Manivel, P.; Prabakaran, K.; Krishnakumar, V.; Khan, F.N.; Maiyalagan, T. Thiourea-Mediated Regioselective Synthesis of Symmetrical and Unsymmetrical Diversified Thioethers. *Ind. Eng. Chem. Res.* **2014**, *53*, 7866–7870. [CrossRef]
44. Niu, H.; Xia, C.; Qu, G.; Wu, S.; Jiang, Y.; Jin, X.; Guo, H. Microwave-Promoted "One-Pot" Synthesis of 4-Nitrobenzylthioinosine Analogues Using Thiourea as a Sulfur Precursor. *Chem. Asian J.* **2012**, *7*, 45–49. [CrossRef] [PubMed]
45. Firouzabadi, H.; Iranpoor, N.; Gholinejad, M. One-Pot Thioetherification of Aryl Halides Using Thiourea and Alkyl Bromides Catalyzed by Copper(I) Iodide Free from Foul-Smelling Thiols in Wet Polyethylene Glycol (PEG 200). *Adv. Synth. Catal.* **2010**, *352*, 119–124. [CrossRef]
46. Mondal, J.; Modak, A.; Dutta, A.; Basu, S.; Jha, S.N.; Bhattacharyya, D.; Bhaumik, A. One-pot thioetherification of aryl halides with thiourea and benzylbromide in water catalyzed by Cu-grafted furfural imine-functionalized mesoporous SBA-15. *Chem. Commun.* **2012**, *48*, 8000–8002. [CrossRef] [PubMed]
47. Ma, X.; Yu, L.; Su, C.; Yang, Y.; Li, H.; Xu, Q. Efficient Generation of C-S Bonds via a By-Product-Promoted Selective Coupling of Alcohols, Organic Halides, and Thiourea. *Adv. Synth. Catal.* **2017**, *359*, 1649–1655. [CrossRef]
48. Abbasi, M.; Khalifeh, R. One-pot odourless synthesis of thioesters via in situ generation of thiobenzoic acids using benzoic anhydrides and thiourea. *Beilstein J. Org. Chem.* **2015**, *11*, 1265–1273. [CrossRef] [PubMed]
49. Swain, S.P.; Chou, Y.; Hou, D. Thioesterifications Free of Activating Agent and Thiol: A Three-Component Reaction of Carboxylic Acids, Thioureas, and Michael Acceptors. *Adv. Synth. Catal.* **2015**, *357*, 2644–2650. [CrossRef]
50. Bezgubenko, L.V.; Pipko, S.E.; Sinitsa, A.D. Dichlorothiophosphoric acid and dichlorothiophosphate anion as thionating agents in the synthesis of thioamides. *Russ. J. Gen. Chem.* **2008**, *78*, 1341–1344. [CrossRef]

51. Meltzer, R.I.; Lewis, A.D.; King, J.A. Antitubercular Substances. IV. Thioamides. *J. Am. Chem. Soc.* **1955**, *77*, 4062–4066. [CrossRef]
52. Perregaard, J.; Lawesson, S.O. Studies on Organophosphoous Compounds. XI.* Oxidation of Aromatic Compounds with Sulfur in Hexamethylphosphoric Triamide (HMPA). A New Method for Preparation of *N,N*-Dimethylthiocarboxamides. *Acta Chem. Scand. B* **1975**, *29*, 604–608.

8

# Understanding the Exceptional Properties of Nitroacetamides in Water: A Computational Model including the Solvent

**Giovanni La Penna [1,*,†] and Fabrizio Machetti [2,*,‡]**

1 Istituto di Chimica dei Composti Organometallici (ICCOM), Consiglio Nazionale delle Ricerche (CNR), via Madonna Del Piano 10, I-50019 Sesto Fiorentino, Firenze, Italy
2 Istituto di Chimica dei Composti Organometallici (ICCOM), Consiglio Nazionale delle Ricerche (CNR), c/o Dipartimento di Chimica "Ugo Schiff" via Della Lastruccia 13, I-50019 Sesto Fiorentino, Firenze, Italy
* Correspondence: glapenna@iccom.cnr.it (G.L.P.); fabrizio.machetti@unifi.it or fabrizio.machetti@cnr.it (F.M.)

† G.L.P. is associated to the Istituto Nazionale di Fisica Nucleare (INFN), Section of Roma-Tor Vergata, via della Ricerca Scientifica 1, I-00133 Roma, Italy.
‡ Dedicated to Prof. Francesco De Sarlo on the occasion of his 80th birthday.

Academic Editor: Michal Szostak

**Abstract:** Proton transfer in water involving C–H bonds is a challenge and nitro compounds have been studied for many years as good examples. The effect of substituents on acidity of protons geminal to the nitro group is exploited here with new p$K_a$ measurements and electronic structure models, the latter including explicit water environment. Substituents with the amide moiety display an exceptional combination of acidity and solubility in water. In order to find a rationale for the unexpected p$K_a$ changes in the (ZZ$'$)NCO- substituents, we measured and modeled the p$K_a$ with Z=Z$'$=H and Z=Z$'$=methyl. The dominant contribution to the observed p$K_a$ can be understood with advanced computational experiments, where the geminal proton is smoothly moved to the solvent bath. These models, mostly based on density-functional theory (DFT), include the explicit solvent (water) and statistical thermal fluctuations. As a first approximation, the change of p$K_a$ can be correlated with the average energy difference between the two tautomeric forms (*aci* and *nitro*, respectively). The contribution of the solvent molecules interacting with the solute to the proton transfer mechanism is made evident.

**Keywords:** amides; carbanions; C–H acidity; nitro-aci tautomerism; molecular dynamics; density-functional theory

---

## 1. Introduction

Nitro compounds are useful reagents in synthetic organic chemistry [1]. They are precursors of dipoles in 1,3-dipolar cycloaddition [2–6], a source of carbon nucleophiles in conjugated additions [7,8] and nitro aldol (Henry) reaction [9], and a substrate in Nef reaction [10,11]. In all of these reactions, C–H protons geminal to the nitro group are involved. Because of the presence of the nitro group, the above C–H protons show a higher degree of acidity (compound **4**, Table 1) compared with the C–H protons of an aliphatic chain. This feature is due to the ability of the nitro group to stabilize the carbanion in the form of the nitronate anion. The species involved in the nitro compound acidity are depicted in Figure 1 for primary nitro compounds.

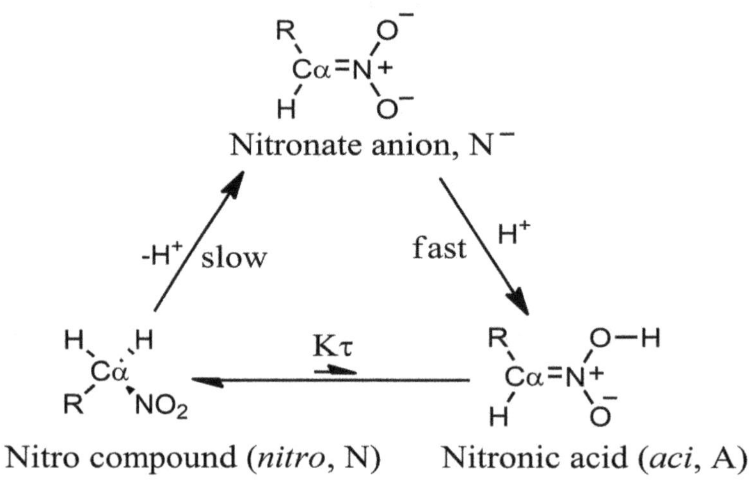

**Figure 1.** Schematic picture of species involved in the acid-base equilibria of nitro compounds.

An interesting aspect of nitro compounds is their lower proton extraction rate from C$\alpha$ than that expected from the acidity (Figure 1). This aspect is due to the required conformational rearrangement of the C$\alpha$ atom (from $sp^3$ to $sp^2$) to delocalize the negative charge of the carbanion to the nitro group [12]. The p$K_a$ of nitronic acid, as it can be derived by kinetic experiments [12], is about 3.5. The issue of the unusual acidity of nitro compounds with labile C–H bonds in a geminal position has been the object of experimental and modeling studies for a long time [12–14].

During our work on condensation of nitro compounds with alkenes or alkynes, we became interested in mechanistic aspects of this reaction [6,15,16]. We envisaged that acid-base properties of the substrates could be involved. The acidity of nitro compounds is enhanced by electron withdrawing groups such as esters and ketones in geminal position, resembling carboxylic acid in acid strength (compound **5** vs. compounds **6**–**8**, Table 1) [12]. Intramolecular interactions, including hydrogen bonds, stabilize to different extents the species involved.

Therefore, in this work, we complete the list of ionization constants for some nitro compounds, including the nitroacetamides **1**–**3** (Figure 2), which are the major focus of our study because of the exceptional combination of acidity and solubility of compound **1**.

**Table 1.** Apparent ionization constants of primary nitro compounds.

| Compound | R | p$K_a$ [a] | p$K_a$ [b] |
|---|---|---|---|
| **1** | $H_2NCO$ | **5.39**; 5.18 [17] | 6.20 |
| **2** | $(CH_3)HNCO$ | 5.46 [16] | 6.75 |
| **3** | $(CH_3)_2NCO$ | **7.23** | 5.99 |
| **4** | H | 10.7 [18] | 10.2 |
| **5** | $CH_3$ | 8.57 [19] | 8.49 |
| **6** | $CH_3CO$ | 5.10 [20] | 5.40 |
| **7** | PhCO | 5.19 [21] | 5.37 |
| **8** | $CH_3OCO$ | **5.70**; 5.68 [16]; 5.56 [12] | 5.73 |

[a] Data reported in the literature and measured (boldface) in this work; [b] Data obtained by ACDLabs available from SciFinder™.

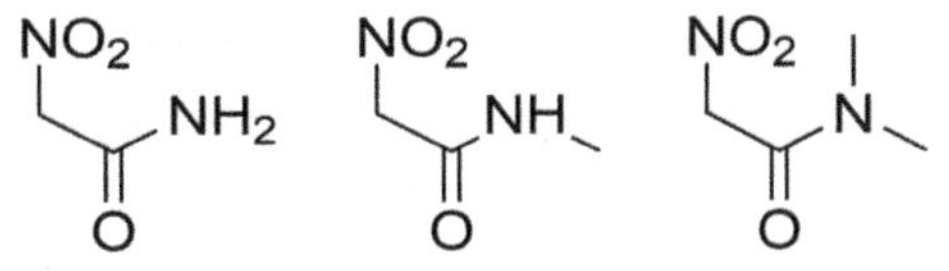

**Figure 2.** Nitroacetamides studied in this work.

Unexpectedly, nitroacetamides **1–3** show significant change in p$K_a$ values by replacing N-CH3 methyl groups in **3** with protons (compounds **2** and **1**, Table 1). As we show with computational models, those values cannot be easily explained with stabilization factors on nitronate ions. In addition, the prediction of p$K_a$ using a popular software [22], available from the SciFinder$^{TM}$ database, does not completely agree with the experimental data (Table 1, last column).

To provide a rationale for the effect of amide derivatives on the acidity of C–H bonds in geminal position to the nitro group, we present in this work an original model where, in addition to electronic and steric intramolecular effects, the role of the water solvent is included. Electronic effects are included using density-functional theory (DFT) with exchange functional described as in the Perdew–Burke–Ernzerhof (PBE) approximation [23], when dynamical methods are used [24,25], or in the Becke three-parameter Lee–Yang–Parr (B3LYP) approximation [26], when static (or single point) calculations are performed. The models, compared to quantum mechanics/molecular mechanics (QM/MM) techniques [27], allow the study of subtle effects due to charge separation during the addressed reaction [28].

It is found that the solvent exerts an essential effect that opens to a new design strategy for further enhancing this important type of acidity.

## 2. Results and Discussion

Following the analysis first reported in Ref. [19], the "apparent" ionization constant $K'$ is a function of the ionization constants of the two tautomeric forms, respectively *aci* and *nitro* (see Figure 1):

$$K' = \frac{[N^-][H_3O^+]}{[A]+[N]}. \tag{1}$$

Manipulating the equation above, the apparent $K'$ constant can be expressed in terms of the equilibrium constant between the two tautomeric forms $K_\tau$:

$$K' = \frac{K_N}{(K_{\o}+1)} \simeq K_N \tag{2}$$

where $K_N$ is the ionization constant of the nitro form (that is the most stable at room conditions) and $K_\tau$ = [A]/[N]. The low ratio between *aci* and *nitro* forms ($K_\tau << 1$) at room conditions in water solution prevents the species from showing the larger acidity of the *aci* form compared to the *nitro*. The former is more acidic because the C–H bond is always stronger than the O–H bond. However, the enhanced chemical properties of rare species present with very low statistical weight in the sample are evident in the measured apparent ionization constant. The stronger acidity of the low-weight *aci* form is evident when the proton exchange between the *aci* form and the *nitro* form is frozen or the kinetics of the *aci* deprotonation can be separated by measured kinetic data [12]. Hereafter, we indicate $K'$ as $K_a$.

The prediction of p$K_a$ for the compounds displayed in Figure 2 is a challenging task also for empirical methods, the latter still the more accurate [29]. The application of a a popular software [22], available from the SciFinder$^{TM}$ database, does not agree with the experimental data (see Introduction above) and our work aims at explaining the disagreement in terms of atomistic models. Theoretical and computational methods achieved significant advancement, but reliable applications are still problematic when protons are released by C atoms, rare species are transiently involved and subtle effects of solvent, especially water, play a role in the thermodynamics of the proton exchange.

From a microscopic point of view, the contribution of rare acidic forms to the average observed property, that is potentially dominated by low-acidic forms, can be explained if the reactive form is trapped within energy barriers. In this case, the conversion from the rare form to the most stable one is slower than the ionization. The average property, provided by the series of sampled microscopic states, slowly converges with sampling.

Indeed, this effect can be achieved in practice with computational models where the model is constrained towards bound states and cannot escape from one chemical configuration to another. Among these models, the tight-binding method forces the sampling of bound states. In this approximation, the sampling of rare chemical species can last for a long time even if in theory the atoms should rapidly change the valence to reach the most stable configuration. Therefore, despite the many limitations of the tight-binding approximation, it is possible to compare the energy of different bound states, while free energy changes are affected by huge errors. In this case, the average energy can be computed in different samples, each mimicking the metastable equilibrium state of the two different and separated tautomeric forms. Another advantage is the possibility to include explicit water molecules in the modeled sample. In this work, we used the self-consistent charge density-functional tight-binding approximation [30] (DFTB, hereafter).

In order to compare the thermodynamic quantities measured by experiments with results of microscopic models, we make the following assumption in the context of the nitro compounds an object of this study. The larger the statistical weight of the *aci* form, the larger the acidity of the sample. The tight-binding approximation can be then used to describe realistic configurations with significant statistical weight for each of the two tautomeric forms. Once this goal is achieved, the proton transfer between the two forms can be described with more detailed computational experiments still including the contribution of the solvation layer. The latter task is accomplished here by adding an external empirical potential to a density-functional theory (DFT) approximation of electron density coupled with molecular dynamics (MD) simulations.

### 2.1. *Tight-Binding Approximation of nitro and aci Forms*

In Table 2, the difference in average energy ($\Delta E_\tau$) at $T = 300$ K and at the water density of bulk water ($\rho_0 = 1$ g/cm$^3$) between the two tautomeric forms is reported.

**Table 2.** Comparison between $\Delta G^0$ ($T = 298$ K, $P = 1$ bar) for ionization of species $RCH_2NO_2$ soluble in water.

| Compound | $\Delta G^0$ | $\Delta E_\tau$ (DFTB) | $\Delta E_\tau$ (DFT) | $\Delta G^0$ |
|---|---|---|---|---|
| **1** | 30.75; 29.55 | −13 | −53.85/−55.70 | 35.37 |
| **2** | 31.15 | −20 | −65.16/−73.74 | 38.51 |
| **3** | **41.25** | −75 | −61.54/−66.73 | 34.17 |
| **4** | 61.04 | −59 | −90.62 | 58.19 |
| **5** | 48.89 | −46 | −60.32 | 48.89 |
| **6** | 29.10 | −24 | −56.44 | 30.81 |
| **7** | 29.61 | −6 | −58.53 | 30.64 |
| **8** | **32.30**; 32.40; 31.71 | −4 | −60.14 | 32.69 |

Column 2 is derived from p$K_a$ in Table 1 (boldface values are obtained in this work). Columns 3–4 ($\Delta E_\tau$) are the difference in average energy between *nitro* and *aci* tautomers for the same species. Column 3 is computed in the explicit solvent DFTB model; column 4 is computed in the mean-field solvent DFT model (see Methods for details). Column 5 are values computed from p$K_a$ obtained with SciFinder™ (Table 1). All energy values are in kJ/mol. DFTB averages are computed at $T = 300$ K and with water bulk density at $T = 300$ K and $P = 1$ bar. In these conditions, root-mean square error on DFTB energy is in the range 105–140 kJ/mol.

These data are compared with the measured ionization free energy change and with the same energy difference computed with an accurate DFT approximation that allows geometry optimization in an implicit model of the water solvent. The final column is the $\Delta G_0$ derived from p$K_a$ values predicted with an empirical method provided by the SciFinder$^{TM}$ database. According to a comparison between different prediction methods [29], the ACDLabs [22] method is one of the best performing.

The approximate DFTB model of the electronic structure and the low statistics are not expected to provide agreement between the measured free energy changes (column 2) and the computed energy difference between ionized and neutral species (not shown here). However, it can be noticed that the most acidic species (less positive free energy of ionization, column 2) display the lowest difference in energy of the *aci* tautomeric form (column 3). With the exception of **6**, the series of substituents displays the correct order for both $\Delta G^0$ and $\Delta E_{\mathcal{T}}$. This rough correlation indicates that the contribution to the apparent acidity due to substituent R can be ascribed to the increasing statistical weight of the *aci* form, the latter characterized by large acidity.

The ACDLabs empirical prediction, though it is excellent for the compounds that are presumably tabulated (**4–8**), fails in predicting the high acidity of compound **1** and the decrease of acidity of **3** with respect to **1**.

Intramolecular interactions have only a partial role in determining the average energy difference between the two tautomeric forms. This is shown by the values of $\Delta E_{\mathcal{T}}$ energy difference computed with the more accurate DFT method (column 4 in Table 2). The values are larger in absolute value than the corresponding DFTB estimate, even though they follow approximately the same ordering, with the smallest absolute values corresponding to the most acidic compounds. The range displayed for the amide derivatives is due to the choice of different structures as initial configurations for the geometry optimization. For instance, the lowest energy *nitro* structure corresponds to an open extended *all-trans* structure, where there is no interaction between the nitro and the amine group. The *aci* form is, compared to this extended structure, at the highest energy. On the other hand, when the *nitro* compound forms intramolecular interactions that favor a closed structure, the *aci* form is at a lower energy. However, in both cases, the energy difference is larger than when the calculation is performed with a less accurate model, but includes the solvent layer explicitly. Therefore, the inclusion of explicit solvent makes the energy landscape flatter than in the case of a polarizable continuum model for the solvent.

The interplay between intramolecular interactions and interactions with solvent molecules is shown by simulations in the explicit solvent.

In Figure 3, the time evolution of the N-C$\alpha$-C$\beta$-N$\gamma$ dihedral angle is displayed for all the three simulation stages (*nitro*, *aci* and ionized forms) for **1** and **3**, performed in the DFTB model. The dihedral angle displays for both compounds large fluctuations when in the *nitro* form because of the $sp^3$ configuration of C$\alpha$. After the displacement of $\alpha$ H to the nitro O atom (in the *aci* form) and then into the bulk water (ionized form), the molecules are sealed into, respectively, E and Z configurations for **1** and **3**. Despite the conformational freezing, keeping the *aci* form in the E configuration, the intramolecular H–N$\cdots$H$_N$-O$_N$-N hydrogen bond is not observed. Also in the ionized form, the O$_N$ atoms of the nitro group strongly interact with water molecules in the solvent (see below).

Despite the absence of stable intramolecular hydrogen bonds, there are significant intramolecular interactions in certain compounds. For instance, there is a high persistence of the intramolecular interaction between the N–O bond and the amide H atom when R = $NH_2CO$ (**1**). This interaction keeps the *aci* form sealed in the E conformation, the latter more hindered to water access than the *aci* form of other compounds (see Table 3 and Figure 4 discussed below). The N–O$\cdots$H–N interaction displays an angle smaller than 135°, thus being not classified as an hydrogen bond, but rather a strong electrostatic interaction. The proton attached to the nitro group in the *aci* form when R = $NH_2CO$ never interacts with N and O of the amide group. The latter atom is always *anti* to the nitro group with respect to C$\alpha$-C bond. As a consequence of this closed *aci* form, the anion displays always the strong N–O$\cdots$H–N intramolecular electrostatic interaction, while such interaction is not effective in the other substituents.

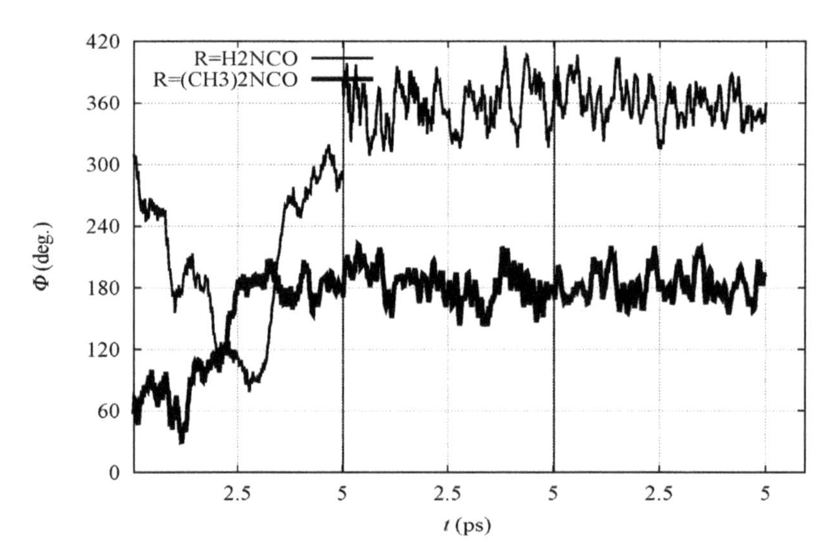


**Figure 3.** Time evolution of N-C$\alpha$-C$\beta$-N$\gamma$ dihedral angle ($\Phi$) with R = $H_2NCO$ (thin line) and $(CH_3)_2NCO$ (thick line) within BO–MD simulations performed with the density-functional tight-binding (DFTB) model. Vertical lines separate the simulation of *nitro* (left), *aci* (middle) and ionized (nitronate) forms (right), respectively.

**Table 3.** Hydrogen bond population (%) for all H-bond donor and acceptor groups in the investigated compounds (residue R, see Table 1).

| Compound | R | *nitro* | *aci* | *anion* |
|---|---|---|---|---|
| **1** | $H_2NCO$ | 287 | 235 | 410 |
| **2** | $(CH_3)HNCO$ | 257 | 258 | 557 |
| **3** | $(CH_3)_2NCO$ | 170 | 282 | 455 |
| **6** | $CH_3CO$ | 175 | 160 | 426 |
| **7** | PhCO | 139 | 249 | 359 |
| **8** | $CH_3OCO$ | 170 | 281 | 437 |

Hydrogen bond is counted when the distance X$\cdots$Y (X donor, Y acceptor) is within 0.3 nm and the X-H$\cdots$Y angle is between 135 and 180°. Percentage is obtained as the sum of occurrence of hydrogen bonds over all the water molecules in the sample, the two O–H bonds in water (when donor), acceptor atoms or X-H donating bond of solute, and the configurations collected at $T$ = 300 K (250 configurations within the simulation time $t$ = 5 ps), finally divided for the number of configurations. Therefore, percentage can be higher than 100.

These observations indicate that the energy of most of the species reported in Table 2 is also strongly modulated by interactions with the water environment, in addition to the electrostatic intramolecular interactions discussed above. The hydrogen bond population of H-bond donor and acceptor groups is reported in Table 3 and in Figure 4.

In this analysis, all the water molecules (213, 210 only when R = PhCO) in the sample are included. From Table 3, it can be observed that the $\alpha$ H atom forms a significant amount of hydrogen bonds with water molecules (C$\alpha$-H$\cdots$Ow, with Ow the O atom in water molecules) when C$\alpha$ is in the *nitro* form. As a term of comparison, when R = $CH_3$ (**5**, data not displayed in Table) the probability for C$\alpha$-H and C$\beta$-H hydrogen bonds with water is, respectively, 23 and 22%.

The C$\alpha$-H$\cdots$Ow hydrogen bond is lost when the C$\alpha$ atom is converted into the $sp^2$ electronic configuration (*aci* and nitronate forms). The highest population C$\alpha$-H$\cdots$Ow hydrogen bond is observed for R = $NH_2CO$ (**1**) in the *nitro* form (81%, see Figure 4).

The probability of any hydrogen bond with water molecules (both donating and accepting solute hydrogen atoms) reported in Table 3 shows that, for **1** and **2**, and when the solute is neutral, the population of hydrogen bonds has its maximum in the *nitro* form. The probability is still high when the proton is moved to the *aci* form. Finally, when the proton is removed from the solute (the negatively charged nitronate form), the probability of hydrogen bonds with the solvent increases, mainly because of the negatively charged nitro group. Looking at the partition of hydrogen bonds (Figure 4), it can be noticed that substituents with at least one N–H bond are particularly efficient in keeping water molecules structured in the first solvation layer around all the regions of the solute, independently of the tautomeric or protonation form.

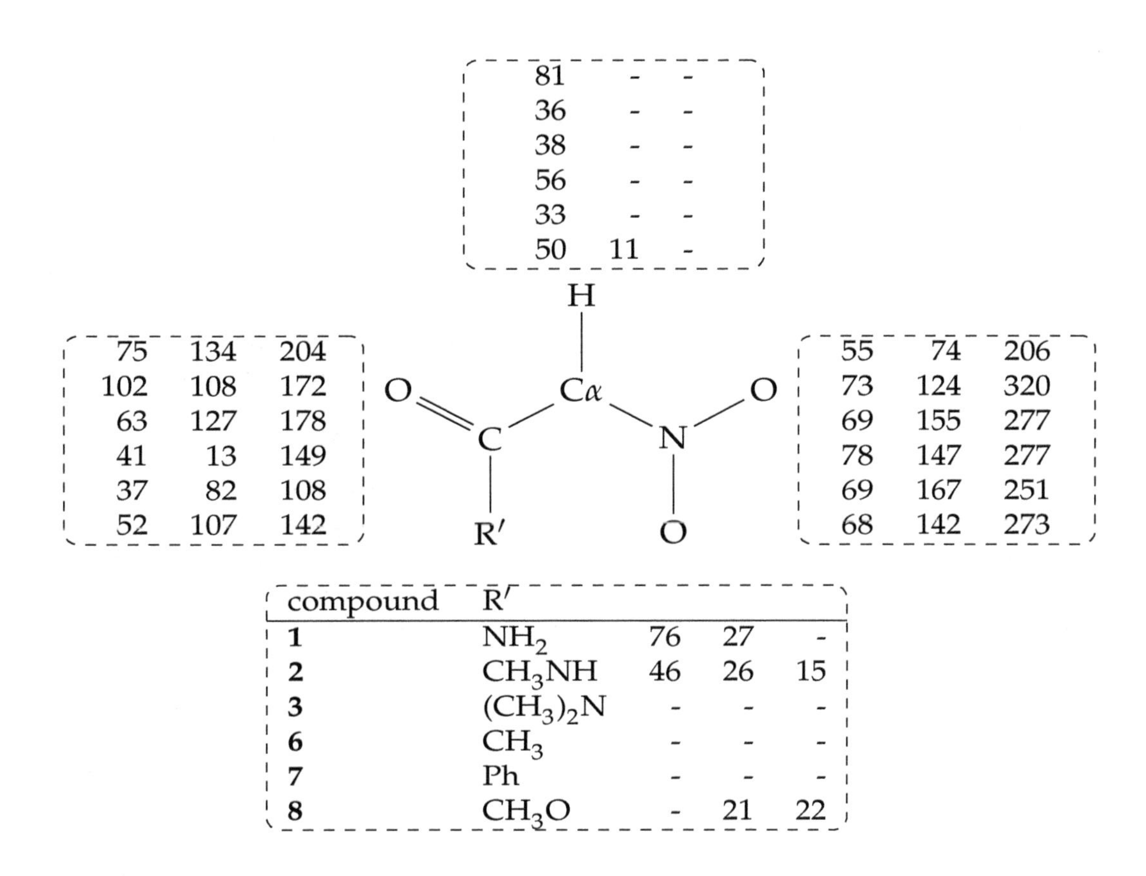


**Figure 4.** Hydrogen bond population (%) measured as in Table 3 and distributed over atomic groups in *nitro* (black, column 1), *aci* (red, column 2), and nitronate (blue, column 3) species. Lines are in the same order of Table 3. The - symbol indicates no hydrogen bond. Hydrogen bonds are counted for atoms: Cα-H (top); CO (left); $NO_2$ (right); R′ (bottom).

The water environment around each molecule has different degrees of basicity and, once it is protonated by the α H extraction, different degrees of acidity towards the solute molecule. To approximately measure acid-base properties of this water environment, α H is extracted from the solute by moving the atom towards the closest water molecule in the solvent bath. In this condition, the water bath (containing a single $H_3O^+$ species) is allowed to give back the proton to the solute. If a short time is provided to the solute, allowing the relaxation of the Cα bond environment, the water bath gives the proton back to other basic groups because, in the tight-binding approximation, the relaxed $sp^2$ Cα atom can not form a new C–H bond. Remarkably, in most of the cases, the group that is able to host the proton provided by the water bath is the nitro group, thus forming the solute in the *aci* form. Therefore, this alchemical process mimics a possible pathway for the proton transfer from position Cα to the nitro O atom, as it is mediated by the water layer around the solute. This simple experiment allows a first exploitation of the mechanism by which the solute can better manifest itself as the more acidic *aci* form. Interestingly, in some cases (R = PhCO), the carbonyl oxygen is able to form a transient covalent bond with the proton provided by the water layer.

In Table 4, the times required to transfer the proton from the water layer to one of the oxygen atoms in the solute, producing either the *aci* or the enolic form, is reported.

**Table 4.** Times ($\tau$) required to transfer the excess proton in the water layer, due to $\alpha$ H extraction from C$\alpha$, to O atoms either in the nitro group or in the carbonyl group (the latter case indicated with an asterisk).

| R | $\tau$ (ps) |
|---|---|
| 1 | 0.28 (*) |
| 2 | 1.92 |
| 3 | - |
| 6 | >5 |
| 7 | 0.14 (*) |
| 8 | >5 |

Each $\alpha$ H extraction to the water layer is performed from a selected configuration displaying an approximately zero or $\pi$ dihedral angle for $\alpha$H'-C$\alpha$-N–O and a water molecule with Ow within 0.2 nm from $\alpha$ H. It must be noticed that, when these two conditions are not fulfilled, in most of the cases, the $\alpha$ proton is rapidly given back to C$\alpha$ because there is no efficient relaxation mechanism for the $H_3O^+$ species formed in the water layer.

The formation of the *aci* form from the reaction between the protonated water environment and the negatively charged form of the solute has different lag-times $\tau$ displayed in Table 4. In some cases (**1** and **7**), the proton is finally bound by the carbonyl oxygen, forming the enolic isomer of the given species. Only in the case of **3**, the proton goes always back to C$\alpha$ because of the strong repulsion between the solute and the close by hydronium species formed by the $\alpha$ H extraction. Therefore, these data show that, for **1**, **2** and **7**, the pathway for proton exchange between C–H bond in the solute and a O–H bond in the solvation water layer, followed by the exchange with the O–H bond in the solute, is easily found.

The large chance of formation of enolic forms in the case of **1** and **7** is an indication of the possibility for enolic form as an intermediate in the slow process of C$\alpha$ deprotonation. A higher probability for enolic form increases the rate for proton release in certain compounds, as observed in the literature [12], because of the $sp^2$ pre-organization of C$\alpha$. In the DFTB model investigated here, the enolic form appears, in the more hydrophilic nitro compounds analyzed here, as a second acidic form of the nitro compound, in addition to the *aci* form. However, in the DFTB model, the mechanism to obtain the enolic form is mediated by the water molecule close to the $\alpha$ H atom that is extracted.

### *2.2. The $\alpha$ H Extraction from the* nitro *Tautomer and Insertion into the aci Tautomer*

The DFT model of the water solution sample circumvents the limitation of the tight-binding model in oversampling bound states. By using an external force that smoothly extracts one of the $\alpha$ H atom away from the C$\alpha$-H bond at room thermal conditions, it is possible to break the C–H bond, keeping the possibility of forming alternative explicit H–O bonds in the first solvent layer of the solute.

The analysis of the change in potential energy along with the $\alpha$ H extraction in the two extreme cases (R = $H_2NCO$ and R = $(CH_3)_2NCO$) is displayed in Figure 5 (see Methods).

The configurations corresponding to some selected points, indicated by letters a–f and A–F, are displayed in Figures 6 and 7, respectively.

It can be noticed that the compact initial structure of **1**, where the electrostatic interaction between the amino and nitro groups is effective, is rapidly lost during equilibration, and extended configurations are sampled in the explicit solvent (Figure 6, panel a). During the application of the external force that drives one of the $\alpha$ H towards the solvent, the more hydrophilic substituent (R = $H_2NCO$, filled circles in Figure 5) displays the increase in potential energy due to the exchange of the C–H bond with a O–H bond (Figure 5c). The potential energy is rapidly decreased ($\sim$150 kJ/mol), producing configurations with the proton confined within the solute and a water molecule in the first solvation layer (Figure 6d, the excess proton is on top-right).

On the other hand, the more hydrophobic substituent (R = $(CH_3)_2NCO$, **3**, circles in Figure 5) displays a fast movement of $\alpha$ H to the closest water molecule (2.7 Å compared to 2.0 of **1**), with a similar increase in potential energy compared to **1**. However, the following relaxation of the charge separation (Figure 7C) does not allow a significant decrease in potential energy. The excess proton (that is visible in panel C on top of the carbonyl group) displays a high energy and the movement of the excess proton away from the first solvation layer does not produce a significant decrease in potential energy (Figure 7D). The oscillation of potential energy (panels e–f and E–F of Figures 6 and 7, respectively) does not allow for the hydrophobic substituent (circles in Figure 5) the dissipation of potential energy that is allowed for the more hydrophilic one (filled circles in the same figure).

For both the substituents, the *aci* form is produced during the forced N–O neutralization process (Figures 6 and 7, panels E–F). Nevertheless, the *aci* form is transient and in rapid exchange with anions displaying hydrogen bonds between the nitro group and water molecules in the first solvation layer.

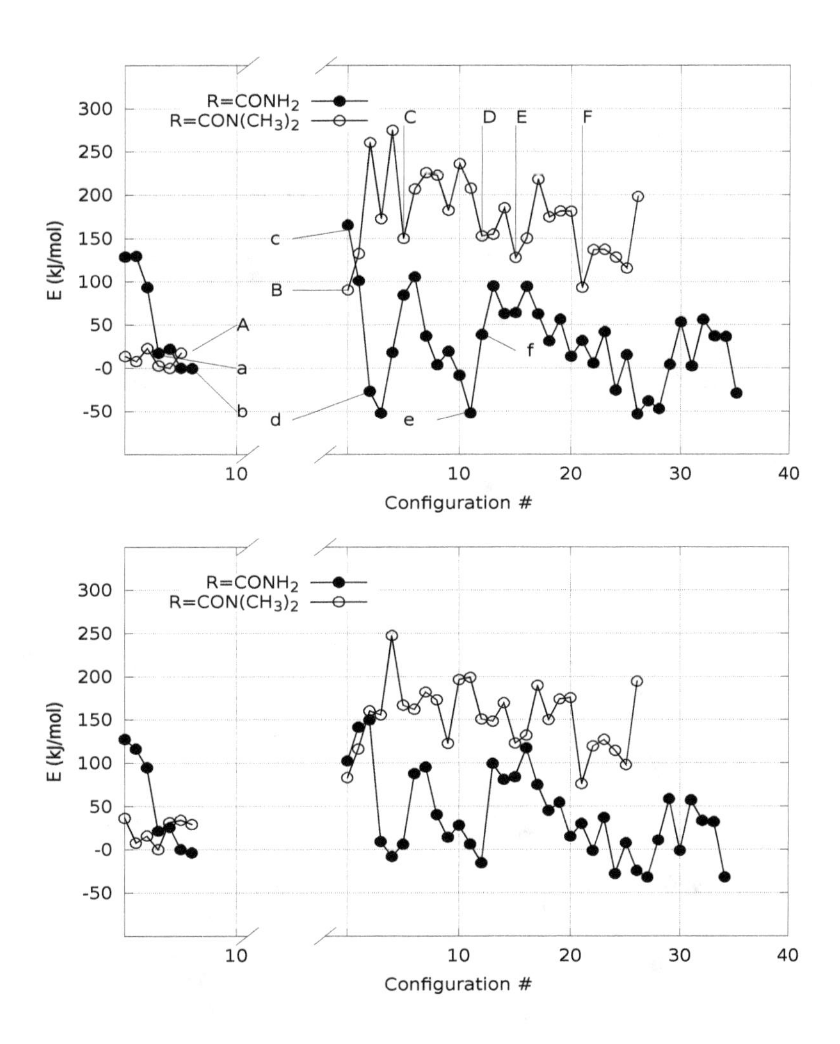


**Figure 5.** Potential energy with R= $H_2NCO$ (filled circles) and $(CH_3)_2NCO$ (circles) along with the extraction of $\alpha$ H from the bond with C$\alpha$ atom. Energy is computed for solute within a distance O(water)-solute atoms of 4 Å (see Methods). Energy reference is the lowest energy obtained in the initial *nitro* form for each compound. The gap in the $x$-axis separates *nitro* (left) from nitronate and *aci* (right) species. Arrows in the left panel indicate the configurations displayed in Figures 6 and 7. The calculation is performed with plane-wave basis-set and PBE exchange functional (left) and localized Gaussian basis-set with hybrid B3LYP exchange functional (right panel). Points indicated with a–f and A–F are are displayed in Figures 6 and 7, respectively. See text for details.

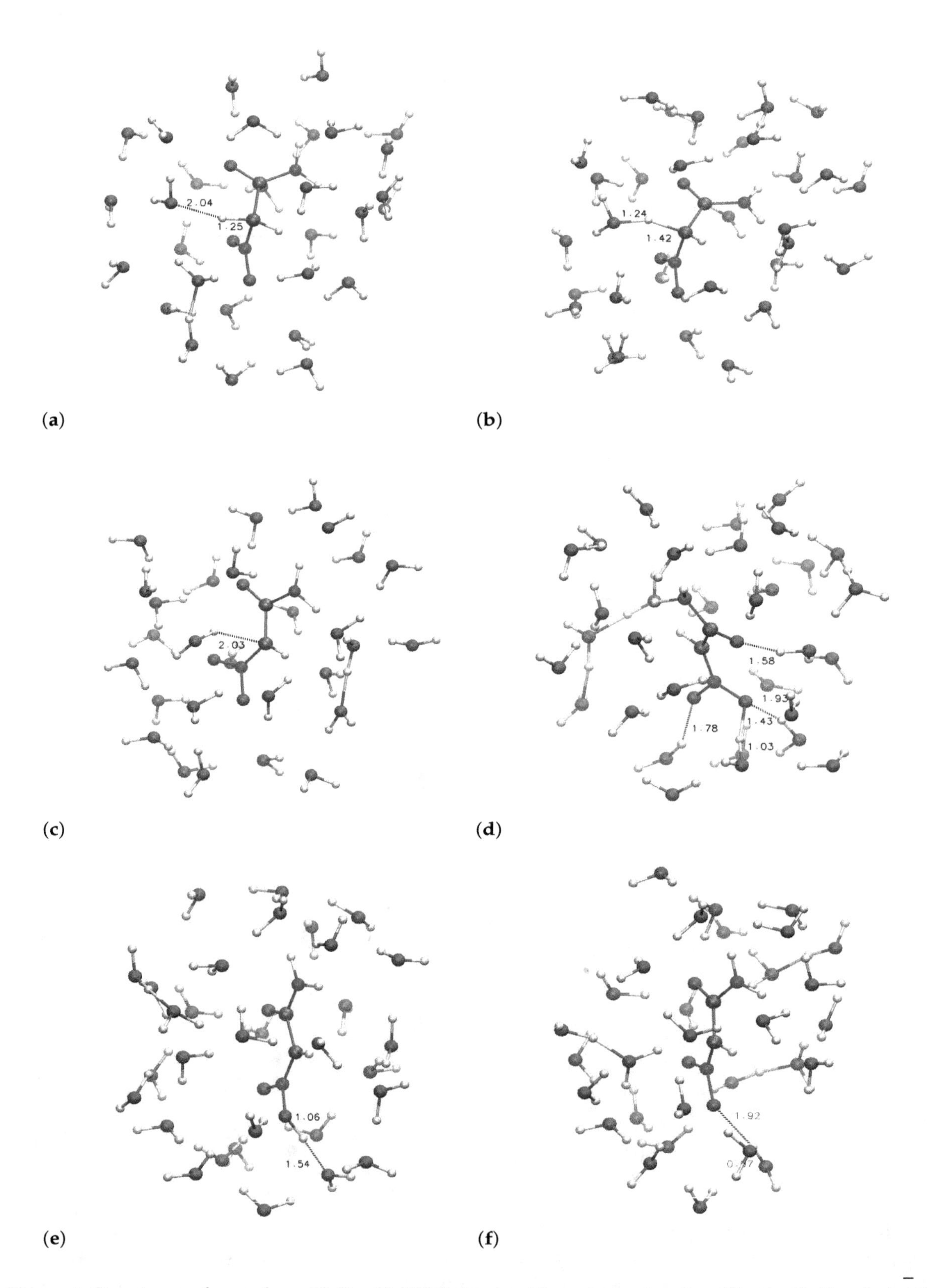


**Figure 6.** Structures of samples with R = $H_2NCO$ along with proton extraction (Figure 5). Panels (**a**–**f**) refer, respectively, to points indicated with a–f in Figure 5. C is gray, N is blue, O is red, H is white. Atomic and bond radii are arbitrary. Some relevant distances are displayed. Explicit bonds are drawn when atoms are closer than 1.6 Å. The VMD [31] program is used for all molecular drawings.

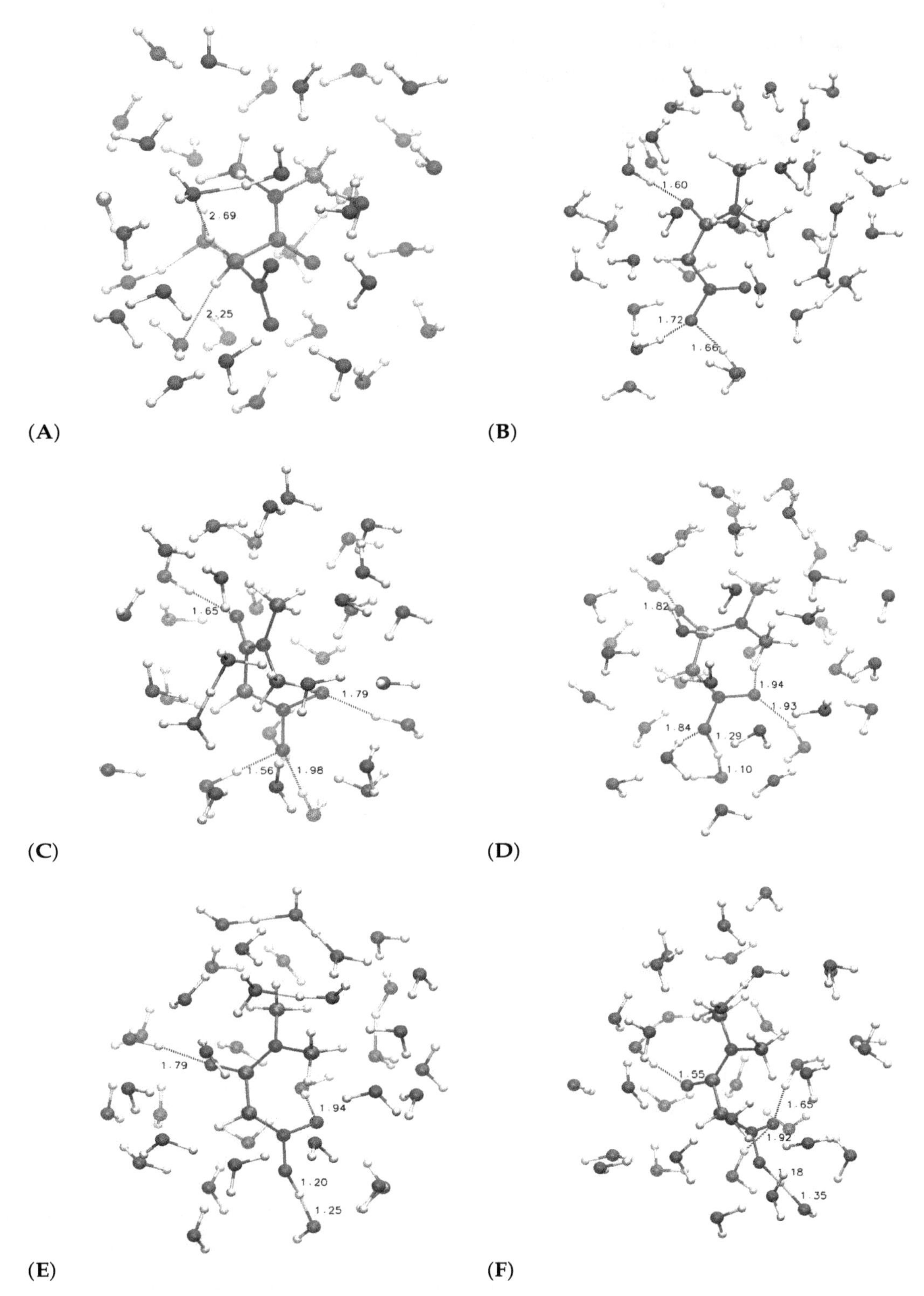


**Figure 7.** The same as Figure 6 when R = $(CH_3)_2NCO$. Panels (**A–F**) refer, respectively, to points indicated with A–F in Figure 5.

## 3. Materials and Methods

### *3.1. Preparation of Nitroacetamides and Determination of Ionization Constants (Apparent $pK_a$)*

Nitroacetamide (**1**) and *N*,*N*-dimethylnitroacetamide (**3**) have been obtained, respectively, by aminolysis from ethyl nitroacetate [32,33] and methyl nitroacetate [34], following previously reported procedures.

Ionization constants of nitro compounds **8**, **1**, and **3** were determined in water by potentiometric titration using a glass electrode (method of partial neutralization). The values of pH were determined with CyberScan510 pH meter produced by Eutech Instruments. Compound **8** was used as reference acid and its $pK_a$ was determined to reproduce published results [12] using our procedure.

The values of $pK_a$ were calculated according to the formula:

$$pK_a = \text{pH} + \text{Log}\,\frac{[\text{HA}]}{[\text{A}]}\,, \tag{3}$$

where [HA] is the concentration of non-dissociated nitroacetamide and [A] is the concentration of its salt.

### 3.1.1. Methyl Nitroacetate (8)

A 0.0100 M (59.5 mg in 50 mL) solution of methyl nitroacetate (**8**) (41.0 mL) was titrated with a 0.100 M solution of sodium hydroxyde. Titration data are reported in Table 5.

**Table 5.** Calculations of the acidity constants of **8** at $T = 23$ °C (Table 1), according to the results of one titration. $pK_a = 5.70$ as the arithmetic mean of all 12 values in the set.

| NaOH (mL) | Log [HA]/[A] | pH | $pK_a$ |
|---|---|---|---|
| 0.3 | 1.10 | 4.59 | 5.69 |
| 0.6 | 0.766 | 4.88 | 5.65 |
| 0.9 | 0.551 | 5.12 | 5.67 |
| 1.2 | 0.383 | 5.3 | 5.68 |
| 1.5 | 0.239 | 5.42 | 5.66 |
| 1.8 | 0.106 | 5.59 | 5.70 |
| 2.1 | −0.0212 | 5.72 | 5.70 |
| 2.4 | −0.150 | 5.84 | 5.69 |
| 2.7 | −0.285 | 5.97 | 5.68 |
| 3.0 | −0.436 | 6.16 | 5.72 |
| 3.3 | −0.615 | 6.33 | 5.71 |
| 3.6 | −0.857 | 6.68 | 5.82 |

### 3.1.2. Nitroacetamide (1)

*Run 1*: a 0.0100 M (103.9 mg in 100 mL) solution of nitroacetamide (**1**) (48.0 mL) was titrated with a 0.100 M solution of sodium hydroxyde at $T = 23$ °C. *Run 2*: the above 0.0100 M solution of nitroacetamide (**1**) (44.0 mL) was titrated with a 0.100 M solution of sodium hydroxyde at $T = 21$ °C. Titration data are reported in Table 6.

**Table 6.** Calculations of the acidity constants of **1** (Table 1), according to the results of two titrations. $pK_a = 5.39$ and 5.38 as the arithmetic mean of all eight values in the set *Run 1* and *Run 2*, respectively.

| NaOH (mL) | *Run 1* | | | *Run 2* | | |
|---|---|---|---|---|---|---|
| | Log [HA]/[A] | pH | $pK_a$ | Log [HA]/[A] | pH | $pK_a$ |
| 0.5 | 0.934 | 4.36 | 5.29 | 0.892 | 4.42 | 5.31 |
| 1.0 | 0.580 | 4.72 | 5.30 | 0.531 | 4.77 | 5.30 |
| 1.5 | 0.342 | 4.96 | 5.30 | 0.286 | 5.02 | 5.31 |
| 2.0 | 0.146 | 5.17 | 5.32 | 0.0792 | 5.26 | 5.34 |
| 2.5 | −0.0362 | 5.37 | 5.33 | −0.119 | 5.48 | 5.36 |
| 3.0 | −0.222 | 5.60 | 5.38 | −0.331 | 5.75 | 5.42 |
| 3.5 | −0.430 | 5.91 | 5.48 | −0.590 | 6.21 | 5.62 |
| 4.0 | −0.699 | 6.42 | 5.72 | - | - | - |

### 3.1.3. *N,N*-Dimethylnitroacetamide (**3**)

*Run 1*: a 0.0100 M (66.1 mg in 50 mL) solution of N,N-Dimethylnitroacetamide (**3**) (43.0 mL) was titrated with a 0.100 M solution of sodium hydroxyde at $T = 25$ °C. *Run 2*: same as *Run 1* at $T = 24$ °C. Titration data are reported in Table 7.

**Table 7.** Calculations of the acidity constants of **3** (Table 1), according to the results of two titrations. $pK_a = 7.25$ and 7.23 as the arithmetic mean of all 8 values in the set *Run 1* and of 13 values in set *Run 2*, respectively.

| NaOH (mL) | *Run 1* | | | NaOH (mL) | *Run 2* | | |
|---|---|---|---|---|---|---|---|
| | Log [HA]/[A] | pH | $pK_a$ | | Log [HA]/[A] | pH | $pK_a$ |
| 0.5 | 0.881 | 6.32 | 7.20 | 0.3 | 1.12 | 6.11 | 7.23 |
| 1.0 | 0.519 | 6.74 | 7.26 | 0.6 | 0.790 | 6.47 | 7.26 |
| 1.5 | 0.271 | 6.99 | 7.26 | 0.9 | 0.577 | 6.69 | 7.27 |
| 2.0 | 0.0607 | 7.2 | 7.26 | 1.2 | 0.412 | 6.86 | 7.27 |
| 2.6 | −0.184 | 7.41 | 7.23 | 1.5 | 0.271 | 6.97 | 7.24 |
| 3.0 | −0.363 | 7.62 | 7.26 | 1.8 | 0.143 | 7.09 | 7.23 |
| 3.5 | −0.641 | 7.91 | 7.27 | 2.1 | 0.0202 | 7.22 | 7.24 |
| 4.0 | −1.12 | 8.41 | 7.29 | 2.4 | −0.101 | 7.34 | 7.24 |
| - | - | - | - | 2.7 | −0.227 | 7.45 | 7.22 |
| - | - | - | - | 3.0 | −0.363 | 7.58 | 7.22 |
| - | - | - | - | 3.3 | −0.519 | 7.73 | 7.21 |
| - | - | - | - | 3.6 | −0.711 | 7.9 | 7.19 |
| - | - | - | - | 3.9 | −0.989 | 8.16 | 7.17 |

### 3.2. *Density Functional Tight-Binding (DFTB) Models*

The final goal of our models is to investigate the mechanism of the reactions described in Figure 1, within a density-functional theory (DFT) approximation of electrons in a system composed by the solute nitro compound and a sample of solvent water molecules. To accomplish this task, we apply in this work implementations of DFT suited for systems of several hundreds of atoms (see the next subsection below). Before applying time-consuming DFT models to systems composed of several hundreds of atoms, we applied, to the same systems, semi-empirical models that are as close as possible to the final DFT models, in order to minimize effects due to the transition from the semi-empirical to the DFT models. Therefore, we performed molecular dynamics (MD) simulations in the Born–Oppenheimer (BO) approximation and at room conditions (BO–MD, hereafter) within a semi-empirical Hamiltonian describing atomic cores and valence electrons. The Hamiltonian of the system was based on the self-consistent charge density-functional tight-binding approximation [30] (DFTB), because geometrical parameters (like distances and angles) of minimal energy conformations are consistent with accurate DFT calculations for a large set of organic molecules, both isolated and in condensed phases. We used the DFTB+ code [35] for these simulations. The valence electrons of each atom are represented as *s* and *p* orbitals.

We built the solute nitro compounds according to standard geometrical parameters and we merged the resulting solute conformation into a snapshot of the sample of water molecules simulated by MD with the TIP3P interaction potential [36]. This sample is a cubic unit cell with the side of 1.8774 nm containing 216 water molecules, in a configuration extracted by the MD simulated trajectory in the NVT (constant number of particles, volume, and temperature statistical ensemble with $T = 300$ K and the fixed density of 0.976 $g/cm^3$. The water molecules with the O atom closer than 1.2 Å from any solute atom were discarded. The number of discarded water molecules was in the range from 6 (R = COPh) to 3 (R = CH3CO). As usual, to minimize finite volume effects, periodic boundary conditions are imposed to the system.

The energy of the system was minimized via the conjugate gradient algorithm for 20 steps, in order to reduce the force initially acting on the atoms. Then, the MD simulation in the NVE (constant number of particles, volume, and energy statistical ensemble was performed for 100 steps, starting with velocities extracted from a Gaussian distribution at $T = 50$ K and with a time-step of 1 fs. During this stage, the temperature never reached values larger than 50 K, indicating the absence of close contacts between atoms. The velocity-verlet algorithm was used to integrate the equations of motion [37]. The MD simulation in the NVT statistical ensemble was then performed continuing the trajectory by using the Nosé–Hoover thermostat [38] at $T = 150$ K for 1000 steps, followed by 5000 steps (5 ps) at $T = 300$ K. A unique effective mass corresponding to a coupling constant of 10 THz was used for the thermostat. The second half of the simulation at $T = 300$ K (2.5 ps) was used for analysis, sampling configurations every 20 fs.

To account for temperature oscillations affecting energy values, the total energy $H$ was corrected for the thermal contribution of $N_{deg}RT(t)/2$, with $N_{deg} = 3N_{at} - 6$, $N_{at}$ the number of atoms in the simulated cell, and $T(t)$ the actual temperature measured in the system at time $t$. Therefore, the corrected total energy $H' = H - N_{deg}RT(t)/2$ was used for computing the average total energy $E = \langle H' \rangle_T$ of each simulated system.

### 3.3. Density Functional Theory (DFT) Models

Car–Parrinello molecular dynamics (CP–MD) simulations [24,25] were performed for models with R = $H_2N$, $(CH_3)HN$ and $(CH_3)_2N$, starting from the final atomic positions and velocities obtained with the corresponding DFTB model at $T$ = 300 K. The parallel version of the Quantum-Espresso package [39], which incorporates Vanderbilt ultra-soft pseudopotentials [40] and the PBE exchange-correlation functional [23], was used in all CP–MD simulations. Electronic wave functions were expanded in plane waves up to an energy cutoff of 25 Ry, while a 250 Ry cutoff was used for the expansion of the augmented charge density in the proximity of the atoms, as required in the ultra-soft pseudopotential scheme. As in the DFTB calculations, periodic boundary conditions were applied in the three directions of space. All calculations were performed under spin-restricted conditions, i.e., with two-electrons effective Kohn–Sham orbitals. The electronic ground state was first calculated using 10 steps of conjugate gradient minimization performed on the dynamic variables representing electrons. Then, the system evolution was followed with CP–MD, using as initial velocities those obtained with the final DFTB configuration. As in the DFTB BO–MD, we used the velocity-Verlet algorithm for integrating the equations of motion, with a time step of 0.121 fs. Empirical dispersive corrections of energy and forces [41,42] were included in the CP–MD simulations to correct for the overestimate of atomic repulsion by the DFT approximation.

We performed CP–MD simulations in the NVT statistical ensemble, with temperature held fixed by a Nosé–Hoover thermostat [38]. The systems were equilibrated for 726 fs (6000 time-steps). After this stage, the simulation was continued in the NPT statistical ensemble. In order to further reduce the overestimate of repulsive forces, possibly producing an unrealistic empty space between solutes and solvent water molecules, a short simulation stage at a pressure slightly larger than room conditions can better settle the water layer around the solute. We performed this step with a short simulation (6000 time-steps) with the same thermostat used in the NVT ensemble and a barostat at $P = 10$ bar [43] with an effective mass of $3/4\ M\pi^2$, $M$ the total mass of the simulation cell. The cell side oscillates around 1.72 nm in all of the simulations. At the end of this NPT stage, we started manipulating the $\alpha$ H atoms with external pulling forces (see below), still in the NPT statistical ensemble. At the end of the H $\alpha$ extraction from C$\alpha$, the sample volume was kept fixed (NVT ensemble).

### 3.4. Pulling $\alpha$H in DFT Models

We performed pulling experiments in order to explore possible pathways for the mechanism of C$\alpha$-H bond breaking, together with the formation of O–H bonds in the water layer around the solute. After extraction of $\alpha$ H from the C–H bond, the $O_N$-H bond formation (with $O_N$ indicating the O atom in the nitro group) was forced in order to achieve the *aci* form of the nitro compound (see Figure 1). To accomplish this pathway, we first applied an external mechanical force on the atoms involved in the C$\alpha$-H bonds. When the extraction of $\alpha$ H into the water sample was achieved, we applied a similar external force to the H atoms in the water sample, potentially binding $O_N$ atoms of the nitro group. With the first pulling experiment, we obtain the nitronate anion from the nitro compound in the *nitro* form; with the second experiment, we obtain the *aci* form of the nitro compound starting from the nitronate anion in contact with a protonated sample of water.

As for the first pulling experiment, we defined a collective variable as the $C\alpha$ coordination number $CN$ according to the equation below [44]:

$$
\begin{aligned}
CN &= \textstyle\sum_{i,j} s_{i,j}, \\
s_{i,j} &= 1 \quad \text{if } r_{i,j} \leq 0, \\
s_{i,j} &= \frac{1-\left(\frac{r_{i,j}}{\sigma}\right)^{6}}{1-\left(\frac{r_{i,j}}{\sigma}\right)^{12}} \quad \text{if } r_{i,j} > 0, \\
r_{i,j} &= |\mathbf{r}_i - \mathbf{r}_j| - d_0 \,,
\end{aligned}
\tag{4}
$$

where the index $j$ runs over the $\alpha$ H (two) atoms of the solute and $i$ indicates $C\alpha$. The actual value of $CN$ can be therefore manipulated by defining an external force as the derivative of an external harmonic potential $U_e = \frac{k}{2}(CN - CN_0(t))^2$. By progressively decreasing $CN_0$ with the simulation time $t$, we allow the smooth release (when the target $CN_0$ value becomes lower than the actual value of $CN$) of one of the two $\alpha$ H atoms. With this procedure and due to the presence of the explicit water molecules in the model, the C–H bond is broken and, when available, a new O–H bond is formed in the water layer around the solute. The parameter $d_0$ in Equation (4) was set to 1.1 Å, while $\sigma$ was 0.2 Å for the first 363 fs and was increased to 0.5 Å for the following 363 fs. The parameter $k$ was set to 1255 kcal/mol in all experiments. The value of $CN_0$ is moved from 2 to 1 at the rate of 1 $CN$ value in 2000 CP–MD steps. The pulling of $\alpha$ H was performed in 726 fs after the equilibration.

As for the second pulling experiment, the index $i$ runs over the $O_N$ atoms of the nitro group, while $j$ runs over all the H atoms not bound to the solute. The $d_0$ parameter was 1.03 Å, the latter the equilibrium distance for $O_N$-$H_N$ measured by DFTB simulations. The $CN_0$ parameter was increased from zero to one, at the same speed and after the same equilibration time of the $\alpha$ H pulling experiment.

### *3.5. Analysis*

We computed a mean-field energy for selected configurations along the pathways sampled with $CN$ manipulations. This calculation is required to correct the energy for contributions due to the periodic boundary conditions and to eliminate thermal fluctuations due to the presence of bulk water around the system of interest (the solute and its hydration layer). From each simulated configuration in the trajectories, we extracted the solute atoms and the water molecules with O atom within 4 Å from any solute atom. For R = $H_2NCO$, the number of extracted water molecules is in the range 25–36. For R = $(CH_3)_2NCO$, the same extraction provides a number of water molecules in the range 28–40 because of the larger size of the solute.

When the $\alpha$ H atom becomes farther than 1.6 Å from any solute atom, then $\alpha$ H is not assumed as part of the solute. Thus, $\alpha$ H becomes part of the solvation layer. When the distance between $\alpha$ H and any water molecule in the layer becomes larger than 1.4 Å, then $\alpha$ H becomes part of the bulk solvent and its energy is estimated according to experimental solvation energy of the proton at room conditions [45]. The number of water molecules in the layer changes from one extracted configuration to another. Therefore, the energy contribution due to the addition or deletion of a number $x$ of water molecules in the layer is computed according to the estimated energy of isolated water molecules and the cohesion energy per water molecule in the layer (see below). These quantities are computed within the same approximations used for the CP–MD trajectories, except as for the following. In the case of energy calculations, the size of the super-cell was chosen as 2.1 nm, i.e., slightly larger than in the CP–MD simulations (1.8774 nm), to achieve better accuracy in total energy. The wavefunction and density energy cut-off were 30 and 300 Ry, respectively. The Makov–Payne correction [46], accounting for the energy contribution of collecting the charge in the given periodic super-cell, was always included in the reported energies. The water environment, i.e., the bulk water around each solvated solute, was modeled as a uniform dielectric medium with relative permittivity of 78.3 (pure water at room conditions). In these calculations, a self-consistent DFT approach based on plane-waves representation of effective monoelectronic (Kohn–Sham) states was used in place of the dynamical extended Lagrangian method used in CP–MD simulations. We used the implicit solvation scheme implemented in the Quantum Espresso code [47]. The energy tolerance for energy change was 0.01 Ry. All the calculations reported in this work are performed with the contribution of plane-waves with $K = 0$ in the super-cell lattice described by the periodic boundary conditions used, i.e., in the $\Gamma$-point approximation of solid state electron density. Since for water layers the energy minimum cannot be achieved, we performed 30 relaxation steps in the conditions reported above. This number of steps has been found as sufficient to relax most of the vibrational stress in the system.

In order to compare the energy of systems composed of different number of atoms, we used the approximation described below. We first calculated the energy of a single water molecule merged in the dielectric, $E_w$, using the same computational conditions of the solvated system (see above). Indicating the different species as in Figure 1, the following equations describe the reaction indicated in the left portion of Figure 1, but including the water layer:

$$\mathrm{N}[n\mathrm{H_2O}] \longrightarrow \mathrm{N}^-[(n-x)\mathrm{H_2O}] + \mathrm{H}^+ + x\mathrm{H_2O} \ . \tag{5}$$

Here, $n$ indicates the water molecules in the solvation layer of the N solute, while $n - x$ the number of water molecules in the layer when the nitronate anion $N^-$ is formed. The energy change due to the addition or deletion of $x$ water molecules ($x$ can be a negative number) to the solvation layer is determined by calculating in one single conformation of R = CONH2 the energy change to increase the size of the layer from 16 to 30 water molecules. The cohesion energy $E_c$ of a single water molecule to the layer is therefore approximated as:

$$E_c = \frac{1}{14}\left[E(n=30) - E(n=16) + 14\,E_w\right] \ , \tag{6}$$

with $E_w$ the energy of the isolated water molecule (see above). We computed this value for a single relaxed configuration of the species R = H2NCO in the *nitro* form. Within this approximation, $E_c = -26.0334$ kJ/mol.

Finally, the value of $-1107$ kJ/mol was used for the solvation energy of $H_3O^+$ [45]. No entropic contribution was taken into account in the calculations reported here, except for the empirical value used for $E(H_3O^+)$.

The final configurations obtained with the DFTB simulations in the *nitro* and *aci* forms were optimized in an implicit model for water with more accurate DFT approximations. These calculations were performed with Gaussian 16 package [48], using the B3LYP [26] hybrid approximation for the exchange functional and with the 6-31++G(d) basis-set. The PCM method [49] for the implicit water solvation was used. All geometries were optimized according to default "optimization" criteria (Gaussian 16 manual [50]).

## 4. Conclusions

The measurement of apparent ionization constant (p$K_a$) for a series of substituted nitromethanes, including the amide moiety (compounds **1–3** in Figure 2), shows the strong effect of hydrophobic and bulky sidechains on the C$\alpha$ acidity. Models including the water molecules interacting with the solute allow for comparing the contribution of intramolecular interactions with that of interactions with structured water layers. This can answer the question about which of these contributions is more efficient to enhance the acidity of the geminal C–H bond.

In this work, we address the inclusion of explicit water molecules in modeling thermodynamic data for this important deprotonation reaction, involving a C–H bond. The reported models, despite the different approximations in the description of ground-state electron density, allow for listing the above observations:

1. The experimental p$K_a$ values approximately follow the statistical weight of the *aci* (more acidic) form as a reactant, with the weight measured by the energy of the *aci* form with respect to the low-energy *nitro* form (Table 2).
2. The extraction of $\alpha$ H from the C$\alpha$-H bond does not occur necessarily when the molecule populates a closed configuration where the ionized form is stabilized by intramolecular hydrogen-bonds. The work required to extract $\alpha$ H is related to the availability of water molecules near the solute, rather than on the internal structure of the solute itself.
3. The hydronium species ($H_3O^+$) formed in the water solvent is different depending on the nitronate species. When the solute is more hydrophilic ($Z=Z^{'}=H$), the presence of a hydronium close to the solute decreases the potential energy. On the other hand, when the solute is more hydrophobic ($Z=Z^{'}=CH_3$), a hydronium species close to the solute does not decrease the energy compared to a hydronium species completely separated by the solute.

These observations indicate that the nature of the R substituent, enhancing the acidity of C$\alpha$-H, should be hydrophilic in order to increase the probability of persistent hydronium species close to the solute. Intramolecular hydrogen bonds and electrostatic interactions enhancing the population of closed configurations do not appear as requirements for the proton release by the C–H bond. The presence of the amide moiety as a substituent in the geminal position to the nitro group greatly enhances the C$\alpha$-H acidity, provided the amide substituent is hydrophilic. The further modification of the amide moiety will be the subject of further studies.

**Author Contributions:** F.M. conceived and designed the experiments; G.L.P. conceived and designed the models; F.M. performed the measurements; G.L.P. applied the molecular models; G.L.P. and F.M. wrote the article together.

**Acknowledgments:** Numerical calculations have been made possible through a CINECA-INFN agreement, providing access to resources on Galileo and Marconi at the CINECA (Consorzio Interuniversitario per il Calcolo Automatico dell'Italia Nord Orientale) computational infrastructure. Luca Guideri is acknowledged for carrying out some preliminary experiments.

## Abbreviations

The following abbreviations are used in this manuscript:

| | |
|---|---|
| DFT | density-functional theory |
| DFTB | density-functional tight-binding |
| MD | molecular dynamics |
| BO | Born–Oppenheimer |
| B3LYP | Becke three-parameter Lee–Yang–Parr exchange-correlation functional |
| PBE | Perdew–Burke–Ernzerhof exchange-correlation functional |
| PCM | polarizable continuum model |

## References

1. Ono, N. *The Nitro Group in Organic Synthesis*; Organic Nitro Chemistry Series; Wiley-VCH: Weinheim, Germany, 2001.
2. Mukaiyama, T.; Hoshino, T. The Reactions of Primary Nitroparaffins with Isocyanates1. *J. Am. Chem. Soc.* **1960**, *82*, 5339–5342. [CrossRef]
3. Basel, Y.; Hassner, A. An Improved Method for Preparation of Nitrile Oxides from Nitroalkanes for In Situ Dipolar Cycloadditions. *Synthesis* **1997**, 309–312. [CrossRef]

4. Nelson, S.D., Jr.; Kasparian, D.J.; Trager, W.F. The Reaction of $\alpha$-Nitro Ketones with the Ketene-Generating Compounds. Synthesis of 3-Acetyl- and 3-Benzoyl-5- Substituted Isoxazoles. *J. Org. Chem.* **1972**, *37*, 2686–2688. [CrossRef]
5. Cecchi, L.; De Sarlo, F.; Machetti, F. Synthesis of 4,5-Dihydroisoxazoles by Condensation of Primary Nitro Compounds with Alkenes by Using a Copper/Base Catalytic System. *Chem. Eur. J.* **2008**, *14*, 7903–7912. [CrossRef] [PubMed]
6. Trogu, E.; Vinattieri, C.; De Sarlo, F.; Machetti, F. Acid-Base Catalysed Condensation Reaction in Water: Isoxazolines and Isoxazoles from Nitroacetates and Dipolarophiles. *Chem. Eur. J.* **2012**, *18*, 2081–2093. [CrossRef] [PubMed]
7. Ballini, R.; Bosica, G.; Fiorini, D.; Palmieri, A.; Petrini, M. Conjugate Additions of Nitroalkanes to Electron-Poor Alkenes: Recent Results. *Chem. Rev.* **2005**, *105*, 933–972. [CrossRef] [PubMed]
8. Ballini, R.; Barboni, L.; Bosica, G.; Fiorini, D.; Palmieri, A. Synthesis of fine chemicals by the conjugate addition of nitroalkanes to electrophilic alkenes. *Pure Appl. Chem.* **2006**, *78*, 1857–1866. [CrossRef]
9. Shvekhgeimer, A.M.G. Aliphatic nitro alcohols. Synthesis, chemical transformations and applications. *Russ. Chem. Rev.* **1998**, *67*, 35–68. [CrossRef]
10. Noland, W.E. The Nef Reaction. *Chem. Rev.* **1955**, *55*, 137–155. [CrossRef]
11. Ballini, R.; Petrini, M. The Nitro to Carbonyl Conversion (Nef Reaction): New Perspectives for a Classical Transformation. *Adv. Synth. Catal.* **2015**, *357*, 2371–2402. [CrossRef]
12. Bernasconi, C.F.; Pérez-Lorenzo, M.; Brown, S.D. Kinetics of the Deprotonation of Methylnitroacetate by Amines: Unusually High Intrinsic Rate Constants for a Nitroalkane. *J. Org. Chem.* **2007**, *72*, 4416–4423. [CrossRef] [PubMed]
13. Bernasconi, C.F.; Ali, M.; Gunter, J.C. Kinetic and Thermodynamic Acidities of Substituted 1-Benzyl-1-methoxy-2-nitroethylenes. Strong Reduction of the Transition State Imbalance Compared to Other Nitroalkanes. *J. Am. Chem. Soc.* **2003**, *125*, 151–157. [CrossRef] [PubMed]
14. Ando, K.; Shimazu, Y.; Seki, N.; Yamataka, H. Kinetic Study of Proton-Transfer Reactions of Phenylnitromethanes. Implication for the Origin of Nitroalkane Anomaly. *J. Org. Chem.* **2011**, *76*, 3937–3945. [CrossRef] [PubMed]
15. Trogu, E.; Cecchi, L.; De Sarlo, F.; Guideri, L.; Ponticelli, F.; Machetti, F. Base- and Copper-Catalysed Condensation of Primary Activated Nitro Compounds with Enolisable Compounds. *Eur. J. Org. Chem.* **2009**, 5971–5978. [CrossRef]
16. Guideri, L.; De Sarlo, F.; Machetti, F. Conjugate Addition versus Cycloaddition/Condensation of Nitro Compounds in Water: Selectivity, Acid-Base Catalysis, and Induction Period. *Chem. Eur. J.* **2013**, *19*, 665–677. [CrossRef] [PubMed]
17. Horst, G.A.; Mortimer, J.K. Fluoronitroaliphatics. I. The Effect of $\alpha$ Fluorine on the Acidities of Substituted Nitrometanes. *J. Am. Chem. Soc.* **1966**, *88*, 4761–4763. [CrossRef]
18. Bernasconi, C.F.; Panda, M.; Stronach, M.W. Kinetics of Reversible Carbon Deprotonation of 2-Nitroethanol and 2-Nitro-1,3-propanediol by Hydroxide Ion, Water, Amines, and Carboxylate Ions. A Normal Brönsted $\alpha$ Despite an Imbalanced Transition State. *J. Am. Chem. Soc.* **1995**, *117*, 9206–9212. [CrossRef]
19. Turnbull, D.; Maron, S.H. The Ionization Constants of Aci and Nitro Forms of Some Nitroparaffins. *J. Am. Chem. Soc.* **1943**, *65*, 212–218. [CrossRef]
20. Pearson, R.G.; Dillon, R.L. Rates of Ionization of Pseudo Acids. IV. Relation between Rates and Equilibria. *J. Am. Chem. Soc.* **1953**, *75*, 2439–2443. [CrossRef]
21. Pearson, R.G.; Anderson, D.H.; Alt, L.L. Mechanism of the Hydrolytic Cleavage of Carbon-Carbon Bonds. III. Hydrolysis of $\alpha$-Nitro and $\alpha$-Sulfonyl Ketones. *J. Am. Chem. Soc.* **1955**, *77*, 527–529. [CrossRef]
22. *ACD/Labs Percepta Platform: Insight-Driven Decision Support for Teams that Design and Synthesize New Chemical Entities*; ACD/Structure Elucidator, version 15.01; Advanced Chemistry Development, Inc.: Toronto, ON, Canada, 2015.
23. Perdew, J.P.; Burke, K.; Ernzerhof, M. Generalized Gradient Approximation Made Simple. *Phys. Rev. Lett.* **1996**, *77*, 3865–3868. [CrossRef] [PubMed]
24. Car, R.; Parrinello, M. Unified Approach for Molecular Dynamics and Density-Functional Theory. *Phys. Rev. Lett.* **1985**, *55*, 2471–2474. [CrossRef] [PubMed]
25. Marx, D.; Hutter, J. *Ab Initio Molecular Dynamics: Basic Theory and Advanced Methods*; Cambridge University Press: Cambridge, UK, 2009.

26. Becke, A.D. Density-functional Thermochemistry. III. The Role of Exact Exchange. *J. Chem. Phys.* **1993**, *98*, 5648–5652. [CrossRef]
27. Dreyer, J.; Brancato, G.; Ippoliti, E.; Genna, V.; De Vivo, M.; Carloni, P.; Rothlisberger, U. First Principles Methods in Biology: From Continuum Models to Hybrid Ab initio Quantum Mechanics/Molecular Mechanics. In *Simulating Enzyme Reactivity: Computational Methods in Enzyme Catalysis*; The Royal Society of Chemistry: London, UK, 2017; pp. 294–339. [CrossRef]
28. La Penna, G.; Andreussi, O. When water plays an active role in electronic structure. Insights from first-principles molecular dynamics simulations of biological systems. In *Computational Methods to Study the Structure and Dynamics of Biomolecules and Biomolecular Processes*, 2nd ed.; Liwo, A.J., Ed.; Springer series in bio- and neurosystems; Springer-Verlag: Berlin/Heidelberg, Germany, 2019; Volume 1. [CrossRef]
29. Liao, C.; Nicklaus, M.C. Comparison of Nine Programs Predicting pKa Values of Pharmaceutical Substances. *J. Chem. Inf. Model.* **2009**, *49*, 2801–2812. [CrossRef] [PubMed]
30. Elstner, M.; Porezag, D.; Jungnickel, G.; Elsner, J.; Haugk, M.; Frauenheim, T.; Suhai, S.; Seifert, G. Self-consistent-charge Density-functional Tight-binding Method for Simulations of Complex Materials Properties. *Phys. Rev. B* **1998**, *58*, 7260–7268. [CrossRef]
31. Humphrey, W.; Dalke, A.; Schulten, K. VMD visual molecular dynamics. *J. Mol. Graph.* **1996**, *14*, 33–38. [CrossRef]
32. Cecchi, L.; De Sarlo, F.; Machetti, F. 1,4-Diazabicyclo[2,2,2]octane (DABCO) as an Efficient Reagent for the Synthesis of Isoxazole Derivatives from Primary Nitro Compounds and Dipolarophiles: The Role of the Base. *Eur. J. Org. Chem.* **2006**, 4852–4860. [CrossRef]
33. Biagiotti, G.; Cicchi, S.; De Sarlo, F.; Machetti, F. Reactivity of [60]Fullerene with Primary Nitro Compounds: Addition or Catalysed Condensation to Isoxazolo[60]fullerenes. *Eur. J. Org. Chem.* **2014**, 7906–7915. [CrossRef]
34. Ciommer, B.; Frenking, G.; Schwarz, H. Massenspektrometrische Untersuchung von Stickstoffverbindungen, XXXI, Experimentelle und theoretische Untersuchungen zur dissoziativen Ionisierung von $\alpha$-nitro- und $\alpha$-halogen substituierten Acetamiden. Pseudo-einstufige Zerfallsprozesse von Radikalkationen in der Gasphase. *Chem. Ber.* **1981**, *114*, 1503–1519. [CrossRef]
35. Aradi, B.; Hourahine, B.; Frauenheim, T. DFTB+, a Sparse Matrix-Based Implementation of the DFTB Method. *J. Phys. Chem. A* **2007**, *111*, 5678–5684. [CrossRef]
36. Jorgensen, W.L.; Chandrasekhar, J.; Madura, J.D.; Impey, R.W.; Klein, M.J. Comparison of simple potential functions for simulating liquid water. *J. Chem. Phys.* **1983**, *79*, 926–935. [CrossRef]
37. Frenkel, D.; Smit, B. *Understanding Molecular Simulation*; Academic Press: San Diego, CA, USA, 1996.
38. Nosé, S. A Molecular Dynamics Method for Simulations in the Canonical Ensemble. *Mol. Phys.* **1984**, *52*, 255–268. [CrossRef]
39. Giannozzi, P.; Baroni, S.; Bonini, N.; Calandra, M.; Car, R.; Cavazzoni, C.; Ceresoli, D.; Chiarotti, G.L.; Cococcioni, M.; Dabo, I.; et al. QUANTUM ESPRESSO: A Modular and Open-Source Software Project for Quantum Simulations of Materials. *J. Phys. Condens. Matter* **2009**, *21*, 395502. [CrossRef] [PubMed]
40. Vanderbilt, D. Soft Self-Consistent Pseudopotentials in a Generalized Eigenvalue Formalism. *Phys. Rev. B* **1990**, *41*, 7892–7895. [CrossRef]
41. Grimme, S. Semiempirical GGA-Type Density Functional Constructed with a Long-Range Dispersion Correction. *J. Comput. Chem.* **2006**, *27*, 1787–1799. [CrossRef] [PubMed]
42. Barone, V.; Casarin, M.; Forrer, D.; Pavone, M.; Sambi, M.; Vittadini, A. Role and Effective Treatment of Dispersive Forces in Materials: Polyethylene and Graphite Crystals As Test Cases. *J. Comput. Chem.* **2009**, *30*, 934–939. [CrossRef] [PubMed]
43. Parrinello, M.; Rahman, A. Polymorphic Transitions in Single Crystals: A New Molecular Dynamics Method. *J. Appl. Phys.* **1981**, *52*, 7182–7190. [CrossRef]
44. Barducci, A.; Chelli, R.; Procacci, P.; Schettino, V.; Gervasio, F.L.; Parrinello, M. Metadynamics simulation of prion protein: $\beta$-structure stability and the early stages of misfolding. *J. Am. Chem. Soc.* **2006**, *128*, 2705–2710. [CrossRef]
45. Raffa, D.F.; Rickard, G.A.; Rauk, A. Ab Initio Modelling of the Structure and Redox Behaviour of Copper(I) Bound to a His-His Model Peptide: Relevance to the $\beta$-Amyloid Peptide of Alzheimer's Disease. *J. Biol. Inorg. Chem.* **2007**, *12*, 147–164. [CrossRef]

46. Makov, G.; Payne, M.C. Periodic Boundary Conditions in *Ab Initio* Calculations. *Phys. Rev. B* **1995**, *51*, 4014. [CrossRef]
47. Andreussi, O.; Dabo, I.; Marzari, N. Revised Self-Consistent Continuum Solvation in Electronic-Structure Calculations. *J. Chem. Phys.* **2012**, *136*, 064102. [CrossRef] [PubMed]
48. Frisch, M.J.; Trucks, G.W.; Schlegel, H.B.; Scuseria, G.E.; Robb, M.A.; Cheeseman, J.R.; Scalmani, G.; Barone, V.; Petersson, G.A.; Nakatsuji, H.; et al. *Gaussian 16, Revision A.03*; Gaussian Inc.: Wallingford, CT, USA, 2016.
49. Tomasi, J.; Mennucci, B.; Cammi, R. Quantum mechanical continuum solvation models. *Chem. Rev.* **2005**, *105*, 2999–3093. [CrossRef] [PubMed]
50. *Gaussian 16 On-Line Manual*; Gaussian, Inc.: Wallingford, CT, USA, 2016. Available online: http://www.Gaussian.com/keywords (accessed on 25 October 2018).

# 9

# Heteroatom Substitution at Amide Nitrogen—Resonance Reduction and HERON Reactions of Anomeric Amides

**Stephen A. Glover * and Adam A. Rosser**

Department of Chemistry, School of Science and Technology, University of New England, Armidale, NSW 2351, Australia; arosser3@une.edu.au
* Correspondence: sglover@une.edu.au

**Abstract:** This review describes how resonance in amides is greatly affected upon substitution at nitrogen by two electronegative atoms. Nitrogen becomes strongly pyramidal and resonance stabilisation, evaluated computationally, can be reduced to as little as 50% that of *N,N*-dimethylacetamide. However, this occurs without significant twisting about the amide bond, which is borne out both experimentally and theoretically. In certain configurations, reduced resonance and pronounced anomeric effects between heteroatom substituents are instrumental in driving the HERON (Heteroatom Rearrangement On Nitrogen) reaction, in which the more electronegative atom migrates from nitrogen to the carbonyl carbon in concert with heterolysis of the amide bond, to generate acyl derivatives and heteroatom-substituted nitrenes. In other cases the anomeric effect facilitates $S_N1$ and $S_N2$ reactivity at the amide nitrogen.

**Keywords:** amide resonance; anomeric effect; HERON reaction; pyramidal amides; physical organic chemistry; reaction mechanism

---

## 1. Introduction

Amides are prevalent in a range of molecules such as peptides, proteins, lactams, and many synthetic polymers [1]. Generically, they are composed of both a carbonyl and an amino functional group, joined by a single bond between the carbon and nitrogen. The contemporary understanding of the resonance interaction between the nitrogen and the carbonyl in amides is that of an interaction between the lowest unoccupied molecular orbital (LUMO) of the carbonyl, $\pi^*_{C=O}$, and the highest occupied molecular orbital (HOMO) of the amide nitrogen ($N2p_z$) (Figure 1). This molecular orbital model highlights the small contribution of the carbonyl oxygen to the LUMO, which indicates that limited charge transfer to oxygen occurs, in line with the resonance model presented in Figure 2, which signifies that charge at oxygen is similar to that in polarized ketones or aldehydes and nitrogen lone pair density is transferred to electron deficient carbon rather than to oxygen. The major factor in the geometry of amides, and the restricted rotation about the amide bond, is the strong $\pi$-overlap between the nitrogen lone pair and the $C2p_z$ component of the LUMO, which dominates the $\pi^*_{C=O}$ orbital, on account of the polarisation in the $\pi_{C=O}$ [2,3].

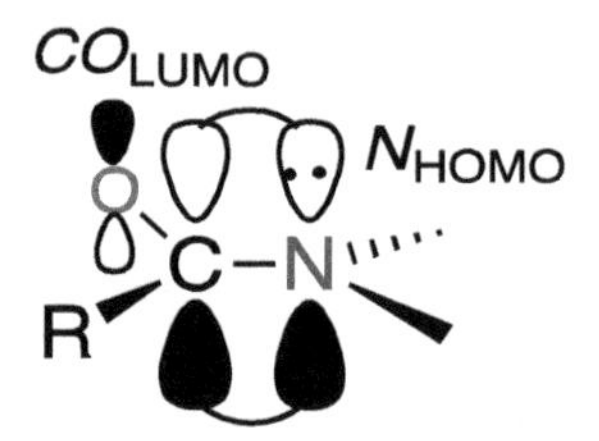


**Figure 1.** The interaction between the nitrogen highest occupied molecular orbital (HOMO) and the carbonyl lowest unoccupied molecular orbital (LUMO) in amides.

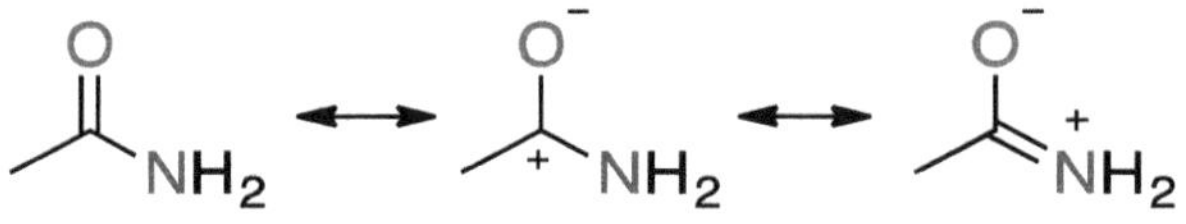

**Figure 2.** Resonance hybrid contributions in amides, showing that the majority of charge transfer occurs between nitrogen and carbon.

**1**

**2**

a: X=Cl, Y=OR
b: X=OAc, Y=OR
c: X=Y=OR
d: X=OR, Y=$NR_2$
e: X=OR, Y=SR
f: X=Cl, Y=$NR_2$
g: X=OR, Y=$O^-$
h: X=OR, Y=N
i: X=$NR_2$, Y=SR
j: X=Cl, Y=SR

**3**

a: R=H
b: R=Me

**4**

a: R=H
b: R=Me

Amide resonance can be diminished by limiting the overlap between the nitrogen lone pair, $n_N$, and $\pi^*_{C=O}$ orbitals. In most instances, this can occur by twisting the amino group about the N–C(O) bond and/or pyramidalising the nitrogen, which amounts to introducing "s" character into the $N2p_z$ orbital.

Computational modelling of N–C(O) rotation and nitrogen pyramidalisation in *N,N*-dimethylacetamide **1**, at the B3LYP/6-31G(d) level, illustrates the energetic changes (Figure 3) [4]. Distortion of the amide linkage can be quantified by Winkler–Dunitz parameters $\chi$ and $\tau$, where $\chi = 60°$ for a fully pyramidal nitrogen, and $\tau = 90°$ for a completely twisted amide where the lone pair orbital on nitrogen is orthogonal to the $C2p_z$ orbital [5,6]. Deformation from the non-twisted, $sp^2$ planar ground state (Figure 3a) through rehybridisation of nitrogen to $sp^3$ (Figure 3b) is accompanied by an increase in energy (~27 kJ $mol^{-1}$), but the majority of amide resonance remains intact. A much larger increase in energy (~100 kJ $mol^{-1}$) results from twisting the N–C(O) bond through 90° whilst maintaining $sp^2$ planarity at nitrogen (Figure 3d), and as the nitrogen is allowed to relax to $sp^3$ hybridisation (Figure 3c), a fully twisted amide devoid of resonance is obtained. The loss of ~31 kJ $mol^{-1}$ in this final step is indicative of the concomitant twisting and pyramidalisation observed in a variety of twisted amides.

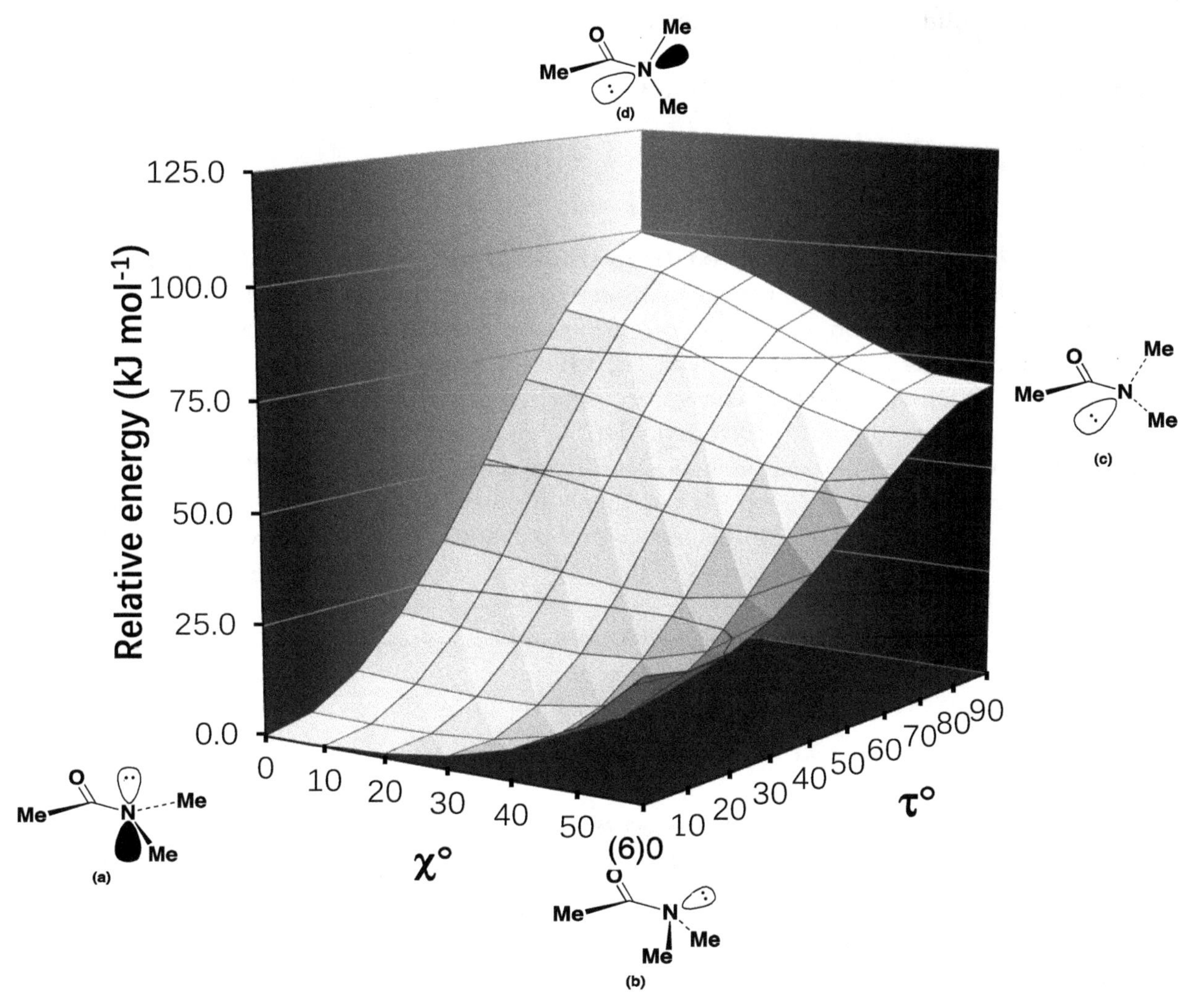


**Figure 3.** A B3LYP/6-31G(d) generated energy surface for the deformation of the amide moiety in *N*,*N*-dimethylacetamide 1 [4]. $\chi$ and $\tau$ are Winkler–Dunitz parameters for pyramidalisation and twist, respectively [5,6]. (**a**) The ground state planar structure for *N*,*N*-dimethylacetamide; (**b**) untwisted system with $sp^3$ hybridised nitrogen; (**c**) 90° rotation about N–C(O), whilst maintaining $sp^3$ hybridisation at nitrogen; and (**d**) the fully twisted moiety, 90° N–C(O) rotation and $sp^2$ hybridised nitrogen.

Typically, amides exhibit restricted rotation about the N–C(O) bond, in the order of 67–84 kJ $mol^{-1}$, along a sigmoid pathway with little change in energy upon moderate pyramidalisation ($\chi$ = 0–40°) and minor twisting ($\tau$ = 0–20°). Clearly, rotation without pyramidalisation is energetically unfavourable and many examples of twisted amides are testament to this. The shift to $sp^3$ hybridisation at the amidic nitrogen is clearly demonstrated when twisting of the N–C(O) bond is geometrically enforced by tricyclic and bicyclic bridged lactams [7–13]. Kirby's "most twisted amide" 1-aza-2-adamantanone, synthesised in 1998 [10,12], and Tani and Stoltz's 2-quinuclidone, synthesised in 2006 [14], exemplify the fully twisted amide, geometrically and chemically. Intramolecular steric hindrance is another source of non-planar twisted amides [15,16], as exemplified by the thioglycolurils [17,18] and other systems. Ring strain in nontwisted amides, such as 1-acylaziridines [19,20] and *N*-acyl-7-azabicyclo[2.2.1]heptanes [21–24] can result in pyramidality at the amide nitrogen, despite retaining a noticeable $n_N$–$\pi^*_{C=O}$ interaction.

The structural changes accompanying the loss of amide resonance include the lengthening of the N–C(O) bond and minor shortening of the (N)C=O bond. Comparing the fully twisted 1-aza-2-adamantanone to an analogous unstrained tertiary δ-lactam, the N–C(O) bond shortens from 1.475 Å to 1.352 Å, and the (N)C=O bond lengthens from 1.196 Å to 1.233 Å [12,25]. Spectroscopically and chemically, the amide carbonyl trends towards ketonic behaviour at large twist angles. In tricyclic 1-aza-2-adamantanone, the carbonyl carbon $^{13}$C NMR resonance is at 200.0 ppm and the IR carbonyl vibrational frequency (1732 $cm^{-1}$) is significantly higher than regular amides (1680 $cm^{-1}$) [10,12].

## 2. Properties of Anomeric Amides

### 2.1. Structural Properties

Another way in which amides may be dispossessed of their planarity and resonance is through bisheteroatom-substitution by electronegative heteroatoms at the amide nitrogen in **2**. We named this class 'anomeric amides' on account of the pronounced anomeric effects that can and do occur between the heteroatoms [26–35]. However, the physical, theoretical, and chemical properties of various congeners differ from those of conventional primary, secondary, and tertiary alkylamides.

Electronegative atoms demand an electron density redistribution which is facilitated by a shift in the hybridisation at nitrogen towards $sp^3$, in accordance with Bent's rule [36,37]. Reduced $n_N$–$\pi^*_{C=O}$ overlap due to pyramidalisation (Figure 4) together with the increased '2s' character of the lone pair on nitrogen results in structural, electronic, spectroscopic, and chemical differences in comparison to traditional amides. Their unique properties of have been reviewed in recent years [31,35].

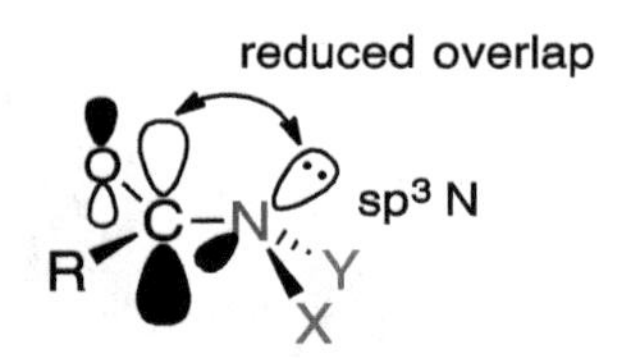


**Figure 4.** Reduced $n_N$–$\pi^*_{C=O}$ overlap due to pyramidalisation.

Much of our interest into, and indeed the discovery of, anomeric amide chemistry emanates from our investigations of the biological activity, structure, and reactivity of *N*-acyloxy-*N*-alkoxyamides (NAA's) **2b** [31,38–49] a class of anomeric amides. Readily synthesised by treatment of *N*-alkoxy-*N*-chloroamides **2a**, themselves a form of anomeric amide, with silver or sodium carboxylate salts in anhydrous solvents [45,46,48,49], NAA's are direct-acting mutagens which react with nucleophilic centres in DNA [31,39–46,49,50]. Additionally, they react with a variety of nucleophiles to produce other congeners, including, reactive anomeric amides in the form of *N*-alkoxy-*N*-aminoamides **2d** [44,47,51] and *N*-alkoxy-*N*-thioalkylamides **2e** [52]. We have also encountered several anomeric reactive intermediates through reactions of *N*-acyloxy-*N*-alkoxyamides: reaction of **2b** with base generates *N*-alkoxy-*N*-hydroxamate anions **2g** [45] and reaction with azide ultimately generates 1-acyl-1-alkoxydiazenes, which are aminonitrenes **2h** [53,54]. We have generated *N,N*-dialkoxyamides **2c** in related studies by solvolysis of *N*-alkoxy-*N*-chloroamides **2a** in aqueous alcohols and through the reaction of hydroxamic esters with hypervalent iodine reagents in appropriate alcohols [30,55,56]. Other anomeric amides of theoretical interest to us are *N*-amino-*N*-chloroamides **2f** [34], as well as *N*-amino-*N*-thioalkylamides **2i** and *N*-chloro-*N*-thioalkyamides **2j**.

The impact of heteroatom substitution at amide nitrogen is clearly demonstrated by comparing *N,N*-dimethylacetamide **1** to *N*-methoxy-*N*-methylacetamide **3b** and *N,N*-dimethoxyacetamide **4b**. In 1996, we reported on the B3LYP/6-31G(d) theoretical properties of the corresponding formamides **3a** and **4a** [27]. The substitution of hydrogen by methoxyl in **3a** led to an increase in N–C(O) bond length from 1.362 Å to 1.380 Å and a second substitution at nitrogen in **4a** resulted in a similar increase to 1.396 Å; the carbonyl bond contracted marginally by 0.006 Å. Nitrogen in *N*-methoxy-*N*-methylformamide and *N,N*-dimethoxyformamide becomes distinctly pyramidal with average angles at nitrogen of ~114°. With similar degrees of pyramidality ($\chi_N$), the increased N–C(O) bond length in *N,N*-dimethoxyformamide could not solely be attributed to deformation at nitrogen. Energetic lowering of the lone pair electrons is also responsible for reduced overlap with the adjacent $C2p_z$ orbital. This is dramatically exemplified by the rotational barriers for *N,N*-dimethylformamide,

*N*-methoxy-*N*-methylformamide, and *N*,*N*-dimethoxyformamide, which were computed to be of the order of 75, 67, and 29 kJ mol$^{-1}$, respectively [27].

Deformation energy surfaces for *N*-methoxy-*N*-methylacetamide **3b** and *N*,*N*-dimethoxyacetamide **4b**, for comparison with that of *N*,*N*-dimethylacetamide **1**, are depicted in Figure 5. In contrast to *N*,*N*-dimethylacetamide (Figure 3), the lowest energy forms clearly deviate from planarity at nitrogen with $\chi_0$ in the region of 40° and 50°, respectively. In both structures, the highest point corresponding to $\chi = 0°$, $\tau = 90°$ is metastable, and the planar fully twisted, and therefore nonconjugated forms, relax to fully pyramidal conformations ($\tau = 90°$, $\chi = 60°$), the energy of which reflects the B3LYP/6-31G(d) barriers to amide isomerisation in each case, which are approximately 67 and 44 kJ mol$^{-1}$, respectively. While the energy lowering for *N*-methoxyacetamide is modest, the attachment of two electronegative oxygens to the amide nitrogen radically lowers the isomerisation barrier by some 29 to 33 kJ mol$^{-1}$. Inversion barriers at nitrogen are low on account of the gain in resonance stabilisation in the planar form, though the barrier is higher for *N*,*N*-dimethoxyacetamide where the resonance capability would be less and a six π-electron repulsive effect would operate; planarisation in the hydroxamic ester is less costly than in *N*,*N*-dimethoxyacetamide (~6.3 vs. 14.6 kJ mol$^{-1}$, respectively).

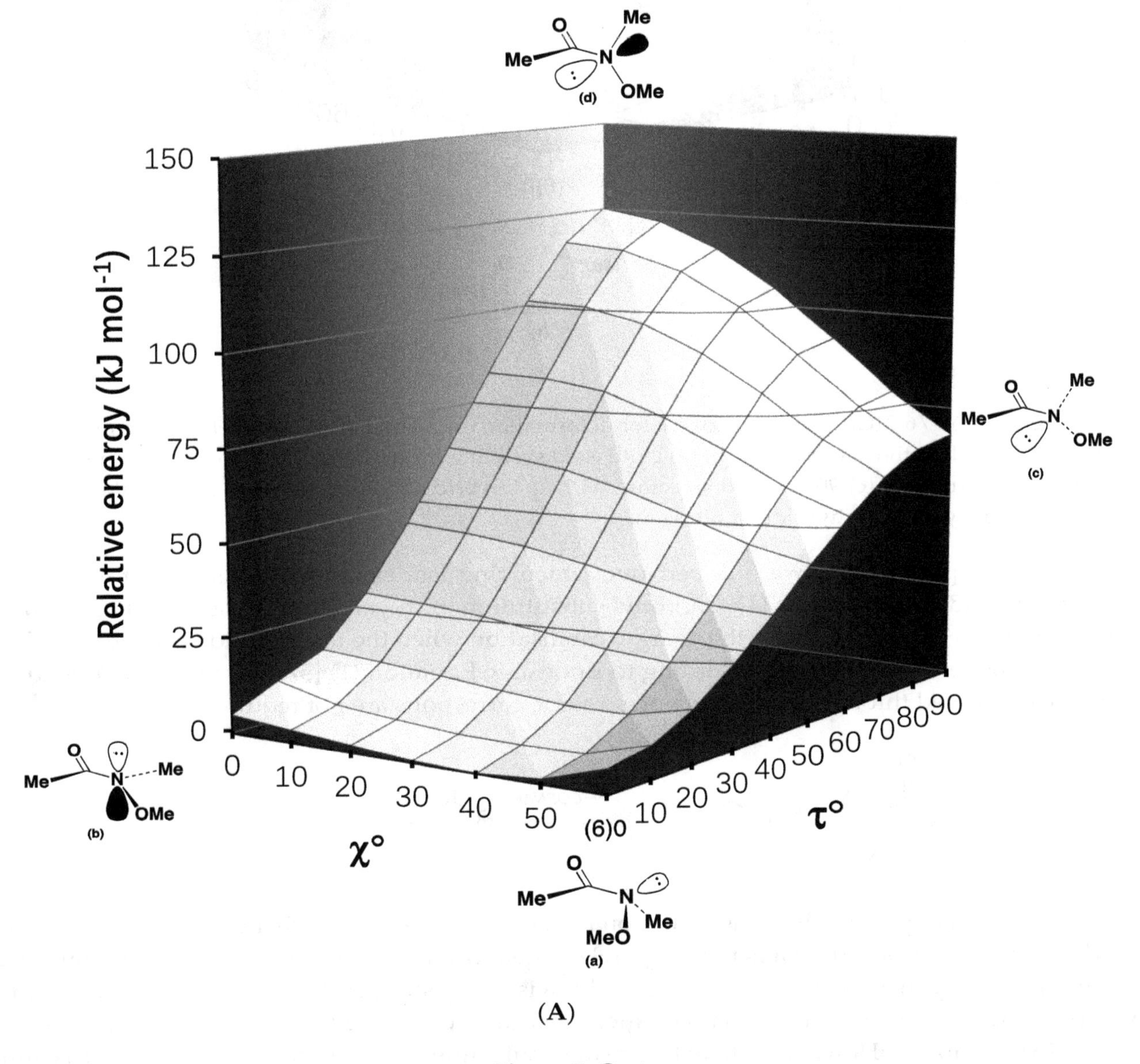


(A)

**Figure 5.** *Cont.*

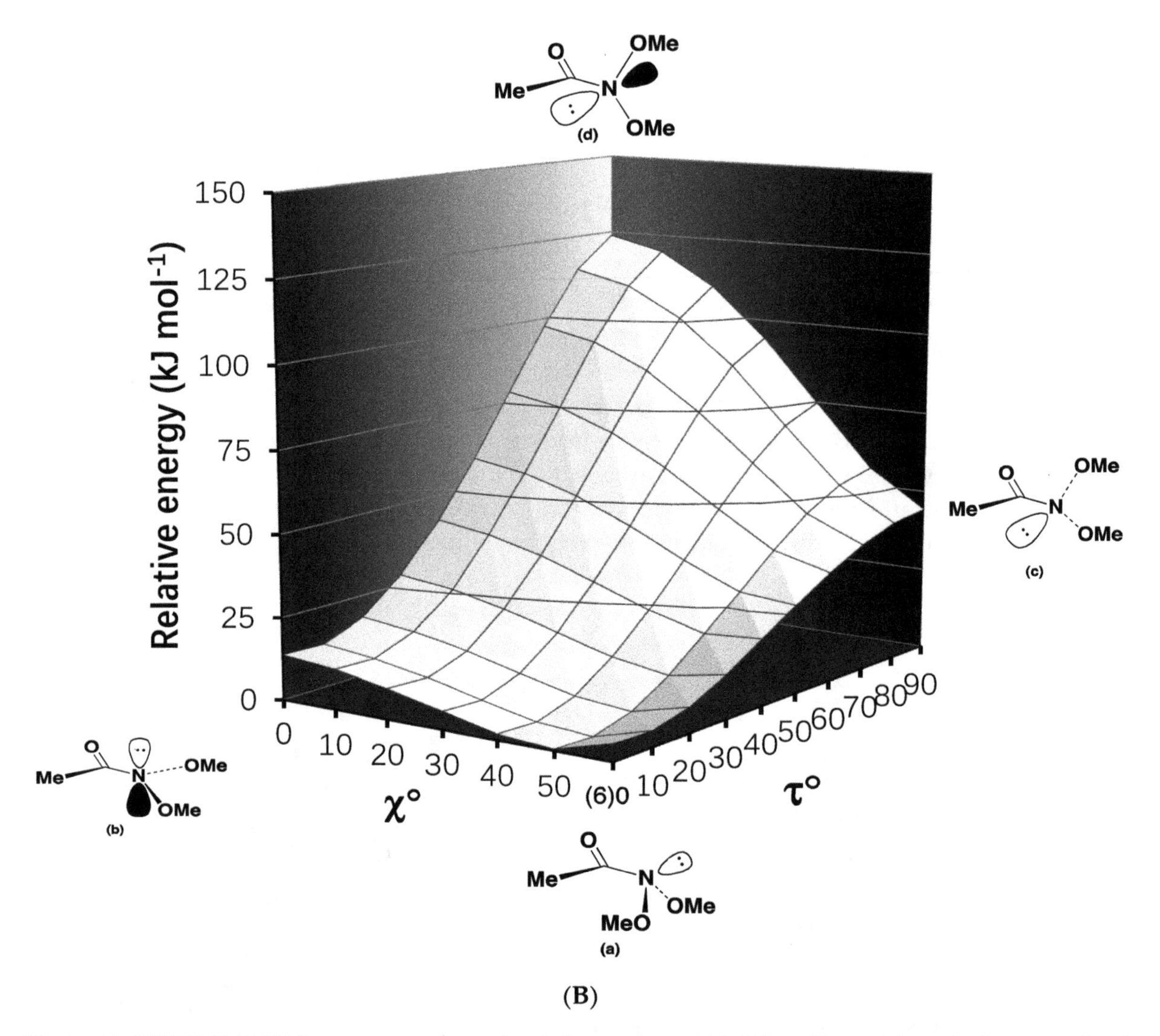


**(B)**

**Figure 5.** B3LYP/6-31G(d) energy surfaces for deformation of (**A**) *N*-methoxy-*N*-methylacetamide **3b** and (**B**) *N,N*-dimethoxyacetamide **4b**. (**a**) Untwisted system with $sp^3$ hybridised nitrogen; (**b**) planar untwisted structure; (**c**) 90° rotated structure with $sp^3$ hybridisation at nitrogen; and (**d**) the fully twisted moiety with $sp^2$ hybridised nitrogen.

In a recent publication, we outlined two concurring isodesmic methods for estimating the amidicity of amides and lactams [4]. Carbonyl substitution, nitrogen atom replacement (COSNAR), developed by Greenberg evaluates the energy stabilisation when the target amide is generated from corresponding ketone and amine according to isodesmic Equation (1) [57–59]. Steric or substituent effects are conserved throughout the reaction so steric corrections are not required.

$$\mathrm{RC(O)CH(X)Y} + \mathrm{RCH_2N(X)Y} \xrightarrow[\mathrm{RE} = \Delta E_{COSNAR}]{\Delta E_{COSNAR}} \mathrm{RC(O)N(X)Y} + \mathrm{RCH_2CH(X)Y} \quad (1)$$

In the second approach, the transamidation method (TA), the energy change is determined when *N,N*-dimethylacetamide **1** transfers the carbonyl oxygen to a target amine according to Equation (2). The limiting energy increase for formation of fully twisted, unstrained 1-aza-2-adamantanone by the corresponding reaction with 1-azaadamantane, constitutes complete loss of resonance stabilisation ($\Delta E_{TA}$ = 76.0 kJ $mol^{-1}$). However, where heteroatom substituents are present at nitrogen, $\Delta E_{TA}$ must be corrected for any additional inductive destabilisation of the carbonyl ($\Delta E_{\text{ind}}$) in the absence of resonance, which we obtain isodemically from reactions such as Equation (3) [33].

(2)

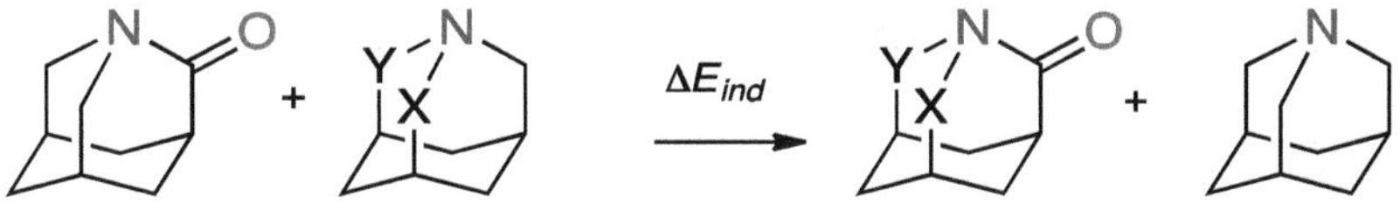

(3)

By the TA method, the residual resonance, $\text{RE}_{\text{TA}}$, is given by Equation (4).

$$\text{RE}_{\text{TA}} = -76.0 \text{ kJ mol}^{-1} + (\Delta E_{\text{TA}} - \Delta E_{\text{ind}}) \quad (4)$$

In both the TA and COSNAR approaches, zero point energies largely cancel and meaningful results are obtained without the need for frequency calculations [58,60]. RE by both methods, the negative of the traditional representation of resonance stabilisation energy, should correlate [4,30,33,34,61,62] and RE as a percentage of $-76.0$ kJ mol$^{-1}$ (or $-77.5$ kJ mol$^{-1}$ in the case of COSNAR) yields the amidicity relative to *N,N*-dimethylacetamide **1** (by definition 100%).

The resonance energy and amidicity of *N*-methoxy-*N*-methylacetamide **3b** has been determined with COSNAR ($-62.1$ kJ mol$^{-1}$, 80% amidicity) and TA ($-61.25$ kJ mol$^{-1}$, 81% amidicity) and accords nicely with the lowest rotational barrier from Figure 5a of 67.5 kJ mol$^{-1}$. In contrast, for the bisoxyl-substituted acetamide, the $\text{RE}_{\text{COSNAR}}$ and $\text{RE}_{\text{TA}}$ were determined at $-35.9$ kJ mol$^{-1}$, respectively just 47% or 46% that of *N,N*-dimethylacetamide [33]. From Figure 5b the rotational barrier was 44.4 kJ mol$^{-1}$ and therefore most of the barrier can be accounted for by loss of resonance.

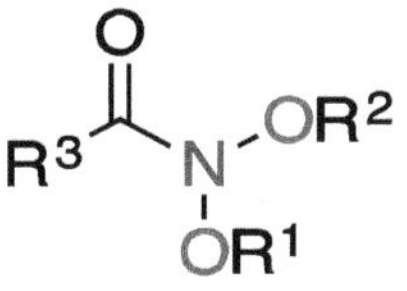

**5** a: $R^1$=Me, $R^2$=Et, $R^3$=4-$NO_2C_6H_4$
b: $R^1$=Me, $R^2$=4-$NO_2C_6H_4CH_2$, $R^3$=Ph

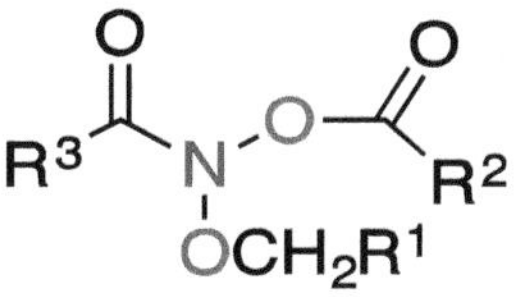

**6** a: $R^1$= $R^2$=$R^3$=4-$Bu^tC_6H_4$
b: $R^1$=4-$Bu^tC_6H_4$, $R^2$=$R^3$=Ph

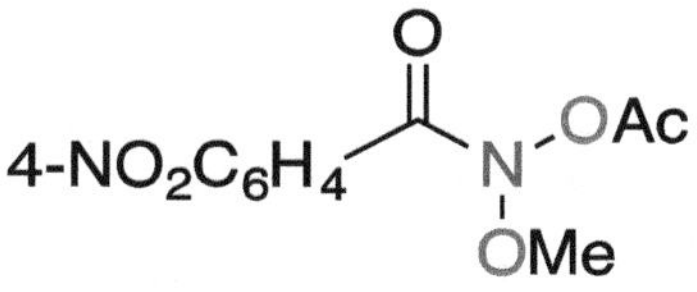

**7**

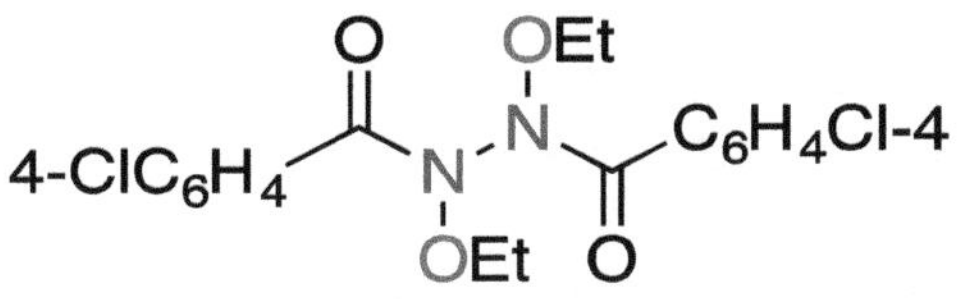

**8**

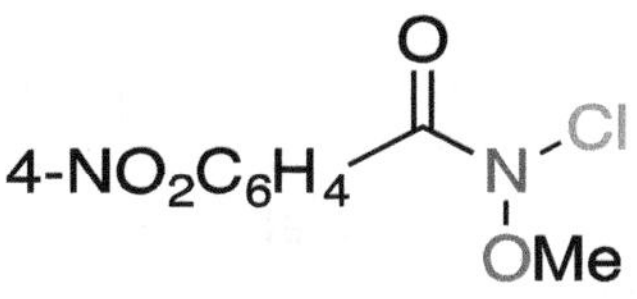

**9**

The unusual structure of a number of stable anomeric amides (**5**–**9**) has been confirmed by X-ray crystallography (Figure 6) and relevant structural parameters of these are given in Table 1. The X-ray structures of anomeric amides **5**–**9** provide clear evidence of reduced amide resonance. The data in Table 1 shows that as the combined electron demands of X and Y increase, N–C(O) bond length and pyramidalisation at nitrogen both increase. The reported average N–C(O) bond length in acyclic amides is 1.359 Å (median 1.353 Å), generated from Cambridge Structural Database (CSD) [29,63], which is significantly shorter than the average 1.418 Å from these seven X-ray structures. While the (N)C=O bond (average 1.207 Å) contracts slightly compared to simple acyclic amides (1.23 Å), there is little to no correlation seen between change in (N)C=O bond length and degree of lone pair dislocation; this may be attributed to bias towards carbon in the LUMO (Figure 1) [2,26,35,64]. Deviation from the usual $sp^2$ hybridisation at the amide nitrogen ($\chi_N = 0$) is significant and in line with the electron demands of substituents. *N*-acyloxy-*N*-alkoxyamides **6a**, **6b**, and **7** ($\chi_N$ = 65.33°, 65.62° and 59.7°, respectively) are pyramidalised at nitrogen to the extent of, and beyond, what is expected for a pure $sp^3$ hybridisation. The 4-nitrobenzamide **7** is less pyramidal than the benzamides **6a** and **6b** presumably on account of greater positive charge at the carbonyl carbon and attendant increase in nitrogen lone pair attraction. Both *N,N*-dialkoxyamides **5a** and **5b** ($\chi_N$ = 58.3° and 55.6°, respectively) are also strongly pyramidalised. For Shtamburg's *N*-alkoxy-*N*-chloroamide **9**, X-ray data reveals a small $\chi_N$ of 52.5°, similar to *N,N*-dialkoxyamides **5a** and **5b**. Both amide nitrogens in hydrazine **34** are the least pyramidalised with $\chi_{N1}$ and $\chi_{N2}$ of 47° and 49°, respectively.

Despite high degrees of pyramidalisation in this set of anomeric amides, there is minimal twist about the N–C(O) bond ($\tau$ = 6.7–15.5°). This indicates that lone pair orbital overlap with the carbonyl $C2p_z$ orbital, though clearly less effective on electronic and geometric grounds, remains a stabilizing influence (Table 1). As can be seen in the B3LYP/6-31G(d) deformation surface for *N,N*-dimethoxyacetamide **4b** (Figure 5b), the strongly pyramidal structure requires twist angles ($\tau$) beyond 20° before there is significant loss of stabilization.

**Table 1.** Selected structural data for anomeric amides **5**–**9** from X-ray structures (Figure 6).

| Structure | N–C(O)/Å | (N)C=O/Å | $\theta$/° | $\chi_N$/° | $\tau$/° | Anomeric Twist [1]/° |
|---|---|---|---|---|---|---|
| **5a** [33] | 1.409 | 1.206 | 331.2 | 58.3 | 6.7 | **C10-O3-N1-O2: 95.9**<br>C8-O2-N1-O3: −114.1 |
| **5b** [33] | 1.421 | 1.211 | 334.5 | 55.6 | 13.9 | **C8-O2-N1-O3: −101.6**<br>C15-O3-N1-O2: −63.8 |
| **6a** [29] | 1.439 | 1.205 | 323.5 | 65.3 | 15.5 | **C18-O3-N1-O1: 96.2**<br>C7-O1-N1-O3: −141.6 |
| **6b** [29] | 1.441 | 1.207 | 324.1 | 65.6 | 13.9 | **C14-O3-N1-O1: 96.7**<br>C7-O1-N1-O3: −137.6 |
| **7** [65] | 1.411 | 1.203 | 330.3 | 59.7 | −14.0 | **C8-O1-N1-O5: −91.8**<br>C9-O5-N1-O1: 116.2 |
| **8**(N1) [32] | 1.412 | 1.213 | 343.2 | 47.3 | −8.5 | LP (N2)-N2 -N1-O2: 47.3 |
| **8**(N2) | 1.410 | 1.207 | 341.1 | 48.9 | −11.4 | **LP (N1)-N1-N2-O3: 178.6** |
| **9** [65] | 1.408 | 1.204 | 337.5 | 52.5 | −13.3 | **C8-O2-N1-Cl1: −84.2** |

[1] Anomeric alignments in bold-face.

Pyramidal nitrogens and corresponding anomeric interactions have also been observed in the structures of a number of urea and carbamate analogues of anomeric amides (**10**–**14**) studied by Shtamburg and coworkers and selected structural data are presented in Table 2 [66–69]. Where substituent electronic effects are largely similar as in **11b** and **13**, it can be deduced that the stronger conjugative effect of the $\alpha$-nitrogen lone pair in the urea **11b**, relative to the $\alpha$-oxygen in the carbamate **13**, results in significantly less amide resonance interaction and, hence, significant increase in pyramidality at the amide nitrogen. Comparison of the ONCl structures **9** and **10a** and **b**, again indicates greater pyramidality in the ureas where competing acyl nitrogen resonance reduces the anomeric amide resonance interaction.

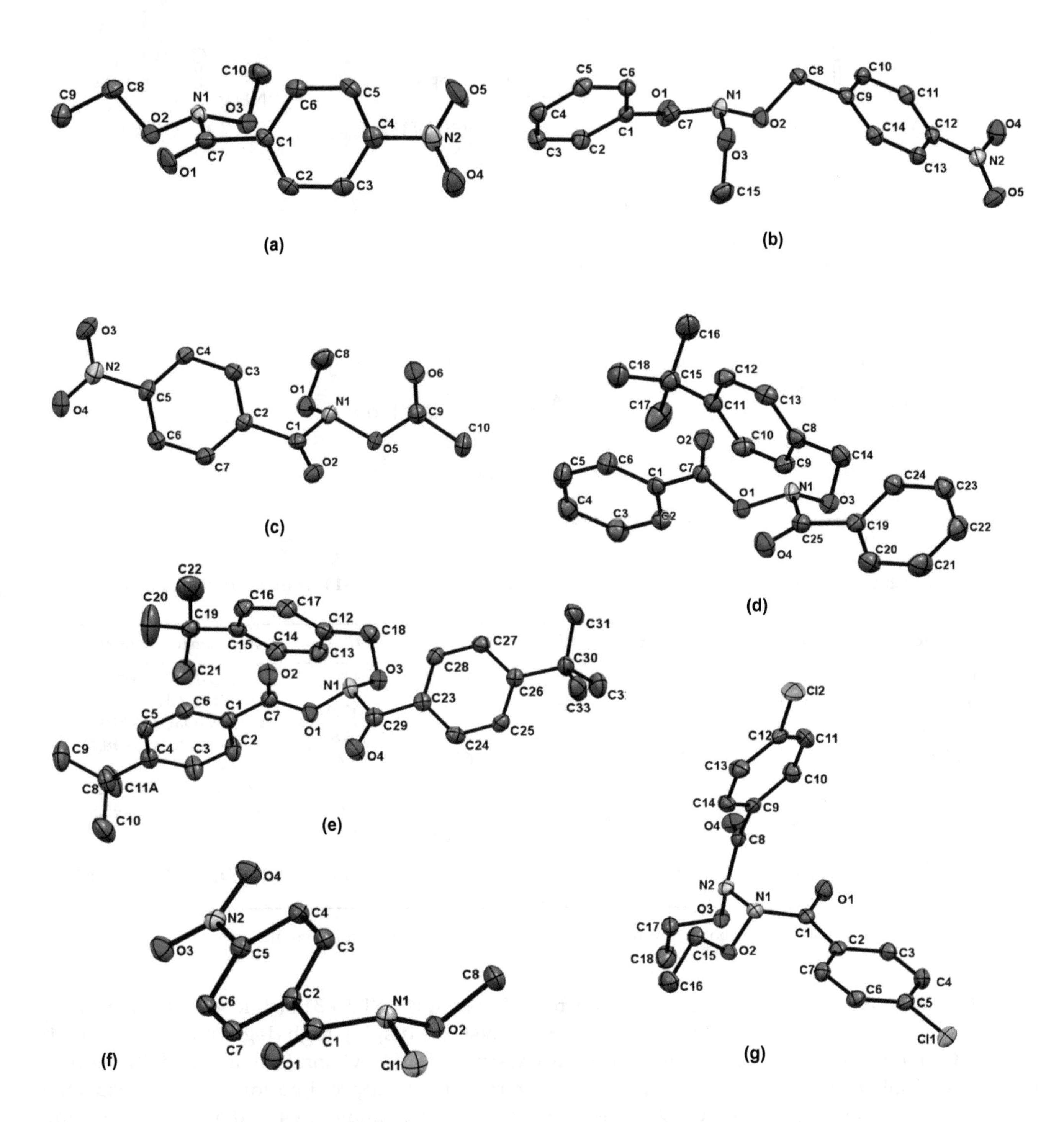


**Figure 6.** X-ray structures for (**a**) *N*-ethoxy-*N*-methoxy-4-nitrobenzamide **5a** [33], (**b**) *N*-methoxy-*N*-(4-nitrobenzyloxy)benzamide **5b** [33], (**c**) *N*-acetoxy-*N*-methoxy-4-nitrobenzamide **7** [65], (**d**) *N*-benzoyloxy-*N*-(4-*tert*-butylbenzyloxy) benzamide **6b** [29], (**e**) *N*-(4-*tert*-butylbenzoyloxy)-*N*-(4-*tert*-butylbenzyloxy)-4-*tert*-butylbenzamide **6a** [29], (**f**) *N*-chloro-*N*-methoxy-4-nitrobenzamide **9** [65], and (**g**) *N*,*N'*-4-chlorobenzoyl-*N*,*N'*-diethoxyhydrazine **8** [32].

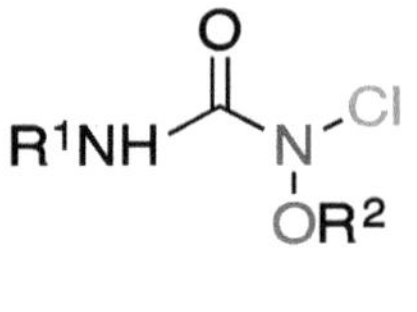

**10**

a: $R^1$=H, $R^2$=Me
b: $R^1$= 4-$NO_2C_6H_4$, $R^2$=$C(Me)_2CO_2Me$

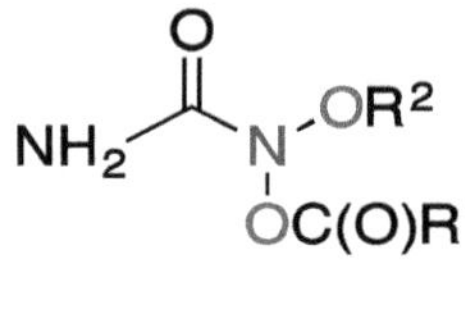

**11**

a: $R^1$=Me, $R^2$=Et
b: $R^1$= 4-$ClC_6H_4$, $R^2$=Bu

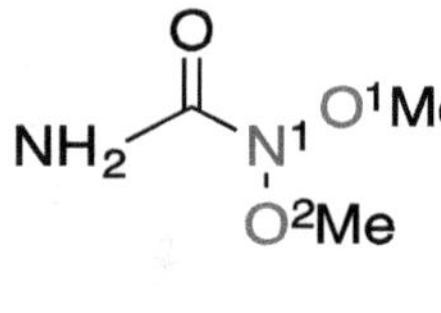

**12**

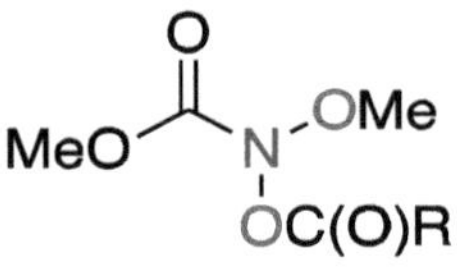

**13**

a: R=4-$ClC_6H_4$

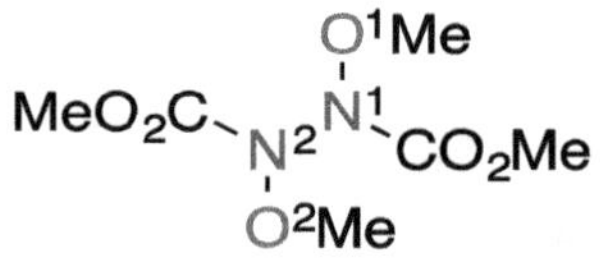

**14**

**Table 2.** Selected structural data for ureas **10–12** and carbamates **13–14** from X-ray structures.

| Structure | N–C(O)/Å | (N)C=O/Å | θ/° | $\chi_N$/° | τ/° | Anomeric Twist [2]/° |
|---|---|---|---|---|---|---|
| **10a** [66] | 1.443 [1] | 1.226 | 329.1 | 59.9 | 8.2 | **C-O-*N*-Cl: −90.9** |
| **10b** [66] | 1.472 [1] | 1.210 | 325.8 | 61.9 | −13.4 | **C-O-*N*-Cl: −100.1** |
| **11a** [67] | 1.426 [1] | 1.222 | 333.6 | −57.1 | −6.8 | **C-O-*N*-Oacyl: −104.0** |
| **11b** [68] | 1.441 [1] | 1.233 | 323.7 | 64.6 | 11.8 | **C-O-*N*-Oacyl: −98.2** |
| **12** [68] | 1.438 [1] | 1.220 | 331.8 | 57.4 | 0.8 | **C-O2-N1-O1: −89.3**<br>C-O1-N1-O2: 55.2 |
| **13** [67] | 1.424 | 1.198 | 334.2 | −56.3 | 2.9 | **C-O-*N*-Oacyl: −95.5** |
| **14** [69] (N1)/(N2) [3] | 1.408 | 1.194 | 340.0 | 59.8 | −9.4 | **LP (N1)-N1-N2-O: 189.2** |

[1] Anomeric amide bond; [2] Anomeric alignments in bold-face; [3] Equivalent nitrogens.

These and other anomeric amides have been modelled at B3LYP/6-31G(d) level in the simplified acetamide system and ground state models of **15a–d** systems display high degrees of pyramidality, little N–C(O) twist, long N–C(O) bonds, and slightly shortened (N)C=O bonds, in line with the electron demands of substituents and, where applicable, are reasonable approximations of their respective X-ray structure counterparts (Figure 7, Table 3). ONS and NNCl analogues, **15d** and **15e**, have only been generated as intermediates in reactions but their theoretical structures are in line with those of anomeric systems **15a–d**; N–C(O) bond lengths and pyramidality at nitrogen ($\chi_N$) are broadly in line with the gross electronegativity of substituents at nitrogen. Sulphur, with its low electronegativity, results in a less pyramidal nitrogen, while the NNCl system **15e** is completely planar (and untwisted) for steric reasons. Nonetheless, its N–C(O) bond is comparatively long. The carbamate and the urea **16a** ($\chi_N$ = 46.3°) and **16b** ($\chi_N$ = 49.6°) are both more pyramidal than the corresponding acetamide **15c** ($\chi_N$ = 41.8°), presumably as a result of competing resonance from the α-oxygen and α-nitrogen lone pairs.

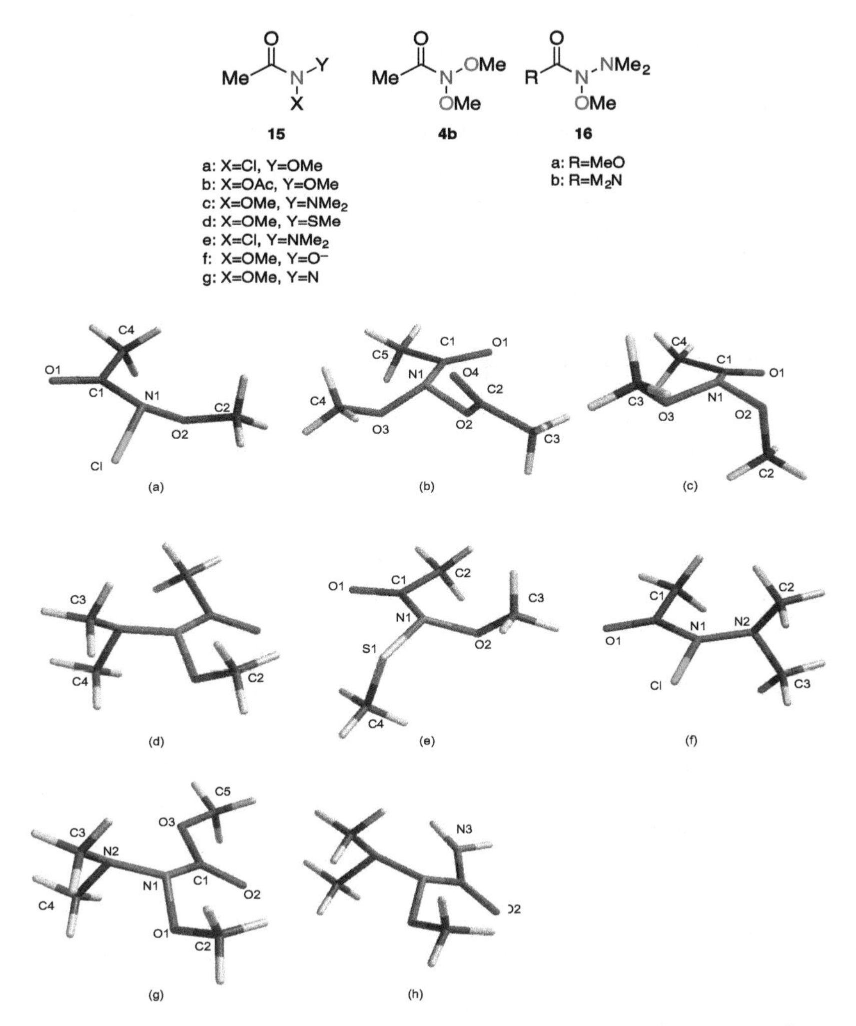


**Figure 7.** B3LYP/6-31G(d) lowest energy ground state conformers of model anomeric acetamides (**a**) *N*-chloro-*N*-methoxyacetamide **15a**, (**b**) *N*-acetoxy-*N*-methoxyacetamide **15b**, (**c**) *N*,*N*-dimethoxyacetamide **4b**, (**d**) *N*-methoxy-*N*-dimethylaminoacetamide **15c**, (**e**) *N*-methoxy-*N*-methylthiylacetamide **15d**, (**f**) *N*-chloro-*N*-dimethylaminocetamide **15e**, (**g**) *O*-methyl *N*-methoxy-*N*-dimethylaminoacetamide **16a**, and (**h**) *N*-methoxy-*N*-dimethylaminourea **16b**.

**Table 3.** Selected structural data for B3LYP/6-31G(d) lowest energy conformers of model acetamides **4b**, **15a–e**, and **16a**,**b**.

| Structure | N–C(O)/Å | (N)C=O/Å | (N–X,N–Y)/Å | θ/° | $\chi_N$/° | τ/° | Anomeric Twist Angles [1]/° |
|---|---|---|---|---|---|---|---|
| **15a** ONCl [34] | 1.432 | 1.207 | Cl:1.787<br>O2:1.389 | 337.6 | 52.3 | −5.3 | **C2-O2-N1-Cl: 88.8** |
| **15b** ONOAc [61] | 1.429 | 1.209 | O2:1.423<br>O3:1.395 | 332.1 | 58.0 | 2.3 | **C4-O3-N1-O2: 101.1** |
| **4b** ONO [33] | 1.417 | 1.212 | O2:1.387<br>O3:1.412 | 342.9 | 48.1 | 8.5 | C2-O2-N1-O3: 66.6<br>**C3-O3-N1-O2: 83.8** |
| **15c** NNO [62] | 1.404 | 1.217 | O1:1.430<br>N2:1.387 | 346.5 | 41.8 | 5.4 | **LP (N2)-N2-N1-O1: 190** |
| **15d** ONS [61] | 1.408 | 1.215 | S1:1.717<br>O2:1.420 | 352.4 | 31.7 | −4.6 | C4-S1-N1-O2: −79<br>**C3-O2-N1-S1: −86.6** |
| **15e** NNCl [34] | 1.414 | 1.209 | Cl:1.820<br>N2:1.351 | 360.0 | 0 | 0 | **LP(N2)N2-N1-Cl: 180** |
| **16a** NNO [61] | 1.406 | 1.212 | O1:1.404<br>N2:1.383 | 342 | 46.3 | 0 | **LP (N2)-N2-N1-O1: 169.4** |
| **16b** NNO | 1.428 | 1.217 | O1:1.397<br>N2:1.428 | 340.0 | 49.6 | −1.7 | **LP (N2)-N2-N1-O1: 167.5** |

[1] Anomeric alignment in bold face.

*2.2. Resonance Energies and Amidicities*

The resonance energy and the amidicity of anomeric amides **4b** and **15a–e** have been calculated at the B3LYP/6-31G(d) level by both the TA and COSNAR methods and by the COSNAR method using dispersion corrected M06/6-311++G(d,p). $\Delta E_{\mathbf{COSNAR}}$ (Equation (1)), reaction energies ($\Delta E_{\mathrm{TA}}$) (Equation (2)), inductive destabilization corrections ($\Delta E_{\mathbf{ind}}$) (Equation (3)), resultant $\mathrm{RE}_{\mathbf{TA}}$ (Equation (4)) together with COSNAR, and TA amidicities are presented in Table 4.

**Table 4.** B3LYP/6-31G(d), B3LYP/6-311++G(d,p), and M06/6-311++G(d,p) derived resonance energies and amidicities of model anomeric acetamides **4b**, **15a–e**, and **16a**.

| Amide (R = Me) | $\Delta E_{\mathrm{COSNAR}}$ [1] /kJ mol$^{-1}$ | $\Delta E_{\mathrm{TA}}$ /kJ mol$^{-1}$ | $\Delta E_{\mathrm{ind}}$ /kJ mol$^{-1}$ | $\mathrm{RE}_{\mathrm{TA}}$ [1,2] /kJ mol$^{-1}$ |
|---|---|---|---|---|
| **15a** (ONCl) [34] | −29.6(38) | 69.9 | 23.4 | −29.5(39) |
| **15a** [3] | −27.2 (36) | | | |
| **15a** [4] | −34.4(45) | | | |
| **15b** (ONOAc) [61] | −39.7(52) | 65.3 | 29.7 | −40.5(53) |
| **15b** [4] | −39.5 (52) | | | |
| **4b** (ONO) [35] | −36.0(47) | 58.2 | 18.0 | −36.0(47) |
| **4b** [4] | −39.5(53) | | - | |
| **15c** (ONN) [62] | −52.3(69) | 35.1 | 10.0 | −51.0(67) |
| **15c** [4] | −55.7(73) | | | |
| **15d** (ONS) [61] | −48.6(64) | 26.8 | 5.0 | −48.6(64) |
| **15d** [4] | −47.3(62) | | | |
| **15e** (NNCl) [34] | −28.7(38) | 64.5 | 17.6 | −29.0(37) |
| **15e** [3] | −28.6(38) | | | |
| **15e** [4] | −48.7(64) | | | |
| **16a** (ONN) [61] | −47.7(63) | 31.8 | 14.6 | −50.6(67) |
| **1** | −75.9(100) | | | |
| **1** [4] [34] | −75.9(100) | | | |

[1] Amidicity (%) in parentheses; [2] From Equation (4); [3] At B3LYP/6-311++G(d,p);[4] M06/6-311++G(d,p).

The TA and COSNAR methodologies give almost identical resonance energies for all the anomeric amides (**4b** and **15a–e**) at the B3LYP/6-31G(d) level. $\Delta E_{\mathbf{COSNAR}}$ determined at M06 with the expanded basis set yields very similar results to B3LYP/6-31G(d) for *N,N*-dimethylacetamide **1** and the anomeric amides with the exception of **15e**, where resonance is computed to be worth 49 kJ mol$^{-1}$, just over 60% that of *N,N*-dimethylacetamide and substantially higher than that predicted from B3LYP/6-311++G(d,p), which was identical to the B3LYP/6-31G(d) value (29 kJ mol$^{-1}$) [34]. The RE parity for **15a** and **15e** at B3LYP is no longer observed with M06, in line with the lower overall electronegativity of nitrogen and chlorine. These results impute the necessity for inclusion of dispersion corrections in treatments of molecules where anomeric interactions are likely to have a pronounced influence.

It is clear that resonance in anomeric amides is impaired, broadly in line with the gross electronegativity or negative inductive effect of the atoms/groups bonded to nitrogen. Interestingly, the ONN and ONS acetamides, **15c** and **15d**, have very similar resonance energies despite the much lower electronegativity of sulphur. From Bent's rule, the opposite effect would be expected, since s-character in the lone pair orbital should decrease with decreasing electronegativity of X. This is likely to be a manifestation of the role of orbital size since the influence of second period elements is significant; nitrogen increases p-character in the bond to sulphur to effect better overlap with the larger, 3p orbital of sulphur. Consequently, the nitrogen lone pair gains more "s" character relative to the amide nitrogen in the NNO acetamide **15c**, resulting in less amide resonance [36,37,70]. Comparing the M06 values, ONCl and ONO systems have about half the resonance of *N,N*-dimethylacetamide while NNO, NNCl, and ONS systems are computed to preserve some 60 to 70% of the resonance of *N,N*-dimethylacetamide.

*2.3. The Anomeric Effect*

In addition to their reduced amide resonance, anomeric amides are exemplars of XNY systems featuring an anomeric effect [26,35]. There are two possible anomeric interactions designated as $n_X$–$\sigma^*_{NY}$, and $n_Y$–$\sigma^*_{NX}$ and where X and Y are different electronegative atoms, one of these interactions will be favoured over the other (Figure 8). By analogy with anomeric carbon centres [26,35,71–73], the relative electronegativities of heteroatoms X and Y at nitrogen and the relative sizes of interacting orbitals contribute to the strength of an anomeric interaction. Heteroatoms Y and X directly influence the relative energies of $n_Y$ and $\sigma^*_{NX}$, which, in turn, affect the net stability gain for the lone pair electrons (Figure 9a).

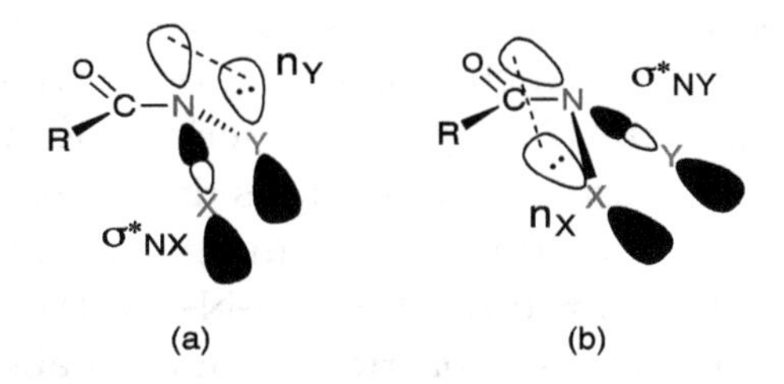


**Figure 8.** Two possible anomeric interactions in anomeric amides (**a**) $n_Y$–$\sigma^*_{NX}$ and (**b**) $n_X$–$\sigma^*_{NY}$.

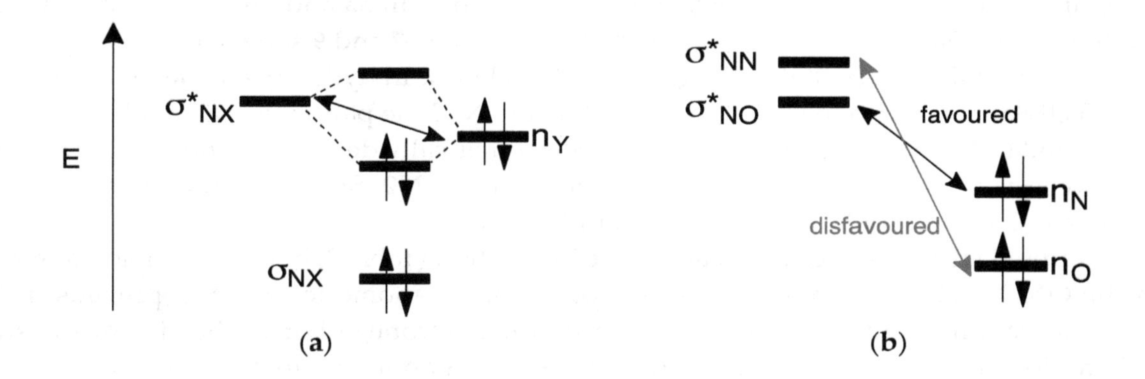


**Figure 9.** (**a**) Stabilisation of a lone pair through a $n_Y$–$\sigma^*_{NX}$ anomeric interaction and (**b**) the energetics of the anomeric effect in **15c**.

As the electronegativity of X and Y increases by going across the p-block row on the periodic table, $\sigma^*_{NX}$ (and $\sigma_{NX}$) decreases in energy as does $n_Y$. Additionally, as X decreases in electronegativity by going down a p-block group, $\sigma^*_{NX}$ decreases in energy due to reduced orbital overlap [26,35,36,71,72]. An optimal anomeric effect can be achieved when Y is an early p-block element and X is a p-block element to the right of Y on the periodic table; more specifically an anomeric stabilisation will be greater when the energy gap between $n_Y$ and $\sigma^*_{NX}$ is lower (Figure 9b) [74,75]. In the unusual case of anomeric amides, where the nitrogen may range between planar $sp^2$ and pyramidal $sp^3$ hybridisation, the geometry of the central nitrogen atom in XNY systems also plays a role, as pyramidal nitrogen is more conducive to edge-on $n_Y$ and $\sigma^*_{NX}$ overlap than is planar, $sp^2$ hybridised nitrogen (Figure 10a) [26,35].

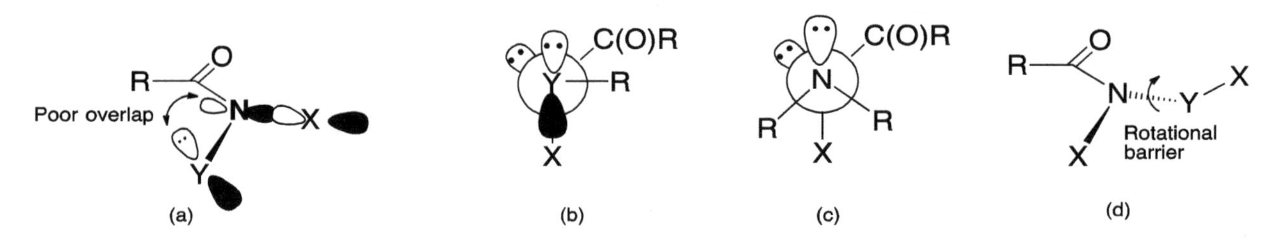


**Figure 10.** (**a**) $sp^2$ hybridised nitrogen hinders overlap with heteroatom Y in an $n_Y$–$\sigma^*_{NX}$ system; (**b**) optimum conformation in an $n_Y$–$\sigma^*_{NX}$ anomeric interaction; (**c**) optimum conformation in an $n_N$–$\sigma^*_{NX}$ anomeric interaction; and (**d**) restricted rotation about the N–Y bond in an $n_Y$–$\sigma^*_{NX}$ anomeric interaction.

Similar to XCY configurations, in an XNY system, a stabilising $n_Y$–$\sigma^*_{NX}$ anomeric effect, for example, can cause the amide to adopt a gauche conformation in which the lone pair of Y is coplanar with the vicinal to C–X bond (Figure 10b) [71,74]. In anomeric amides, when $n_Y$ is divalent oxygen or sulphur, an R-Y-N-X dihedral angle close to |90°| aligns the p-type lone pair on those atoms with the vicinal $\sigma^*_{NX}$ (Figure 10b). Where the donor is $n_N$, an optimum anomeric effect would need the lone pair on nitrogen to be antiperiplanar to X, LP(N)-N-N-X = 180° (Figure 10c). Consequences of the $n_Y$–$\sigma^*_{NX}$ interaction include an increased barrier to rotation about the N–Y bond (Figure 10d), a contraction of the N–Y bond, and an extension of the N–X bond as electrons from Y populate the $\sigma^*_{NX}$ orbital.

In each of the X-ray structures in Figure 6 there is clear evidence of these anomeric interactions (Table 1, bold-face torsion angles). On the basis of energetics, the expected anomeric interactions are $n_O$–$\sigma^*_{NOAc}$ in **6a**, **6b**, and **7**, $n_N$–$\sigma^*_{NO}$ in **8**, and $n_O$–$\sigma^*_{NCl}$ in **9**. In **5a** and **5b** one $n_O$–$\sigma^*_{NO}$ would be expected to prevail. Dihedral angles about the N–O bonds in **5**–**7** and **9** show a preferred p-type lone pair alignment with the adjacent $\sigma^*_{NO}$, $\sigma^*_{NOAc}$, or $\sigma^*_{NCl}$ bond. In hydrazine **8**, the lone pair on N1 is almost perfectly aligned with the N2–O3 bond while the N2 lone pair makes an angle of only 47° to the N1–O2 bond (1.403 Å), which is shorter than the anomerically destabilised bond N2–O3 (1.411 Å). The structure is asymmetrical with a donor N1 and recipient N2. Similar anomeric interactions are observable in ureas **10**–**12** and carbamates **13** and **14** (Table 2).

All computed structures exhibit an anomeric interaction (Table 3, bold-face torsion angles) and, besides the ONO system where one $n_O$–$\sigma^*_{NO}$ prevails, an anomeric $n_O$–$\sigma^*_{NX}$ prevails in ONCl, ONOAc, and ONS in line with expectations based on electronegativity. The size of the sulphur p-orbital and lower energy of the $\sigma^*_{NS}$ renders an $n_S$–$\sigma^*_{NO}$ anomeric interaction less likely than a $n_O$–$\sigma^*_{NS}$ stabilisation. Where nitrogen is present, the $n_N$–$\sigma^*_{NO}$ and $n_N$–$\sigma^*_{NCl}$ interactions are clearly in evidence.

The anomeric effects not only dictate stereochemistry at nitrogen but, as will be seen, combined with reduced amide resonance, they have a profound impact upon spectroscopic properties and the reactivity of anomeric amides.

### *2.4. Spectroscopic Properties of Anomeric Amides*

Spectroscopic properties of anomeric amides are strongly influenced by reduced resonance due to electronegativity of substituents at nitrogen. Infrared and $^{13}C$ NMR data for a diverse range of ONCl, ONOAcyl [26,31,35], and ONO systems [35,55] have been reported as well as for a number of ONN *N,N′*-dialkoxy-*N,N′*-diacylhydrazines [76–78]. Representative infrared carbonyl stretch frequencies in solution for stable anomeric amides (Table 5) are significantly higher (1700 to 1750 $cm^{-1}$) [32,47,55,79–81] than those of their precursor hydroxamic esters (1650 to 1700 $cm^{-1}$) and primary (1690 $cm^{-1}$), secondary (1665 to 1700 $cm^{-1}$), and tertiary (1630 to 1670 $cm^{-1}$) alkylamides [26,35,82]. While there is a slight tightening of the (N)C=O bond, the increase in $\nu_{C=O}$ has been attributed primarily to electronic destabilisation of single-bond resonance form of the carbonyl as electron density is pulled towards the electronegative heteroatoms on nitrogen, resulting in a more ketonic carbonyl bond [2,64]. Likewise, anomeric amides exhibit more ketonic carbonyl $^{13}$C chemical shifts ($CDCl_3$), with downfield shifts of approximately 8.0 ppm from their hydroxamic ester precursors. Deshielding of the carbonyl carbon is an expected consequence of the electron-withdrawing substituents. However, like acid chlorides and anhydrides, the carbonyls resonate upfield of ketones, relative to which the electron density at the carbonyl is increased on account of greater double bond character.

**Table 5.** Typical spectroscopic data for stable anomeric alkanamides and arylamides (RC(O)NXY), and their hydroxamic ester precursors.

| System | X | Y | R | Amide $\nu$/cm$^{-1}$ ($\delta^{13}$C) | Hydroxamic Ester $\nu$/cm$^{-1}$ ($\delta^{13}$C) |
|---|---|---|---|---|---|
| **2a** [79] | Cl | OBu | Me | 1740 [1](175.3) | 1678 (167.9) |
| **2a** [35] | Cl | OBu | Ph | 1719 (174.2) | 1654 (165.7) |
| **2b** [79] | OAc | OBu | Me | 1746 (176.2) | 1678 (167.9) |
| **2b** [80] | OAc | OBu | Ph | 1732 (173.9) | 1654 (165.7) |
| **2c** [81] | OBu | OBu | Me | 1707 (174.1) | 1678[1] (167.9) |
| **2c** [55] | OMe | OMe | Ph | 1711 (174.3) | 1683 (166.4) |
| **2d** [76] | 4-MeBnONAc | 4-MeBnO | Me | 1734/1700 (171.3) | 1693 (168.0) |
| **2d** [76] | BzNOEt | OEt | Ph | 1708 (170.0) | 1685 (166.5) |

[1] Neat.

Common primary, secondary, and tertiary amides have significant *cis–trans* isomerisation barriers for rotation about the N–C(O) bond, due to the stabilising effect of amide resonance in their ground state [2]. Restricted rotation through lone pair overlap with the adjacent carbonyl $C2p_z$ orbital, as illustrated in Figure 1, often results in observation of different chemical shifts of *cis* and *trans* conformers in their $^1$H NMR spectra, from which barriers to isomerisation can be deduced [83,84]. The computed surface for *N,N*-dimethylacetamide in Figure 3 indicates that the barrier (difference between the completely planar and the fully rotated-pyramidal forms) is of the order of 71–75 kJ mol$^{-1}$. Hydroxamic esters have slightly less resonance and amidicity, and the rotation barrier for *N*-methoxy-*N*-methylacetamide from Figure 5a is approximately 67 kJ mol$^{-1}$. Many hydroxamic esters have broadened $^1$H NMR signals at ambient temperatures. Anomeric amides with lower resonance should have lower *cis–trans* isomerisation barriers as exemplified in the model *N,N*-dimethoxyacetamide in Figure 5b. Accordingly, the isomerisation barrier in bisoxyl-substituted amides **2b** and **2c** are too low to be measured by usual dynamic NMR methods. All proton signals of *N,N*-dimethoxy-4-toluamide remained sharp down to 180 K (in $d_4$-methanol) and the $^1$H NMR signals for *N*-acetoxy-*N*-benzyloxybenzamide remained isochronous down to 190 K (in $d_8$-toluene) [26]. The barrier for *N*-benzyloxy-*N*-chlorobenzamide **18** could likewise not be determined by dynamic NMR [35].

Figure 11 illustrates the dramatic difference between $^1$H NMR spectra of *N*-butoxyacetamide and its *N*-chloro- or *N*-acetoxyl derivatives as a consequence of reduced resonance in the anomeric structures, which clearly have much lower isomerization barriers. At room temperature, the anomeric amides are in the fast exchange region as opposed to the slow exchange in the hydroxamic ester. Earlier theoretical studies by Glover and Rauk support this assertion [27,28]. For example, a comparison of formamide **17a**, *N*-methoxyformamide **17b**, and *N*-chloro-*N*-methoxyformamide **17d**, calculated at the B3LYP/6-31G(d) level, showed little reduction in the *cis–trans* isomerisation barrier of 73.2 kJ mol$^{-1}$ in formamide, to 67–75 kJ mol$^{-1}$ in *N*-methoxyformamide. However, the introduction of a second electronegative heteroatom, Cl, reduced the barrier to only 32.2 kJ mol$^{-1}$, indicating that monosubstitution alone is insufficient to impact upon the isomerisation barrier [27]. A theoretical isomerisation barrier for *N*-methoxy-*N*-(dimethylamino)formamide, which, by analogy with typical NNO systems, should retain ~70% the resonance in *N,N*-dimethylacetamide, was computed to be higher at 52.7 kJ mol$^{-1}$ [28]. Such an amide isomerisation barrier in the hydrazine, *N,N′*-diacetyl-*N,N′*-di(4-chlorobenzyloxy)hydrazine **20d**, was measurable at $\Delta G^{\ddagger}_{278}$ = 54.0 kJ mol$^{-1}$, in line with this theoretical barrier [27].

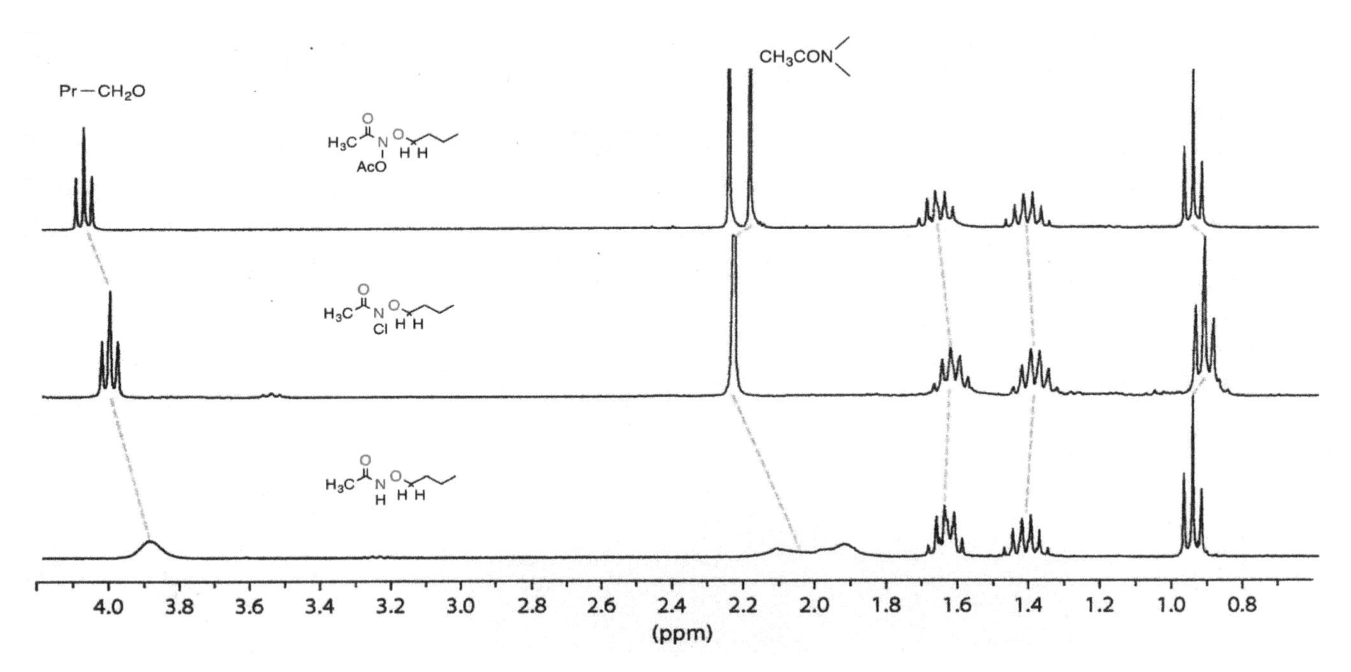


**Figure 11.** 300 MHz $^1$H NMR spectra (300K, $CDCl_3$) of *N*-butoxyacetamide, and anomeric amides *N*-butoxy-*N*-chloroacetamide and *N*-acetoxy-*N*-butoxyacetamide.

The intrinsic barrier to inversion at nitrogen in bisheteroatom-substituted amides is expected to be much lower than those of analogous amines. The transition state for inversion in anomeric amides is expected to be planar, where the nitrogen lone pair can interact with the carbonyl $2p_Z$ orbital generating stabilisation [35]. For *N,N*-dimethoxyacetamide (Figure 5b) this barrier is represented by the energy of structure (b), while the difference in energy between fully twisted structures (d) and (c) represents the much larger barrier to inversion in the corresponding anomeric amine. Several theoretical estimates put these barriers in anomeric amides at about 10 kJ mol$^{-1}$ [28].

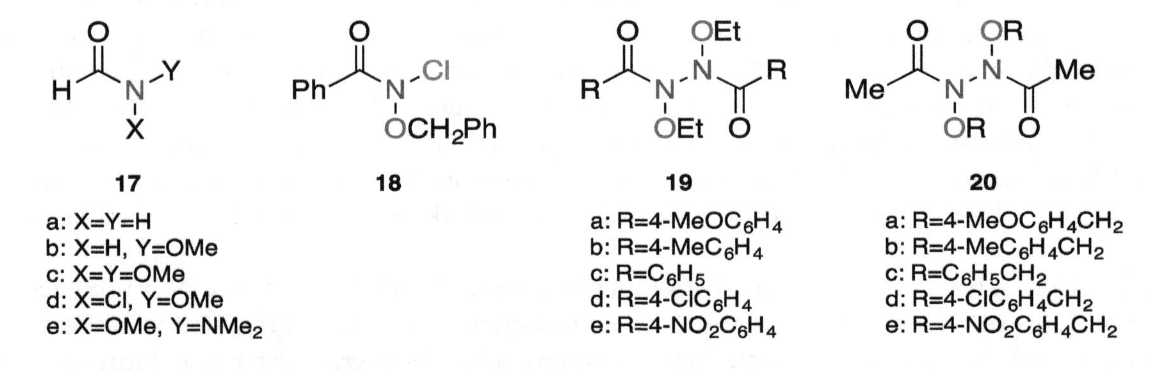


In anomeric systems, strong $n_Y$–$\sigma^*_{NX}$ interactions should increase the bond order of the N–Y bond, which should impose barriers to rotation about those bonds. For *N*-chloro-*N*-methoxyformamide **17d**, the N–O and N–C(O) rotation barriers have been estimated at the B3LYP/6-31G(d) level at 44.7 and 29.2 kJ mol$^{-1}$, respectively [27], while the theoretical barrier to rotation about the N–N bond in *N*-methoxy-*N*-dimethylaminoformamide **17e** was computed at ~60 kJ mol$^{-1}$ [28]. Experimental measurements for both anomerically induced barriers have been made. An N–O anomeric rotational barrier of $\Delta G^‡ = 43$ kJ mol$^{-1}$ has been determined for *N*-chloro-*N*-benzloxybenzamide **18**, where the benzylic methylenes become diastereotopic at 217 K in $d_8$-toluene, but as no amide isomerisation could be detected, that barrier must be significantly lower [26]. An anomerically induced rotational barrier for N–N′ bond in a number of *N,N*-diacyl-*N,N*-dialkoxyhydrazines **19a–e** and **20a–e** has been measured in dynamic $^1$H NMR studies [32]. Methylene signals of *N,N′*-diethoxy **19a–e** and *N,N′*-dibenzyloxy groups **20a–e**, which were diastereotopic at room temperature, as a consequence of restricted rotation about the N–N′ bond, coalesced at higher temperatures ($T_c$ = 316–346 K) from which rotational free energy barriers of the order of 60−70 kJ mol$^{-1}$ could be determined [32], which compare favourably with the theoretically calculated *N–N′* rotational barrier for **17e** of 60 kJ mol$^{-1}$ [28].

## 3. Reactivity of Anomeric Amides

The reduced resonance and attendant pyramidalisation at the amide nitrogen together with anomeric properties of these unusual molecules results in a plethora of amide reactivity, some known, and now better understood, and others that are unique to anomeric amides. The destabilisation of the amide bond, coupled with the substitution pattern, facilitates reactivity at the amide nitrogen in which the amides are usually transformed from one anomeric amide form to another. Moreover, it can induce a novel process, known as the HERON reaction (Named at the Third Heron Island Conference on Reactive Intermediates and Unusual Molecules, Heron Island 1994.), in which the amide bond is broken to form acyl derivatives and heteroatom-stabilised nitrenes. This reaction is facilitated by weakened amide resonance, but is driven by $n_Y$–$\sigma^*_{NX}$ anomeric destabilisation of the N–X bond.

### 3.1. *Reactivity at the Amide Nitrogen*

Due to their characteristic, diminished amide resonance and anomeric destabilisation, this class of amide has been shown to undergo $S_N2$ reaction at nitrogen, and elimination of *N*-substituents leading to $S_N1$-type processes. Several congeners undergo thermolytic homolysis to give alkoxyamidyl radicals.

#### 3.1.1. $S_N2$ Reactions

In XNY systems, a moderate $n_Y$–$\sigma^*_{NX}$ negative hyperconjugation leads, through neighbouring group participation, to weakening of the N–X bond, which can encourage $S_N2$ reactions at nitrogen [26,31,35]. The increased electrophilicity of nitrogen in *N*-acyloxy-*N*-alkoxyamides **21** leaves them vulnerable to attack by arylamines (Scheme 1 i) [42,44,47,51], azide (Scheme 1 ii) [54], hydroxide (Scheme 1 iii) [45], and thiols (Scheme 1 iv) [52], the outcomes from which are anomerically substituted intermediates **22–24** and **26** that ultimately may undergo HERON reactions. Furthermore, *N*-acyloxy-*N*-alkoxyamides **21** themselves may be synthesised in $S_N2$ reactions between *N*-alkoxy-*N*-chloroamides **25** and sodium carboxylates (Scheme 1 vi) [43,45,46,48,51,80,85]. *N*-alkoxy-*N*-chloroamides **25** also react bimolecularly with azide generating reactive *N*-alkoxy-*N*-azidoamides **26** (Scheme 1 v) [54].

Reactions of **21** systems with amines and thiols have been modelled at the AM1, HF/6-31G(d), and pBP/DN* levels, which reveal significant charge separation in the transition states and alkoxynitrenium ion character (Figure 12) [86]. These reactions should be favoured by electron-donor substituents on the nucleophile and electron-acceptor substituents on the acyloxyl group.

The $S_N2$ reaction of *N*-methylaniline with a wide range of *N*-acyloxy-*N*-alkoxyamides **21** has been studied (Scheme 1 i), and relative rate constants, Arrhenius activation energies, and entropies of activation are in accord with a transition state with significant charge separation [31,44,51,87]. $E_A$'s are of the order of 40 to 60 kJ mol$^{-1}$. Entropies of activation (−90–160 J K$^{-1}$ mol$^{-1}$) are more negative than found in $S_N2$ reactions of alkyl halides, owing to a greater degree of solvation in the charge separated transition state [88]. In addition, the rates for reactions i, iii, and iv with a series of NAA's bearing *N*-*p*-substituted benzoyloxyl leaving groups correlated with Hammett $\sigma$ constants with positive slope (i, $\varrho$ = 1.7, iii, $\varrho$ = 0.6, and iv, $\varrho$ = 1.1) [44,45,52]. In addition, a series of anilines reacted bimolecularly and rate constants correlate with Hammett $\sigma^+$ constants ($\varrho$ = −0.9) [44]. In most respects, the $S_N2$ reactions are electronically and geometrically reminiscent of those at carbon centres and are accelerated by electron-donor groups on the nucleophile and electron-withdrawing groups on the leaving group. The amide carbonyl facilitates $S_N2$ reactivity in line with enhanced reactivity in phenacyl bromides [89]. In particular, bimolecular reaction rates are radically impeded with branching $\alpha$ to the carbonyl [50], which is analogous to the resistance to $S_N2$ reactions of $\alpha$-halo ketones bearing substituents at the $\alpha'$ position [90,91].

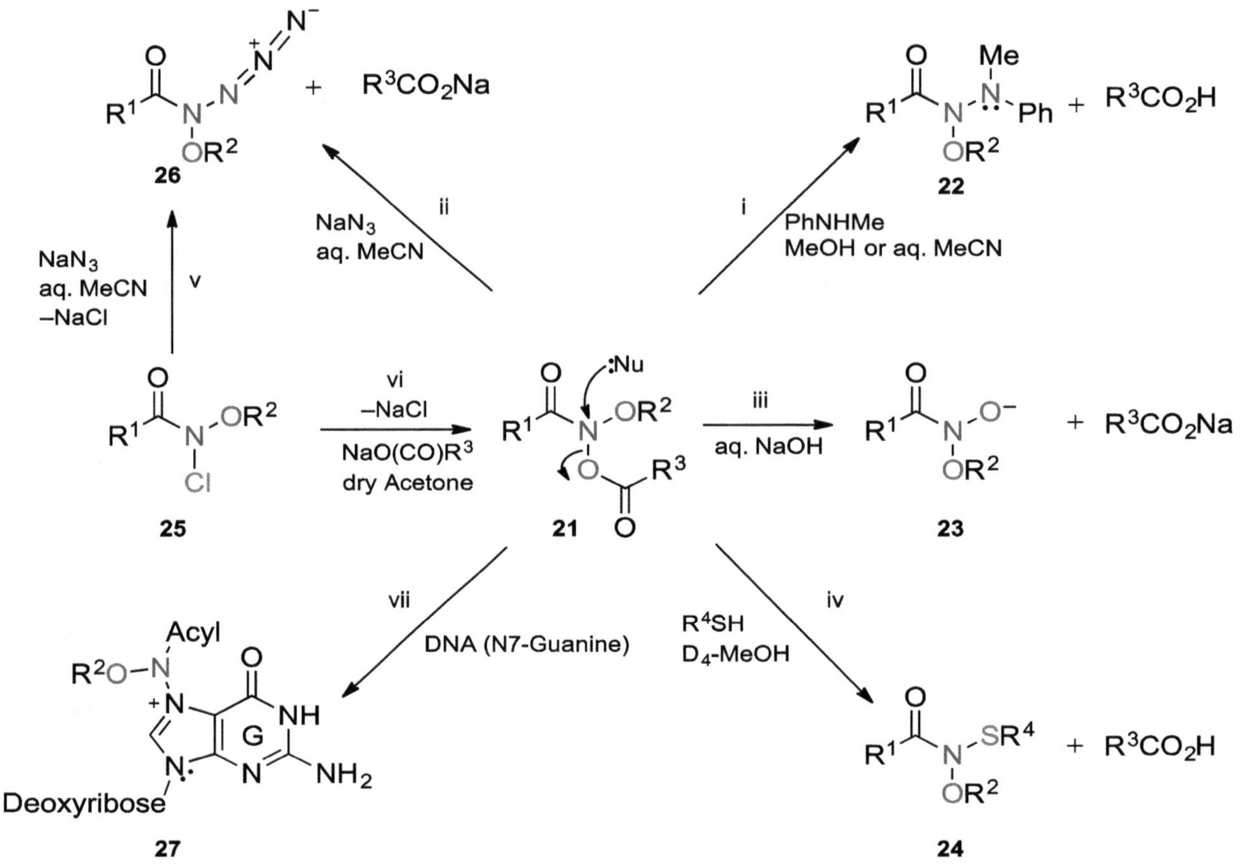

**Scheme 1.** $S_N2$ reactions of *N*-acyloxy-*N*-alkoxyamides **21** with: (i) arylamines; (ii) azide; (iii) hydroxide; (iv) thiols; (vii) DNA and reaction of *N*-alkoxy-*N*-chloroamides **25** with (v) azide and (vi) sodium carboxylates.

(**a**) (**b**)

**Figure 12.** HF/6-31G* predicted charge separation in transition states for reactions of *N*-formyloxy-*N*-methoxyformamide with (**a**) ammonia and (**b**) methanethiol.

Anomeric substitution at nitrogen in *N*-acyloxy-*N*-alkoxyamides **21** renders this class of amides as direct-acting mutagens. Mutagenicity towards *S. typhimurium* in the Ames reverse mutation assay does not require premetabolic activation [92,93]. Our DNA damage studies on plasmid DNA at physiological pH, as well as extensive structure–activity relationships [31,38–41,43,45,46,48–50,79,80], point to binding of NAA's **21**, intact, into the major groove of DNA, where an $S_N2$ reaction occurs at the most nucleophilic centre, the electron-rich N7 in guanine (Scheme 1 vii) [94–97]. All three side chains ($R^1$, $R^2$, and $R^3$) of **21** have an impact upon both DNA damage profiles as well as mutagenicity levels. An $S_N1$ mechanism, yielding electrophilic *N*-acyl-*N*-alkoxynitrenium ions, was ruled out since only $R^1$ and $R^2$ would influence binding and reactivity. Moreover, mutagenic activity is radically reduced when there is branching $\alpha$ to the carbonyl in parallel with the impaired $S_N2$ reactivity [40,50,51]. The mutagenic activity of *N*-acyloxy-*N*-alkoxyamides **21** has been used recently to show how hydrophobicity and intercalating side chains impact upon DNA binding [38,39].

### 3.1.2. Elimination Reactions

In an anomeric amide where $n_Y$–$\sigma^*_{NX}$ is a strong interaction, where X has a high electron affinity and Y is a strong electron donor, polarisation can lead to elimination of $X^-$, leaving a Y-stabilised nitrenium ion **28** (Scheme 2) [98]. The stronger the anomeric effect, the more readily the elimination is expected to occur. In the case of *N*-alkoxy-*N*-chloroamides **25**, elimination can be facilitated by Lewis acid complexation with X and by the use of polar solvents [35]. For example, treatment of *N*-chloro-*N*-(2-phenylethyloxy)- **29a** and *N*-chloro-*N*-(3-phenylpropyloxy)amides **29b** with silver tetrafluoroborate in ether, initiates a ring closing reaction to form *N*-acyl-1*H*-3,4-dihydro-2,1-benzoxazines **30a** and *N*-acyl-1,3,4,5-tetrahydrobenzoxazepines **30b**, respectively, via chlorine elimination to form nitrenium ions (Scheme 3). *N*-acyl-*N*-alkoxynitrenium ions are strongly stabilised by delocalisation of the positive charge onto oxygen [98,99]. This methodology has been used widely since its discovery in 1984 by Glover [100,101] and Kikugawa [102–108]. In addition, treatment of *N*-alkoxy-*N*-chloroamides **25** with silver carboxylates in diethyl ether, allows the nitrenium ion to be scavenged by carboxylate in a versatile reaction which has been used to synthesise a range of *N*-acyloxy-*N*-alkoxyamides **21** [48,80].

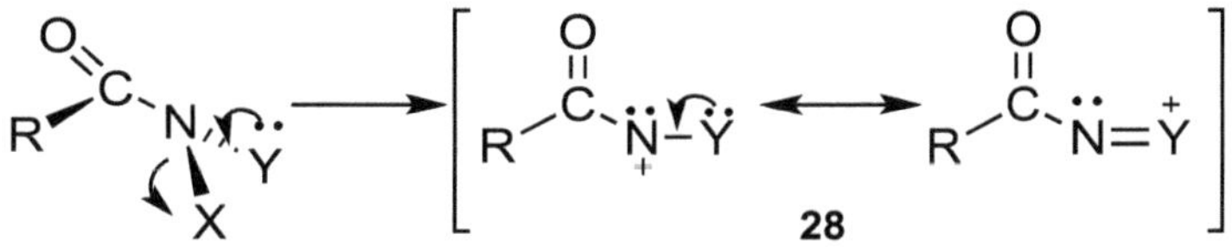

**Scheme 2.** Nitrenium ion formation by elimination of X due to a strong $n_Y$–$\sigma^*_{NX}$ interaction.

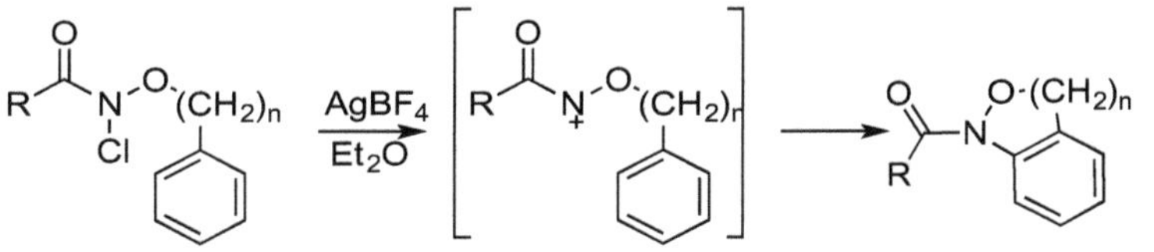

**Scheme 3.** Cyclisation by silver ion catalysed elimination reactions.

Elimination of chloride in the alcoholysis of **25** to give nitrenium ion provided a synthetic pathway to *N,N*-dialkoxyamides **5** (Scheme 4) [55,56]. We recently reported a more versatile synthesis effected by PIFA oxidation of hydroxamic esters **31** in appropriate alcohol, which proceeds through a reactive phenylbistrifluoroacetate derivative **32** [30,55]. Similar hypervalent iodine oxidations have been used in nitrenium ion cyclisations onto aromatic rings [105].

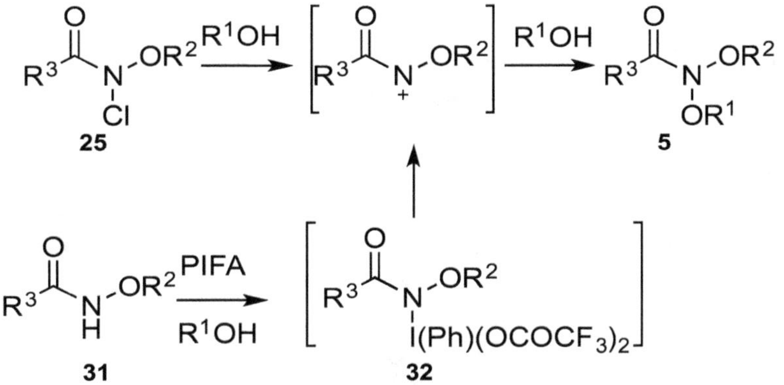

**Scheme 4.** *N*-Acyl-*N*-alkoxynitrenium ion mediated syntheses of *N,N*-dialkoxyamides **5**.

*N*-Alkoxy-*N*-benzoylnitrenium ions **34** are generated through $A_{A1}1$ acid-catalysed solvolysis of *N*-acetoxy-*N*-alkoxybenzamides **33** (Scheme 5) [46,48,49]. Acetoxyl, upon protonation with a catalytic amount of mineral acid, is eliminated from *N*-acetoxy-*N*-alkoxybenzamides **33** and the nitrenium ions are trapped by water to form *N*-alkoxyhydroxamic acids **35**. The anomeric **35** undergoes secondary reactions to form a range of products.

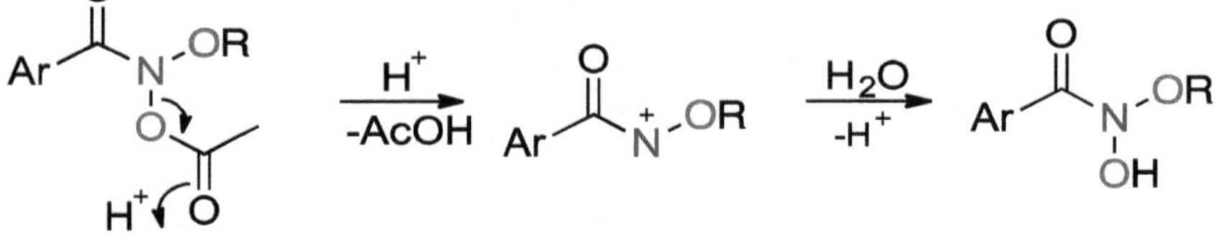

**Scheme 5.** Acid catalysed hydrolysis of *N*-acetoxy-*N*-butoxybenzamides.

The elimination reactions of *N*-alkoxy-*N*-chloroamides **25** and the acid-catalysed solvolysis reactions of **33**, both of which proceed through intermediacy of *N*-acyl-*N*-alkoxynitrenium ions, can better be re-evaluated in terms of anomeric destabilisation in combination with their reduced amidicities.

### 3.2. *The HERON Reaction*

#### 3.2.1. HERON Reactions of *N*-Amino-*N*-Alkoxymides

A novel reaction of suitably constituted anomeric amides is the HERON (Heteroatom Rearrangement On Nitrogen) reaction [56,109–113]. In such amides, when the X heteroatom of an anomerically destabilised N–X bond is a poor leaving group, the amide can undergo a concerted rearrangement involving the migration of X to the carbonyl carbon and the ejection of a Y-stabilised nitrene (Scheme 6).

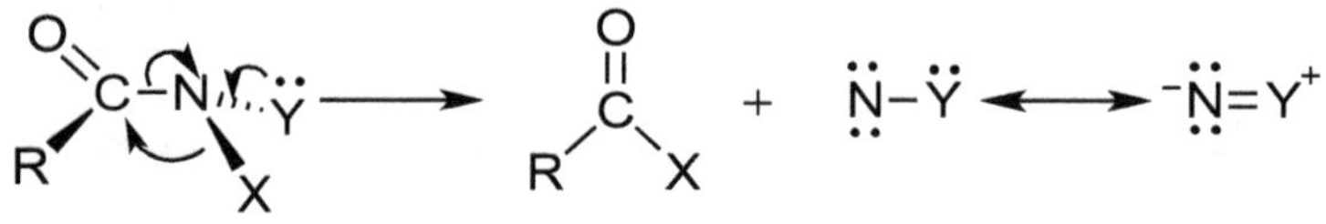

**Scheme 6.** The heteroatom rearrangement on nitrogen (HERON) reaction of an anomeric amide.

The HERON reaction was discovered by Glover and Campbell during research into $S_N2$ reactivity of *N*-acyloxy-*N*-alkoxyamides **21**, specifically the reaction between *N*-acetoxy-*N*-butoxybenzamide and *N*-methylaniline according to Scheme 1 i [47,111]. In a polar solvent such as methanol, *N*-methylaniline attacks the amide nitrogen, replacing the acetoxyl side chain to form an unstable intermediate, *N*-butoxy-*N*-(*N*′-methylanilino)benzamide **36**, which undergoes the HERON reaction to form butyl benzoate **37**, and an aminonitrene, 1-methyl-1-phenyldiazene **38** (Scheme 7 i). Aminonitrenes are highly reactive intermediates with a singlet ground state, which persist long enough under reaction conditions to dimerise to tetrazenes [114–117], in this case, **39** (Scheme 7 ii) [44,47]. *N*,*N*′-Diacyl-*N*,*N*′-dialkoxyhydrazines **40**, the only forms of *N*-alkoxy-*N*-aminoamides **2d** to have been isolated, undergo tandem HERON reactions to form two equivalents of ester **41** and a molecule of nitrogen; the *N*-acyl-*N*-alkoxyaminonitrenes **42**, formed in this HERON reaction, rapidly undergo a second rearrangement to form a molecule of nitrogen and ester before dimerisation of the *N*-acyl-*N*-alkoxyaminonitrene can occur (Scheme 8) [56,76]. Step ii can also be regarded as a HERON process, driven by a high energy electron pair on 1,1-diazene, a charge separated form of aminonitrene. Barton and coworkers studied the decomposition at about the same time, and both groups established the operation of three-centre mechanisms using asymmetric hydrazines [77]. In addition, Barton found its concerted nature facilitated the formation of a range of highly hindered esters and, recently,

Zhang has utilised the reaction to generate hindered esters from *N*,*N*′-dialkoxy-*N*,*N*′-diacylhydrazines, synthesised through *N*-bromosuccinimide (NBS) oxidation of hydroxamic esters [118].

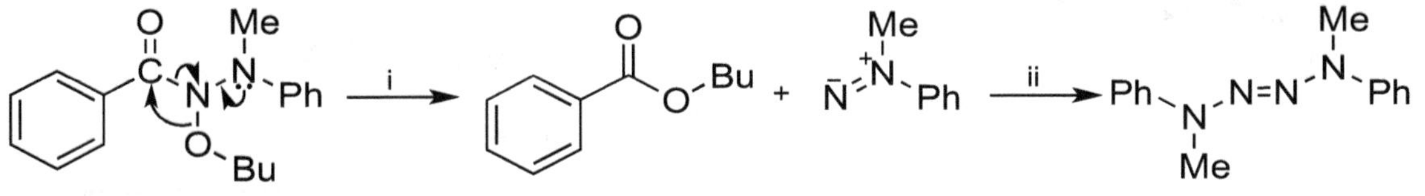

**Scheme 7.** The first HERON reaction of *N*-butoxy-*N*-(*N*′-methylanilino)benzamide **36**.

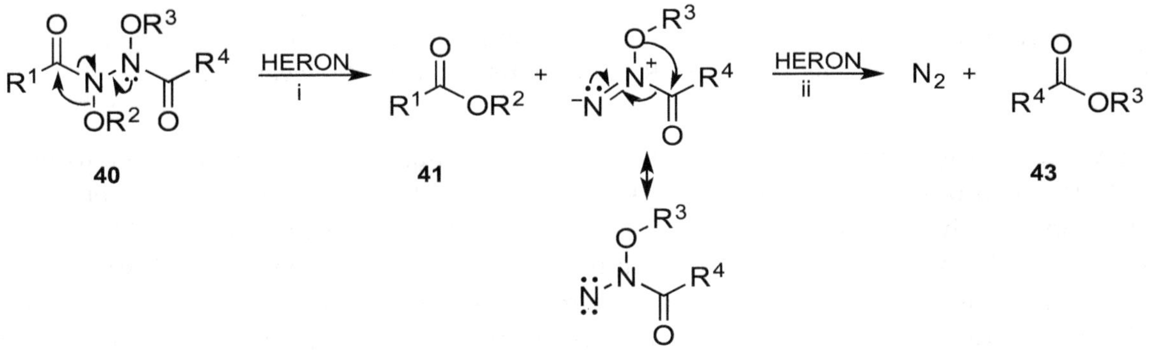

**Scheme 8.** Tandem HERON reactions of *N*,*N*′-diacyl-*N*,*N*′-dialkoxyhydrazines **40**.

3.2.2. Theoretical and Experimental Validation of the HERON Reaction

The HERON reaction of **22** and **40** has been modelled and validated computationally. Initially, AM1 modelling predicted that the three-centre reaction in the first HERON (Scheme 7 i) had an energy barrier of 184 kJ mol$^{-1}$, while that of the second step was very low (25 kJ mol$^{-1}$) [56,77]. An extensive AM1 study of HERON reactions of *N*-amino-*N*-alkoxyacetamides predicted a similar barrier of 159 kJ mol$^{-1}$ in the gas phase, but a lower barrier of 126 kJ mol$^{-1}$ in solution [111]. The same study also predicted lower barriers with electron-donor groups on the amino nitrogen. In a more rigorous study at the B3LYP/6-31G(d) level [76], Glover et al. modelled the HERON reaction of *N*-methoxy-*N*-dimethylaminoformamide **17e**, a model representative of *N*-(*N*′-methylanilino)-*N*-butoxybenzamide **36**, and *N*,*N*′- diacyl-*N*,*N*′-dialkoxyhydrazines **40**, for which HERON reactions had been experimentally observed. It was confirmed that the $n_N$–$\sigma^*_{NO}$ anomeric destabilisation resulted in migration of the methoxyl group with an activation barrier of 90 kJ mol$^{-1}$ and the reaction was exothermic by 23 kJ mol$^{-1}$. Similar diasteromeric transition states were located, but the transition state accessible from the lowest energy (*syn*) conformer of **17e** was found to be that depicted in Figure 13a. Importantly, modelling showed that amide resonance in the transition state was largely lost, as migration occurs in a plane perpendicular to the carbonyl, twisting the nitrogen lone pair away from alignment with the $\pi^*_{C=O}$ orbital. Anomeric destabilisation, however, remained along the reaction coordinate, driving the reaction forward. The N–C(O) bond is largely intact at the transition state but breaks as the O2–C1 bond forms in an internal, $S_N2$-like reaction at the amide carbon. Significantly, a tetrahedral intermediate is avoided by this process. Subsequent high level calculations on the decomposition of anomeric hydrazines by Tomson and Hall, yielded similar energetics for the HERON process [119].

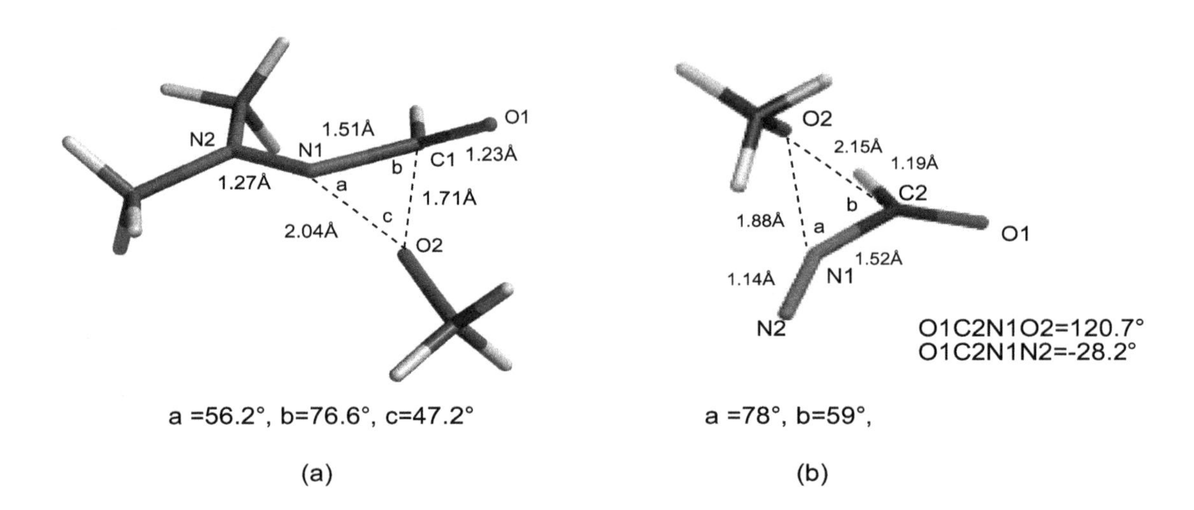


**Figure 13.** Twisted HERON transition states of (**a**) *N*-methoxy-*N*-dimethylaminoformamide **17e** and (**b**) 1-formyl-1-methoxydiazene **44** ($R^1$ = H, $R^2$ = Me) at B3LYP/6-31G(d) level [53,76].

Analysis of charge separation in the B3LYP/6-31G(d) transition state revealed a partial positive charge of +0.5 on amino group, partial negative charge of −0.3 on the migrating methoxyl group and little change in charge at the carbonyl. This indicated that HERON in these NNO systems could be assisted by polar solvents, electron-donating groups on the stationary amino substituent, and electron-withdrawing groups on the migrating oxygen substituent. The activation barriers and charge separation in the transition state were validated experimentally by Arrhenius studies and Hammett correlations from thermal decomposition of a range of substituted hydrazines **19a–e** and **20a–e** in mesitylene (Table 6) [76]. In **19**, the donor ability of $n_N$ is increased by electron-rich aroyl groups, leading to enhanced reaction rates and a negative Hammett $\sigma^+$ correlation ($\varrho$ = −0.35, $R^2$ = 0.978) (Figure 14a). However, acceptor benzyloxy substituents in **20** facilitate the migration, leading to a positive Hammett σ-correlation ($\varrho$ = 1.02, $R^2$ = 0.911) (Figure 14b). Rate constants were lower in **20** on account of a negative impact at the donor nitrogen. Donor groups, on the other hand, have little impact on the carbonyl in **19**.

**Table 6.** Hammett reaction constants, Arrhenius activation energies, entropies of activation, and rate constants for HERON decomposition of **19a–e** and **20a–e** [76].

| Series | System | $E_A$/kJ mol$^{-1}$ | $\Delta S^‡$/J K$^{-1}$ mol$^{-1}$ | $10^6 \cdot k_{298}$/s$^{-1}$ |
|---|---|---|---|---|
| Series **19**: $\sigma^+$ reaction constant $\varrho$ = −0.35 [1] | **19a** | 99.2 (4.3) | −21.2 (13) | 5.56 |
| | **19b** | 100.7 (8.2) | −19.6 (25) | 3.72 |
| | **19c** | 100.4 (5.1) | −23.8 (15) | 2.52 |
| | **19d** | 107.7 (1.7) | 0.9 (5) | 2.50 |
| | **19e** | 104.0 (1.7) | −15.4 (5) | 1.57 |
| | | | | $10^8 \cdot k_{298}$/s$^{-1}$ |
| Series **20**: σ reaction constant $\varrho$ = +1.02 [1] | **20a** | 111.4 (0.9) | −21.1 (2) | 4.02 |
| | **20b** | 125.9 (1.0) | 24.5 (3) | 2.76 |
| | **20c** | 114.1 (3.9) | −8.7 (11) | 6.21 |
| | **20d** | 125.1 (8.5) | 27.4 (24) | 5.44 |
| | **20e** | 98.8 (0.3) | −46.4 (1) | 31.9 |

[1] Hammett correlations using $k_{298K}$.

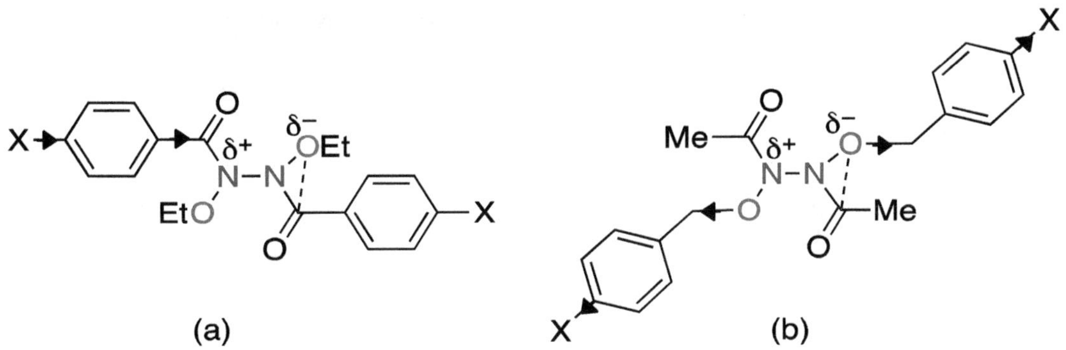

**Figure 14.** Influence on the HERON transition state of (**a**) electron-rich benzoyl groups in **19** and (**b**) electron-deficient benzyloxy groups in **20**.

3.2.3. HERON Reaction of 1-Acyl-1-Alkoxydiazenes

The second step in the thermal decomposition of *N,N′*-dialkoxy-*N,N′*-diacylhydrazines **40**, **19**, and **20** has also been modelled at both B3LYP/6-31G(d) and CCSD(T)//B3P86 level using *N*-formyl-*N*-methoxydiazene, and was found to have an extremely small $E_A$ of between 5 and 12 kJ mol$^{-1}$ and to be highly exothermic ($\Delta E = -400$ kJ mol$^{-1}$) [53,119]. We encountered this process from the reaction of *N*-acyloxy-*N*-alkoxyamides **21** with azide (Scheme 1 ii), which generates ester and two molecules of nitrogen [54]. The reaction of *N*-acyloxy-*N*-alkoxyamides with azide was originally conceived in an attempt to trap and determine the lifetimes of *N*-acyl-*N*-alkoxynitrenium ions, by analogy with the determination of lifetimes of arylnitrenium ions in water [120,121]. However, *N*-Alkoxy-*N*-azidoamides **26** are highly unstable intermediates, losing nitrogen to generate 1-acyl-1-alkoxydiazenes **44**, which react further to nitrogen and ester Scheme 9. The transition state for this reaction of *N*-formyl-*N*-methoxydiazene, modelled at the B3LYP/6-31G(d) level, is depicted in Figure 13b. Once again, the methoxyl group migrates in a plane orthogonal to the N1-C2-O1 plane in an earlier transition state with little N–C(O) bond cleavage, which occurs in concert with O2–C2 bond formation, again avoiding a tetrahedral intermediate. The reaction of azide with *N*-chloro-*N*-alkoxyamides **25** proved to be an excellent means of generating highly hindered esters [54]—not surprisingly, in light of the very low $E_A$ and extreme exothermicity born out of the entropically favourable generation of two highly stable molecules (methyl formate and nitrogen) [53,119]. Overall, the decomposition of *N*-azido-*N*-methoxyformamide to two molecules of nitrogen and methylformate was computed to be exothermic by some 575 kJ mol$^{-1}$ [53]. Yields of esters prepared by this method are given in Table 7.

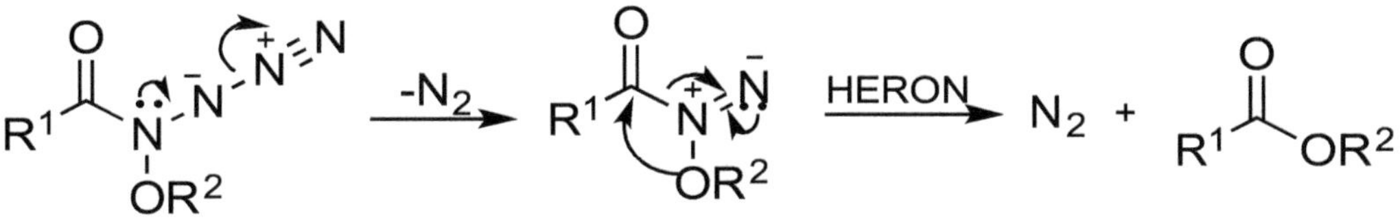

**Scheme 9.** Decomposition and HERON reaction of *N*-alkoxy-*N*-azidoamides **26**.

**Table 7.** Ester **41** ($R^1COOR^2$) formation from the reaction of *N*-alkoxy-*N*-chloroamides **25** ($R^1CONClOR^2$) with sodium azide in aqueous acetonitrile.

| R | R′ | Isolated Crude Yield/% |
|---|---|---|
| Ph | $(CH_3)_3C$ | 87 |
| $(CH_3)_3C$ | $(CH_3)_3C$ | 30 |
| 1-adamantyl | $(CH_3)_3C$ | 82 |
| $(CH_3)_3C$ | cyclohexyl | 97 |
| Ph | $(CH_3)_2CH$ | 92 |
| Ph | $PhCH_2$ | 93 |
| $CH_3$ | $PhCH_2$ | 92 |
| p-$NO_2C_6H_4$ | Et | 94 |
| Ph | Et | 94 |

Both Barton and recently Zhang have shown that the thermal decomposition of *N,N′*-dialkoxy-*N,N′*-diacylhydrazines is a source of hindered esters. The avoidance of a tetrahedral intermediate in both HERON steps of these reactions, and which is a limiting structure in Fisher esterification, is critical. This, and the clear role of anomeric substitution at nitrogen in both reducing amidicity, as well as promoting the rearrangement, are paramount.

### 3.2.4. HERON Reactions of Anionic Systems

Hydroxide substitution of the acyloxyl side chain from a range of *N*-acyloxy-*N*-alkoxybenzamides **21**, at room temperature in aqueous acetonitrile, generated esters and their hydrolysis products [45]. Rate data and crossover experiments pointed to an $S_N2$ reaction at nitrogen and the intramolecular nature of the reaction, implicating a HERON process. The hydroxamic acid intermediates **45** from initial attack (Scheme 1 iii) would be converted to their conjugate base **23** under basic reaction conditions, generating a strong $n_{O^-}$–$\sigma^*_{NO}$ anomeric destabilisation of the N–O bond (Scheme 10). A HERON migration of the alkoxyl side chain to the carbonyl carbon results in the formation of alkyl benzoate and the ejection of the nitric oxide anion (Scheme 10, $R^1$ = Ph). This route to esters from hydroxamic acids in equilibrium with **23** was earlier invoked to explain non-crossover ester formation in the $A_{Al}1$ solvolysis of *N*-acyloxy-*N*-alkoxyamides at low acid concentrations (Scheme 5) [45,46,48,49].

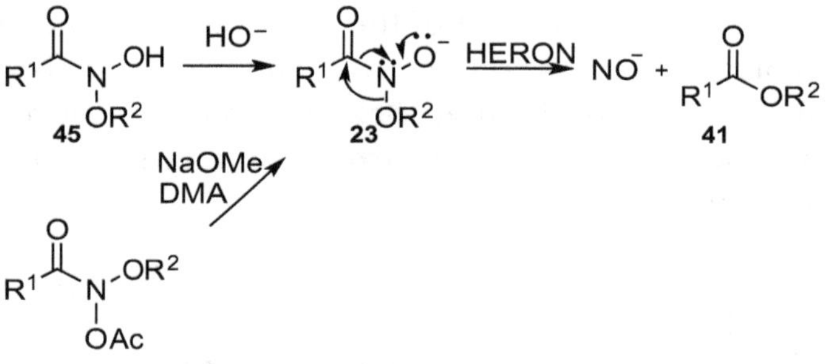

**Scheme 10.** HERON reaction of hydroxamate anion **23**.

The HERON reaction was also invoked by Shtamburg and coworkers to account for formation of ethyl benzoate **41** ($R^1$ = Ph, $R^2$ = Et) when *N*-acetoxy-*N*-ethoxybenzamide **46** ($R^1$ = Ph, $R^2$ = Et) was treated with methoxide in aprotic media. Anion **23** ($R^1$ = Ph, $R^2$ = Et) leading to the HERON reaction was generated by methoxide addition at the ester carbonyl (Scheme 10), while methoxide attack at the amide carbonyl lead to the formation of methyl benzoate [122].

Dialkyl azadicarboxylates, widely used in the Mitsonobu reaction, decompose vigorously with methoxide in methanol [123,124]. **47a** afforded methyl isopropyl carbonate **49a** and isopropyl formate **50a** in a 1:1 ratio by $^1$H NMR, and diethyl azadicarboxylate **47b** behaved similarly, though volatile ethyl formate **50b** was less prevalent in the reaction mixture (Scheme 11) [112,125]. Since the nitrogens in **47** are the overwhelming contributors to the LUMO of azadicarboxylates [112], the most probable route to these products is methoxide addition at nitrogen and a facile HERON reaction of the anionic adducts **48**. Calculations based on the HERON reaction of dimethyl azodicarboxylate **5c**, gave an $E_A$ of 27 kJ mol$^{-1}$ and exothermicity of 59 kJ mol$^{-1}$, at the B3LYP/6-31G*//HF/6-31G(d) level [112].

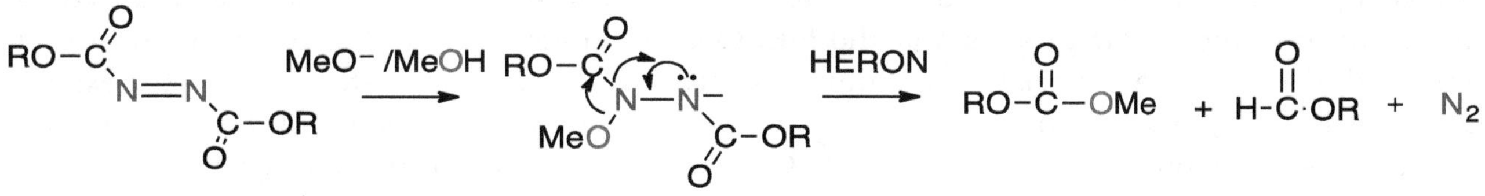

a: R = *i*-Pr
b: R = Et
c: R = Me

**Scheme 11.** Amide anion induced HERON reaction in reactions of azadicarboxylates with methoxide.

### 3.2.5. HERON Reactions of *N*-Alkoxy-*N*-Aminocarbamates

By analogy with the HERON reactions of *N*-acyloxy-*N*-alkoxyamides, several carbamates have been shown to undergo a similar reaction (Scheme 12). *N*-Acetoxy-*O*-alkyl-*N*-benzyloxycarbamates **51a–c** and *N*-methylaniline reacted bimolecularly in [D$_4$]-methanol to produce the corresponding carbonates **53** and tetrazene **39**, presumably through HERON reaction of the *N*-methylanilino intermediate **52** [126].

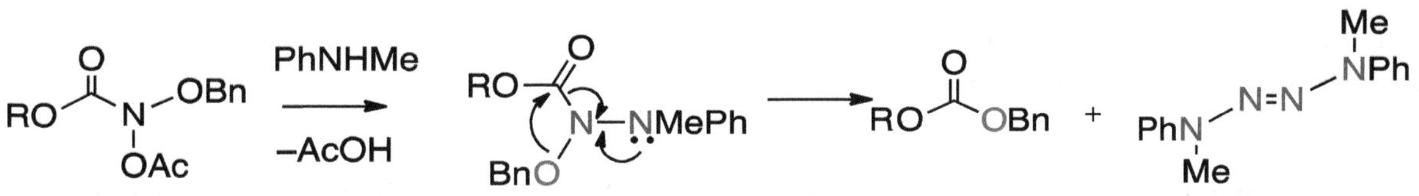

a: R = Me
b: R = Et
c: R = Bn

**Scheme 12.** HERON reactions of *N*-alkoxy-*N*-aminocarbamates.

### 3.2.6. HERON Reactions of *N*-Acyloxy-*N*-Alkoxyamides

Based on RE's of models **15c** and **16a** in Table 4, the amidicity of NNO systems such as *N*-alkoxy-*N*-aminoamides **2d** or *N*-alkoxy-*N*-aminocarbamates such as **52**, is likely to be reduced by modest amounts (60–70% that of *N*,*N*-dimethylacetamide), yet these systems undergo HERON reactions at room temperature, as do the 1,1-diazenes **2h** and hydroxamates **2g**. However, all have in common a strong anomeric destabilisation of the N–O bond through high energy electron pairs on the donor atom, $n_Y$, and high electronegativity of oxygen. $n_O$–$\sigma^*_{NO}$ systems such as *N*-acyloxy-*N*-alkoxyamides **2b** and *N*,*N*-dialkoxyamides **2c**, on the other hand, should have lower RE's (amidicities approximately 50% that of *N*,*N*-dimethylacetamide), yet they are thermally stable at room temperature on account of a weaker $n_O$–$\sigma^*_{NO}$ anomeric interaction. However, at elevated temperatures, *N*-acyloxy-*N*-alkoxyamides **2b** also undergo HERON reactivity [85].

A tandem mass spectrometric analysis by electrospray ionisation, used in characterising mutagenic *N*-acyloxy-*N*-alkoxyamides **21** (Scheme 13), produced, in addition to sodiated parent compound **54**, three characteristic sodiated ions, a sodiated alkoxyamidyl radical, **55**, which was generally most prevalent, a sodiated anhydride, **56**, and a minor cation due to sodiated ester, **57**, which was absent in the ionisation of aliphatic amides. While the fragments formed in parallel with each product ion, **58–60**, were undetectable, under these conditions the source of anhydrides must be an intramolecular process and the HERON rearrangement at elevated temperature was implicated. Furthermore, the relative amounts of sodiated anhydride and ester reflected the expected bias towards an $n_O$–$\sigma^*_{NOAc}$ anomeric stabilisation with attendant weakening of the N–OAc, rather than the N–OR, bond [112]. A B3LYP/6-31G(d) computational study of migration tendencies in *N*-formyloxy-*N*-methoxyformamide predicted high $E_A$'s for migration of both formyloxyl and methoxyl in the gas phase (162 and 182 kJ mol$^{-1}$, respectively) [112]. While acyloxyl, rather than alkoxyl, migration would be energetically more favourable, in solution at room temperature and particularly in polar media, the HERON reaction would not be competitive with $S_N2$ and $S_N1$ reactions at nitrogen.

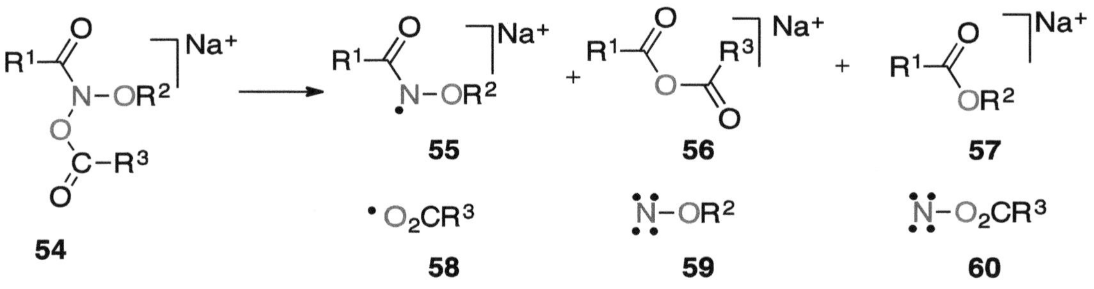

**Scheme 13.** Fragments from collision induced electrospray ionisation-mass spectrometric (ES-MS) analysis of *N*-acyloxy-*N*-alkoxyamides **21**.

However, in toluene at 90 °C, HERON reactions of **21** have been detected in competition with a homolytic decomposition pathway, and analysis of the complex reaction mixtures provides further support for the driving force behind the HERON process (Scheme 14) [85]. Homolysis of the N–OAc bond in **61** gave relatively long-lived alkoxyamidyl radicals **62**, which in solvent cage reactions with product radicals generated dioxazole **63**, or upon escaping the solvent cage dimerise to hydrazines leading to the expected thermal decomposition esters. The HERON reaction of **61** generates anhydrides **65** and alkoxynitrenes **66**. Anhydrides, which in the case of symmetrical benzoic anhydride from **61a** and mixed benzoyl heptanoyl anhydride from **61b** were relatively stable, react further to give esters **68** and **69** with alcohols **67** generated in the reaction mixture, the source of which was the alkoxynitrenes, the other HERON product. Critical evidence for the HERON process derived from products of alkoxynitrenes **66**, which (1) could be trapped by oxygen; (2) dimerised to hyponitrites; or, (3) underwent characteristic rearrangements leading aldehydes, nitriles, and alcohols. Competition between the HERON and homolytic reaction pathways was evident in a comparison of products from **61c–e**. The polarity of these HERON transition states would require a build-up of positive charge on the donor alkoxyl oxygen, $n_O$. This would be stabilised by electron-donor *para* substituents on the benzyloxy group, but destabilised by electron-withdrawing *para* substituents. In accord with this, **61c** and **61e** generated dioxazole and esters. Little dioxazole was formed from **61d** in which the methoxyl group would lower the energy of the HERON transition state, but would have little impact on the non-polar transition state for homolysis of the N–OAc bond (Figure 15).

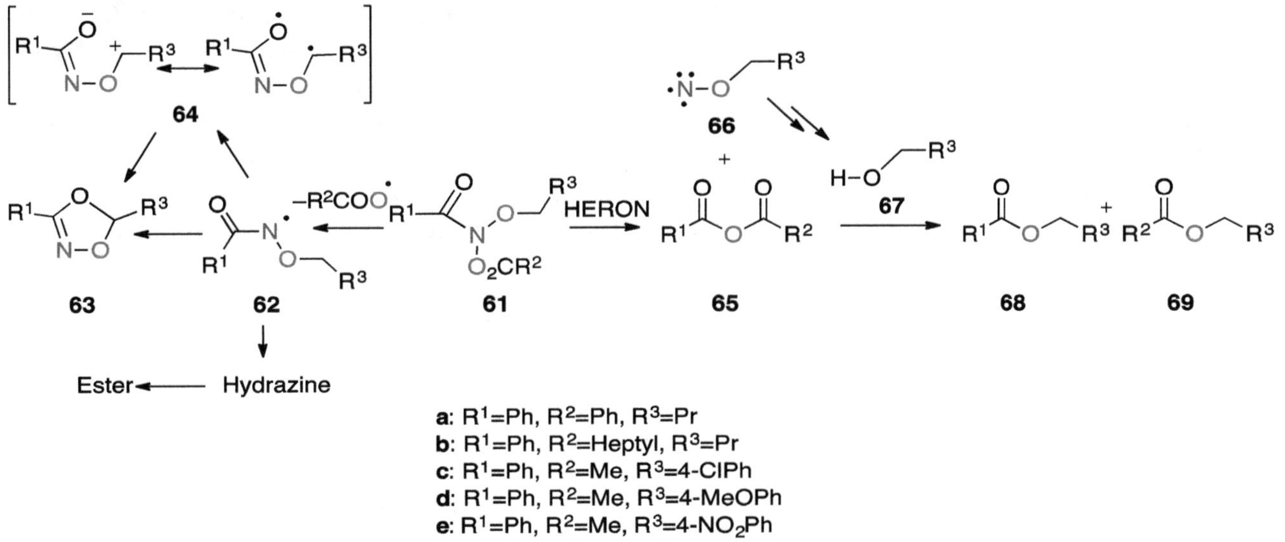

a: $R^1$=Ph, $R^2$=Ph, $R^3$=Pr
b: $R^1$=Ph, $R^2$=Heptyl, $R^3$=Pr
c: $R^1$=Ph, $R^2$=Me, $R^3$=4-ClPh
d: $R^1$=Ph, $R^2$=Me, $R^3$=4-MeOPh
e: $R^1$=Ph, $R^2$=Me, $R^3$=4-$NO_2$Ph

**Scheme 14.** HERON and radical decomposition pathways for *N*-acyloxy-*N*-alkoxyamides in toluene at 90 °C.

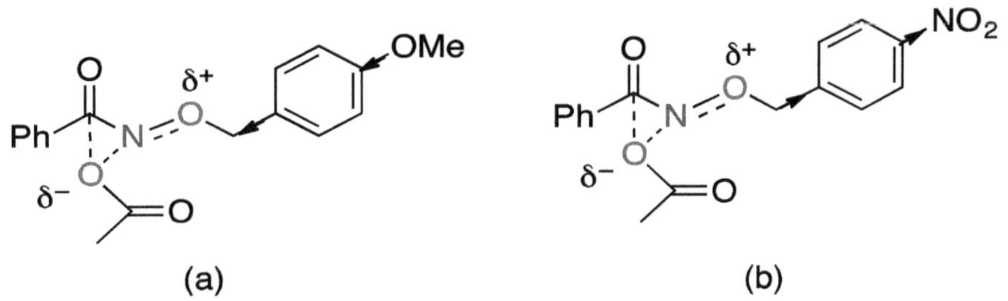

**Figure 15.** (**a**) Stabilisation of the HERON transition state for *N*-acyloxy-*N*-alkoxyamides by *para* methoxyl group and (**b**) destabilisation by a *para* nitro group.

3.2.7. HERON Reactions of *N,N*-Dialkoxyamides

Like *N*-acyloxy-*N*-alkoxyamides **2b**, *N,N*-dialkoxyamides **2c** possess low amidicity and they are thermally unstable, but require higher temperatures (typically 155 °C in mesitylene). However, their reaction proceeds exclusively by homolysis. Secondary products from alkoxynitrenes, which would be produced by the HERON pathway, were not observed for acyclic *N,N*-dialkoxyamides. Rather, they produce alkoxyamidyl radicals, which dimerise to *N,N'*-dialkoxy-*N,N'*-diacylhydrazines and ultimately esters. In addition, they can be trapped by hydrogen donors and solvent derived radicals [55].

On the other hand, room temperature HERON reactivity was found to occur exclusively in several alicyclic ONO systems [30]. Cyclic *N,N*-dialkoxyamides, *N*-butoxy-3(2*H*)-benzisoxazolone **73**, *N*-butoxyisoxazolidin-3-one **74**, and *N*-butoxytetrahydro-2*H*-1,2-oxazin-3-one **75,** can be synthesised by PIFA oxidation of the salicamide **70**, $\beta^-$ and $\gamma$-hydroxyhydroxamic esters, **71** and **72**, respectively, by analogy with the synthesis of acyclic *N,N*-dialkoxyamides (Scheme 4) [55]. Only *N*-butoxy-3(2*H*)-benzisoxazolone **73** is stable; *N*-butoxyisoxazolidin-3-one **74** and *N*-butoxytetrahydro-2*H*-1,2-oxazin-3-one **75** both react at room temperature and the reactions can be monitored by $^1$H NMR and mass spectrometry (Scheme 15). The $\gamma$-oxazinolactam undergoes quantitative ring opening to a diastereomeric mixture of the stable hyponitrite **76**, which must arise from dimerization of alkoxynitrene **77**, a known reaction of alkoxynitrenes. The $\delta$-oxazinolactam, on the other hand, undergoes a quantitative ring contraction to the $\gamma$-valerolactone **78** with production of butoxynitrene **79**. Both are clearly HERON reactions [30].

The $E_A$ for both HERON reactions must be radically lower than that for acyclic *N,N*-dialkoxyamides. Analysis of the B3LYP/6-31G(d) optimised ground state structures of *N*-methoxy-$\gamma$-oxazinolactam **81** (Figure 16b) and *N*-methoxy-$\delta$-oxazinolactam **82** (Figure 16c) provide insight into this unusual difference in reactivity. Firstly, the model $\gamma$-oxazinolactam is strongly pyramidal at nitrogen ($\chi_N = 64.6°$) and significantly twisted ($\tau = -36°$), with attendant loss of amide character. The N–C(O) bond is very long compared to *N,N*-dimethoxyacetamide **4b** (1.417 Å, Table 3). The COSNAR and TA resonance energies for **81** are $-20$ and $-19$ kJ mol$^{-1}$, respectively, translating to amidicities of only 26% and 25%. Torsion angles close to 90° indicate that the *endo* and *exo* oxygen lone pairs are ideally aligned for maximum $n_O$–$\sigma^*_{NO}$ stabilisation. Low amidicity and a strong stereoelectronic effect would favour either reaction but, clearly, ring opening would be more favourable than ring contraction, which would give a highly strained $\beta$-lactone.

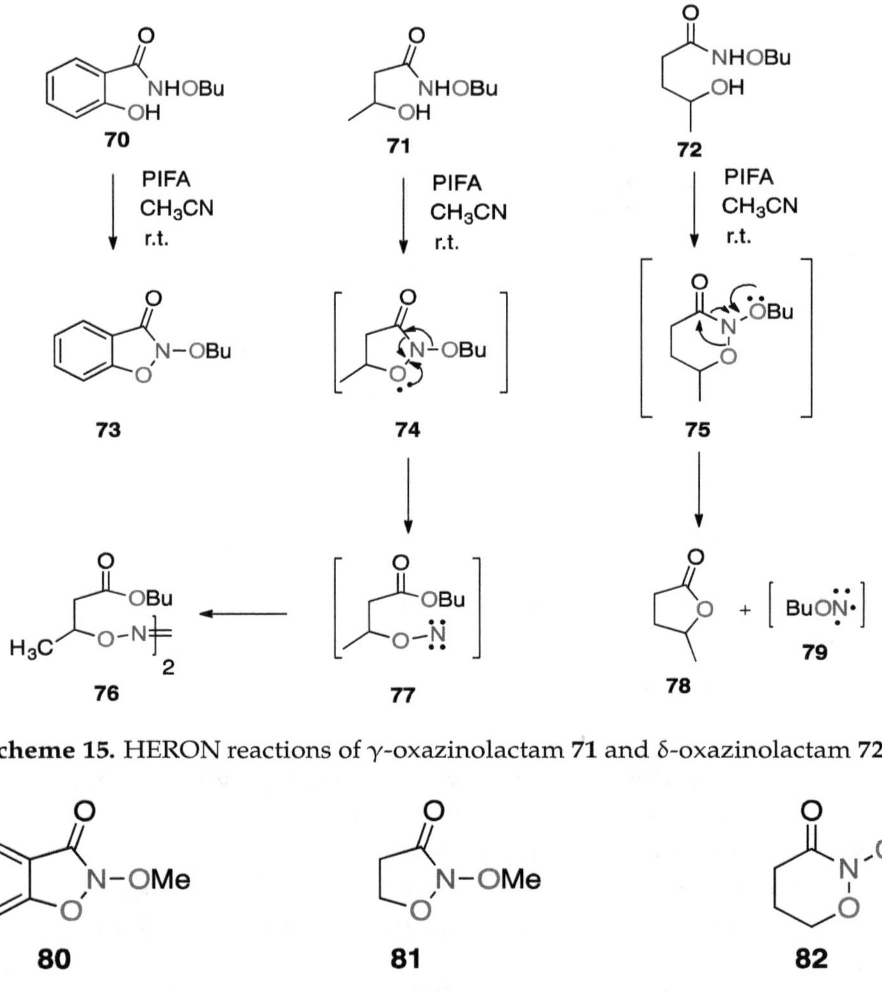

**Scheme 15.** HERON reactions of $\gamma$-oxazinolactam **71** and $\delta$-oxazinolactam **72**.

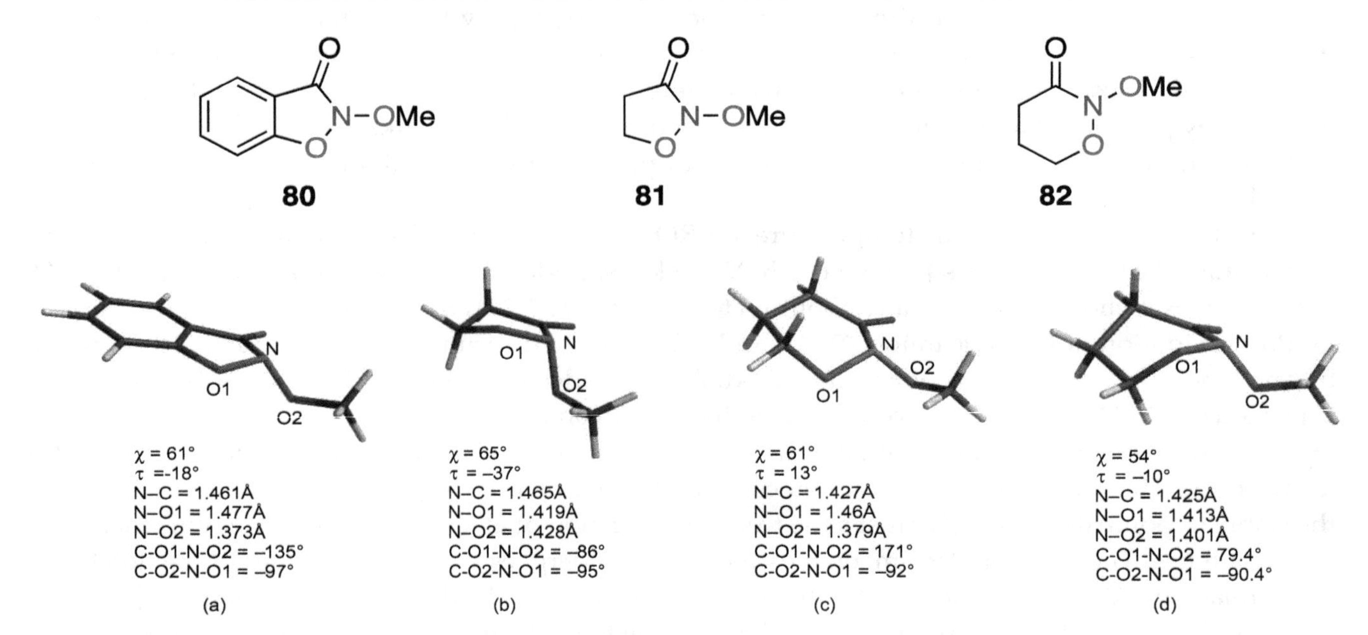


**Figure 16.** Ground state structures of models (**a**) *N*-methoxybenzisoxazolone **80**, (**b**) *N*-methoxy-$\gamma$-oxazinolactam **81**, and (**c**) chair and (**d**) boat *N*-methoxy-$\delta$-oxazinolactam **82**.

The B3LYP/6-31G(d) structure of the model δ-oxazinolactam **82** ($\chi_N = 61°$ and $\tau = 13°$) matches the experimental ($^1$H NMR) chair conformation of **75** which has a chiral nitrogen. It is also more pyramidal at nitrogen and slightly more twisted than the alicyclic *N,N*-dimethoxyacetamide **4b** ($\chi_N = 48°$ and $\tau = 9°$, Figure 7c, Table 3), but its resonance and amidicity ($RE_{\mathbf{COSNAR}} = -38$ kJ mol$^{-1}$, amidicity 49%, and $RE_{\mathbf{TA}} = -37$ kJ mol$^{-1}$, amidicity 47%) is almost identical to the open chain form. However, only an $n_{O(exo)}$–$\sigma^*_{NO(endo)}$ anomeric alignment is evident; with a torsion angle of 171°, the $n_{O(endo)}$–$\sigma^*_{NO(exo)}$ interaction is completely switched off. The *endo* N–O bond is nearly 0.1 Å longer than the *exo* N–O bond (the *endo* bond is marginally shorter than the *exo* bond by 0.01 Å in γ-oxazinolactam and N–O bonds differ by 0.02 Å in *N,N*-dimethoxyacetamide). Ring opening and ring contraction are computed to have about the same $E_A$ and $\Delta H^{\ddagger}$ at B3LYP/6-31G(d) [30]. It is evident that the ring contraction of the δ-oxazinolactam to γ-butyrolactone is largely driven by a strong, conformationally imposed anomeric effect, a remarkable impact of anomeric substitution at an amide nitrogen [30]. While the computed transition state for ring opening of model *N*-methoxy-δ-oxazinolactam is marginally lower in energy, the required $n_{O(endo)}$–$\sigma^*_{NO(exo)}$ is only accessible from the boat conformation of the δ-oxazinolactam (Figure 16d). Experimentally, accessing this conformation in **75** by nitrogen inversion could be energetically unfavourable, owing to steric hindrance between the axial 4-methyl and *N*-butoxy groups. However, even in the boat conformation, the strong $n_{O(exo)}$–$\sigma^*_{NO(endo)}$, which favours ring contraction is still evident.

The γ-lactam in benzisoxazolone is stable at room temperature, though model **80** has a moderately suitable anomeric alignment for migration of O2. However, the $n_{O(endo)}$–$\sigma^*_{NO(exo)}$ interaction is probably weakened by conjugation of the $O_{endo}$ p-type lone pair onto the aromatic ring. Ring contraction driven by a favourable $n_{O(exo)}$–$\sigma^*_{NO(endo)}$ interaction would be disfavoured.

### 3.2.8. *N*-Alkoxy-*N*-Alkylthiylamides

The combination of sulphur and oxygen attachment to amide nitrogen in *N*-methoxy-*N*-methylthiylacetamide **15d** results in a similar reduction in amide resonance to that of *N*-methoxy-*N*-dimethylaminoacetamide **15c**, namely about 64% vs. 67% (Table 4). However, the reaction of *N*-acyloxy-*N*-alkoxyamides **21** with biological thiols, glutathione, and methyl and ethyl esters of cysteine, which resulted in $S_N2$ displacement of carboxylate, produced exclusively hydroxamic esters **83** and disulphides **84** [52] (Scheme 16). The driving force for HERON reactivity is not only reduced resonance, but the anomeric effect. Instead of HERON reactions, the intermediate *N*-alkoxy-*N*-alkylthiylamides **24** undergo an $S_N2$ reaction at sulphur by thiol. The distinction between reactivity modes for NNO and SNO systems lies in the $n_S$–$\sigma^*_{NO}$ anomeric interaction, which is much weaker than the $n_N$–$\sigma^*_{NO}$ of *N*-alkoxy-*N*-aminoamides.

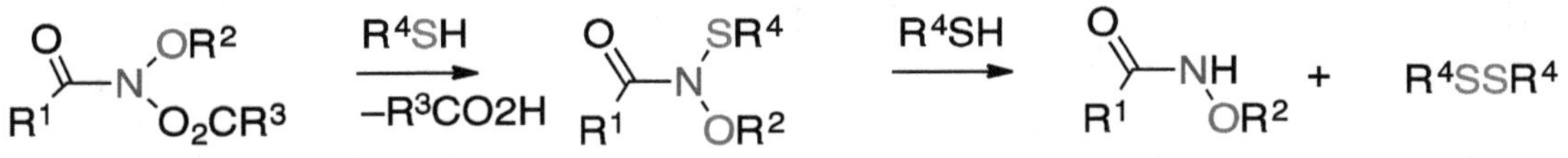

$R^1C(O)N(OR^2)(O_2CR^3)$ (**21**) $\xrightarrow[-R^3CO_2H]{R^4SH}$ $R^1C(O)N(SR^4)(OR^2)$ (**24**) $\xrightarrow{R^4SH}$ $R^1C(O)NH(OR^2)$ (**83**) + $R^4SSR^4$ (**84**)

**Scheme 16.** Reaction of *N*-acyloxy-*N*-alkoxyamides with alkylthiols.

### 3.3. *Driving Force for the HERON Reaction*

The most accessible transition states for HERON migration of methoxyl in a number of model anomeric amides, *N*-methoxy-*N*-dimethylaminoacetamide **15c**, *N,N*-dimethoxyacetamide **4b**, *N*-acetoxy-*N*-methoxyacetamide **15b**, and *O*-methyl-*N*-methoxy-*N*-dimethylaminocarbamate **16a**, ring opening of *N*-methoxy-γ-oxazinolactam **81**, and ring contracting of *N*-methoxy-δ-oxazinolactam **82**, each of which represents a class of neutral anomeric amides known to undergo HERON reactions, as well as for methoxyl migration in *N*-methoxy-*N*-methylthiylacetamide **15d**, *N*-methoxyacetohydroxamate **15f**, and 1-acetyl-1-methoxydiazene **15g**, have been derived at the

B3LYP/6-31G(d) level as part of several studies [30,61,62]. Transition state geometries of **4b**,**15b–d**,**f** and **g**, **16a**, **81**, and **82** are presented in Figure 17.

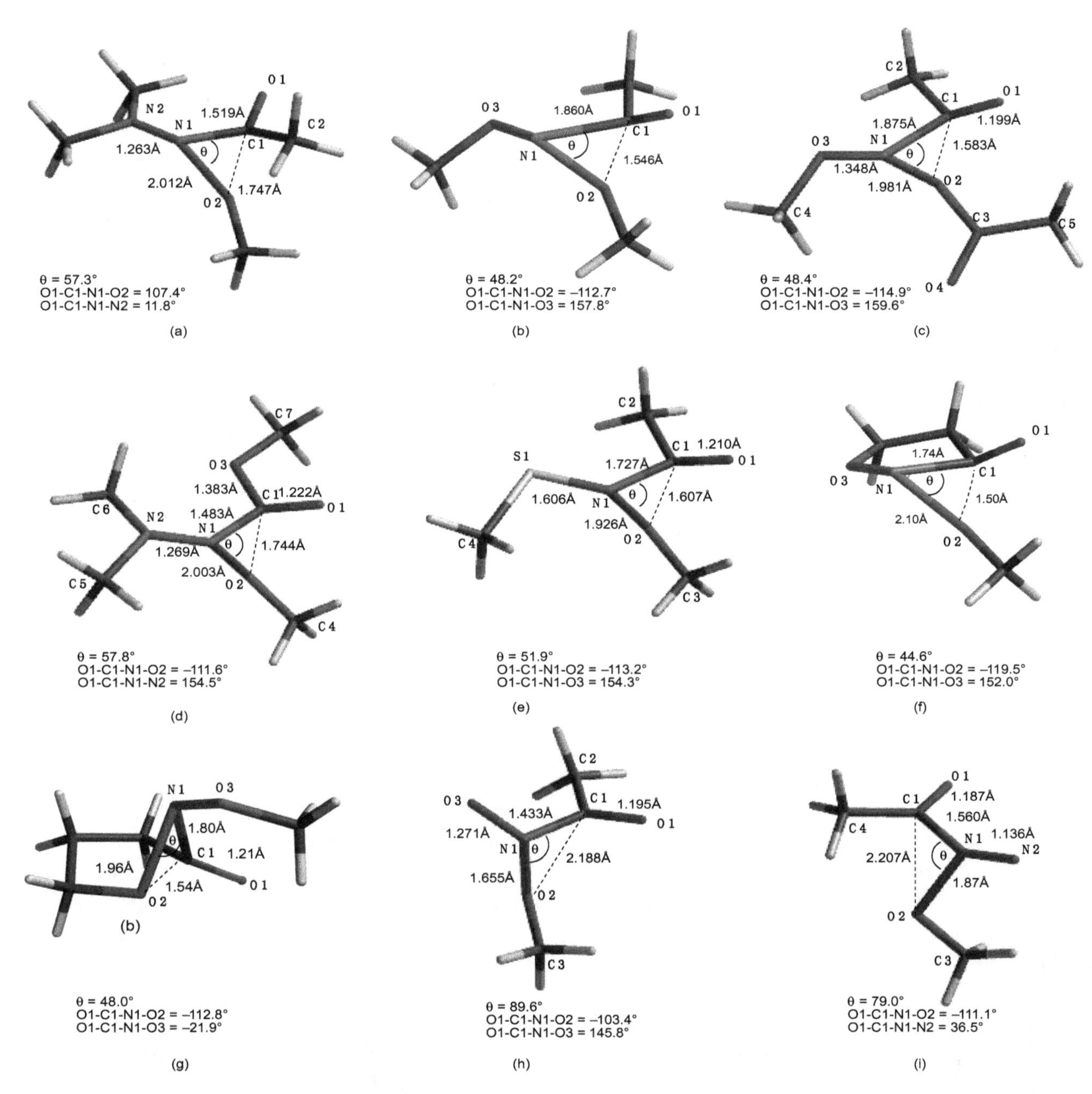


**Figure 17.** B3LYP/6-31G(d) optimised transition states of favoured HERON reactions of **(a)** *N*-methoxy-*N*-dimethylaminoacetamide **15c**, **(b)** *N,N*-dimethoxyacetamide **4b**, **(c)** *N*-acetoxy-*N*-methoxyamide **15b**, **(d)** *O*-methyl-*N*-methoxy-*N*-dimethylaminocarbamate **16a**, **(e)** *N*-methoxy-*N*-methylthiylacetamide **15d**, **(f)** *N*-methoxy-γ-oxazinolactam **81** (ring opening), **(g)** *N*-methoxy-δ-oxazinolactam **82** (ring contraction), **(h)** *N*-methoxyacetohydroxamate **15f** and **(i)** 1-acetyl-1-methoxydiazene **15g**.

In all the transition state complexes, the migrating oxygen does so in a plane largely orthogonal to the N1–C1–O1 plane and the donor atom $n_Y$, driving the migration, is largely in the N1–C1–O1 plane. As a consequence, the amide nitrogen lone pair lies close to the plane and amide resonance is largely lost in the transition state. The resonance energy, RE is therefore a component of the overall $E_A$, the balance being the energy required for the rearrangment under anomeric assistance, $E_{rearr}$, and must reflect the relative nature of the $n_X$–$\sigma^*_{NO}$ driving force. It is therefore possible to approximate the influence of the anomeric substituents on the resonance interaction on the one hand, and on the migration process, on the other. Table 8 gives the $E_A$, $RE_{TA}$, and net $E_{rearr}$ data for these transition states.

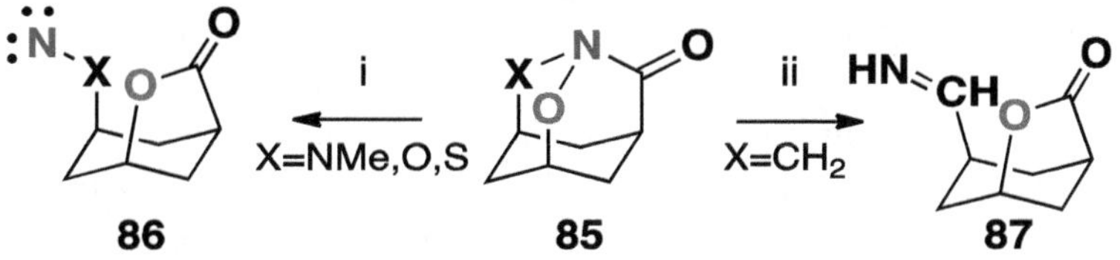

**Scheme 17.** HERON migrations of alkoxyl in the absence of resonance.

**Table 8.** B3LYP/6-31G(d) activation energies ($E_A$), resonance energies ($RE_{TA}$), and rearrangement energies ($E_{rearr}$) for HERON reactions of model anomeric amides and tricyclic anomeric amides; all values in kJ mol$^{-1}$.

| Migratory Mode | $E_A$ | XNY [1] | $RE_{TA}$ [2] | $E_{rearr}$ |
|---|---|---|---|---|
| MeO from *anti* **15c** | 95.0 | ONN | −51.5 | 43.5 |
| MeO from **4b** | 156.5 | ONO | −35.9 | 120.5 |
| MeO from *syn* **16a** | 92.4 | ONN | −50.7 | 41.8 |
| AcO from *syn* **15b** | 181.0 | AcONO | −40.5 | 140.5 |
| MeO from *syn* **15b** | 207.9 | ONOAc | −40.5 | 167.4 |
| AcO from *anti* **15b** | 181.5 | AcONO | −40.5 | 141.0 |
| MeO from *anti* **15b** | 182.8 | ONOAc | −40.5 | 142.3 |
| MeO from *anti* **15d** | 174.1 | ONS | −48.6 | 125.4 |
| MeO from **15f** [3] | 48.0 | ONO$^-$ | −32.4 | 15.6 |
| MeO from **15g** | 8.8 | ONNitrene | −14.8 | −6.0 |
| Ring opening **81** | 113.0 | ONO | −19.2 | 93.8 |
| Ring contraction **82** | 145.2 | ONO | −36.7 | 108.4 |
| Ring opening of **82** | 136.4 | ONO | −36.7 | 99.7 |
| Scheme 17 i X=NMe | 59.0 | ONN | 0.0 | 59.0 |
| Scheme 17 i X=O | 146.9 | ONO | 0.0 | 146.9 |
| Scheme 17 i X=S | 152.7 | ONS | 0.0 | 152.7 |
| Scheme 17 ii X=$CH_2$ | 242.7 | $ONCH_2$ | 0.0 | 242.7 |
| Scheme 18 i | 133.5 | AcONO | 0.0 | 133.5 |
| Scheme 18 ii | 161.5 | ONOAc | 0.0 | 161.5 |

[1] $n_Y$ donor to $\sigma^*_{NX}$; [2] Transamidation data (COSNAR data very similar); [3] HF/6-31G(d) values.

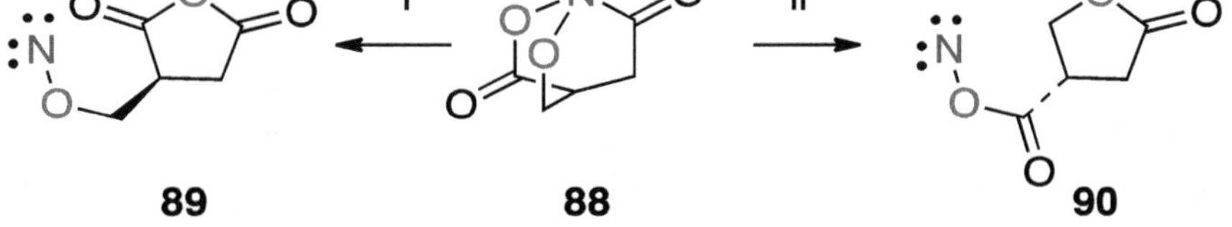

**Scheme 18.** HERON reactions of *N*-acyloxy-*N*-alkoxyamide in absence of resonance.

Methoxyl migration in the NNO systems **15c** and **16a** ($E_A$'s 95 and 92 kJ mol$_-^1$ respectively) have similar RE's of around −50 kJ mol$^{-1}$, and therefore $E_{rearr}$ of approximately 40 kJ mol$^{-1}$. The change in charge on the carbonyl carbon in the HERON transition state is negligible [76], so replacement of methyl by methoxyl has little bearing on the rearrangement energies. ONO in **4b** and ONOAc in **15b** migrations have much higher $E_A$'s despite the lower amide resonance energy. The large difference lies in the $E_{rearr}$ which is nearly 100 kJ mol$^{-1}$ less favourable and a reflection of the relative efficacy of the $n_N$–$\sigma^*_{NO}$ vs. the weaker $n_O$–$\sigma^*_{NO}$ anomeric interaction. The difference of about 80 kJ mol$^{-1}$ between methoxy migration in the ONS **15d** and ONN acetamides **15c** lies, again, in the much weaker n –$\sigma^*_{NO}$ anomeric effect. This can be accounted for by size mismatch due the larger 3p orbitals of the sulphur. Both HERON reactions of the cyclic forms of *N*,*N*-dialkoxyamides **81** and **82** have substantially lower overall $E_A$'s than *N,N*-dimethoxyacetamide **4b**. After RE has been taken into account, $E_{rearr}$ values indicate that in the cyclic forms, reorganisation to the HERON transition state is easier. Better stereoelectronic control in the cyclic system accounts for this.

The RE of the 1,1-diazene **15g**, is computed to be about the same as the overall $E_A$ for its HERON reaction. This is a very early transition state in keeping with the exothermicity of the process. $E_{rearr}$ is essentially zero on account of the high energy electron-pair on the amino nitrene and the $E_A$ is essentially equivalent to the RE that must be sacrificed. Likewise, the hydroxamate **15f** bears a very high energy electron pair on the anomeric donor oxygen. While the resonance energy is similar to that of *N,N*-dimethoxyacetamide, the $E_{rearr}$ is small. This too is a very early transition state.

Overall, the decreasing order of $E_{rearr}$ of XNY systems based on the deconvolution in Table 8, which can be regarded as the order of decreasing effectiveness of an $n_Y$–$\sigma^*_{NX}$ interaction and destabilisation of the N–OMe bond, is AcONO > ONS ~ONO > ONN > $ONO^-$ > ONNitrene. The order of $RE_{TA}$ is ONN > ONS > AcONO > ONO > $ONO^-$ > ONNitrene and the overall activation energies decrease in the order AcONO > ONS > ONO > ONN > $ONO^-$ > ONNitrene. Clearly, the dominant influence in HERON reactivity is the strength of the anomeric effect rather than the decrease in amidicity.

Rearrangement energies, $E_{rearr}$'s, obtained by the deconvolution method represent activation energies in the absence of resonance. Relative $E_{rearr}$'s can be compared to relative $E_A$'s for intramolecular rearrangement in fully twisted amides, heteroatom-substituted 1-aza-2-adamantanones **85** (Scheme 17) and 2-quinuclidone **88** (Scheme 18) and the $E_A$'s are also presented in Table 8. $\Delta E_A$'s relative to migration of oxygen (Scheme 17i) in **85b**, for **85a**, and **85c** are $-88$ and 5.8 kJ mol$^{-1}$, respectively, which correlate well with the respective differences in $E_{rearr}$ of the respective acetamides, namely $-77$ and 5 kJ mol$^{-1}$. Similarly, the difference in the ease of migration of alkoxyl to form **90** and acyloxyl to form **89** in **88** (Scheme 18i,ii) is 28 kJ mol$^{-1}$ and the difference between the corresponding $E_{rearr}$ form the *syn* conformer of **15b** is 27 kJ mol$^{-1}$.

No transition state can be found for migration of alkoxyl group in hydroxamic ester **3b**. However, **85d** can be rearranged to lactone **87** (Scheme 17ii) with concomitant rearrangement of the nitrene product to the imine. The difference between $E_A$ for this process and the rearrangement of **85a** is 184 kJ mol$^{-1}$. If this translates to the difference in $E_{rearr}$ between twisted *N*-methoxy-*N*-methylacetamide **3b** and *N*-methoxy-*N*-dimethylaminoacetamide **15c**, for which $E_{rearr}$ is 43.5 kJ mol$^{-1}$, the rearrangement of **3** to methyl acetate and methylnitrene would have an activation energy of about 233 kJ mol$^{-1}$ from twisted **3b** or, adding in the RE for **3b** of 62 kJ mol$^{-1}$, 295 kJ mol$^{-1}$ from conjugated **3**. It is clear that hydroxamic esters do not rearrange to esters and alkylnitrenes and the role of anomeric (bisheteroatom) substitution in the HERON reaction is vital.

## 4. Conclusions

In this review we have outlined the theoretical and structural properties and the reactivity of the class of anomeric amides. Much of the data from this, and several other groups, is relatively recent and, while several compilations on the subject have appeared in the literature, a focus on the perturbation of the amide structure is befitting this special edition. The fairly recent accrual of structural data from our laboratory and that of Shtamburg has provided experimental verification of the unusual properties of bisheteroatom-substituted amides. Coupled with extensive computational results, the effect of heteroatoms at nitrogen on amide resonance and conformation at the amide nitrogen can be better understood and predicted, in particular the role played by electronegativity of the bonded atoms and orbital interactions. Ensuing from the work is a clearer understanding of the energetic consequences of distortion of the amide linkage. It is clear that spectroscopic properties of various congeners are dictated by two principle concepts: the first is impaired resonance owing to change in hybridisation bought about by orbital interactions that have their foundation in Bent's, and more recently Alabugin's theories of orbital interactions. Of particular importance is the reassignment of 2s character to the amide nitrogen lone pair orbital, as consequence of both electronegativity and orbital overlap considerations. Secondly, the participating atoms in this purturbation of regular amide bonding possess intrinsic orbital interactions by virtue of both their electronegativity and their lone pairs. The anomeric interaction bought about by $n_Y$–$\sigma^*_{NX}$ overlap is pronounced in anomeric amides. What is clear is that the electronegativity induces a shift to less resonance through pyramidalisation and lowering the energy of the amide nitrogen lone pair and the anomeric interaction is better served by this shift to $sp^3$ character. Stronger electronegativity serves to reduce resonance as well as to promote the anomeric interaction.

Regarding reactivity, it is abundantly clear that both resonance impairment and strength of the anomeric effect are definitive, but the anomeric effect dominates in promoting both $S_N2$ and elimination reactions at the amide nitrogen. Most significant is the anomeric driving force for the HERON reaction, notably in systems such as *N*-amino-*N*-alkoxyamides (NNO systems) where the energetics for anomeric weakening of the N–O bond are optimised, and in cyclic systems such as the *N*-alkoxy-γ- and δ-oxazinolactams where conformation and configuration appear to favour and enhance anomeric overlap. The HERON reaction is unique to this class of amides and has no equivalence in the literature. It operates in the opposite sense to the well-known Curtius, Hoffman, and Lossen rearrangements in which an acyl substituent migrates from the carbonyl to the nitrogen. Finally our studies of the HERON and those of Barton and coworkers, shed light on the well-known decomposition reactions of *N,N′*-dialkoxy-*N,N′*-diacylhydrazines. The HERON mechanism that operates in these decompositions and in those of the 1-acyl-1-alkoxydizenes is critical to the synthesis of highly hindered esters.

## References


1. Greenberg, A.; Breneman, C.M.; Liebman, J.F. *The Amide Linkage: Structural Significance in Chemistry, Biochemistry and Materials Science*; John Wiley & Sons, Inc.: Hoboken, NJ, USA, 2003.
2. Wiberg, K.B. Origin of the Amide Rotational Barrier. In *The Amide Linkage. Structural Significance in Chemistry, Biochemistry and Materials Science*; Greenberg, A., Breneman, C.M., Liebman, J.F., Eds.; John Wiley & Sons, Inc.: New York, NY, USA, 2003; p. 33.
3. Wiberg, K.B.; Breneman, C.M. Resonance interactions in acyclic systems. 3. Formamide internal rotation revisited. Charge and energy redistribution along the C–N bond rotational pathway. *J. Am. Chem. Soc.* **1992**, *114*, 831–840. [CrossRef]
4. Glover, S.A.; Rosser, A.A. Reliable determination of amidicity in acyclic amides and lactams. *J. Org. Chem.* **2012**, *77*, 5492–5502. [CrossRef] [PubMed]
5. Dunitz, J.D. *X-ray Analysis and Structure of Organic Molecules*; Cornell University Press: London, UK, 1979.
6. Winkler, F.K.; Dunitz, J.D. The non-planar amide group. *J. Mol. Biol.* **1971**, *59*, 169–182. [CrossRef]
7. Szostak, M.; Aube, J. Chemistry of bridged lactams and related heterocycles. *Chem. Rev.* **2013**, *113*, 5701–5765. [CrossRef] [PubMed]
8. Szostak, M.; Aube, J. Medium-bridged lactams: A new class of non-planar amides. *Org. Biomol. Chem.* **2011**, *9*, 27–35. [CrossRef] [PubMed]
9. Kirby, A.J.; Komarov, I.V.; Feeder, N. Spontaneous, millisecond formation of a twisted amide from the amino acid, and the crystal structure of a tetrahedral intermediate. *J. Am. Chem. Soc.* **1998**, *120*, 7101–7102. [CrossRef]
10. Kirby, A.J.; Komarov, I.V.; Feeder, N. Synthesis, structure and reactions of the most twisted amide. *J. Chem. Soc. Perk. Trans.* **2001**, *2*, 522–529. [CrossRef]
11. Kirby, A.J.; Komarov, I.V.; Kowski, K.; Rademacher, P. Distortion of the amide bond in amides and lactams. Photoelectron-spectrum and electronic structure of 3,5,7-trimethyl-1-azaadamantan-2-one, the most twisted amide. *J. Chem. Soc. Perk. Trans.* **1999**, *2*, 1313–1316. [CrossRef]
12. Kirby, A.J.; Komarov, I.V.; Wothers, P.D.; Feeder, N. The most twisted amide: Structure and reactions. *Angew. Chem. Int. Ed.* **1998**, *37*, 785–786. [CrossRef]

13. Morgan, J.; Greenberg, A. Novel bridgehead bicyclic lactams: (a) Molecules predicted to have *O*-protonated and *N*-protonated tautomers of comparable stability; (b) hyperstable lactams and their *O*-protonated tautomers. *J. Chem. Thermodyn.* **2014**, *733*, 206–212. [CrossRef]
14. Tani, K.; Stoltz, B.M. Synthesis and structural analysis of 2-quinuclidonium tetrafluoroborate. *Nature* **2006**, *441*, 731–734. [CrossRef] [PubMed]
15. Yamada, S. Chemistry of highly twisted amides. *Rev. Heteroat. Chem.* **1998**, *19*, 203–236.
16. Yamada, S. Sterically Hindered Twisted Amides. In *The Amide Linkage: Structural Aspects in Chemistry, Biochemistry, and Material Science*; Greenberg, A., Breneman, C.M., Liebman, J.F., Eds.; John Wiley & Sons, Inc.: Hoboken, NJ, USA, 2000; pp. 215–246.
17. Yamada, S. Structure and reactivity of an extremely twisted amide. *Angew. Chem.* **1993**, *105*, 1128–1130. [CrossRef]
18. Cow, C.N.; Britten, J.F.; Harrison, P.H.M. X-ray crystal structure of 1,6-diacetyl-3,4,7,8-tetramethyl-2, 5-dithioglycoluril a highly twisted acetamide. *J. Chem. Soc. Chem. Commun.* **1998**, 1147–1148. [CrossRef]
19. Anet, F.A.L.; Osyani, J.M. Nuclear magnetic resonance spectra and nitrogen inversion in 1-acylaziridinies. *J. Am. Chem. Soc.* **1967**, *89*, 352–356. [CrossRef]
20. Boggs, G.R.; Gerig, J.T. Nitrogen inversion in *N*-Benzoylaziridines. *J. Org. Chem.* **1969**, *34*, 1484–1486. [CrossRef]
21. Ohwada, T.; Achiwa, T.; Okamoto, I.; Shudo, K.; Yamaguchi, K. On the planarity of amide nitrogen. Intrinsic pyramidal nitrogen of *N*-acyl-7-azabicyclo 2.2.1 heptanes. *Tetrahedron Lett.* **1998**, *39*, 865–868. [CrossRef]
22. Otani, Y.; Nagae, O.; Naruse, Y.; Inagaki, S.; Ohno, M.; Yamaguchi, K.; Yamamoto, G.; Uchiyama, M.; Ohwada, T. An Evaluation of amide group planarity in 7-azabicyclo[2.2.1]heptan amides. Low amide bond rotation barrier in solution. *J. Am. Chem. Soc.* **2003**, *125*, 15191–15199. [CrossRef] [PubMed]
23. Otani, Y.; Futaki, S.; Kiwada, T.; Sugiura, Y.; Ohwada, T. Synthesis of non-planar peptides bearing the 7-azabicyclo[2.2.2]heptane skeleton, and possible self-organized structures. *Pept. Sci.* **2005**, *41*, 173–174.
24. Ohwada, T.; Ishikawa, S.; Mine, Y.; Inami, K.; Yanagimoto, T.; Karaki, F.; Kabasawa, Y.; Otani, Y.; Mochizuki, M. 7-Azabicyclo[2.2.2]heptane as a structural motif to block mutagenicity of nitrosomines. *Biorg. Med. Chem.* **2011**, *19*, 2726–2741. [CrossRef] [PubMed]
25. Allen, F.H.; Kennard, O.; Watson, D.G.; Brammer, L.; Orpen, A.G.; Taylor, R. Tables of bond lengths determined by x-ray and neutron diffraction. Part 1. Bond lengths in organic compounds. *J. Chem. Soc. Perkin Trans. 2* **1987**, *12*, S1–S19. [CrossRef]
26. Glover, S.A. Anomeric amides—Structure, properties and reactivity. *Tetrahedron* **1998**, *54*, 7229–7272. [CrossRef]
27. Rauk, A.; Glover, S.A. A computational investigation of the stereoisomerism in bisheteroatom-substituted amides. *J. Org. Chem.* **1996**, *61*, 2337–2345. [CrossRef]
28. Glover, S.; Rauk, A. Conformational stereochemistry of the HERON amide, *N*-methoxy-*N*-dimethylaminoformamide: A theoretical study. *J. Org. Chem.* **1999**, *64*, 2340–2345. [CrossRef]
29. Gillson, A.-M.E.; Glover, S.A.; Tucker, D.J.; Turner, P. Crystal structures and properties of mutagenic *N*-acyloxy-*N*-alkoxyamides—"Most pyramidal" acyclic amides. *Org. Biomol. Chem.* **2003**, *1*, 3430–3437. [CrossRef] [PubMed]
30. Glover, S.A.; Rosser, A.A.; Taherpour, A.; Greatrex, B.W. Formation and HERON reactivity of cyclic *N,N*-dialkoxyamides. *Aust. J. Chem.* **2014**, *67*, 507–520. [CrossRef]
31. Glover, S.A. *N*-Acyloxy-*N*-alkoxyamide—Structure, properties, reactivity and biological activity. *Adv. Phys. Org. Chem.* **2007**, *42*, 35–123.
32. Glover, S.A.; Mo, G.; Rauk, A.; Tucker, D.J.; Turner, P. Structure, conformation, anomeric effects and rotational barriers in the HERON amides, *N,N′*-diacyl-*N,N′*-dialkoxyhydrazines. *J. Chem. Soc. Perkin Trans. 2* **1999**, 2053–2058. [CrossRef]
33. Glover, S.A.; White, J.M.; Rosser, A.A.; Digianantonio, K.M. Structures of *N,N*-Dialkoxyamides—Pyramidal anomeric amides with low amidicity. *J. Org. Chem.* **2011**, *76*, 9757–9763. [CrossRef] [PubMed]
34. Glover, S.A.; Rosser, A.A.; Spence, R.M. Studies of the structure, amidicity and reactivity of *N*-chlorohydroxamic esters and *N*-chloro-β, β-dialkylhydrazides: Anomeric amides with low resonance energies. *Aust. J. Chem.* **2014**, *67*, 1344–1352. [CrossRef]
35. Glover, S.A. *N*-Heteroatom-Substituted Hydroxamic Esters. In *The Chemistry of Hydroxylamines, Oximes and Hydroxamic*; Rappoport, Z., Liebman, J.F., Eds.; Wiley: Chichester, UK, 2009; pp. 839–923.

36. Alabugin, I.V.; Bresch, S.; Manoharan, M. Hybridization trends for main group elements and expanding the Bent's rule beyond carbon: More than electronegativity. *J. Phys. Chem. A* **2014**, *118*, 3663–3677. [CrossRef] [PubMed]
37. Bent, H.A. An appraisal of valence-bond structures and hybridization in compounds of the first-row elements. *Chem. Rev.* **1961**, *61*, 275–311. [CrossRef]
38. Glover, S.A.; Schumacher, R.R. The effect of hydrophobicity upon the direct mutagenicity of *N*-acyloxy-*N*-alkoxyamides—Bilinear dependence upon LogP. *Mutat. Res.* **2016**, *795*, 41–50. [CrossRef] [PubMed]
39. Banks, T.M.; Clay, S.F.; Glover, S.A.; Schumacher, R.R. Mutagenicity of *N*-acyloxy-*N*-alkoxyamides as an indicator of DNA intercalation Part 1: Evidence for naphthalene as a DNA intercalator. *Org. Biomol. Chem.* **2016**, *14*, 3699–3714. [CrossRef] [PubMed]
40. Glover, S.A.; Schumacher, R.R.; Bonin, A.M.; Fransson, L.E. Steric effects upon the direct mutagenicity of *N*-acyloxy-*N*-alkoxyamides—Probes for drug-DNA interactions. *Mutat. Res.* **2011**, *722*, 32–38. [CrossRef] [PubMed]
41. Andrews, L.E.; Banks, T.M.; Bonin, A.M.; Clay, S.F.; Gillson, A.-M.E.; Glover, S.A. Mutagenic *N*-acyloxy-*N*-alkoxyamides—Probes for drug—DNA interactions. *Aust. J. Chem.* **2004**, *57*, 377–381. [CrossRef]
42. Banks, T.M.; Bonin, A.M.; Glover, S.A.; Prakash, A.S. Mutagenicity and DNA damage studies of *N*-acyloxy-*N*-alkoxyamides—The role of electrophilic nitrogen. *Org. Biomol. Chem.* **2003**, *1*, 2238–2246. [CrossRef] [PubMed]
43. Bonin, A.M.; Banks, T.M.; Campbell, J.J.; Glover, S.A.; Hammond, G.P.; Prakash, A.S.; Rowbottom, C.A. Mutagenicity of electrophilic *N*-acyloxy-*N*-alkoxyamides. *Mutat. Res.* **2001**, *494*, 115–134. [CrossRef]
44. Campbell, J.J.; Glover, S.A. (Synopsis) Bimolecular reactions of mutagenic *N*-(acyloxy)-*N*-alkoxybenzamides with aromatic amines. *J. Chem. Res.* **1999**, *8*, 474–475. [CrossRef]
45. Glover, S.A.; Hammond, G.P.; Bonin, A.M. A comparison of the reactivity and mutagenicity of *N*-benzoyloxy-*N*-benzyloxybenzamides. *J. Org. Chem.* **1998**, *63*, 9684–9689. [CrossRef]
46. Bonin, A.M.; Glover, S.A.; Hammond, G.P. Reactive intermediates from the solvolysis of mutagenic *O*-alkyl *N*-acetoxybenzohydroxamates. *J. Chem. Soc. Perkin Trans. 2* **1994**, *6*, 1173–1180. [CrossRef]
47. Campbell, J.J.; Glover, S.A. Bimolecular reactions of mutagenic *N*-acetoxy-*N*-alkoxybenzamides and *N*-methylaniline. *J. Chem. Soc. Perk. Trans. 2* **1992**, *10*, 1661–1663. [CrossRef]
48. Campbell, J.J.; Glover, S.A.; Hammond, G.P.; Rowbottom, C.A. Evidence for the formation of nitrenium ions in the acid-catalysed solvolysis of mutagenic *N*-acetoxy-*N*-alkoxybenzamides. *J. Chem. Soc. Perkin Trans. 2* **1991**, 2067–2079. [CrossRef]
49. Campbell, J.J.; Glover, S.A.; Rowbottom, C.A. Solvolysis and mutagenesis of *N*-acetoxy-*N*-alkoxybenzamides—Evidence for nitrenium ion formation. *Tetrahedron Lett.* **1990**, *31*, 5377–5380. [CrossRef]
50. Andrews, L.E.; Bonin, A.M.; Fransson, L.E.; Gillson, A.-M.E.; Glover, S.A. The role of steric effects in the direct mutagenesis of *N*-acyloxy-*N*-alkoxyamides. *Mutat. Res.* **2006**, *605*, 51–62. [CrossRef] [PubMed]
51. Cavanagh, K.L.; Glover, S.A.; Price, H.L.; Schumacher, R.R. $S_N2$ Substitution reactions at the amide nitrogen in the anomeric mutagens, *N*-acyloxy-*N*-alkoxyamides. *Aust. J. Chem.* **2009**, *62*, 700–710. [CrossRef]
52. Glover, S.A.; Adams, M. Reaction of *N*-acyloxy-*N*-alkoxyamides with biological thiols. *Aust. J. Chem.* **2011**, *64*, 443–453. [CrossRef]
53. Glover, S.A.; Rauk, A. A computational investigation of the structure of the novel anomeric amide *N*-azido-*N*-methoxyformamide and its concerted decomposition to methyl formate and nitrogen. *J. Chem. Soc. Perk. Trans. 2* **2002**, *0*, 1740–1746. [CrossRef]
54. Glover, S.A.; Mo, G. Hindered ester formation by $S_N2$ azidation of *N*-acetoxy-*N*-alkoxyamides and *N*-alkoxy-*N*-chloroamides—Novel application of HERON rearrangements. *J. Chem. Soc. Perk. Trans. 2* **2002**, *10*, 1728–1739. [CrossRef]
55. Digianantonio, K.M.; Glover, S.A.; Johns, J.P.; Rosser, A.A. Synthesis and thermal decomposition of *N,N*-dialkoxyamides. *Org. Biomol. Chem.* **2011**, *9*, 4116–4126. [CrossRef] [PubMed]
56. Buccigross, J.M.; Glover, S.A.; Hammond, G.P. Decomposition of *N,N*-diacyl-*N,N*-dialkoxyhydrazines revisited. *Aust. J. Chem.* **1995**, *48*, 353–361.
57. Greenberg, A. The Amide Linkage as a Ligand: Its Properties and the Role of Distortion. In *The Amide Linkage. Structural Significance in Chemistry, Biochemistry and Materials Science*; Greenberg, A., Breneman, C.M., Liebman, J.F., Eds.; John Wiley & Sons, Inc.: New York, NY, USA, 2003; pp. 47–83.

58. Greenberg, A.; Moore, D.T.; DuBois, T.D. Small and medium-sized bridgehead lactams: A systematic ab initio molecular orbital study. *J. Am. Chem. Soc.* **1996**, *118*, 8658–8668. [CrossRef]
59. Greenberg, A.; Venanzi, C.A. Structures and enrgetics of two bridgehead lactams and their *N*- and *O*-protonated froms: An ab initio molecular orbital study. *J. Am. Chem. Soc.* **1993**, *115*, 6951–6957. [CrossRef]
60. Hehre, W.J.; Radom, L.; Schleyer, P.; Pople, J.A. *Ab Initio Molecular Orbital Theory*; John Wiley & Sons: New York, NY, USA, 1986.
61. Glover, S.A.; Rosser, A.A. The role of substituents in the HERON reaction of anomeric amides. *Can. J. Chem.* **2016**, *94*, 1169–1180. [CrossRef]
62. Glover, S.A.; Rosser, A.A. HERON reactions of anomeric amides: Understanding the driving force. *J. Phys. Org. Chem.* **2015**, *28*, 215–222. [CrossRef]
63. Allen, F.H. The cambridge structural database: A quarter of a million crystal structures and rising. *Acta Crystallogr. Sect. B Struct. Sci.* **2002**, *58*, 380–388. [CrossRef]
64. Wiberg, K.B.; Rablen, P.R.; Rush, D.J.; Keith, T.A. Amides. 3. Experimental and theoretical studies of the effect of the medium on the rotational barriers for *N*,*N*-dimethylformamide and *N*,*N*-dimethylacetamide. *J. Am. Chem. Soc.* **1995**, *117*, 4261–4270. [CrossRef]
65. Shtamburg, V.G.; Tsygankov, A.T.; Shishkin, O.V.; Zubatyuk, R.I.; Uspensky, B.V.; Shtamburg, V.V.; Mazepa, A.V.; Kostyanovsky, R.G. The properties and structure of *N*-chloro-*N*-methoxy-4-nitrobenzamide. *Mendeleev Commun.* **2012**, *22*, 164–166. [CrossRef]
66. Shtamburg, V.G.; Shishkin, O.V.; Zubatyuk, R.I.; Kravchenko, S.V.; Tsygankov, A.V.; Mazepa, A.V.; Klots, E.A.; Kostyanovsky, R.G. *N*-Chloro-*N*-alkoxyureas: Synthesis, structure and properties. *Mendeleev Commun.* **2006**, *16*, 323–325. [CrossRef]
67. Shishkin, O.V.; Zubatyuk, R.I.; Shtamburg, V.G.; Tsygankov, A.V.; Klots, E.A.; Mazepa, A.V.; Kostyanovsky, R.G. Pyramidal amide nitrogen in *N*-acyloxy-*N*-alkoxyureas and *N*-acyloxy-*N*-alkoxycarbamates. *Mendeleev Commun.* **2006**, *16*, 222–223. [CrossRef]
68. Shtamburg, V.G.; Shishkin, O.V.; Zubatyuk, R.I.; Kravchenko, S.V.; Tsygankov, A.V.; Shtamburg, V.V.; Distanov, V.B.; Kostyanovsky, R.G. Synthesis, structure and properties of *N*-alkoxy-*N*-(1-pyridinium)urea salts, *N*-alkoxy-*N*-acyloxyureas and *N*,*N*-dialkoxyureas. *Mendeleev Commun.* **2007**, *17*, 178–180. [CrossRef]
69. Shtamburg, V.G.; Shishkin, O.V.; Zubatyuk, R.I.; Shtamburg, V.V.; Tsygankov, A.V.; Mazepa, A.V.; Kadorkina, G.K.; Kostanovsky, R.G. Synthesis and structure of *N*-alkoxyhydrazines and *N*-alkoxy-*N*′,*N*′,*N*′-trialkylhydrazinium salts. *Mendeleev Commun.* **2013**, *23*, 289–291. [CrossRef]
70. Alabugin, I.V.; Bresch, S.; dos Passos Gomez, G. Orbital hybridization: A key electronic factor in control of structure and reactivity. *J. Phys. Org. Chem.* **2015**, *28*, 147–162. [CrossRef]
71. Alabugin, I.V. *Stereoelectronic Effects: A Bridge between Structure and Reactivity*, 1st ed.; John Wiley & Sons: Chichester, UK, 2016.
72. Alabugin, I.V.; Zeidan, T.A. Stereoelectronic effects and general trends in hyperconjugative acceptor ability of $\sigma$ bonds. *J. Am. Chem. Soc.* **2002**, *124*, 3175–3185. [CrossRef] [PubMed]
73. Rauk, A. *Orbital Interaction Theory of Organic Chemistry*; John Wiley & Sons, Inc.: New York, NY, USA, 1994; p. 102.
74. Eliel, E.L.; Wilen, S.H.; Mander, L.N. *Stereochemistry of Organic Compounds*; John Wiley & Sons, Inc: New York, NY, USA, 1994; pp. 753–1191.
75. Epiotis, N.D.; Cherry, W.R.; Shaik, S.; Yates, R.L.; Bernardi, F. *Structural Theory of Organic Chemistry*; Springer: Berlin, Germany, 1977.
76. Glover, S.A.; Mo, G.; Rauk, A. HERON rearrangement of *N*,*N*′-diacyl-*N*,*N*′-dialkoxyhydrazines—A theoretical and experimental study. *Tetrahedron* **1999**, *55*, 3413–3426. [CrossRef]
77. De Almeida, M.V.; Barton, D.H.R.; Bytheway, I.; Ferriera, J.A.; Hall, M.B.; Liu, W.; Taylor, D.K.; Thomson, L. Preparation and thermal decomposition of *N*,*N*′-diacyl-*N*,*N*′-dialkoxyhydrazines: Synthetic applications and mechanitic insights. *J. Am. Chem. Soc.* **1995**, *117*, 4870–4874. [CrossRef]
78. Cooley, J.H.; Mosher, M.W.; Khan, M.A. Preparation and reactions of *N*,*N*′-diacyl-*N*,*N*′-dialkoxyhydrazines. *J. Am. Chem. Soc.* **1968**, *90*, 1867–1871. [CrossRef]
79. Banks, T.M. Reactivity, Mutagenicity and DNA Damage of *N*-Acyloxy-N-Alkoxyamides. Ph.D. Thesis, University of New England, Armidale, Australia, 2003.
80. Gerdes, R.G.; Glover, S.A.; Ten Have, J.F.; Rowbottom, C.A. *N*-Acetoxy-*N*-alkoxyamides—A new class of nitrenium ion precursors which are mutagenic. *Tetrahedron Lett.* **1989**, *30*, 2649–2652. [CrossRef]

81. Taherpour, A. University of New England, Armidale, New South Wales, Australia. Personal communication, 2012.
82. Williams, D.H.; Fleming, I. *Spectroscopic Methods in Organic Chemistry*, 3rd ed.; McGraw Hill Ltd.: Maidenhead, UK, 1980.
83. Jackman, L.M. Rotation about Partial Double Bonds in Organic Molecules. In *Dynamic Nuclear Magnetic Resonance Spectroscopy*; Jackman, L.M., Cotton, F.A., Eds.; Academic Press: Cambridge, MA, USA, 1975.
84. Günther, H. *NMR Spectroscopy—Basic Principles, Concepts, and Applications in Chemistry*, 2nd ed.; John Wiley & Sons Ltd.: Chichester, UK, 1995.
85. Johns, J.P.; van Losenoord, A.; Mary, C.; Garcia, P.; Pankhurst, D.S.; Rosser, A.A.; Glover, S.A. Thermal decomposition of *N*-acyloxy-*N*-alkoxyamides—A new HERON reaction. *Aust. J. Chem.* **2010**, *63*, 1717–1729. [CrossRef]
86. Glover, S.A. $S_N2$ reactions at amide nitrogen—Theoretical models for reactions of mutagenic *N*-acyloxy-*N*-alkoxyamides with bionucleophiles. *Arkivoc* **2002**, *2001*, 143–160. [CrossRef]
87. Campbell, J.J.; Glover, S.A. (Microfiche) Bimolecular reactions of mutagenic *N*-(acyloxy)-*N*-alkoxybenzamides with aromatic amines. *J. Chem. Res.* **1999**, *8*, 2075–2096.
88. Isaacs, N.S. *Physical Organic Chemistry*, 2nd ed.; Longman Scientific and Technical: New York, NY, USA, 1995; p. 422.
89. Forster, W.; Laird, R.M. The mechanism of alkylation reactions. Part 1. The effect of ubstituents on the reaction of phenacyl bromide with pyridine in methanol. *J. Chem. Soc. Perk. Trans. 2* **1982**, *2*, 135–138. [CrossRef]
90. De Kimpe, N.; Verhé, R. *The Chemistry of α-Haloketones, α-Haloaldehydes and α-Haloimines*; John Wiley & Sons: Chichester, UK, 1988; p. 38.
91. Thorpe, J.W.; Warkentin, J. Stereochemical and steric effects in nucleophilic substitution of α-halo ketones. *Can. J. Chem.* **1973**, *51*, 927–935. [CrossRef]
92. Maron, D.M.; Ames, B.N. Revised methods for salmonella mutagenicity tests. *Mutat. Res.* **1983**, *113*, 173–215. [CrossRef]
93. Mortelmans, K.; Zeiger, E. The ames salmonella/microsome mutagenicity assay. *Mutat. Res.* **2000**, *455*, 29–60. [CrossRef]
94. Pullman, A.; Pullman, B. Electrostatic effect of macromolecular structure on the biochemical reactivity of the nucleic acids. Significance for chemical carcinogenesis. *Int. J. Quantum Chem. Symposia* **1980**, *18*, 245–259. [CrossRef]
95. Kohn, K.W.; Hartley, J.A.; Mattes, W.B. Mechanism of sequence selective alkylation of guanine-N7 positions by nitrogen mustards. *Nucleic Acids Res.* **1987**, *15*, 10531–10544. [CrossRef] [PubMed]
96. Warpehoski, M.A.; Hurley, L.H. Sequence selectivity of DNA covalent modification. *Chem. Res. Toxicol.* **1988**, *1*, 315–333. [CrossRef] [PubMed]
97. Prakash, A.S.; Denny, W.A.; Gourdie, T.A.; Valu, K.K.; Woodgate, P.D.; Wakelin, L.P.G. DNA-directed alkylating ligands as potential antitumor agents: Sequence specificity of alkylation by aniline mustards. *Biochemistry* **1990**, *29*, 9799–9807. [CrossRef] [PubMed]
98. Glover, S.A.; Scott, A.P. MNDO properties of heteroatom and phenyl substituted nitrenium ions. *Tetrahedron* **1989**, *45*, 1763–1776. [CrossRef]
99. Schroeder, D.; Grandinetti, F.; Hrusak, J.; Schwarz, H. Experimental and ab initio MO studies on [H2,N,O]+ ions in the gas phase: Characterization of the isomers H2NO+, HNOH+ and NOH2+ and the mechanism of unimolecular dehydrogenation of [H2,N,O]+. *J. Phys. Chem.* **1992**, *96*, 4841–4845. [CrossRef]
100. Glover, S.A.; Goosen, A.; McCleland, C.W.; Schoonraad, J.L. *N*-alkoxy-*N*-acylnitrenium ions as possible intermediates in intramolecular aromatic substitution: Novel formation of *N*-acyl-3,4-dihydro-1*H*-2,1-benzoxazines and *N*-acyl-4,5-dihydro-1*H*-2,1-benzoxazepine. *J. Chem. Soc. Perkin Trans. 1* **1984**, 2255–2260. [CrossRef]
101. Glover, S.A.; Rowbottom, C.A.; Scott, A.P.; Schoonraad, J.L. Alkoxynitrenium ion cyclisations: Evidence for difference mechanisms in the formation of benzoxazines and benzoxazepines. *Tetrahedron* **1990**, *46*, 7247–7262. [CrossRef]
102. Kikugawa, Y.; Kawase, M. Electrophilic aromatic substitution with a nitrenium ion generated from *N*-chloro-*N*-methoxyamides. Application to the synthesis of 1-methoxy-2-oxindoles. *J. Am. Chem. Soc.* **1984**, *106*, 5728–5729. [CrossRef]

103. Kawase, M.; Kitamura, T.; Kikugawa, Y. Electrophilic aromatic substitution with *N*-methoxy-*N*-acylnitrenium ions generated from *N*-chloro-*N*-methoxy amides: Syntheses of nitrogen heterocyclic compounds bearing a *N*-methoxy amide group. *J. Org. Chem.* **1989**, *54*, 3394–3403. [CrossRef]
104. Kikugawa, Y.; Shimada, M.; Kato, M.; Sakamoto, T. A new synthesis of *N*-alkoxy-2-ethoxyarylacetamides from *N*-alkoxy-*N*-chloroarylacetamides with triethylamine in ethanol. *Chem. Pharm. Bull.* **1993**, *41*, 2192–2194. [CrossRef]
105. Miyazawa, E.; Sakamoto, T.; Kikugawa, Y. Syntheis of spirodienones by intramolecular ipso-cyclization of *N*-methoxy-(4-halegenophenyl)amides using [hydroxy(tosyloxy)iodo]benzene in trifluoroethanol. *J. Org. Chem.* **2003**, *68*, 5429–5432. [CrossRef] [PubMed]
106. Kikugawa, Y.; Kawase, M.; Miyake, Y.; Sakamoto, T.; Shimada, M. A convenient synthesis of eupolauramine. *Tetrahedron Lett.* **1988**, *29*, 4297–4298. [CrossRef]
107. Kikugawa, Y.; Shimado, M.; Matsumoto, K. Cyclization with nitrenium ions generated from *N*-methoxy- or *N*-allyloxy-*N*-chloroamides with anhydrous zinc acetate. Synthesis of *N*-hydroxy- and *N*-methoxynitrogen heterocyclic compounds. *Heterocycles* **1994**, *37*, 293–301. [CrossRef]
108. Kikugawa, Y. Uses of hydroxamic acids and *N*-alkoxyimidoyl halides in organic synthesis. *Rev. Heteroat. Chem.* **1996**, *15*, 263–299.
109. Greenstein, G. *The Merck Index: An Encyclopedia of Chemicals, Drugs, and Biologicals*, 14nd ed.; O'Neil, M.J., Heckelman, P.E., Koch, C.B., Roman, K.J., Eds.; Emerald Group Publishing Limited: Whitehouse Station, NJ, USA, 2006; p. 40.
110. The Heron Reaction—Merck Index Named Reaction Index. Available online: https://www.rsc.org/Merck-Index/reaction/r197/ (accessed on 30 October 2018).
111. Buccigross, J.M.; Glover, S.A. Molecular orbital studies of N to C migrations in *N,N*-bisheteroatom-substituted amides—HERON rearrangements. *J. Chem. Soc. Perk. Trans. 2* **1995**, 595–603. [CrossRef]
112. Glover, S.A.; Rauk, A.; Buccigross, J.M.; Campbell, J.J.; Hammond, G.P.; Mo, G.; Andrews, L.E.; Gillson, A.-M.E. Review: The HERON reaction: Origin, theoretical background and prevalence. *Can. J. Chem.* **2005**, *83*, 1492–1509. [CrossRef]
113. Glover, S.A. Development of the HERON reaction: A historical account. *Aust. J. Chem.* **2017**, *70*, 344–361. [CrossRef]
114. Hinsberg, W.D., III; Dervan, P.B. Synthesis and direct spectroscopic observation of a 1,1-dialkyldiazene. Infrared and electronic spectrum of *N*-(2,2,6,6-tetramethylpiperidyl)nitrene. *J. Am. Chem. Soc.* **1978**, *100*, 1608–1610. [CrossRef]
115. Hinsberg, W.D., III; Dervan, P.B. Kinetics of the thermal decomposition of a 1,1-dialkyldiazene, *N*-(2,2,6,6-tetramethylpiperidyl)nitrene. *J. Am. Chem. Soc.* **1979**, *101*, 6142–6144. [CrossRef]
116. Hinsberg, W.D., III; Schultz, P.G.; Dervan, P.B. Direct studies of 1,1-diazenes. Syntheses, infrared and electronic spectra, and kinetics of the thermal decomposition of *N*-(2,2,6,6-tetramethylpiperidyl)nitrene and *N*-(2,2,5,5-tetramethylpyrrolidyl)nitrene. *J. Am. Chem. Soc.* **1982**, *104*, 766–773. [CrossRef]
117. Schultz, P.G.; Dervan, P.B. Synthesis and direct spectroscopic observation of *N*-(2,2,5,5-tetramethylpyrrolidyl)nitrene. Comparison of five- and six-membered cyclic 1,1-dialkyldiazenes. *J. Am. Chem. Soc.* **1980**, *102*, 878–880. [CrossRef]
118. Zhang, N.; Yang, R.; Zhang-Negrerie, D.; Du, Y.; Zhao, K. Direct conversion of *N*-alkoxyamides to carboxylic esters through tandem NBS-mediated oxidative homocoupling and thermal denitrogenation. *J. Org. Chem.* **2014**, *78*, 8705–8711. [CrossRef] [PubMed]
119. Thomson, L.M.; Hall, M.B. Theoretical study of the thermal decomposition of *N,N'*-diacyl-*N,N'*-dialkoxyhydrazines: A comparison of HF, MP2, and DFT. *J. Phys. Chem. A* **2000**, *104*, 6247–6252. [CrossRef]
120. Novak, M.; Glover, S.A. Generation and trapping of the 4-biphenylyloxenium ion by water and azide: Comparisons with the 4-biphenylylnitrenium ion. *J. Am. Chem. Soc.* **2004**, *126*, 7748–7749. [CrossRef] [PubMed]
121. Novak, M.; Rajagopal, S. *N*-arylnitrenium ions. *Adv. Phys. Org. Chem.* **2001**, *36*, 167–254. [CrossRef]
122. Shtamburg, V.G.; Klots, E.A.; Pleshkova, A.P.; Avramenko, V.I.; Ivonin, S.P.; Tsygankov, A.V.; Kostyanovsky, R.G. Geminal systems. 50. Synthesis and alcoholysis of *N*-acyloxy-*N*-alkoxy derivatives of ureas, carbamates, and benzamides. *Russ. Chem. Bull.* **2003**, *52*, 2251–2260. [CrossRef]
123. Crawford, R.J.; Raap, R. The Synthesis and reaction of *N,N*-dicarboalkoxy-*N,N*-dialkoxyhydrazines and some observations on carbalkoxylium ions. *J. Org. Chem.* **1963**, *28*, 2419–2424. [CrossRef]

124. Fahr, E.; Lind, H. Chemisry of a-carbonyl azo compounds. *Angew. Chem.* **1966**, *5*, 372–384. [CrossRef]
125. Mo, G. Properties and Reactions of Anomeric Amides. Ph.D. Thesis, University of New England, Armidale, Australia, 1999.
126. Cavanagh, K.L.; Glover, S.A. Heron reactivity of *N*-acetoxycarbamates with anilines. Unpublished work.

# 10

# Weinreb Amides as Directing Groups for Transition Metal-Catalyzed C-H Functionalizations

**Jagadeesh Kalepu and Lukasz T. Pilarski** *

Department of Chemistry-BMC, Uppsala University, BOX 576, 75-123 Uppsala, Sweden; jagadeesh.kalepu@kemi.uu.se
* Correspondence: lukasz.pilarski@kemi.uu.se

Academic Editor: Michal Szostak

**Abstract:** Weinreb amides are a privileged, multi-functional group with well-established utility in classical synthesis. Recently, several studies have demonstrated the use of Weinreb amides as interesting substrates in transition metal-catalyzed C-H functionalization reactions. Herein, we review this part of the literature, including the metal catalysts, transformations explored so far and specific insights from mechanistic studies.

**Keywords:** C-H functionalization; directing groups; amides; transition metals; catalysis

---

## 1. Introduction

The pursuit of efficient methods for the direct, catalytic substitution of otherwise inert C-H bonds in organic molecules has become a major area of research focus in recent years [1]. Tremendous advances have been made in the development of previously impossible transformations [2–5], mechanistic understanding [6–9], milder and safer protocols [10], and selectivity [11–14]. The ubiquity of C-H bonds in organic molecules makes their regioselective activation and substitution particularly attractive, but also challenging. To this end, the use of directing groups–parts of an organic substrate that can coordinate to and position a metal center over the desired C-H bond–has met with enormous success [15]. The use of *ortho*-directing groups especially has provided the basis of many new homogeneous catalytic C-H functionalization reactions [16]. Directing groups able to deliver *meta* [17,18] and *para* [19] selectivity in C-H functionalization catalysis have also been described, although the generality of that approach is still some way off in the future.

Ideally, directing groups should not require separate and/or laborious installation/removal or otherwise be inutile in later steps of a synthesis. It is preferable that they should either be part of the desired target compound or that they could be converted to another useful group once their role during the C-H functionalization step is over. A plethora of strategies to realize the latter objective has been pursued. The invention of various removable or modifiable [20–22], traceless [23], or otherwise transient [24–29] directing groups has formed a sizeable category in and of itself within the field of C-H functionalization catalysis.

Amides are an incomparably important class of compounds. The development of methods for their selective synthesis and derivatization is, therefore, a key pursuit in synthetic methodology. That the amide group can serve as a ligand for transition metals enables various new approaches to this via catalytic C-H functionalization [30]. Most commonly, the amide group has been called upon to direct the catalytic substitution of neighboring Ar-H bonds but, as discussed below, they also enable $C(sp^3)$-H bond manipulation.

*N*-methoxy-*N*-methyl amides (**1**, Figure 1), or Weinreb amides [31], are a valuable branch of the amide family. They provide the 'textbook' route to mono-addition products (especially ketones and aldehydes) via nucleophilic attack on the carbonyl group. Such attacks give rise to stable five-membered tetrahedral cyclic intermediates (**2**) to which a second addition ("over-addition") is precluded. Weinreb amides may be prepared with ease from carboxylic acids or their chlorides, esters, aldehydes or ketones [32].

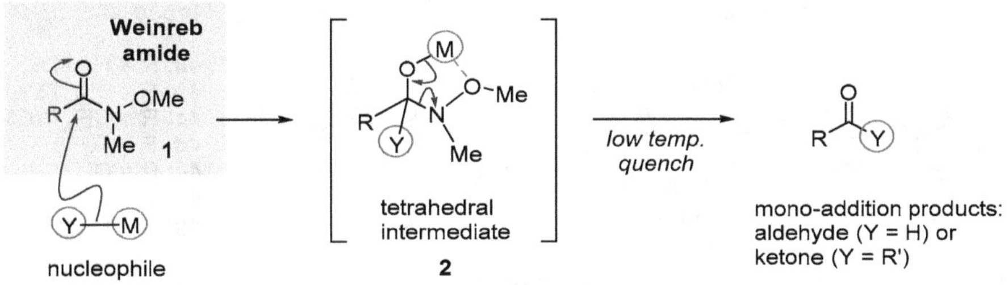

**Figure 1.** The 'textbook' application of Weinreb amides: generation of mono-addition products resulting from nucleophilic attack on their carbonyl groups.

That Weinreb amides can also steer the regioselectivity of transition metal-catalyzed C-H functionalizations (Figure 2) qualifies them as noteworthy multi-functional directing groups. Remarkably, and despite the various synthetic advantages of this (for example that it can obviate the need for lithiation strategies and make available previously impossible reactions), Weinreb amides have only recently attracted attention as substrates for C-H functionalization. In part, this is due to the amide oxygen's weaker coordination ability to most transition metal centers. The latter has presented a challenge to the development of their use in C-H functionalization reactions, although many carbonyl-directed reactions are now known [33].

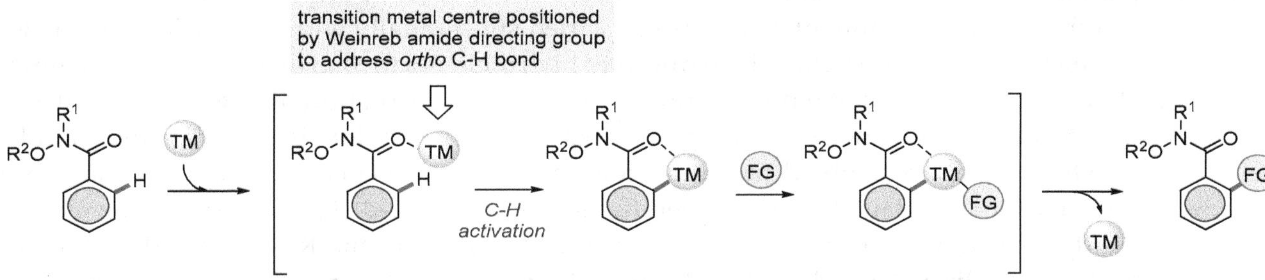

**Figure 2.** A generic representation of a Weinreb amide-directed catalytic C-H functionalization (TM = transition metal).

This review describes progress in the use of Weinreb amides as directing groups in catalytic C-H functionalization. A variety of reactions falls under this category. We have chosen to group these according to the transition metal center responsible for the C-H functionalization catalysis, rather than the overall transformation, in order to maximize the ease of comparison between researchers working in similar areas, and to track the different rates at which progress has occurred and insights in to the underlying mechanisms. Overwhelmingly, the focus of the studies reviewed herein falls on the utility of Weinreb amides as directing groups for the C-H functionalization step. Therefore, whilst several publications describe subsequent manipulation of the Weinreb amide group, this is usually to illustrate the possibility, rather than a key development. We have opted therefore not to include many examples of the latter; we take the possibility of *a posteriori* conversion of Weinreb amides using conventional approaches (e.g., to ketones or aldehydes), for the most part, to be a safe assumption.

## 2. Ru-catalyzed Reactions

The versatility of Ru, as well as its considerably lower price compared to other 2nd and 3rd-row transition metals, make it an appealing candidate around which to develop economical C-H functionalization methodology [34,35]. Moreover, considerable advances have been made in Ru-catalyzed C-H functionalization directed by weakly coordinating groups, including amides [33].

In 2013, Ackermann and co-workers described the Weinreb amide-directed C-H *ortho*-oxygenation of arenes **3** using a Ru(II)-based system [36]. $[RuCl_2(p\text{-cymene})]_2$ served as the catalyst precursor and $PhI(OAc)_2$ as the most effective oxidant. Representative results from the study are shown in Scheme 1a.

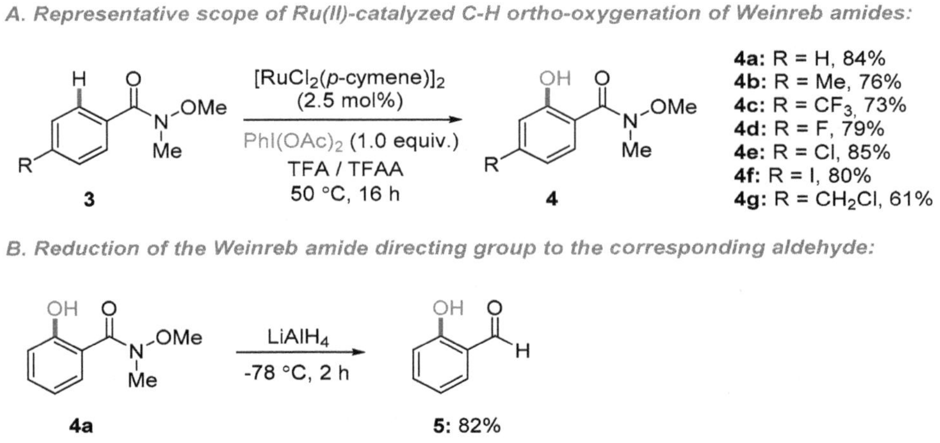

*A. Representative scope of Ru(II)-catalyzed C-H ortho-oxygenation of Weinreb amides:*

**3** → **4**: $[RuCl_2(p\text{-cymene})]_2$ (2.5 mol%), $PhI(OAc)_2$ (1.0 equiv.), TFA / TFAA, 50 °C, 16 h

**4a**: R = H, 84%
**4b**: R = Me, 76%
**4c**: R = $CF_3$, 73%
**4d**: R = F, 79%
**4e**: R = Cl, 85%
**4f**: R = I, 80%
**4g**: R = $CH_2Cl$, 61%

*B. Reduction of the Weinreb amide directing group to the corresponding aldehyde:*

**4a** → **5**: 82%; $LiAlH_4$, -78 °C, 2 h

**Scheme 1.** Ru(II)-catalyzed C-H oxidation of arenes directed by a Weinreb amide group. (**a**) Representative scope of the reaction with respect to arene substituents; (**b**) Reduction of the Weinreb amide to reveal aldehyde functionality.

The reaction showed a high selectivity for mono-oxygenated products **4**, a preference for electron-rich substrates in competition experiments, and a kinetic isotope effect (KIE) of $k_H/k_D = 3.0$ [6]. The authors proposed that an irreversible C-H activation event was a key step *en route* to the products. It is notable that the *N*-alkyl substituent could also be varied substantially without any loss of reaction efficiency. The Weinreb amide group could be reduced in high yield (**4a** to **5**) to reveal the corresponding aldehyde (Scheme 1b). A single Weinreb amide substrate was also shown to work as part of a study by Jeganmohan and co-workers on the *ortho*-directed C-H benzoyloxylation of various amides. The reaction system closely resembled that reported above by Ackermann, except for the use of $(NH_4)S_2O_8$ as the terminal oxidant, higher temperatures and use of 1,2-dichloroethane as solvent [37].

Subsequently, Das and Kapur produced a report demonstrating the Ru-catalyzed olefination of various amides, including Weinreb amide-decorated arenes [38]. The protocol closely resembles the Pd-catalyzed oxidative Heck reaction [39] (also known as the Fujiwara-Moritani reaction) and other related Ru-catalyzed C-H alkenylations [34]. However, for the majority of their entries (selected examples are shown in Scheme 2a), the authors observed cleavage of the Weinreb amide N-O bond; *N*-methyl amides were obtained as the main products. It is instructive to consider the proposed mechanism, an adapted version of which is shown in Scheme 2b. Presumably, the N-O bond serves as an oxidant with $Cu(OAc)_2{\cdot}H_2O$ as the carboxylate source to facilitate repeated C-H functionalization [40] The use of similar "internal oxidant" strategies in C-H functionalization, including using N-methoxy amides, has been recently reviewed by Cui and co-workers [41]. Exceptions wherein the N-O bond was preserved presumably resulted from Cu(II) (or Cu(III) species arising via disproportionation) outcompeting N-O as an internal oxidant. The importance of such examples is under-appreciated, in our view. That an exogenous oxidant may divert reactivity away from damaging a group under otherwise identical conditions is a key aspect of modulating functional group tolerance; a factor that will govern the extent to which C-H activation–and other catalytic methods–gain acceptance as generalizable routes to construct molecular complexity.

Das and Kapur reported that five-membered cyclic amides of the type **7** ($n$ = 0) typically underwent N-O bond cleavage, affording ring-opened products. To prevent this, and thereby retain the Weinreb amide functionality, they subsequently sought to show that N-O bond cleavage could be prevented through judicious substrate design [42]. Increasing the Weinreb amide size to six- and seven-membered rings (**7**, $n$ = 1 or 2, respectively) rendered the N-O bond cleavage energetically unfavorable. Under otherwise unchanged conditions, the oxidative C-H olefination afforded products **8** (Scheme 3), restoring Cu(II) to its role as the terminal oxidant. Further manipulation of the cyclic Weinreb amide moieties in **8**, for example in their conversion to the corresponding aldehyde or ketones, was demonstrated to proceed in high yield.

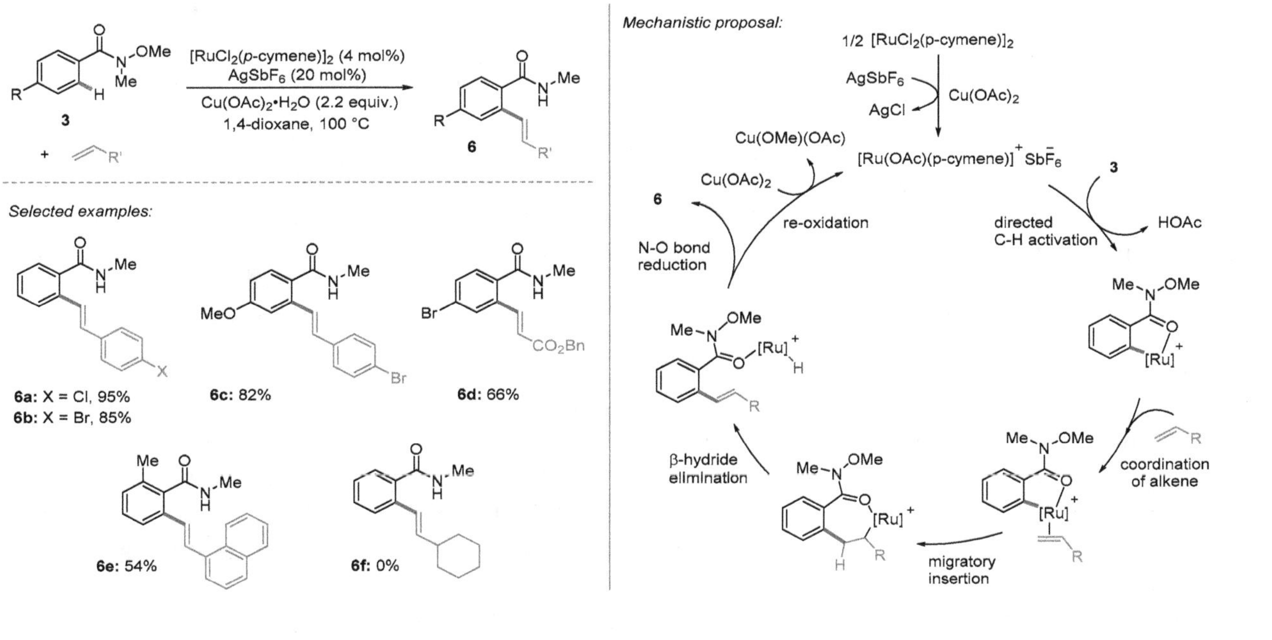

(a) (b)

**Scheme 2.** Ru(II)-catalyzed oxidative C-H *ortho*-alkenylation of Weinreb amides; selected products and proposed mechanism.

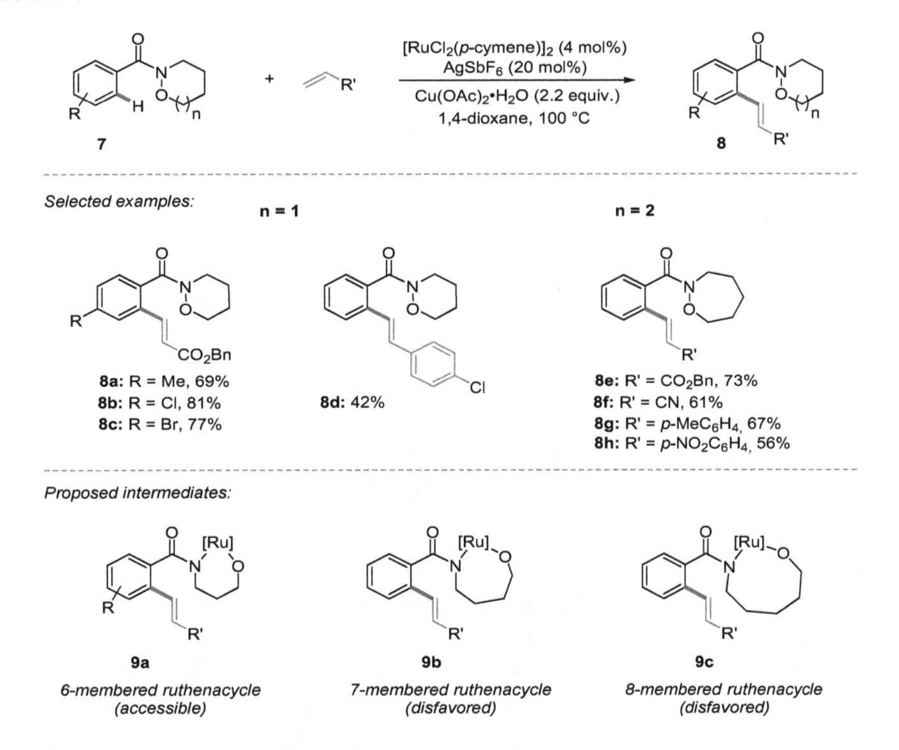


**Scheme 3.** Ru-catalyzed C-H olefination using directed by cyclic Weinreb amides without N-O bond cleavage. Putative intermediates **9a–c** are suggested to arise from the insertion of Ru into the substrate N-O bond.

An explanation for the greater relative stability of the 6- and 7-membered cyclic Weinreb amide groups in this protocol was proposed: In the course of the reaction, oxidative addition of the Ru center to the N-O bond would furnish intermediates of type **9**. Of these, only the formation of **9a** would be favorable; the 7- and 8-membered ruthenacycles (**9b** and **9c**, respectively) are presumably too high in energy to be accessed.

## 3. Co-catalyzed Reactions

The expense and eventual scarcity of second- (4d) and third-row (5d) transition metals has, in recent years, motivated a shift towards exploring the potential of their more abundant first-row (3d) cousins for catalytic C-H functionalization [43]. Cobalt stands out in this context as a comparatively cheap, abundant option that comes with a proven track record in broad areas of homogenous catalysis, including several industrially important reactions [44–46].

In 2013, Yoshino and Matsunaga disclosed the first C-H *ortho*-directed functionalization catalyzed by Cp*Co(III) species [47,48]. In a subsequent study on the Co-catalyzed C-H allylation of aryl purines and benzamides using allylic alcohol, they demonstrated the new reaction also on Weinreb amide **3a** [49]. Here, the bench-stable complex Cp*Co(CO)$I_2$, which was previously shown by Matsunaga and Kanai to be highly efficient for indole C2-H amidation [50], served as the pre-catalyst (Scheme 4). Hexafluoroisopropanol (HFIP) [51,52] was needed for the C-H allylation, as were acetate salts. The authors interpreted the latter as an indication that the C-H activation proceeded via a Concerted-Metalation-Deprotonation (CMD) mechanism [53], as has been elucidated in more detail for second row transition metals [8,40,54]. The moderate yield of product **9a** in this reaction was attributed to the weaker coordinating ability of the Weinreb amide group compared to its simpler benzamide relatives or, indeed, to that of the purine $N_{sp}{}^2$ centers.

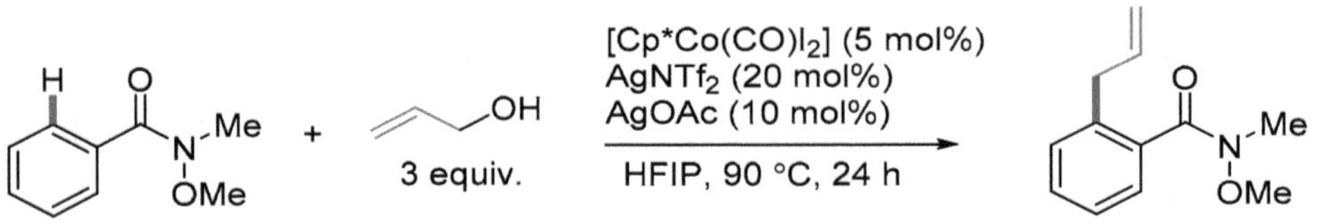

3a + OH (3 equiv.) — [Cp*Co(CO)$I_2$] (5 mol%), $AgNTf_2$ (20 mol%), AgOAc (10 mol%); HFIP, 90 °C, 24 h → **9a**, 44%

**Scheme 4.** An initial example of the Weinreb amide-directed Co(III)-catalyzed *ortho*-C-H allylation.

In a later study, the group published a considerably expanded range of Co-catalyzed C-H functionalizations using aromatic Weinreb amides (Scheme 5) [55]. These included C-H allylation using allylic carbonates (Scheme 5A), C-H alkenylation under oxidative conditions (Scheme 5B), C-H iodination using *N*-iodosuccinimide (Scheme 5C) [56] and C-H amidation using dioxazolones (Scheme 5D). These systems relied on similar combinations of [Cp*Co(CO)$I_2$] pre-catalyst, a cationic silver additive ($AgNTf_2$ or $AgSbF_6$) and AgOAc. Unlike in the Ru-catalyzed case (see Scheme 2), the N-O bond of the Weinreb amide moiety was preserved. Competition and kinetic isotope exchange experiments showed that the C-H activation event in the case of the allylation reaction was both rate-limiting and all but irreversible, supplying further evidence for the CMD pathway Yoshino and Matsunaga proposed in their earlier study [49]. Their proposed mechanism for the Co(III)-catalyzed C-H allylation is shown in Figure 3. In related recent work, Whiteoak and Hamilton have elucidated the mechanistic details governing the oxidative Co-catalyzed C-H alkenylation of amides, and specifically whether and why alkyl or alkenyl products are obtained depending on the choice of alkene substrate [57].

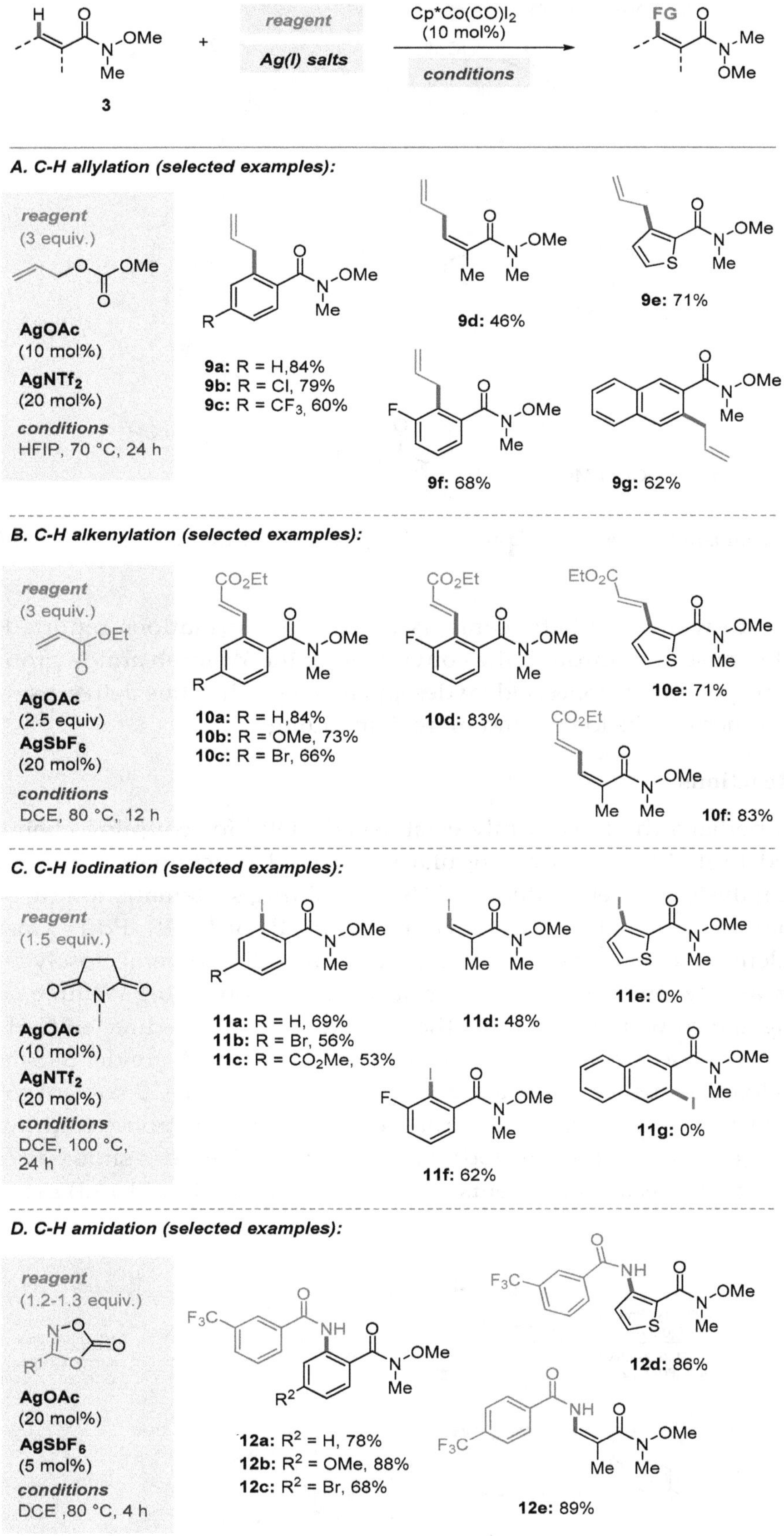

reagent + Ag(I) salts → $Cp^*Co(CO)I_2$ (10 mol%), conditions

**A. C-H allylation (selected examples):**

reagent (3 equiv.); AgOAc (10 mol%); $AgNTf_2$ (20 mol%); conditions HFIP, 70 °C, 24 h

**9a:** R = H, 84%
**9b:** R = Cl, 79%
**9c:** R = $CF_3$, 60%
**9d:** 46%
**9e:** 71%
**9f:** 68%
**9g:** 62%

**B. C-H alkenylation (selected examples):**

reagent (3 equiv.); AgOAc (2.5 equiv); $AgSbF_6$ (20 mol%); conditions DCE, 80 °C, 12 h

**10a:** R = H, 84%
**10b:** R = OMe, 73%
**10c:** R = Br, 66%
**10d:** 83%
**10e:** 71%
**10f:** 83%

**C. C-H iodination (selected examples):**

reagent (1.5 equiv.); AgOAc (10 mol%); $AgNTf_2$ (20 mol%); conditions DCE, 100 °C, 24 h

**11a:** R = H, 69%
**11b:** R = Br, 56%
**11c:** R = $CO_2Me$, 53%
**11d:** 48%
**11e:** 0%
**11f:** 62%
**11g:** 0%

**D. C-H amidation (selected examples):**

reagent (1.2-1.3 equiv.); AgOAc (20 mol%); $AgSbF_6$ (5 mol%); conditions DCE, 80 °C, 4 h

**12a:** $R^2$ = H, 78%
**12b:** $R^2$ = OMe, 88%
**12c:** $R^2$ = Br, 68%
**12d:** 86%
**12e:** 89%

**Scheme 5.** Selected examples of C-H transformations enabled by a Co(III)-catalyzed system.

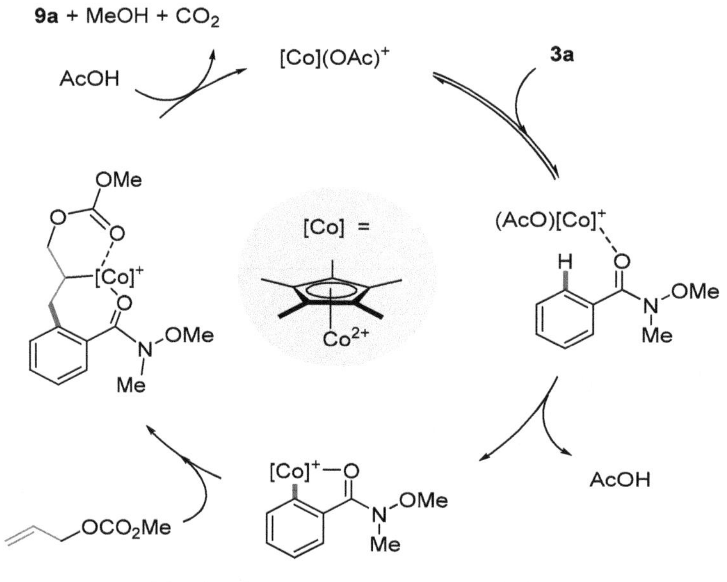

**Figure 3.** A mechanism proposed by Yoshino and Matsunaga for the Co(III)-catalyzed C-H allylation of Weinreb amides.

The relevance to wider synthetic contexts of the transformations reported by Yoshino and Matsunaga was demonstrated through the conversion of the Weinreb amide group in their various products to the corresponding ketones, aldehydes and alkenes. This thus delivers several motifs whose preparation might otherwise be longer and more demanding.

## 4. Pd-catalyzed Reactions

Palladium is perhaps the most firmly established metal for coupling chemistry [58–62], and this has translated to its high level of popularity for C-H functionalization reactions [16,63–66]. Many Pd-based catalysts are well-defined [67,68], permitting systematic tuning of their properties, and are understood, most usually, to work via Pd(0)/Pd(II) or Pd(II)/Pd(IV) manifolds [62,64,66]. Despite the prevalence of many *ortho*-directing groups, including several closely related amides [30], there have been relatively few examples of Pd-catalyzed reactions using Weinreb amides.

In 2015, Wang and co-workers reported the first Weinreb amide-directed C-H functionalization catalyzed by Pd. The reaction took place between either aryl Weinreb amides **3** (Scheme 6a) or benzyl amides **13** and iodoarenes using $Pd(OAc)_2$ as the pre-catalyst in DCE solvent [69]. Both Weinreb types showed high tolerance towards electron-donating and electron-withdrawing substituents, as well as halogens (e.g., the bromide in **15d**). However, benzyl amides showed considerably worse mono-selectivity, except if *meta* substituents were present (e.g., **15c** and **15d**) presumably to impart steric hindrance.

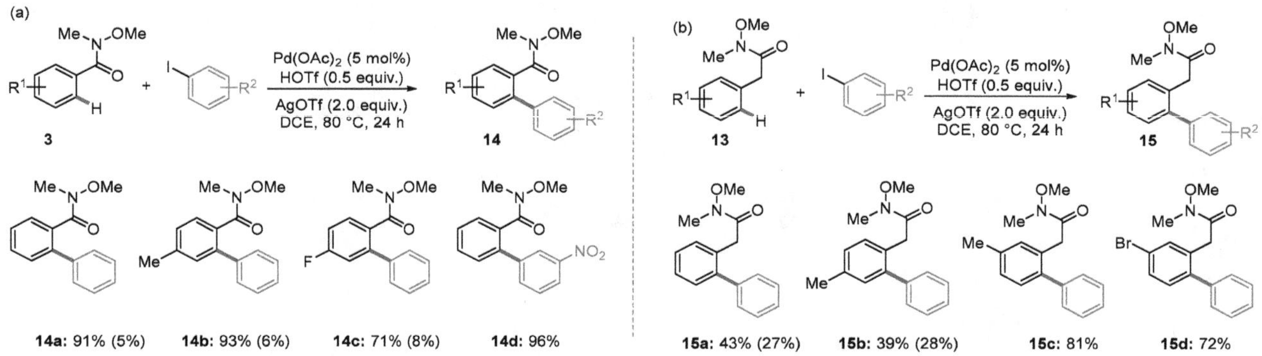

**Scheme 6.** Pd-catalyzed C-H arylation of (**a**) aryl and (**b**) benzyl amides. Yields in parentheses refer to the amount of di-arylated products detected.

Figure 4 shows one of the proposed mechanisms, adapted for the specific case of substrate **3a**. A kinetic isotope effect (KIE) of 1.1 suggested that the C-H activation step itself (forming **16** from **3**) was not rate-determining. Oxidative addition of the aryl iodide to **16** was postulated to form **17** from which the product **14a** is formed via reductive elimination. The authors reported that AgOTf, as well as acting as an oxidant, was crucial for the reaction and hypothesized that in situ generated $Pd(OTf)_2$ is a key aspect of the catalytic cycle. (We note that the various roles Ag plays in Pd-catalyzed reactions have recently been reviewed by Pérez-Temprano and co-workers [70].)

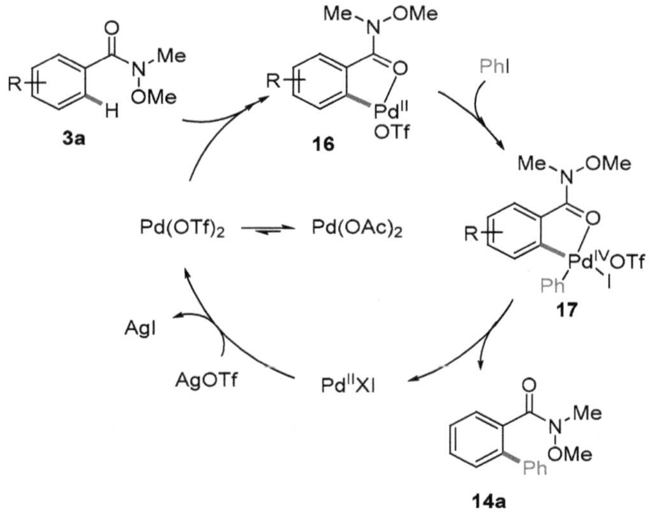

**Figure 4.** Mechanism proposed by Wang and co-workers for the C-H *ortho*-arylation of Weinreb amides proceeding via Pd(IV) intermediates.

To the best of our knowledge, the only other example of a Pd-catalyzed Weinreb amide-directed C-H arylation comes from a single entry in a study by Bhanage and workers on the use of anilines as arylating reagents for N-methoxybenzamides (Scheme 7) [71]. Conditions were largely analogous to those described for Wang's reaction above, with the principal exception that aniline underwent in situ oxidation using $^t$BuONO to generate the electrophilic coupling partner. However, the yield for the Weinreb amide specifically was low, indicating that although this approach holds some promise, it is harder to optimize for the Weinreb amide compared to the using more conventional coupling partners.

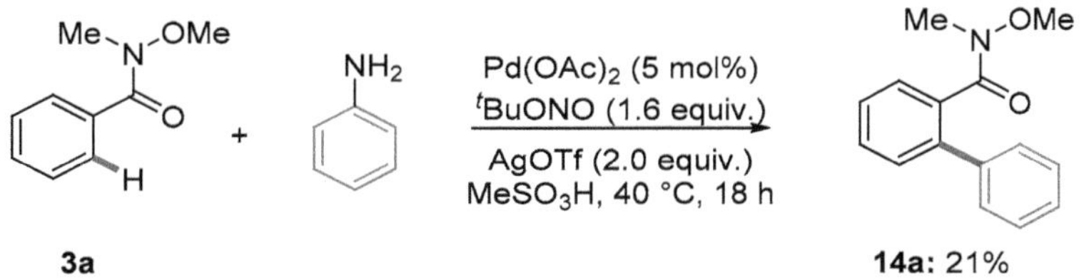

**Scheme 7.** Bhanage's C-H arylation using aniline as the electrophile source.

Also in 2015, Yu and co-workers described conditions for the efficient alkenylation and acetoxylation of Weinreb amides (Scheme 8) [72], overcoming the additional entropic penalty imposed by the extra methylene unit present in the directing group (and thus the challenge of forming seven-membered palladacyclic intermediates). In keeping with related protocols reported by Yu and co-workers, the presence of Ac-Gly-OH was found to be crucial to facilitate the C-H functionalization step [73].

Yu and co-workers found the olefination reaction to have moderate to excellent yields, exclusively mono C-H olefination with *ortho* and *meta*-substituted benzyl Weinreb amides, but reduced selectivity if the substituents were positioned *para* with respect to the directing group (Scheme 8).

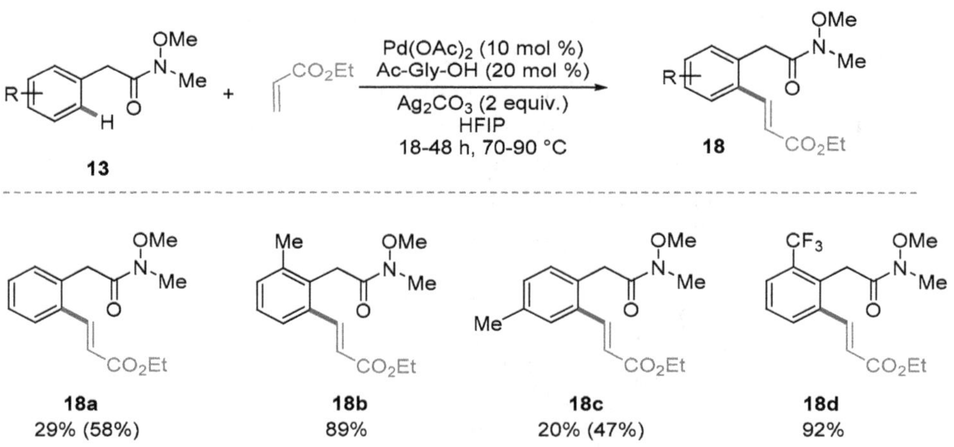

Scheme 8. Yu and co-workers' protocol for the oxidative *ortho* C-H acetoxylation of benzylic Weinreb amides. Yields in parentheses refer to the amount of di-olefinated products detected.

The acetoxylation of benzyl amides (Scheme 9) used $PhI(OAc)_2$ as the oxidant in the presence of either $Ac_2O$ or tert-butyl peroxyacetate ($MeC(O)OO^tBu$). Yields and tolerance towards various substituents were good (examples **19a–c**). Furthermore, it was possible to expand the scope by an extra methylene unit (i.e., the reaction can proceed with good efficiency via an 8-membered palladacycle) to give product **19d**.

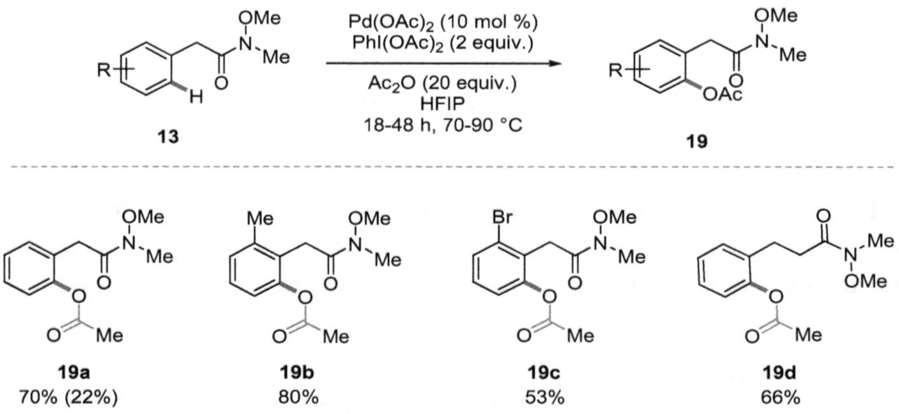

Scheme 9. Yu and co-workers' protocol for the oxidative *ortho* C-H acetoxylation of benzylic Weinreb amides. Yields in parentheses refer to the amount of di-acetoxylated products detected.

The introduction of halogens to organic substrates is an inherently valuable pursuit in methodology development, owing the great versatility of C-halogen bonds in synthesis [74]. In 2017, the group of Kapur disclosed the Pd-catalyzed C-H halogenation (iodination, bromination and chlorination) of Weinreb amides (Scheme 10) [75]. The reaction relied on a combination of $Pd(OAc)_2$ and $Cu(OTf)_2$ with N-halo-succinimide (NXS; X = I, Br or Cl) as the halogen source. The reaction worked well with a variety of substituents, except for nitro groups, which switch off the reaction from either of the *ortho* or *para* position.

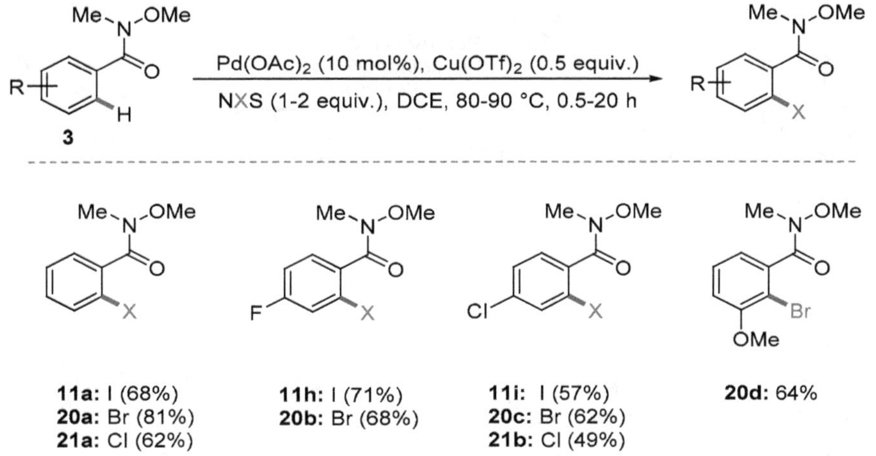
Me N OMe
O
R
H
3
Pd(OAc)2 (10 mol%), Cu(OTf)2 (0.5 equiv.)
NXS (1-2 equiv.), DCE, 80-90 °C, 0.5-20 h
X
F
Cl
Br
OMe
11a: I (68%)
20a: Br (81%)
21a: Cl (62%)
11h: I (71%)
20b: Br (68%)
11i: I (57%)
20c: Br (62%)
21b: Cl (49%)
20d: 64%

**Scheme 10.** Kapur and co-workers' reaction for the *ortho* C-H halogenation of aromatic Weinreb amides.

Mechanistic studies suggested that the C-H activation step was irreversible, but not rate-limiting. The catalytic cycle (Figure 5) was proposed to proceed via a route closely related to that of the corresponding arylation reaction (Figure 4, above): carbonyl directed cyclometallation to give **22** precedes oxidative addition into NXS, the latter of which produces Pd(IV) species **23** (a bimetallic Pd(III) species [76–79]—not shown—is also possible). Thereafter, reductive elimination releases the desired product.

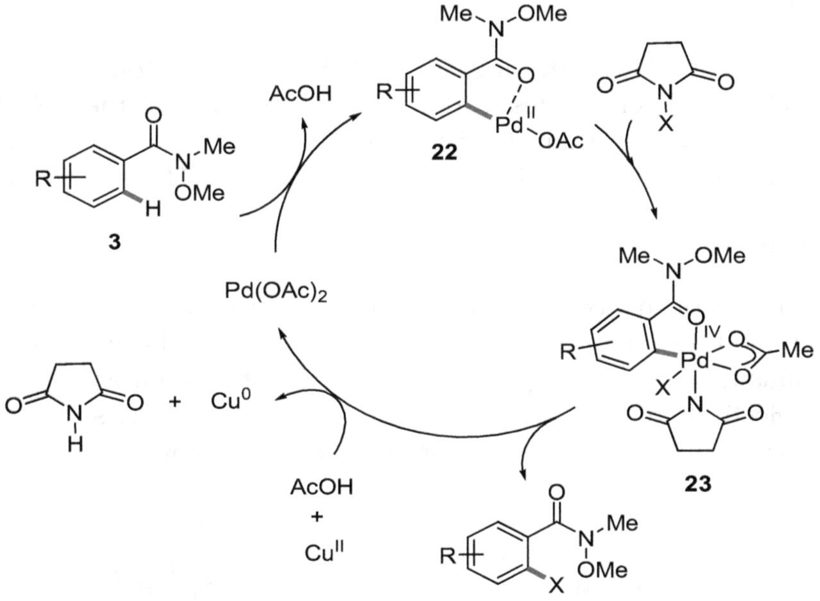
AcOH
22
3
Pd(OAc)2
+
Cu0
23
AcOH
+
CuII

**Figure 5.** An adapted version of the mechanism proposed by Kapur and co-workers for the Pd-catalyzed *ortho* C-H halogenation of aromatic Weinreb amides. X = $Cl^-$, $Br^-$ or $I^-$.

2018 saw Yu and co-workers extend the utility of the Weinreb amide directing group further [80]. This time, the Yu group used it as the basis for directing Pd-catalyzed $C_{sp}{}^3$-H arylation of alkyl groups (Scheme 11). The discovery of general and efficient methods that allow selective substitution of C(sp$^3$)-H at transition metal centers is a long-standing challenge. For Pd, progress has been notably rapid in in the past few years [63,81–83], and its various aspects have been reviewed recently [65,84,85].

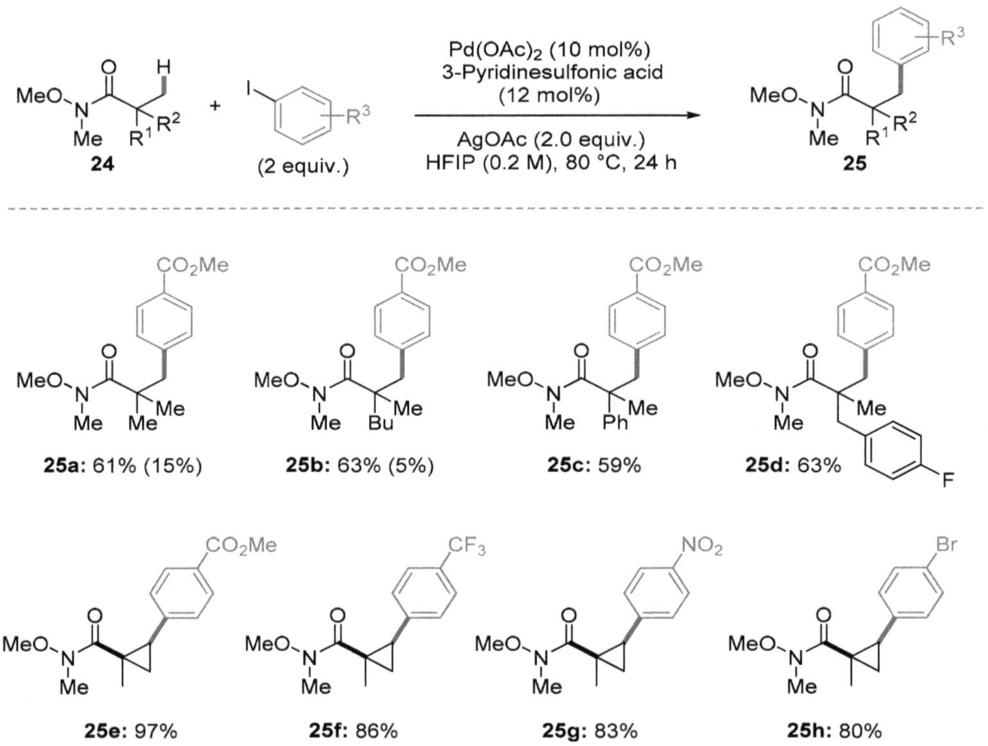

**Scheme 11.** Selected scope from Yu and co-workers' Pd-catalyzed C($sp^3$)-H arylation protocol enabled by Weinreb amide directing groups.

Yu and co-workers found that the inclusion of 3-pyridinesulfonic acid was crucial to the reaction, both Weinreb and other amides. Computational studies at the DFT level revealed that 3-pyridinesulfonic acid stabilizes cationic Pd intermediates during the reaction and that it promotes the dissociation of acetate ligands, which is required for the C($sp^3$)-H bond cleavage to occur [80].

## 5. Rh-catalyzed Reactions

The first two decades of this century have seen an explosion of interest in the use of Rh(III) catalysts for selective C-H functionalization. Substantial progress has been made in both scope [86,87] and attendant mechanistic understanding [7,88–90]. As part of their efforts to answer key questions relating to the regioselectivity of Rh(III)-catalyzed transformations, Rovis and co-workers reported in 2013 an intramolecular alkene hydroarylation directed by Weinreb amide **26** in excellent yield (Scheme 12) [91].

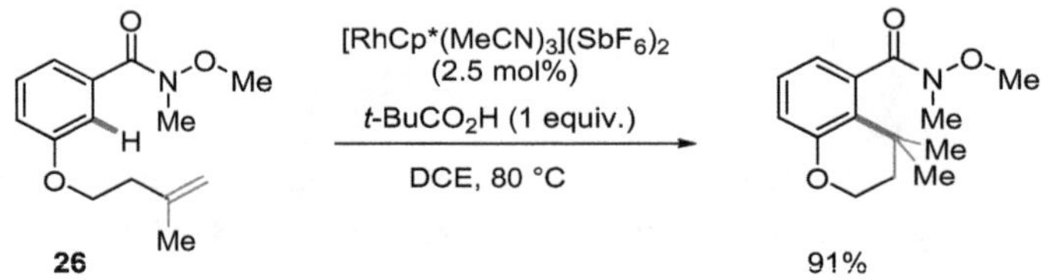

**Scheme 12.** A Rh(III)-catalyzed, Weinreb amide-directed alkene hydroarylation.

Wang and co-workers reported a closely-related but more elaborate system, involving $Cu(OAc)_2$ as a catalytic oxidant regenerated under air to retain olefin functionality at the end of a cycle coupling aromatic Weinreb amides with alkenes (Scheme 13) [92]. An ample scope demonstrated the reaction's tolerance towards various functional groups, including halogens, various *ortho*-substituents and a range of electron-donating and electron-withdrawing groups. Competition experiments revealed that electron-rich arenes reacted faster than their electron-poor counterparts.

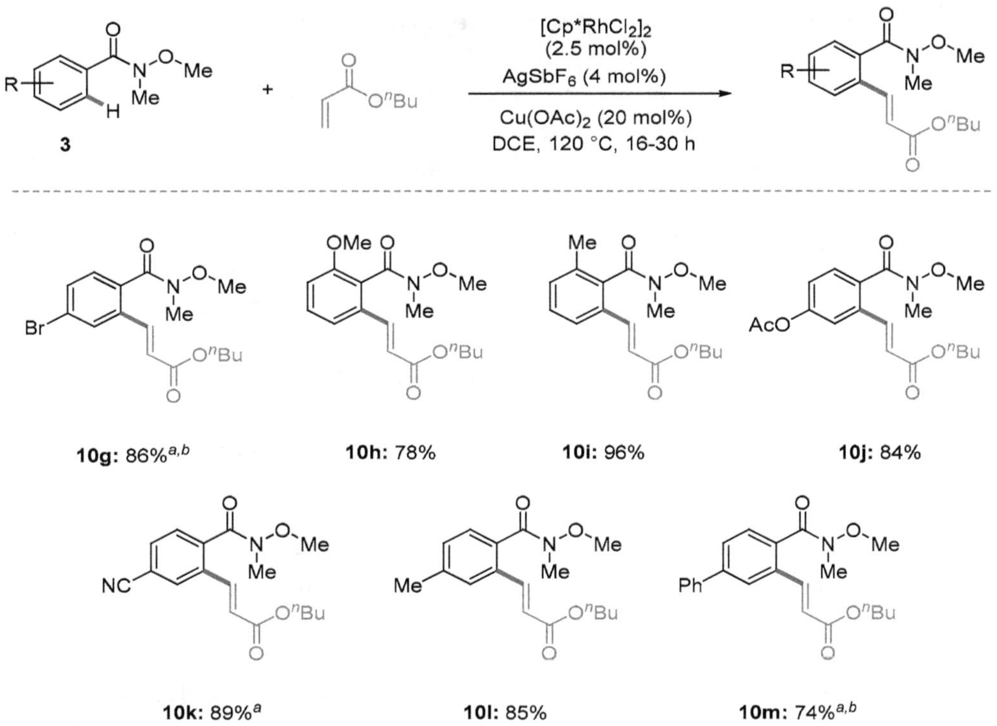

**Scheme 13.** Selected examples from Wang and co-workers' oxidative Rh-catalyzed C-H alkenylation directed by Weinreb amides. [a] $[Cp^*RhCl_2]_2$ (2.5 mol%) and $AgSbF_6$ (10 mol%) loadings were used. [b] A reaction temperature of 130 °C was used.

The mechanism proposed by Wang and co-workers (Figure 6) involved coordination of the Rh(III) center to the carbonyl moiety of the Weinreb amide, insertion of the alkene to the rhodacyclic intermediate (**27**), $\beta$-hydride elimination and regeneration of the active catalyst by $Cu(OAc)_2$.

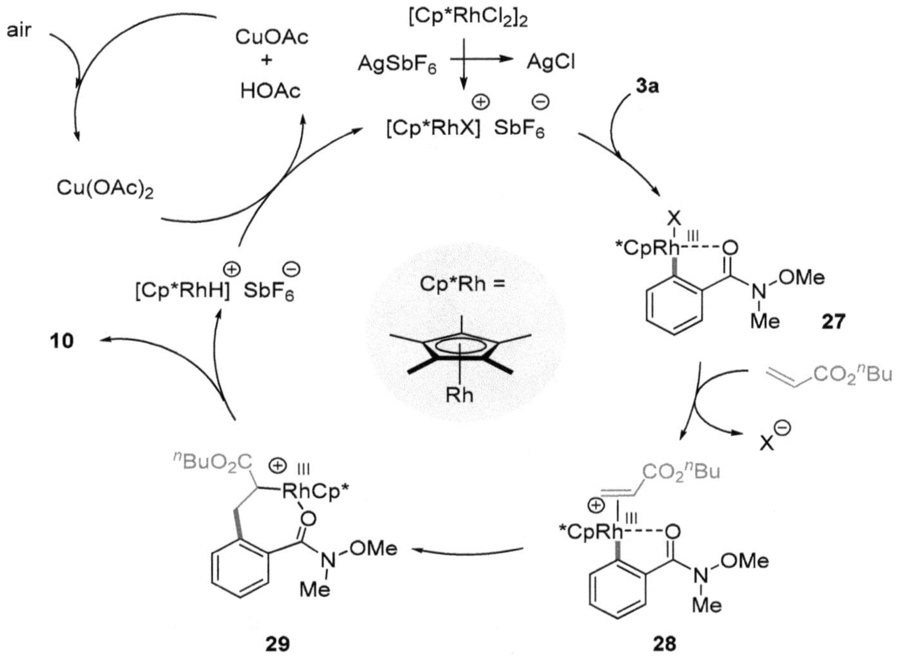

**Figure 6.** An adapted version of the mechanism proposed by Wang and co-workers for the oxidative C-H alkenylation of aromatic Weinreb amides catalyzed by Rh(III).

Whilst, strictly speaking, it represents a slight departure from the current focus on the Weinreb amide group, it is noteworthy that a very closely related class of substrates, has been used extensively wherein their C(O)NH-OMe bond as an "internal oxidant" for transition metal centers. Thus, C-H functionalization reactions may be performed without exogenous oxidants. This strategy has worked with a number of transition metals, Rh(I/III) catalytic cycles have dominated in this area [41].

In 2018, Qin and co-workers disclosed a near identical set of conditions to those used by Wang above, but using ethenesulfonyl fluoride (ESF) as the alkene coupling partner and 1,4-dioxane as the solvent [93]. However, the Qin group further observed that increasing the loading of $AgSbF_6$ to 1 equivalent favored the formation of cyclic lactones. Residual water introduced from the hygroscopic $AgSbF_6$ was proposed to promote the in situ hydrolysis of the Weinreb amide group to account for the lactone formation, though mechanistic work proved inconclusive. In switching away from Weinreb amides to N-methoxybenazmides (ArC(O)NH-OMe), Qin and co-workers were able to cause a similar oxidative cyclization involving insertion of the ESF double bond into the amide N-H unit.

Recent years have witnessed the rediscovery and subsequent explosion of interest in functionalizations enabled by photoredox catalysis. The photoexcitation of transition metal complexes serves as a greener and more economically viable method of generating radical intermediates able to initiate a wide range of valuable reactions [94–98]. One application of this is the replacement of terminal oxidants with lighter loadings of a photocatalyst whose oxidative power is obtained via photoexcitation. Rueping and co-workers applied this strategy to demonstrate that the oxidative alkenylation of aromatic amides, including Weinreb amides, is viable using a manifold analogous to that described by Wang's group (see above), but driven by a visible light-regenerated Ru-based photocatalyst, $[Ru(bpy)_3][PF_6]_2$ (Scheme 14) [99,100]. The protocol is notable not just for its efficiency and high functional group tolerance, but also for the fact that it could be extended to other amides as well as a variety of olefin substrates (various groups $R^2$).

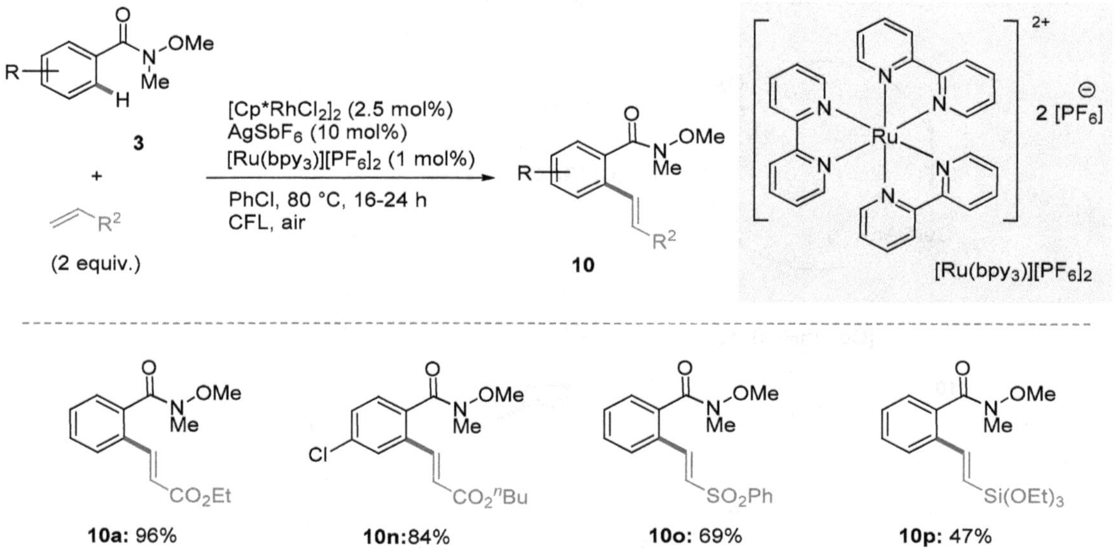

**Scheme 14.** Rh-catalyzed olefination of aromatic Weinreb amides enabled by Ru photocatalysis.

The catalytic cycle proposed by Rueping and co-workers closely resembles to that of Wang: *ortho*-rhodation of **3** to give **29** coordination and insertion of the olefin (complex **30** to **31**) and $\beta$-hydride elimination to form the hydride intermediate **32** and the product (Figure 7). Reductive loss of the hydride and re-oxidation to Rh(III), however, is mediated by the photo-excited $[Ru(bpy)3]^{2+\cdot}$ species, of which only a 1 mol% loading is required.

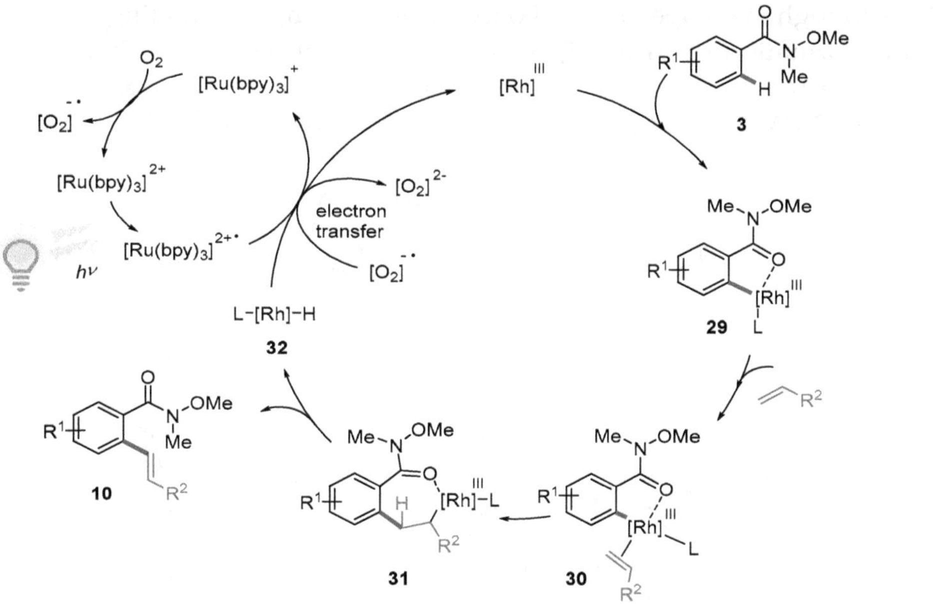

**Figure 7.** Mechanism proposed by Rueping and co-workers for the oxidative C-H olefination of aromatic Weinreb amides catalyzed by Rh(III) in the presence of a Ru photocatalyst.

## 6. Ir-catalyzed Reactions

Boronates rank amongst the most versatile of groups; they may be converted to an extraordinarily broad range of functionality through a many different mechanisms [101–103]. Moreover, recent years have seen boronate-based methodology emerge as the basis of automated synthesis, which holds enormous potential to streamline the synthesis of many complex (hetero)aromatic and olefinic molecules [104–107]. Such advantages render especially important methods that allow the regioselective introduction of boronates to organic substrates. Amongst these, Ir-catalyzed C-H borylation ranks as one of the mildest and most enabling, as it is amenable to regiocontrol through sterics [108,109], directing groups (both to the *ortho* [110–115] and *para* [116] positions) and/or the inherent electronic properties [117–119] of a substrate.

Krska, Maleczka and Smith have demonstrated [120] that Weinreb amides rank amongst groups that competently direct Ir catalysts towards *ortho*-C-H borylation (Scheme 15). Although only a single entry using a Weinreb amide was reported in their Communication, the success of the reaction (a high yield and regioselectivity) was representative, suggesting that other advantages of the method, such as the high functional group tolerance, could easily be paired with the Weinreb amide functionality. Conditions for this transformation deviated little from those commonly used for Ir-catalyzed C-H borylation more generally: $[Ir(\mu\text{-}OMe)(COD)]_2$ as the pre-catalyst with dtbtpy as a ligand and $B_2(pin)_2$ as the boron source. The corresponding mono-borylated product, **33** was reported to form in 84% yield.

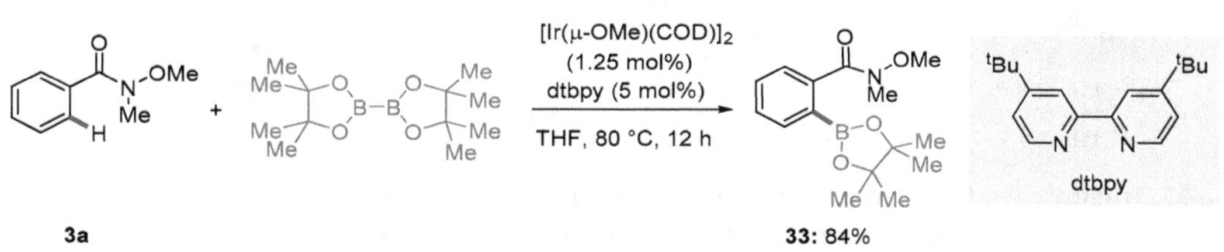

**Scheme 15.** Ir-catalyzed *ortho*-C-H borylation of an arene using the Weinreb amide as a directing group.

Enantioselective C-H functionalization marries the benefits of directly substituting a C-H bond with with establishing a new stereocenter in the product. This pursuit continues to inspire a growing body of research [121]. In 2015, Yamamoto and Shirai described an Ir-catalyzed protocol for the asymmetric intermolecular hydroarylation of arenes directed by oxygen based directing groups [122]. Although the scope included only a single example exploiting a Weinreb amide directing group (the transformation of **34** to **35**), this entry was representative both in terms of yield and selectivity (Scheme 16). Enantioselectivity was induced using the bidentate bis(phosphoramidite) ligand (*R*,*R*)-*S*-Me-BIPAM ligand (**36**) and the N-O bond was preserved in the product.

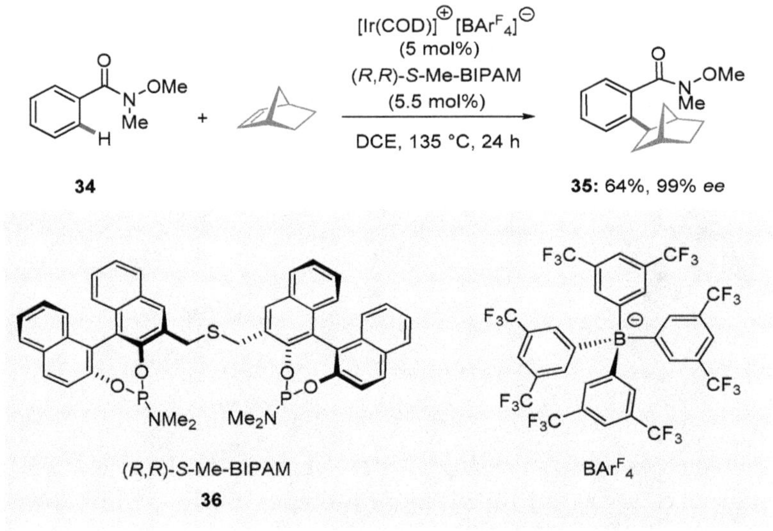

**Scheme 16.** Ir-catalyzed enantioselective C-H hydroarylation of an olefin, directed by a Weinreb amide.

Most recently, the group of Martín-Matute has developed an Ir(III)-catalyzed C-H *ortho*-iodination of various amides [123]. The scope includes a strong focus on Weinreb amides. Their reaction used $[Cp^*Ir(H_2O)_3][SO_4]$ as the catalyst precursor, and *N*-iodo-succinimide (NIS) as the halogenating reagent in the presence of trifluoroacetic acid. Selected examples are shown below (Scheme 17). It is notable that Weinreb amides returned only the mono-iodinated products **11**; a result only tertiary amides gave (primary and secondary amides gave at least small amounts of di-iodination). Whilst the substrate scope included a broad range of functional groups, substituents positioned *ortho* to the directing group at the outset restricted reactivity, presumably by imposing prohibitive amounts of steric hindrance.

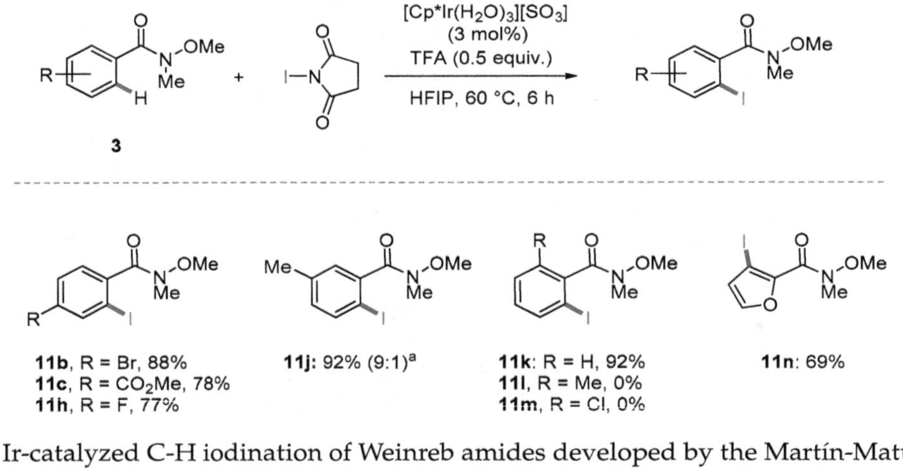

**Scheme 17.** Ir-catalyzed C-H iodination of Weinreb amides developed by the Martín-Matute group. [a] Ratio of isomers is indicated in parenthesis; the major isomer is shown.

The authors also performed a robustness screen [124,125] to identify the reaction's tolerance towards various functional groups. To this end, various small molecules with different functional groups were added to the reaction mixture to see how the reaction would be affected. Tolerance towards several common functional groups, including ketones amides, carboxylic acids and alkyl halides proved excellent. Aldehydes and alkenes returned less satisfactory results, however. Mechanistically, the reaction was proposed to follow a pathway closely related to that of Martín-Matute and co-workers' recently published Ir-catalyzed C-H *ortho*-iodination of aryl carboxylic acids [126]. The acid additive is understood to play a dual role: 1) activation of NIS via protonation of its carbonyl group and 2) by encouraging the dissociation of the iodinated product from the Ir center. Indeed, Martín-Matute and co-workers found in their optimization study on amides that lowering the amount of acid additive favored the formation of di-iodinated products for non-tertiary amides.

## 7. Conclusions

The Weinreb amide is a privileged functional group in organic synthesis that enables otherwise impossible transformations. Recent developments have seen C-H functionalization methodology greatly expand the range of chemical contexts in which its advantages may be exploited. Challenges remain, however. Presently, Weinreb amides are rarely explored as a substrate class in their own right; they are most often presented as specialized examples in studies describing a more general scope, typically amidst other amides. Thus, it is usual that only the simplest examples of Weinreb amides are demonstrated to work under newly developed reaction conditions. This is at least a little unfair since, as some of the examples described above demonstrate, Weinreb amides may return different results to other amides, for example by virtue of the reactivity of their N-O bond. Weinreb amides might, in this sense, be considered hitherto as "sleeper" substrates.

Despite this, it is evident that Weinreb amides can direct a wide range of C-H functionalization reactions catalyzed by transition metals. C-H functionalization, and catalytic methodology in general, is undergoing a shift of emphasis towards the use of less "endangered" [127,128], especially first-row transition elements for catalysis [43]. However, their use with Weinreb amides is rare. For example, to the best of our knowledge, no Weinreb amide-directed C-H functionalization reactions catalyzed by Mn, Fe or Ni have yet been reported.

Finally, it is unfortunate that the preponderance of conditions used for Weinreb amide directed C-H functionalization involve toxic halogenated solvents. Increasing legislative pressure is being brought to bear on this problem, with a particular focus against 1,2-dichloroethane (DCE) [129], which features in many of the reactions discussed above. We look forward to the use of Weinreb amides under greener conditions [130–132].

**Acknowledgments:** We thank the Wenner-Gren Foundation for funding (JK).

## References

1. Crabtree, R.H.; Lei, A. Introduction: CH Activation. *Chem. Rev.* **2017**, *117*, 8481–8482. [CrossRef] [PubMed]
2. Yang, Y.; Lan, J.; You, J. Oxidative C-H/C-H Coupling Reactions between Two (Hetero)arenes. *Chem. Rev.* **2017**, *117*, 8787–8863. [CrossRef] [PubMed]
3. Park, Y.; Kim, Y.; Chang, S. Transition Metal-Catalyzed C–H Amination: Scope, Mechanism, and Applications. *Chem. Rev.* **2017**, *117*, 9247–9301. [CrossRef] [PubMed]
4. Murakami, K.; Yamada, S.; Kaneda, T.; Itami, K. C–H Functionalization of Azines. *Chem. Rev.* **2017**, *117*, 9302–9332. [CrossRef] [PubMed]
5. Sandtorv, A.H. Transition Metal-Catalyzed C-H Activation of Indoles. *ACS Catal.* **2015**, *357*, 2403–2435. [CrossRef]
6. Simmons, E.M.; Hartwig, J.F. On the Interpretation of Deuterium Kinetic Isotope Effects in C-H Bond Functionalizations by Transition-Metal Complexes. *Angew. Chem. Int. Ed. Engl.* **2012**, *51*, 3066–3072. [CrossRef] [PubMed]
7. Qi, X.; Li, Y.; Bai, R.; Lan, Y. Mechanism of Rhodium-Catalyzed C–H Functionalization: Advances in Theoretical Investigation. *Acc. Chem. Res.* **2017**, *50*, 2799–2808. [CrossRef] [PubMed]
8. Davies, D.L.; Macgregor, S.A.; McMullin, C.L. Computational Studies of Carboxylate-Assisted C–H Activation and Functionalization at Group 8–10 Transition Metal Centers. *Chem. Rev.* **2017**, *117*, 8649–8709. [CrossRef] [PubMed]

9. Balcells, D.; Clot, E.; Eisenstein, O. C–H Bond Activation in Transition Metal Species from a Computational Perspective. *Chem. Rev.* **2010**, *110*, 749–823. [CrossRef] [PubMed]
10. Gensch, T.; Hopkinson, M.N.; Glorius, F.; Wencel-Delord, J. Mild metal-catalyzed C-H activation: Examples and concepts. *Chem. Soc. Rev.* **2016**, *45*, 2900–2936. [CrossRef] [PubMed]
11. Hartwig, J.F.; Larsen, M.A. Undirected, Homogeneous C–H Bond Functionalization: Challenges and Opportunities. *ACS Cent. Sci.* **2016**, *2*, 281–292. [CrossRef] [PubMed]
12. Brückl, T.; Baxter, R.D.; Ishihara, Y.; Baran, P.S. Innate and Guided C–H Functionalization Logic. *Acc. Chem. Res.* **2012**, *45*, 826–839. [CrossRef] [PubMed]
13. Ping, L.; Chung, D.S.; Bouffard, J.; Lee, S.G. Transition metal-catalyzed site- and regio-divergent C-H bond functionalization. *Chem. Soc. Rev.* **2017**, *46*, 4299–4328. [CrossRef] [PubMed]
14. Saint-Denis, T.G.; Zhu, R.-Y.; Chen, G.; Wu, Q.-F.; Yu, J.-Q. Enantioselective C(sp3)-H bond activation by chiral transition metal catalysts. *Science* **2018**, *359*, eaao4798. [CrossRef] [PubMed]
15. Sambiagio, C.; Schönbauer, D.; Blieck, R.; Dao-Huy, T.; Pototschnig, G.; Schaaf, P.; Wiesinger, T.; Zia, M.F.; Wencel-Delord, J.; Besset, T.; et al. A comprehensive overview of directing groups applied in metal-catalysed C–H functionalisation chemistry. *Chem. Soc. Rev.* **2018**, *47*, 6603–6743. [CrossRef] [PubMed]
16. Lyons, T.W.; Sanford, M.S. Palladium-Catalyzed Ligand-Directed C—H Functionalization Reactions. *Chem. Rev.* **2010**, *110*, 1147–1169. [CrossRef] [PubMed]
17. Dey, A.; Agasti, S.; Maiti, D. Palladium catalysed meta-C–H functionalization reactions. *Org. Biomol. Chem.* **2016**, *14*, 5440–5453. [CrossRef] [PubMed]
18. Leitch, J.A.; Frost, C.G. Ruthenium-catalysed σ-activation for remote meta-selective C–H functionalisation. *Chem. Soc. Rev.* **2017**, *46*, 7145–7153. [CrossRef] [PubMed]
19. Dey, A.; Maity, S.; Maiti, D. Reaching the south: Metal-catalyzed transformation of the aromatic para-position. *Chem. Commun.* **2016**, *52*, 12398–12414. [CrossRef] [PubMed]
20. Rousseau, G.; Breit, B. Removable Directing Groups in Organic Synthesis and Catalysis. *Angew. Chem. Int. Ed.* **2011**, *50*, 2450–2494. [CrossRef] [PubMed]
21. Yadav, M.R.; Rit, R.K.; Shankar, M.; Sahoo, A.K. Reusable and Removable Directing Groups for C(sp$^2$)—H Bond Functionalization of Arenes. *ASIAN J. Org. Chem.* **2015**, *4*, 846–864. [CrossRef]
22. Zhang, F.; Spring, D.R. Arene C–H functionalisation using a removable/modifiable or a traceless directing group strategy. *Chem. Soc. Rev.* **2014**, *43*, 6906–6919. [CrossRef] [PubMed]
23. Font, M.; Quibell, J.M.; Perry, G.J.P.; Larrosa, I. The use of carboxylic acids as traceless directing groups for regioselective C–H bond functionalisation. *Chem. Commun.* **2017**, *53*, 5584–5597. [CrossRef] [PubMed]
24. Gandeepan, P.; Ackermann, L. Transient Directing Groups for Transformative C–H Activation by Synergistic Metal Catalysis. *Chem* **2018**, *4*, 199–222. [CrossRef]
25. Ihara, H.; Koyanagi, M.; Suginome, M. Anthranilamide: A Simple, Removable ortho-Directing Modifier for Arylboronic Acids Serving also as a Protecting Group in Cross-Coupling Reactions. *Org. Lett.* **2011**, *13*, 2662–2665. [CrossRef] [PubMed]
26. Ihara, H.; Suginome, M. Easily Attachable and Detachable ortho-Directing Agent for Arylboronic Acids in Ruthenium-Catalyzed Aromatic C—H Silylation. *J. Am. Chem. Soc.* **2009**, *131*, 7502–7503. [CrossRef] [PubMed]
27. Ihara, H.; Ueda, A.; Suginome, M. Ruthenium-catalyzed C-H Silylation of Methylboronic Acid Using a Removable α-Directing Modifier on the Boron Atom. *Chem. Lett.* **2011**, *40*, 916–918. [CrossRef]
28. Yamamoto, T.; Ishibashi, A.; Suginome, M. Regioselective Synthesis of o-Benzenediboronic Acids via Ir-Catalyzed o-C–H Borylation Directed by a Pyrazolylaniline-Modified Boronyl Group. *Org. Lett.* **2017**, *19*, 886–889. [CrossRef] [PubMed]
29. Zhao, Q.; Poisson, T.; Pannecoucke, X.; Besset, T. The Transient Directing Group Strategy: A New Trend in Transition-Metal-Catalyzed C–H Bond Functionalization. *Synthesis* **2017**, *49*, 4808–4826. [CrossRef]
30. Zhu, R.-Y.; Farmer, M.E.; Chen, Y.-Q.; Yu, J.-Q. A Simple and Versatile Amide Directing Group for C—H Functionalizations. *Angew. Chem. Int. Ed.* **2016**, *55*, 10578–10599. [CrossRef] [PubMed]
31. Nahm, S.; Weinreb, S.M. N-methoxy-n-methylamides as effective acylating agents. *Tetrahedron Lett.* **1981**, *22*, 3815–3818. [CrossRef]
32. Nowak, M. Weinreb Amides. *Synlett* **2015**, *26*, 561–562. [CrossRef]
33. De Sarkar, S.; Liu, W.; Kozhushkov, S.I.; Ackermann, L. Weakly Coordinating Directing Groups for Ruthenium(II)- Catalyzed C—H Activation. *Adv. Synth. Catal.* **2014**, *356*, 1461–1479. [CrossRef]

34. Arockiam, P.B.; Bruneau, C.; Dixneuf, P.H. Ruthenium(II)-catalyzed C-H bond activation and functionalization. *Chem. Rev.* **2012**, *112*, 5879–5918. [CrossRef] [PubMed]
35. Ruiz, S.; Villuendas, P.; Urriolabeitia, E.P. Ru-catalysed C–H functionalisations as a tool for selective organic synthesis. *Tetrahedron Lett.* **2016**, *57*, 3413–3432. [CrossRef]
36. Yang, F.; Ackermann, L. Ruthenium-Catalyzed C–H Oxygenation on Aryl Weinreb Amides. *Org. Lett.* **2013**, *15*, 718–720. [CrossRef] [PubMed]
37. More, N.Y.; Padala, K.; Jeganmohan, M. Ruthenium-Catalyzed C–H Benzoxylation of tert-Benzamides with Aromatic Acids by Weak Coordination. *J. Org. Chem.* **2017**, *82*, 12691–12700. [CrossRef] [PubMed]
38. Das, R.; Kapur, M. Fujiwara-Moritani Reaction of Weinreb Amides using a Ruthenium-Catalyzed C-H Functionalization Reaction. *Chem. Asian J.* **2015**, *10*, 1505–1512. [CrossRef] [PubMed]
39. Ferreira, E.M.; Zhang, H.; Stoltz, B.M. Oxidative Heck-Type Reactions (Fujiwara–Moritani Reactions). In *The Mizoroki–Heck Reaction*; Oestreich, M., Ed.; Wiley: Chichester, UK, 2009.
40. Ackermann, L. Carboxylate-Assisted Transition-Metal-Catalyzed C−H Bond Functionalizations: Mechanism and Scope. *Chem. Rev.* **2011**, *111*, 1315–1345. [CrossRef] [PubMed]
41. Mo, J.; Wang, L.; Liu, Y.; Cui, X. Transition-Metal-Catalyzed Direct C–H Functionalization under External-Oxidant-Free Conditions. *Synthesis* **2015**, *47*, 439–459. [CrossRef]
42. Das, R.; Kapur, M. Product Control using Substrate Design: Ruthenium-Catalysed Oxidative C-H Olefinations of Cyclic Weinreb Amides. *Chem. Eur. J.* **2016**, *22*, 16986–16990. [CrossRef] [PubMed]
43. Gandeepan, P.; Müller, T.; Zell, D.; Cera, G.; Warratz, S.; Ackermann, L. 3d Transition Metals for C–H Activation. *Chem. Rev.* **2018**. [CrossRef] [PubMed]
44. Moselage, M.; Li, J.; Ackermann, L. Cobalt-Catalyzed C–H Activation. *ACS Catal.* **2016**, *6*, 498–525. [CrossRef]
45. Gao, K.; Yoshikai, N. Low-Valent Cobalt Catalysis: New Opportunities for C–H Functionalization. *Acc. Chem. Res.* **2014**, *47*, 1208–1219. [CrossRef] [PubMed]
46. Yoshikai, N. Development of Cobalt-Catalyzed C–H Bond Functionalization Reactions. *Bull. Chem. Soc. Jpn.* **2014**, *87*, 843–857. [CrossRef]
47. Yoshino, T.; Ikemoto, H.; Matsunaga, S.; Kanai, M. Cp*CoIII-Catalyzed C2-Selective Addition of Indoles to Imines. *Chem. A Eur. J.* **2013**, *19*, 9142–9146. [CrossRef] [PubMed]
48. Yoshino, T.; Ikemoto, H.; Matsunaga, S.; Kanai, M. A Cationic High-Valent Cp*CoIII Complex for the Catalytic Generation of Nucleophilic Organometallic Species: Directed C–H Bond Activation. *Angew. Chem. Int. Ed. Engl.* **2013**, *52*, 2207–2211. [CrossRef] [PubMed]
49. Bunno, Y.; Murakami, N.; Suzuki, Y.; Kanai, M.; Yoshino, T.; Matsunaga, S. Cp*CoIII-Catalyzed Dehydrative C–H Allylation of 6-Arylpurines and Aromatic Amides Using Allyl Alcohols in Fluorinated Alcohols. *Org. Lett.* **2016**, *18*, 2216–2219. [CrossRef] [PubMed]
50. Sun, B.; Yoshino, T.; Matsunaga, S.; Kanai, M. Air-Stable Carbonyl(pentamethylcyclopentadienyl)cobalt Diiodide Complex as a Precursor for Cationic (Pentamethylcyclopentadienyl)cobalt(III) Catalysis: Application for Directed C-2 Selective C-H Amidation of Indoles. *Adv. Synth. Catal.* **2014**, *356*, 1491–1495. [CrossRef]
51. Colomer, I.; Chamberlain, A.E.R.; Haughey, M.B.; Donohoe, T.J. Hexafluoroisopropanol as a highly versatile solvent. *Nat. Rev. Chem.* **2017**, *1*, 0088. [CrossRef]
52. Wencel-Delord, J.; Colobert, F. A remarkable solvent effect of fluorinated alcohols on transition metal catalysed C–H functionalizations. *Org. Chem. Front.* **2016**, *3*, 394–400. [CrossRef]
53. Planas, O.; Chirila, P.G.; Whiteoak, C.J.; Ribas, X. Current Mechanistic Understanding of Cobalt-Catalyzed C–H Functionalization. *Adv. Organomet. Chem.* **2018**, *69*, 209–282. [CrossRef]
54. Gorelsky, S.I.; Lapointe, D.; Fagnou, K. Analysis of the Concerted Metalation-Deprotonation Mechanism in Palladium-Catalyzed Direct Arylation Across a Broad Range of Aromatic Substrates. *J. Am. Chem. Soc.* **2008**, *130*, 10848–10849. [CrossRef] [PubMed]
55. Kawai, K.; Bunno, Y.; Yoshino, T.; Matsunaga, S. Weinreb Amide Directed Versatile C–H Bond Functionalization under (eta(5)-Pentamethylcyclopentadienyl)cobalt(III) Catalysis. *Chem. Eur. J.* **2018**, *24*, 10231–10237. [CrossRef] [PubMed]
56. Kuhl, N.; Schröder, N.; Glorius, F. Rh(III)-Catalyzed Halogenation of Vinylic C–H Bonds: Rapid and General Access to Z-Halo Acrylamides. *Org. Lett.* **2013**, *15*, 3860–3863. [CrossRef] [PubMed]

57. Chirila, P.G.; Adams, J.; Dirjal, A.; Hamilton, A.; Whiteoak, C.J. Cp*Co(III)-Catalyzed Coupling of Benzamides with α,β-Unsaturated Carbonyl Compounds: Preparation of Aliphatic Ketones and Azepinones. *Chem. Eur. J.* **2018**, *24*, 3584–3589. [CrossRef] [PubMed]
58. Johansson Seechurn, C.C.C.; Kitching, M.O.; Colacot, T.J.; Snieckus, V. Palladium-Catalyzed Cross-Coupling: A Historical Contextual Perspective to the 2010 Nobel Prize. *Angew. Chem. Int. Ed. Engl.* **2012**, *51*, 5062–5085. [CrossRef] [PubMed]
59. Wu, X.-F.; Neumann, H.; Beller, M. Synthesis of Heterocycles via Palladium-Catalyzed Carbonylations. *Chem. Rev.* **2013**, *113*, 1–35. [CrossRef] [PubMed]
60. Wang, D.; Weinstein, A.B.; White, P.B.; Stahl, S.S. Ligand-Promoted Palladium-Catalyzed Aerobic Oxidation Reactions. *Chem. Rev.* **2018**, *118*, 2636–2679. [CrossRef] [PubMed]
61. Ruiz-Castillo, P.; Buchwald, S.L. Applications of Palladium-Catalyzed C–N Cross-Coupling Reactions. *Chem. Rev.* **2016**, *116*, 12564–12649. [CrossRef] [PubMed]
62. Sehnal, P.; Taylor, R.J.K.; Fairlamb, I.J.S. Emergence of Palladium(IV) Chemistry in Synthesis and Catalysis. *Chem. Rev.* **2010**, *110*, 824–889. [CrossRef] [PubMed]
63. Timsina, Y.N.; Gupton, B.F.; Ellis, K.C. Palladium-Catalyzed C–H Amination of C($sp^2$) and C($sp^3$)–H Bonds: Mechanism and Scope for N-Based Molecule Synthesis. *ACS Catal.* **2018**, *8*, 5732–5776. [CrossRef]
64. Topczewski, J.J.; Sanford, M.S. Carbon–hydrogen (C–H) bond activation at PdIV: A Frontier in C–H functionalization catalysis. *Chem. Sci.* **2015**, *6*, 70–76. [CrossRef] [PubMed]
65. He, J.; Wasa, M.; Chan, K.S.L.; Shao, Q.; Yu, J.-Q. Palladium-Catalyzed Transformations of Alkyl C–H Bonds. *Chem. Rev.* **2017**, *117*, 8754–8786. [CrossRef] [PubMed]
66. Bonney, K.J.; Schoenebeck, F. Experiment and computation: A combined approach to study the reactivity of palladium complexes in oxidation states 0 to iv. *Chem. Soc. Rev.* **2014**, *43*, 6609–6638. [CrossRef] [PubMed]
67. Valente, C.; Çalimsiz, S.; Hoi, K.H.; Mallik, D.; Sayah, M.; Organ, M.G. The Development of Bulky Palladium NHC Complexes for the Most-Challenging Cross-Coupling Reactions. *Angew. Chem. Int. Ed. Engl.* **2012**, *51*, 3314–3332. [CrossRef] [PubMed]
68. Hazari, N.; Melvin, P.R.; Beromi, M.M. Well-defined nickel and palladium precatalysts for cross-coupling. *Nat. Rev. Chem.* **2017**, *1*, 0025. [CrossRef] [PubMed]
69. Wang, Y.; Zhou, K.; Lan, Q.; Wang, X.-S. Pd(II)-catalyzed C–H arylation of aryl and benzyl Weinreb amides. *Org. Biomol. Chem.* **2015**, *13*, 353–356. [CrossRef] [PubMed]
70. Mudarra, Á.L.; Martínez de Salinas, S.; Pérez-Temprano, M.H. Beyond the traditional roles of Ag in catalysis: The transmetalating ability of organosilver(i) species in Pd-catalysed reactions. *Org. Biomol. Chem.* **2018**, *17*, 1655–1667. [CrossRef] [PubMed]
71. Yedage, S.L.; Bhanage, B.M. Palladium-Catalyzed Deaminative Phenanthridinone Synthesis from Aniline via C–H Bond Activation. *J. Org. Chem.* **2016**, *81*, 4103–4111. [CrossRef] [PubMed]
72. Li, G.; Wan, L.; Zhang, G.; Leow, D.; Spangler, J.; Yu, J.-Q. Pd(II)-Catalyzed C–H Functionalizations Directed by Distal Weakly Coordinating Functional Groups. *J. Am. Chem. Soc.* **2015**, *137*, 4391–4397. [CrossRef] [PubMed]
73. Engle, K.M.; Wang, D.-H.; Yu, J.-Q. Ligand-Accelerated C−H Activation Reactions: Evidence for a Switch of Mechanism. *J. Am. Chem. Soc.* **2010**, *132*, 14137–14151. [CrossRef] [PubMed]
74. Das, R.; Kapur, M. Transition-Metal-Catalyzed Site-Selective C−H Halogenation Reactions. *Asian J. Org. Chem.* **2018**, *7*, 1524–1541. [CrossRef]
75. Das, R.; Kapur, M. Palladium-Catalyzed, ortho-Selective C–H Halogenation of Benzyl Nitriles, Aryl Weinreb Amides, and Anilides. *J. Org. Chem.* **2017**, *82*, 1114–1126. [CrossRef] [PubMed]
76. Powers, D.C.; Lee, E.; Ariafard, A.; Sanford, M.S.; Yates, B.F.; Canty, A.J.; Ritter, T. Connecting Binuclear Pd(III) and Mononuclear Pd(IV) Chemistry by Pd–Pd Bond Cleavage. *J. Am. Chem. Soc.* **2012**, *134*, 12002–12009. [CrossRef] [PubMed]
77. Powers, D.C.; Benitez, D.; Tkatchouk, E.; Goddard, W.A.; Ritter, T. Bimetallic Reductive Elimination from Dinuclear Pd(III) Complexes. *J. Am. Chem. Soc.* **2010**, *132*, 14092–14103. [CrossRef] [PubMed]
78. Powers, D.C.; Ritter, T. Bimetallic Pd(III) complexes in palladium-catalysed carbon–heteroatom bond formation. *Nat. Chem.* **2009**, *1*, 419. [CrossRef]
79. Powers, D.C.; Geibel, M.A.L.; Klein, J.E.M.N.; Ritter, T. Bimetallic Palladium Catalysis: Direct Observation of Pd(III)−Pd(III) Intermediates. *J. Am. Chem. Soc.* **2009**, *131*, 17050–17051. [CrossRef] [PubMed]

80. Park, H.; Chekshin, N.; Shen, P.-X.; Yu, J.-Q. Ligand-Enabled, Palladium-Catalyzed β-C(sp3)–H Arylation of Weinreb Amides. *ACS Catal.* **2018**, *8*, 9292–9297. [CrossRef]
81. Willcox, D.; Chappell, B.G.N.; Hogg, K.F.; Calleja, J.; Smalley, A.P.; Gaunt, M.J. A general catalytic β-C–H carbonylation of aliphatic amines to β-lactams. *Science* **2016**, *354*, 851–857. [CrossRef] [PubMed]
82. Zhu, R.-Y.; Li, Z.-Q.; Park, H.S.; Senanayake, C.H.; Yu, J.-Q. Ligand-Enabled γ-C(sp3)–H Activation of Ketones. *J. Am. Chem. Soc.* **2018**, *140*, 3564–3568. [CrossRef] [PubMed]
83. Zhu, R.-Y.; He, J.; Wang, X.-C.; Yu, J.-Q. Ligand-Promoted Alkylation of C(sp3)–H and C(sp2)–H Bonds. *J. Am. Chem. Soc.* **2014**, *136*, 13194–13197. [CrossRef] [PubMed]
84. Baudoin, O. Ring Construction by Palladium(0)-Catalyzed C(sp3)–H Activation. *Acc. Chem. Res.* **2017**, *50*, 1114–1123. [CrossRef] [PubMed]
85. Baudoin, O. Transition metal-catalyzed arylation of unactivated C(sp3)–H bonds. *Chem. Soc. Rev.* **2011**, *40*, 4902–4911. [CrossRef] [PubMed]
86. Chatani, N.; Rej, S. Rh-Catalyzed Removable Directing Group Assisted sp2 or sp3-C-H Bond Functionalization. *Angew. Chem. Int. Ed.* **2018**. [CrossRef]
87. Colby, D.A.; Bergman, R.G.; Ellman, J.A. Rhodium-Catalyzed C—C Bond Formation via Heteroatom-Directed C—H Bond Activation. *Chem. Rev.* **2010**, *110*, 624–655. [CrossRef] [PubMed]
88. Vásquez-Céspedes, S.; Wang, X.; Glorius, F. Plausible Rh(V) Intermediates in Catalytic C–H Activation Reactions. *ACS Catal.* **2018**, *8*, 242–257. [CrossRef]
89. Davies, D.L.; Ellul, C.E.; Macgregor, S.A.; McMullin, C.L.; Singh, K. Experimental and DFT Studies Explain Solvent Control of C–H Activation and Product Selectivity in the Rh(III)-Catalyzed Formation of Neutral and Cationic Heterocycles. *J. Am. Chem. Soc.* **2015**, *137*, 9659–9669. [CrossRef] [PubMed]
90. Li, J.; Qiu, Z. DFT Studies on the Mechanism of the Rhodium(III)-Catalyzed C–H Activation of N-Phenoxyacetamide. *J. Org. Chem.* **2015**, *80*, 10686–10693. [CrossRef] [PubMed]
91. Davis, T.A.; Hyster, T.K.; Rovis, T. Rhodium(III)-catalyzed intramolecular hydroarylation, amidoarylation, and Heck-type reaction: Three distinct pathways determined by an amide directing group. *Angew. Chem. Int. Ed.* **2013**, *52*, 14181–14185. [CrossRef] [PubMed]
92. Wang, Y.; Li, C.; Li, Y.; Yin, F.; Wang, X.-S. Rhodium-Catalyzed C-H Olefination of Aryl Weinreb Amides. *Adv. Synth. Catal.* **2013**, *355*, 1724–1728. [CrossRef]
93. Wang, S.-M.; Li, C.; Leng, J.; Bukhari, S.N.A.; Qin, H.-L. Rhodium(iii)-catalyzed Oxidative Coupling of N-Methoxybenzamides and Ethenesulfonyl fluoride: A C–H Bond Activation Strategy for the Preparation of 2-Aryl ethenesulfonyl fluorides and Sulfonyl fluoride Substituted γ-Lactams. *Org. Chem. Front.* **2018**, *5*, 1411–1415. [CrossRef]
94. Wang, C.-S.; Dixneuf, P.H.; Soulé, J.-F. Photoredox Catalysis for Building C–C Bonds from C(sp2)–H Bonds. *Chem. Rev.* **2018**, *118*, 7532–7585. [CrossRef] [PubMed]
95. Shaw, M.H.; Twilton, J.; MacMillan, D.W.C. Photoredox Catalysis in Organic Chemistry. *J. Org. Chem.* **2016**, *81*, 6898–6926. [CrossRef] [PubMed]
96. Romero, N.A.; Nicewicz, D.A. Organic Photoredox Catalysis. *Chem. Rev.* **2016**, *116*, 10075–10166. [CrossRef] [PubMed]
97. Lang, X.; Zhao, J.; Chen, X. Cooperative photoredox catalysis. *Chem. Soc. Rev.* **2016**, *45*, 3026–3038. [CrossRef] [PubMed]
98. Prier, C.K.; Rankic, D.A.; MacMillan, D.W.C. Visible Light Photoredox Catalysis with Transition Metal Complexes: Applications in Organic Synthesis. *Chem. Rev.* **2013**, *113*, 5322–5363. [CrossRef] [PubMed]
99. Fabry, D.C.; Rueping, M. Merging Visible Light Photoredox Catalysis with Metal Catalyzed C–H Activations: On the Role of Oxygen and Superoxide Ions as Oxidants. *Acc. Chem. Res.* **2016**, *49*, 1969–1979. [CrossRef] [PubMed]
100. Fabry, D.C.; Zoller, J.; Raja, S.; Rueping, M. Combining rhodium and photoredox catalysis for C-H functionalizations of arenes: Oxidative Heck reactions with visible light. *Angew. Chem. Int. Ed.* **2014**, *53*, 10228–10231. [CrossRef] [PubMed]
101. Fyfe, J.W.B.; Watson, A.J.B. Recent Developments in Organoboron Chemistry: Old Dogs, New Tricks. *Chem* **2017**, *3*, 31–55. [CrossRef]
102. Xu, L.; Zhang, S.; Li, P. Boron-selective reactions as powerful tools for modular synthesis of diverse complex molecules. *Chem. Soc. Rev.* **2015**, *44*, 8848–8858. [CrossRef] [PubMed]

103. Lennox, A.J.J.; Lloyd-Jones, G.C. Selection of boron reagents for Suzuki–Miyaura coupling. *Chem. Soc. Rev.* **2014**, *43*, 412–443. [CrossRef] [PubMed]
104. Lehmann, J.W.; Blair, D.J.; Burke, M.D. Towards the generalized iterative synthesis of small molecules. *Nat. Rev. Chem.* **2018**, *2*, 0115. [CrossRef] [PubMed]
105. Li, J.; Grillo, A.S.; Burke, M.D. From Synthesis to Function via Iterative Assembly of N-Methyliminodiacetic Acid Boronate Building Blocks. *Acc. Chem. Res.* **2015**, *48*, 2297–2307. [CrossRef] [PubMed]
106. Li, J.; Ballmer, S.G.; Gillis, E.P.; Fujii, S.; Schmidt, M.J.; Palazzolo, A.M.E.; Lehmann, J.W.; Morehouse, G.F.; Burke, M.D. Synthesis of many different types of organic small molecules using one automated process. *Science* **2015**, *347*, 1221–1226. [CrossRef] [PubMed]
107. Woerly, E.M.; Roy, J.; Burke, M.D. Synthesis of most polyene natural product motifs using just 12 building blocks and one coupling reaction. *Nat. Chem.* **2014**, *6*, 484. [CrossRef] [PubMed]
108. Hartwig, J.F. Borylation and Silylation of C–H Bonds: A Platform for Diverse C–H Bond Functionalizations. *Acc. Chem. Res.* **2011**, *45*, 864–873. [CrossRef] [PubMed]
109. Mkhalid, I.A.; Barnard, J.H.; Marder, T.B.; Murphy, J.M.; Hartwig, J.F. C-H activation for the construction of C-B bonds. *Chem. Rev.* **2010**, *110*, 890–931. [CrossRef] [PubMed]
110. Smith, M.R.; Bisht, R.; Haldar, C.; Pandey, G.; Dannatt, J.E.; Ghaffari, B.; Maleczka, R.E.; Chattopadhyay, B. Achieving High Ortho Selectivity in Aniline C–H Borylations by Modifying Boron Substituents. *ACS Catal.* **2018**, *8*, 6216–6223. [CrossRef] [PubMed]
111. Li, H.-L.; Kanai, M.; Kuninobu, Y. Iridium/Bipyridine-Catalyzed ortho-Selective C–H Borylation of Phenol and Aniline Derivatives. *Org. Lett.* **2017**, *19*, 5944–5947. [CrossRef] [PubMed]
112. Chattopadhyay, B.; Dannatt, J.E.; Andujar-De Sanctis, I.L.; Gore, K.A.; Maleczka, R.E.; Singleton, D.A.; Smith, M.R. Ir-Catalyzed ortho-Borylation of Phenols Directed by Substrate–Ligand Electrostatic Interactions: A Combined Experimental/in Silico Strategy for Optimizing Weak Interactions. *J. Am. Chem. Soc.* **2017**, *139*, 7864–7871. [CrossRef] [PubMed]
113. Hale, L.V.A.; Emmerson, D.G.; Ling, E.F.; Roering, A.J.; Ringgold, M.A.; Clark, T.B. An ortho-directed C–H borylation/Suzuki coupling sequence in the formation of biphenylbenzylic amines. *Org. Chem. Front.* **2015**, *2*, 661–664. [CrossRef]
114. Crawford, K.M.; Ramseyer, T.R.; Daley, C.J.A.; Clark, T.B. Phosphine-Directed C—H Borylation Reactions: Facile and Selective Access to Ambiphilic Phosphine Boronate Esters. *Angew. Chem.* **2014**, *53*, 7589–7593. [CrossRef] [PubMed]
115. Kawamorita, S.; Ohmiya, H.; Hara, K.; Fukuoka, A.; Sawamura, M. Directed Ortho Borylation of Functionalized Arenes Catalyzed by a Silica-Supported Compact Phosphine—Iridium System. *J. Am. Chem. Soc.* **2009**, *131*, 5058–5059. [CrossRef] [PubMed]
116. Hoque, M.E.; Bisht, R.; Haldar, C.; Chattopadhyay, B. Noncovalent Interactions in Ir-Catalyzed C–H Activation: L-Shaped Ligand for Para-Selective Borylation of Aromatic Esters. *J. Am. Chem. Soc.* **2017**, *139*, 7745–7748. [CrossRef] [PubMed]
117. Tajuddin, H.; Harrisson, P.; Bitterlich, B.; Collings, J.C.; Sim, N.; Batsanov, A.S.; Cheung, M.S.; Kawamorita, S.; Maxwell, A.C.; Shukla, L.; et al. Iridium-catalyzed C–H borylation of quinolines and unsymmetrical 1,2-disubstituted benzenes: Insights into steric and electronic effects on selectivity. *Chem. Sci.* **2012**, *3*, 3505–3515. [CrossRef]
118. Larsen, M.A.; Hartwig, J.F. Iridium-Catalyzed C–H Borylation of Heteroarenes: Scope, Regioselectivity, Application to Late-Stage Functionalization, and Mechanism. *J. Am. Chem. Soc.* **2014**, *136*, 4287–4299. [CrossRef] [PubMed]
119. Hartwig, J.F. Regioselectivity of the borylation of alkanes and arenes. *Chem. Soc. Rev.* **2011**, *40*, 1992–2002. [CrossRef] [PubMed]
120. Ghaffari, B.; Preshlock, S.M.; Plattner, D.L.; Staples, R.J.; Maligres, P.E.; Krska, S.W.; Maleczka, R.E.; Smith, M.R. Silyl Phosphorus and Nitrogen Donor Chelates for Homogeneous Ortho Borylation Catalysis. *J. Am. Chem. Soc.* **2014**, *136*, 14345–14348. [CrossRef] [PubMed]
121. Newton, C.G.; Wang, S.-G.; Oliveira, C.C.; Cramer, N. Catalytic Enantioselective Transformations Involving C–H Bond Cleavage by Transition-Metal Complexes. *Chem. Rev.* **2017**, *117*, 8908–8976. [CrossRef] [PubMed]
122. Shirai, T.; Yamamoto, Y. Cationic Iridium/S-Me-BIPAM-Catalyzed Direct Asymmetric Intermolecular Hydroarylation of Bicycloalkenes. *Angew. Chem. Int. Ed.* **2015**, *54*, 9894–9897. [CrossRef] [PubMed]

123. Erbing, E. Development of New Efficient Iridium-Catalyzed Methods for the Construction of Carbon-Heteroatom Bonds. Ph.D. Thesis, Stockholm University, Stockholm, Sweden, 2018.
124. Collins, K.D.; Glorius, F. A robustness screen for the rapid assessment of chemical reactions. *Nat. Chem.* **2013**, *5*, 597–601. [CrossRef] [PubMed]
125. Gensch, T.; Teders, M.; Glorius, F. Approach to Comparing the Functional Group Tolerance of Reactions. *J. Org. Chem.* **2017**, *82*, 9154–9159. [CrossRef] [PubMed]
126. Erbing, E.; Sanz-Marco, A.; Vázquez-Romero, A.; Malmberg, J.; Johansson, M.J.; Gómez-Bengoa, E.; Martín-Matute, B. Base- and Additive-Free Ir-Catalyzed ortho-Iodination of Benzoic Acids: Scope and Mechanistic Investigations. *ACS Catal.* **2018**, *8*, 920–925. [CrossRef]
127. Egorova, K.S.; Ananikov, V.P. Which Metals are Green for Catalysis? Comparison of the Toxicities of Ni, Cu, Fe, Pd, Pt, Rh, and Au Salts. *Angew. Chem. Int. Ed.* **2016**, *55*, 12150–12162. [CrossRef] [PubMed]
128. Hayler, J.D.; Leahy, D.K.; Simmons, E.M. A Pharmaceutical Industry Perspective on Sustainable Metal Catalysis. *Organometallics* **2019**, *38*, 36–46. [CrossRef]
129. Sherwood, J. European Restrictions on 1,2-Dichloroethane: C-H Activation Research and Development Should Be Liberated and not Limited. *Angew. Chem. Int. Ed.* **2018**, *57*, 14286–14290. [CrossRef] [PubMed]
130. Clarke, C.J.; Tu, W.-C.; Levers, O.; Bröhl, A.; Hallett, J.P. Green and Sustainable Solvents in Chemical Processes. *Chem. Rev.* **2018**, *118*, 747–800. [CrossRef] [PubMed]
131. Wilson, K.L.; Murray, J.; Sneddon, H.F.; Wheelhouse, K.M.P.; Watson, A.J.B. Connecting the Dots: Method Development Using Sustainable Solvents. *Chem* **2017**, *3*, 365–368. [CrossRef]
132. Welton, T. Solvents and sustainable chemistry. *Proc. R. Soc. A Math. Phys. Eng. Sci.* **2015**, *471*. [CrossRef] [PubMed]

11

# Direct Transamidation Reactions: Mechanism and Recent Advances

**Paola Acosta-Guzmán, Alejandra Mateus-Gómez and Diego Gamba-Sánchez ***

Laboratory of Organic Synthesis Bio- and Organocatalysis, Chemistry Department, Universidad de los Andes, Cra. 1 No 18A-12 Q:305, Bogotá 111711, Colombia; pa.acostag@uniandes.edu.co (P.A.-G.); a.mateus@uniandes.edu.co (A.M.-G.)

* Correspondence: da.gamba1361@uniandes.edu.co

Academic Editor: Michal Szostak

**Abstract:** Amides are undeniably some of the most important compounds in Nature and the chemical industry, being present in biomolecules, materials, pharmaceuticals and many other substances. Unfortunately, the traditional synthesis of amides suffers from some important drawbacks, principally the use of stoichiometric activators or the need to use highly reactive carboxylic acid derivatives. In recent years, the transamidation reaction has emerged as a valuable alternative to prepare amides. The reactivity of amides makes their direct reaction with nitrogen nucleophiles difficult; thus, the direct transamidation reaction needs a catalyst in order to activate the amide moiety and to promote the completion of the reaction because equilibrium is established. In this review, we present research on direct transamidation reactions ranging from studies of the mechanism to the recent developments of more applicable and versatile methodologies, emphasizing those reactions involving activation with metal catalysts.

**Keywords:** transamidation; amide; amine; catalyst; catalysis

---

## 1. Introduction

The amide functionality has been recognized as one of the most important functional groups, not only because of its widespread presence in Nature (in proteins, peptides, and alkaloids, among others) [1] but also because of the vast number of synthetic structures bearing this group [2]. It is estimated that approximately 25% of the existing pharmaceuticals contain an amide bond as part of their structures [3] and that approximately 33% of the new drug candidates are "amides" [4], thus making amidation reactions some of the most performed chemical processes in the pharmaceutical industry and in drug discovery activities [5].

Traditional methods to synthesize amides suffer from significant issues, principally the use of stoichiometric amounts of activating reagents with the consequent production of waste or the use of corrosive and troublesome reagents such as acyl chlorides or anhydrides. As a consequence, the ACS Green Chemistry Institute assessed the amide bond formation with good atom economy as one of the biggest challenges for organic chemists [6]. In recent years, amide bond synthesis by nontraditional methods has been reviewed, and some alternatives are available to perform acylations on nitrogen [7–10]. Among those unconventional methods, the transamidation reaction appears to be a useful strategy.

The acyl exchange between an amide and an amine has been known since 1876 with the studies carried out by Flescher [11]; however, the method was pretty limited. For approximately one hundred years, the transamidation reaction was almost unexplored, and only a few successful examples were published [12–15]. The biggest drawback of this method was the use of high temperatures and very long reaction times. The first catalytic transamidation was performed using carbon dioxide [14];

however, the quantity used makes it not properly a catalyst but rather a promoter; additionally, the yields were always lower than 67%. It was not until 1994 that a modern and complete study on a direct transamidation was published by Bertrand and coworkers [16]. Their work was based on the use of aluminum chloride as a promoter, and the method was limited to the use of aliphatic amines since low yields were obtained with aromatic amides. Chemists noticed the importance and the potential of the direct transamidation reaction and developed some useful and general methods performed under the influence of different types of catalysts. Today, heterogeneous, metallic, acidic and basic catalysts are available to perform transamidation reactions. In this review article, we will begin with the mechanistic studies for transamidation of primary and secondary amides, followed by studies of the reaction with tertiary amides and some other mechanistic studies using proline as a catalyst. In the following sections, we will present a set of selected examples of direct transamidation reactions catalyzed by metals as well as some alternative catalysts.

## 2. Mechanistic Studies

In transamidation, an alternative strategy to prepare amides, the direct exchange of the amine moiety in an amide can occur only under the influence of a suitable catalytic or stoichiometric activating agent because of the poor electrophilic nature of amides. Depending on the reaction conditions, the structure of the amide, and the activating agent employed, the reactivity of these amides differs from reaction to reaction; however, it follows the regular pattern that the primary amides are more active than the secondary and tertiary amides. Nevertheless, some activating agents capable of activating secondary amides have been described in the literature with considerable success [10]. The order of reactivity shown before implies that the structure of the amide is the most important constraint in transamidation processes. Another very important factor is the acidic nature of the N-H function, which could hamper the process by simply inactivating the catalytic species. Considering these facts, understanding how this process occurs represents one of the main challenges for the organic chemistry community [8].

In this sense, Stahl and coworkers [17–19] studied the mechanism of the transamidation reaction catalyzed by metals, focusing on the reaction between a secondary amine and a secondary amide (Scheme 1). They evaluated the behavior of nucleophilic alkali-metal amides, Lewis acidic metal complexes, transition metals and different amides. The authors found that metallic complexes of titanium and aluminum could catalyze the transamidation reaction due to their relatively low basicity; furthermore, metallic complexes of titanium [$Ti(NMe_2)_4$] and aluminum [$Al(NMe_2)_6$] could catalyze the transamidation reaction due to an increase in electron density in the metal center. This effect is related to the reduced basicity of the ligand, which increases the Lewis acidic character of the metal.

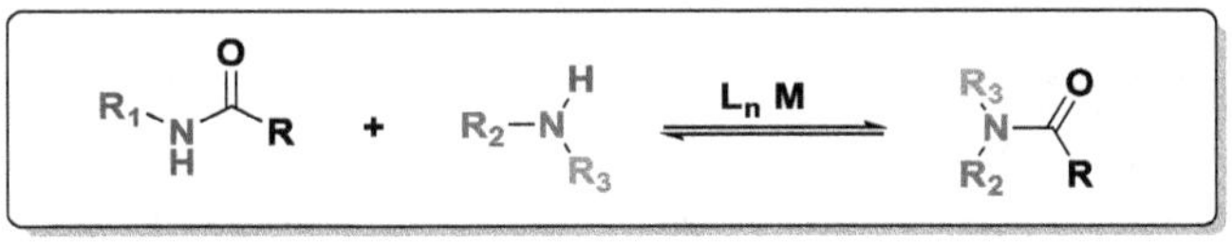

**Scheme 1.** Transamidation of a secondary amide with a secondary amine.

Based on the previous information, the author carried out a complete mechanistic study, which included the determination of the deuterium kinetic isotope effect, the rate law and the identity of the catalyst resting state. With the results obtained in the kinetic studies, they proposed a catalytic mechanism to explain the transamidation process (Scheme 2). They suggested that in the first step, the secondary amide reacts with the precatalyst to yield metal-amidate complex **I**. Once this complex is formed, it enters the catalytic cycle by a bimolecular reaction with a primary amine to generate adduct **II**. After that, a proton transfer between the nitrogen atoms is proposed to reach intermediate **III**. Later, an intramolecular nucleophilic attack of the amido-ligand to the carbonyl of the amide forms metallacycle **IVa**. This intermediate is the most important species in the catalytic cycle because it embodies the bifunctional role of the metal center *via* activation of both the amine and amide substrate.

However, this intermediate can either revert the cycle to complex **III** or continue the catalytic cycle by isomerization to form **IVb**. The isomerization proceeds by transition state **V**, where the metal is linked to both nitrogen atoms, thus changing the nitrogen coordinated to the metal center and providing exchanged intermediate **IVb**. This product is easily converted into **VI**, and finally, the breakdown of intermediate **VI** promoted by a second molecule of amide leads to the formation of the desired amide and the liberation of an amine molecule that is responsible, in most of the cases, for the equilibrium displacements; in other words: it is the driving force of this reaction [18–20].

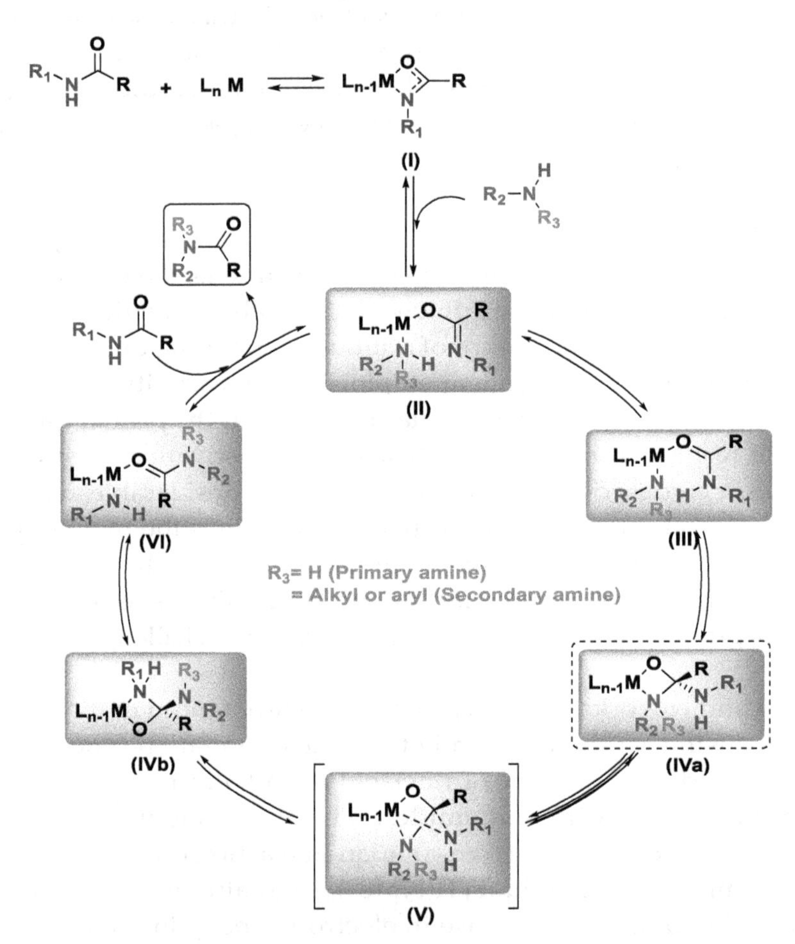


**Scheme 2.** Proposed catalytic cycle for transamidation reactions between primary or secondary amides with primary amines.

Careful analysis of the proposed mechanism suggests that tertiary amides are unsuitable reagents for the transamidation reaction. This is because of the absence of an N-H bond and the consequently different reactivity between these amides and bases. However, successful examples of transamidation reactions with tertiary amides have been described in the literature, suggesting that tertiary amides should follow a different mechanistic pathway. Stahl and coworkers confronted this issue and developed a new study focused on the mechanistic details of the transamidation reaction between tertiary amides and secondary amines employing kinetic, spectroscopic, and computational studies (Scheme 3). According to their results, the first step occurs when the amide interacts with the metal complex to form intermediate **I**, where the simultaneous activation of the electrophile (amide) and the nucleophile (amine) by the metal center is clear. After that, an intramolecular attack of the amido ligand to the coordinated amide gives rise to metal-stabilized tetrahedral intermediate **IIa**, which can interconvert to its isomeric intermediate **IIb** by a simple cleavage and coordination sequence through intermediate III. In the last step, a new amine/amido pair coordinated to the metal **IV** is formed, from which it is possible to obtain the desired product [21,22].

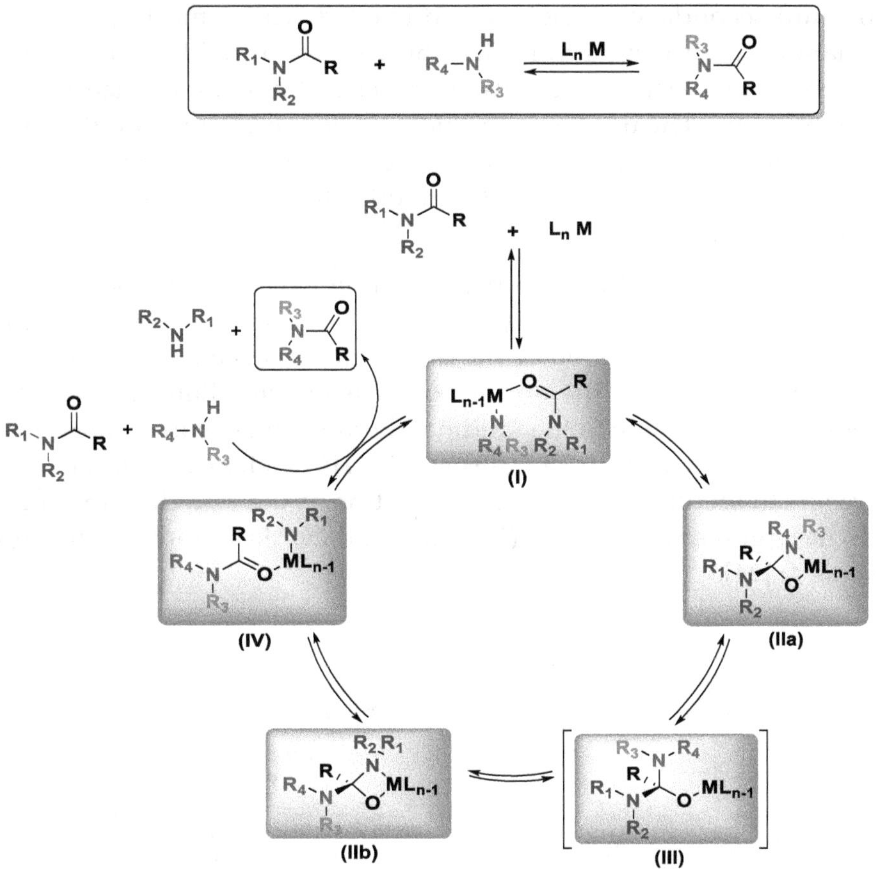

**Scheme 3.** Proposed catalytic cycle for transamidation reaction between tertiary amides with secondary amines.

Comparing the transamidation processes for secondary amides and tertiary amides makes some similarities evident. First, the mechanisms exhibit not only that there are obvious structural similarity between the intermediates but also that the key step in both processes is the intramolecular nucleophilic attack of the amido ligand on a metal-coordinated amide. Regarding the differences between the two transamidation processes, perhaps the most important one is the identity of the catalyst resting state, implying substantial differences in the kinetic properties of both reactions since the transamidation reaction of secondary amides follows first-order kinetics with regards to the metal and amine involved and the amide follows zero-order kinetics. Contrary to that, for the tertiary amide, the rate law is zero-order for the amine, is half- to first-order for the metal and varies from zero-order to saturation behavior for the amide. The transamidation of primary amides is easier than the cases described before because ammonia is produced in the reaction medium, thus making the reaction simpler and usually better yielding. Concerning the mechanisms, each research team describes its own proposal, typically based on the Stahl studies. In our opinion, slight differences between metals should exist, and as a consequence, the reaction mechanism for the transamidation of primary amides should be pretty similar to that described for secondary amides. Additionally, the large number of non-metal-catalyzed transamidation reactions makes the investigation of their mechanisms interesting. Unfortunately, only one computational study on the reaction catalyzed by L-proline has been described; mechanisms with other catalysts remain unexplored.

The organocatalyzed transamidation reaction has been described as a greener methodology compared with the metal-catalyzed version of this reaction. L-Proline has received much attention due to its dual role as a ligand and a catalyst in other reactions. In view of the above perceptions, L-proline was described as a useful catalyst for the transamidation reaction. Adimurthy and coworkers [23] found that various amides react with a variety of amines in the presence of L-proline as a catalyst. The reaction scope included benzylamines with electron-rich and deficient substituents and alkyl aromatic, aliphatic, and secondary amines, which reacted with remarkable ease. Specifically, the reactions of benzylamine with primary amides showed good yields compared to reactions with secondary and tertiary amides. The authors also demonstrated that the reactions have a high degree of functional group tolerance. The mechanistic study carried out to explain the role of L-proline and to disclose why the transamidation can be efficiently catalyzed by proline was reported by Xue and coworkers some years after the reaction was described. They used density functional theory (DFT) calculations to achieve a reasonable proposal.

The transamidation of acetamide with benzylamine was selected as the model system. The calculations revealed that the reactions catalyzed by L-proline follow a stepwise mechanism, which involves a first step where the activation of the amide takes place through a proton transfer from L-proline to the carbonyl group of the amide to form intermediate imidine **I**. Subsequently, a nucleophilic addition of benzylamine to the imidine intermediate occurs to form **II**. After eliminating one molecule of ammonia, a hydrolysis of intermediate **III** produces the target product. This study allowed the researchers to identify the hydrolysis reaction as the rate-determining step (RDS) in the catalytic cycle with the largest energy barrier (26.0, 26.8 and 29.7 kcal/mol in toluene, ethanol and $H_2O$ respectively) (Scheme 4) [24].

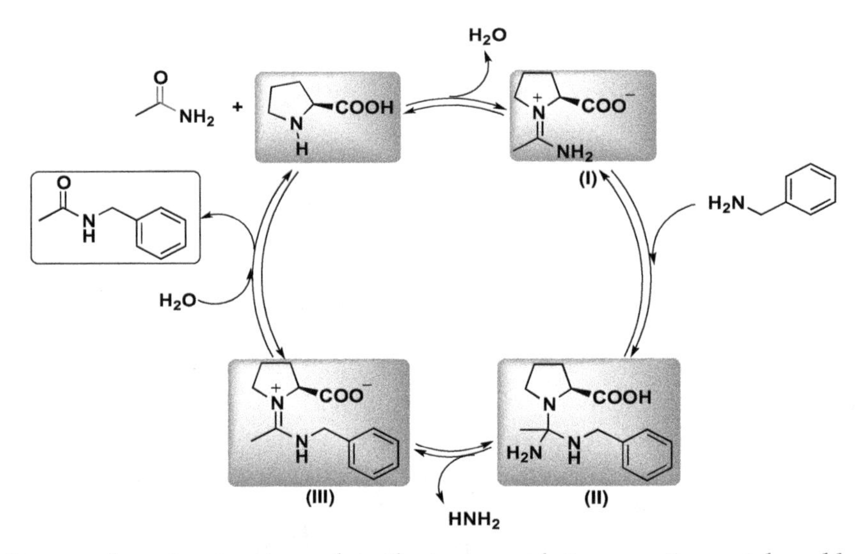


**Scheme 4.** Proposed mechanism to explain the transamidation reaction catalyzed by L-proline.

As already mentioned, the mechanistic studies on the transamidation reaction are quite limited; however, the metal-catalyzed version presumably follows similar mechanisms independent of the metal identity, with the mechanistic pathway depending on the type of amide reacting. Concerning the non-metal-catalyzed version, some examples are described in the literature, but among them, only L-proline has received enough attention that a mechanistic study was carried out. Nevertheless, the summary of mechanistic studies presented in this section is sufficient to understand the principles and particularities of transamidation reactions. In the following chapters, we will present some selected examples of transamidations and discuss their particularities and issues.

## 3. Metal-Catalyzed Direct Transamidations

Since the pioneering work reported by Stahl and coworkers [18–22], several metal catalysts have been described for direct transamidation reactions. Among them, one of the most commonly used is iron(III), particularly in the form of simple Fe(III) salts, supported in clay or even as nanoparticles. The first study was reported in 2014 by Shimizu and coworkers [25]; they used Fe(III) supported in montmorillonite to perform the direct transamidation of primary amides (Scheme 5). The authors showed that the reaction can be performed under solvent-free conditions with a slight excess of amine at 140 °C.

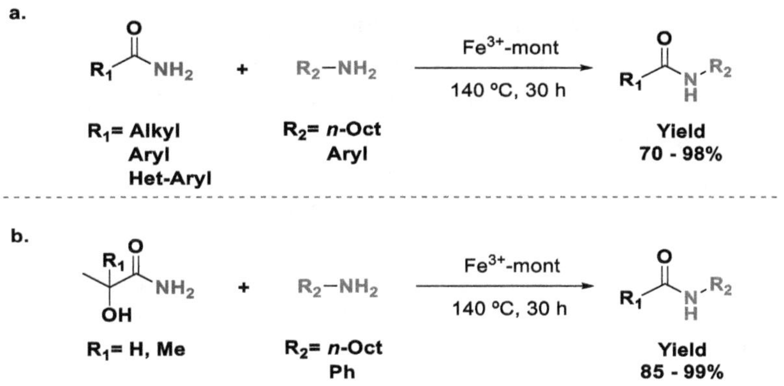

**Scheme 5.** (**a**) Transamidation of amides (alkyl, aryl, Het-aryl) with amines catalyzed by $Fe^{3+}$-mont. (**b**) Transamidations of α-hydroxyamides with amines.

Unfortunately, the scope of the reaction is limited; only primary amides and primary amines can be used (Scheme 5a). Additionally, the method is limited to the use of liquid starting materials since at least one of the two reagents has to be a liquid, thus making this method unsuitable for a large number of reagents. A few examples with free α-hydroxy amides were described (Scheme 5b). Typically, reactions carried out under solvent-free conditions are seen as good examples of green chemistry principles; however, the transamidation reaction has to be seen from another perspective. As we mentioned before, amidation reactions are one of the most performed processes in drug discovery and the pharmaceutical industry, and the structure of amides used in those fields are somehow complex, meaning they are usually polyfunctional compounds with more than 20 carbon atoms or, in some cases, amino acid derivatives. As a consequence, a good method to synthesize amides has to be amenable to all the structures mentioned before, and solid reagents are highly desirable even if the use of a solvent becomes imperative. This has served us as inspiration to develop a more general method for direct transamidations using Fe(III) [26]. Our method is based on the use of simple Fe(III) salts. In fact, we demonstrated that any Fe(III) salt can be used as a catalyst for this reaction; the sole condition is the presence of water, which has a complementary catalytic role in this process. Water can be added or obtained from hydrated salts, the identity of which is only limited by the availability of the reagents. Technical grade toluene is also a suitable solvent. Consequently, this method is one of the cheapest reported in the literature (Scheme 6).

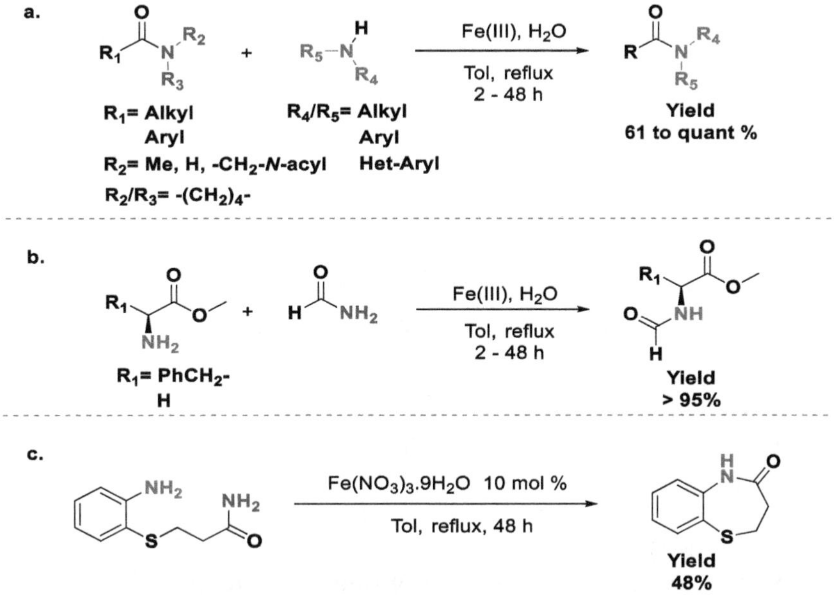

**Scheme 6.** Transamidation catalyzed by Fe(III) and water of (**a**) carboxamides. (**b**) α-amino esters. (**c**) Synthesis of 2,3-dihydro-5*H*-benzo[*b*]-1,4-thiazepin-4-one by intramolecular transamidation.

The reaction scope is very good since the reaction can be performed not only with an excess of amides or amines but also with stoichiometric quantities. The reaction time depends on the amide used, but it is usually shorter with primary amides and longer with secondary and tertiary amides. We also described the transamidation of secondary amides where the liberated amine is not a gas or a volatile liquid and showed that Fe(III)/water is a suitable catalyst in those cases. It should be noted that no previous functionalization of the secondary amide was needed (Scheme 6a). Finally, the reaction was described with some amino acids as the amino source (Scheme 6b), and one intramolecular example was provided (Scheme 6c). A few years later, magnetic nanoparticles were used in direct transamidation reactions, with some advantages and issues. The method described by Thale [27] using $Fe_3O_4$ has the same problem that all transamidations performed under solvent-free conditions show. The loading of the catalyst is also high (20 wt.%), and only primary amines are used as nucleophiles. However, this particular method has some advantages: the catalyst is easily recovered and can be used in several reactions without loss of activity. It is also simple to prepare, and the reaction can be performed with DMF as an example of a tertiary amide. On the other hand, the method described by Heidari and Arefi [28] has a better substrate scope and uses smaller quantities of catalyst. Unfortunately, the preparation process for the catalyst is more complicated and uses simple iron salts as precursors; this is potentially problematic since the authors did not quantify the free salts in their catalyst, and it is possible that some remaining salts are responsible for the transformation, on the other hand if the simple salts are good and suitable catalysts for this reactions, prepare a catalysts that do not provide very different results using them a starting materials consumes time unnecessarily. The recovery of the catalyst was also described, but the identity of the catalytic species is still unclear; both Heidari and Thale attribute the catalytic activity to the metal center.

It is clear that the transamidation reaction is an equilibrium, and as such, when ammonia or gaseous amines are liberated in the reaction media, the reaction is easier. On the other hand, when the liberated amine is a liquid, the reaction becomes more complicated. Stahl and Gellman et al. [29] faced this problem and developed a method applicable to *N*-aryl amides. In this case, $Ti(NMe_2)_4$ and $Sc(OTf)_3$ were suitable catalysts affording good yields and acceptable selectivities to the cross-product. The reaction was performed with nonpolar solvents and at low temperatures. However, the best results were obtained with an excess of aliphatic amines (Scheme 7a); the use of stoichiometric aromatic amines produced mixtures of products (Scheme 7b).

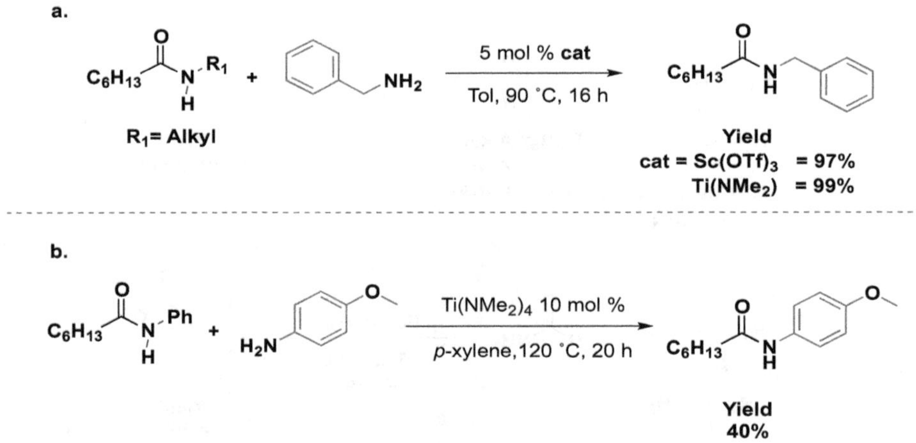

**Scheme 7.** Catalytic transamidation of (**a**) *N*-alkyl heptylamines with benzylamine and (**b**) *N*-aryl amides with aryl amines.

The reactions described before have some common characteristics: they are all performed in nonpolar solvents, and an excess (at least 1 equiv.) of one of the reagents is needed to obtain complete conversions and good yields. One interesting development was the use of $Cu(OAc)_2$ as a catalyst described by Beller and coworkers [30], who used *tert*-amyl alcohol as the solvent and 0.3 excess equivalents of amine. The reaction is limited to the use of primary amides but has a very good substrate scope, as it tolerates free OH groups in the substrate and can be performed with chiral reagents (amides or amines) without racemization. The sole problem is the use of sealed tubes since high pressures are a safety issue in organic chemistry laboratories.

As the reader can notice, several efforts have been made to have more active and general catalysts for transamidations. Simple metal salts have been proven as suitable catalysts for this transformation, however, the use of more developed metal catalysts may have some advantages. In particular, the use of sulfated tungstate proved to be effective with $\alpha,\beta$-unsaturated amides [31]. These substrates can be potentially complicated because of the presence of two reactive sites. Once the amide forms the complex with the metal center, the carbonyl is more electrophilic, but the $\beta$ position suffers a similar activation; as a consequence, Michael additions are competitive reactions. In the case of sulfated tungstate, no Michael product was observed (Scheme 8a). The catalyst was also active with $\alpha$-amino acid esters, as observed in Scheme 8b.

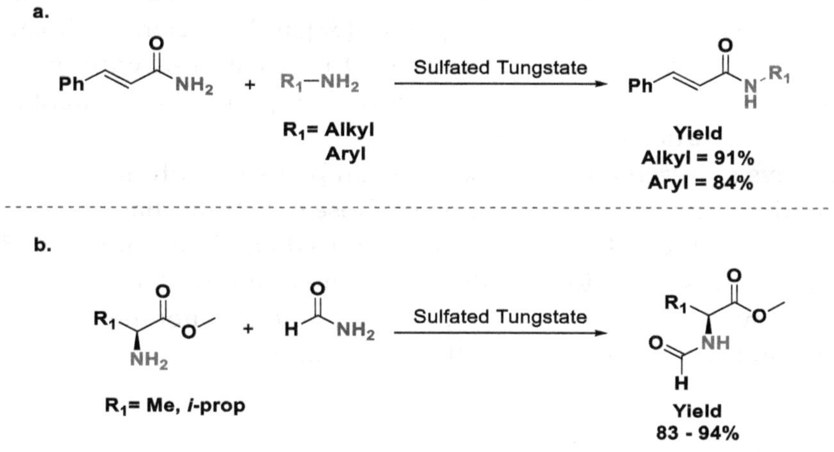

a.

Sulfated Tungstate

$R_1$= Alkyl, Aryl

Yield: Alkyl = 91%, Aryl = 84%

b.

Sulfated Tungstate

$R_1$= Me, *i*-prop

Yield 83 - 94%

**Scheme 8.** Transamidation reactions in the presence of sulfated tungstate of (**a**) cinnamide with alkyl and aryl amines and (**b**) $\alpha$-amino esters with formamide.

The use of tertiary amides is typically exemplified with DMF, and very few examples with other tertiary amides have been described. Fortunately, the use of $Pd(OAc)_2$ with 2,2′-bipyridine (bpy) as the ligand can be used for the transamidation of tertiary amides [32]. This catalyst afforded very good yields with aromatic amines, which are easily oxidized and consequently gave lower yields than their aliphatic partners (Scheme 9). The biggest issue of this method is the use of a large excess of amine: 10 equivalents are required. However, the reaction can be performed on a gram scale with excellent results.

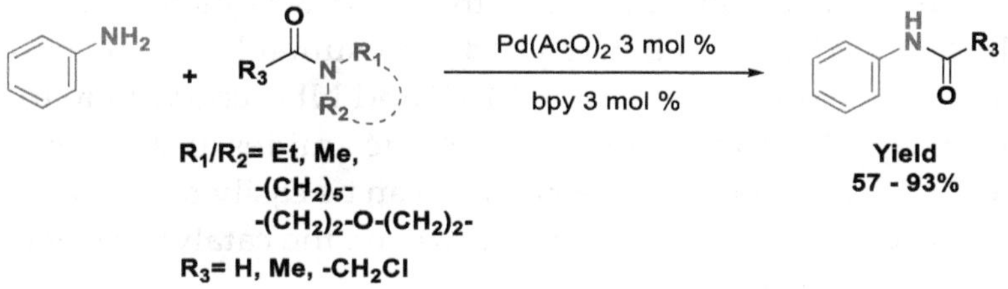

$Pd(AcO)_2$ 3 mol %, bpy 3 mol %

$R_1/R_2$= Et, Me, $-(CH_2)_5-$, $-(CH_2)_2-O-(CH_2)_2-$

$R_3$= H, Me, $-CH_2Cl$

Yield 57 - 93%

**Scheme 9.** Transamidation of DMF derivatives with aniline.

Very recently, Hu and coworkers [33] described a reductive transamidation of tertiary amides (very challenging substrates) with nitrobenzenes promoted by manganese. Presumably, the manganese acts not only as a reducing agent by producing the amine in situ but also as an activating agent for the amide. This was proven by adding some metals usually found as impurities of manganese, such as palladium or nickel. Additionally, this metal has also been described as a useful catalyst in a new method of transamidations, such as a reaction where the amide is activated by the use of BOC or tosyl groups to increase their reactivity. There are some amazing recent publications in this field, but this are not within the scope of this review article. The reaction performed by Hu *et al.* showed very good substrate scope and, to the best of our knowledge, is the sole general method for direct transamidation of tertiary amides.

From the previous examples, it is clear that the transamidation reaction is a powerful alternative in amide synthesis. The use of metal catalysts has shown tremendous growth in the last few years, and many researchers have developed a vast variety of catalysts using metallic centers. We selected the most general and simplest methods to discuss in this review article. Nevertheless, other less important catalysts have to be superficially mentioned, including manganese oxide ($MnO_2$), which was used under solvent-free conditions [34] with a limited substrate scope but with good yields. The use of lanthanides has also been described; in particular, a bimetallic lanthanum alkoxide [35], and immobilized Ce(III) [36] were used as heterogeneous catalysts with the associated advantages, such as easy catalyst recovery and low catalyst charge, but also with the normal issues, such as the catalyst preparation and activation. Some more innovative catalysts have been prepared and used successfully in transamidation. Mesoporous niobium oxide spheres [37] and *N*-heterocyclic carbene ruthenium (II) complexes [38] are some of the most general and easy to use catalysts with the best substrate scope. Very recently, the use of $Co(AcO)_2{\cdot}4H_2O$ was described [39] with an excellent substrate scope, but there are no special considerations to mention here.

Some of the catalysts presented in this section can be used with ureas, with phthalimide and, in very few cases, with thioureas. The activation of those substrates has proven more difficult than amides, even though the C–N bond has similar features to them. In addition, no details are provided about the application of transamidation reactions with ureas and phthalimides since it is outside the scope of this review. Concerning the catalysts, the development of nonmetallic catalysts has received comparable attention and will be discussed in the next chapter.

## 4. Other Catalysts

In an effort to develop safe, environmentally benign, economical, and energy saving chemical reactions, the design and synthesis of recyclable catalysts have become an important objective. In this context, innovative catalysts for transamidations have been developed. One of the most active catalysts is the recyclable polymer-based metallic Lewis acid catalyst developed by Cui and coworkers [40]. This catalyst, which is based on $HfCl_4$/KSF-polyDMAP, is recyclable and was described for transamidations of different amide/amine pairs (Scheme 10). The authors chose $HfCl_4$ because it is active in the direct condensation of carboxylic acids and alcohols and because the Hf(IV) salts are easily manipulated and applicable in large-scale reactions. During the optimization process, the authors tried an HF/KSF system and found it suitable for transamidations with mainly nonpolar or low polarity solvents. Interestingly, this study led them to discover that by using mixtures of solvents such as acetonitrile/water and acetonitrile/$NH_3$, the yield was increased. The use of $HfCl_4$/KSF with polyDMAP or $NH_3$ afforded even better yields; presumably, the reaction rate is increased as a result the change in Lewis acidity when polyDMAP and $NH_3$ coordinate to the metal center (Hf). This catalytic system was applied to a variety of aliphatic amides and benzylamines with excellent results. Additionally, the $HfCl_4$/KSF-polyDMA catalyst can be easily recovered and reused in the same reaction at least five times without loss of activity. However, the catalytic system has some limitations when other aliphatic amines are used.

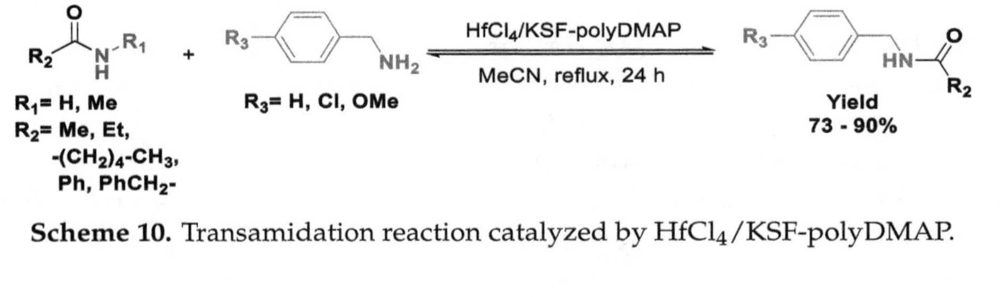

$HfCl_4$/KSF-polyDMAP
MeCN, reflux, 24 h
$R_1$= H, Me
$R_2$= Me, Et, -$(CH_2)_4$-$CH_3$, Ph, $PhCH_2$-
$R_3$= H, Cl, OMe
Yield 73 - 90%

**Scheme 10.** Transamidation reaction catalyzed by $HfCl_4$/KSF-polyDMAP.

The use of stoichiometric boron reagents for amidation reactions has also been reported several times in the literature [8]. Borate esters were used by Sheppard et al. [41] to activate amides in transamidations. They evaluated different boron reagents and found that $B(OMe)_3$ and $B(OCH_2CF_3)_3$ show the best behavior in carboxamidation and transamidation processes. This $B(OCH_2CF_3)_3$-mediated transamidation reaction gave secondary amides in good yields; unfortunately, the main focus of this study was direct carboxamidation, and consequently, very few examples of transamidations were provided. A big issue in this is the use of stoichiometric amounts of the boron reagent. Consequently, the development of new methods that use catalytic amounts of boron reagents is highly desirable.

Nguyen et al. [42] developed the first transamidation using substoichiometric amounts of boron derivatives. In that research, the authors studied boric acid-catalyzed transamidations of amides and phthalimide. Different solvents were tested in an attempt to increase the reaction conversion. As is typical for this kind of reaction, nonpolar solvents such as toluene or xylene showed good results; however, a higher conversion was observed when the reaction was performed under solvent-free conditions. Additionally, the presence of water (1–2 equivalent) helped promote the transamidation process. Different substrates were used successfully and good yields were obtained in most cases. Secondary amines such as morpholine or piperidine proved to be less reactive and higher temperatures were needed in order to obtain acceptable yields. The most important advantage of this method compared with the other reports is the applicability on primary, secondary and tertiary amides (Scheme 11a).

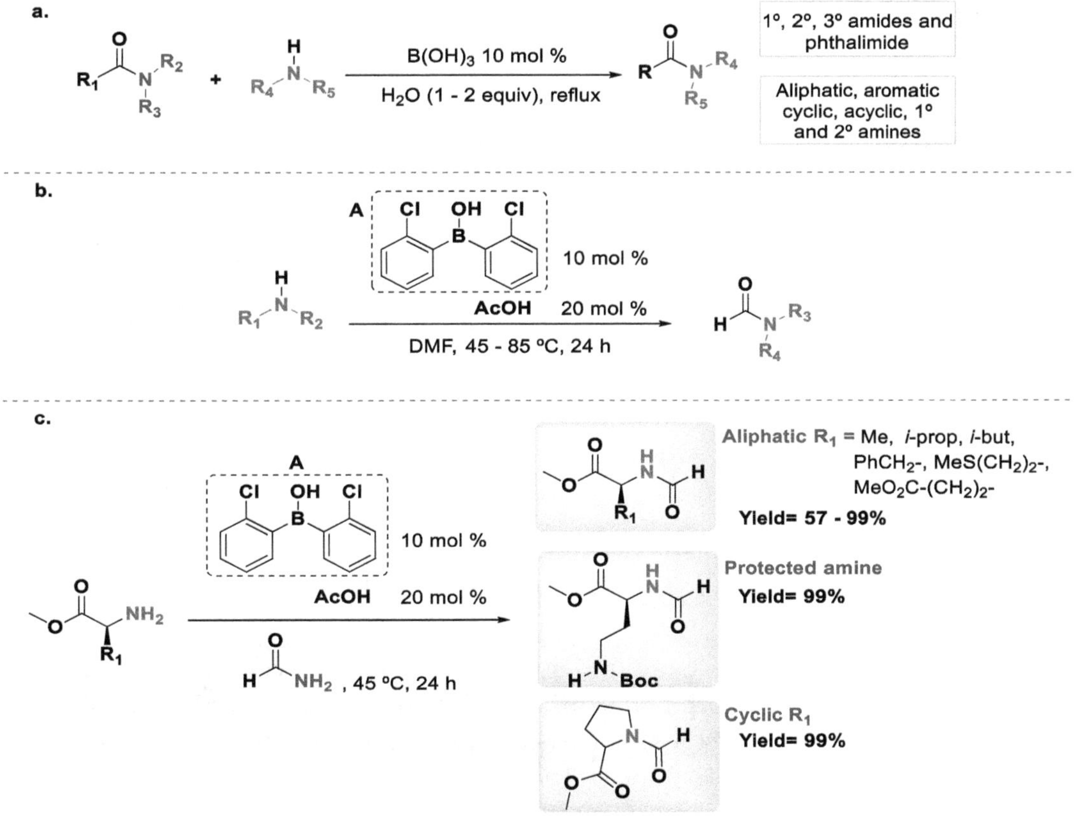

a. $B(OH)_3$ 10 mol %; $H_2O$ (1 - 2 equiv), reflux. 1°, 2°, 3° amides and phthalimide. Aliphatic, aromatic cyclic, acyclic, 1° and 2° amines

b. A 10 mol %; AcOH 20 mol %; DMF, 45 - 85 °C, 24 h

c. A 10 mol %; AcOH 20 mol %; HCONH₂, 45 °C, 24 h. Aliphatic $R_1$ = Me, *i*-prop, *i*-but, $PhCH_2$-, $MeS(CH_2)_2$-, $MeO_2C$-$(CH_2)_2$- Yield= 57 - 99%. Protected amine Yield= 99%. Cyclic $R_1$ Yield= 99%

**Scheme 11.** (**a**) Boron-mediated transamidation of primary, secondary and tertiary amides with different kinds of amines. (**b**) Formylation of amines catalyzed by a boronic acid. (**c**) Transamidation catalyzed by boron between α-amino esters and formamide.

However, the high temperatures needed to carry out this method have limited the reaction to using stable achiral amides. To circumvent this problem, Blanchet and coworker [43] exploited the remarkable efficiency of boronic acids as catalysts for the transamidation of DMF with amines under lower temperatures (Scheme 11b). Many tests were carried out, and the results demonstrated that the combination of 10 mol % of boronic acid **A** and 20 mol % of acetic acid was the best combination to promote the transamidation reactions (Scheme 11b). The authors proposed a cooperation between both acids and that a Lewis acid-assisted Bronsted acid (LBA) catalytic system could be involved. Excellent yields were found in all cases, and the process showed very good behavior with primary and secondary amines as well as α-amino esters. Nonetheless, to perform the reaction with chiral substrates, the DMF solvent was changed to formamide in order to keep racemization at a low level (Scheme 11c).

The activation of primary amides to promote transamidations using catalytic amounts of hydroxylamine was reported by Williams and coworkers [44]. They screened different catalysts and found that the hydroxylamine salt produced the best results. The study, performed in order to compare the differences between base-free and acid salts, demonstrated that the use of salts has a positive effect on the conversion of primary amides into secondary amides. The optimal conditions for this transformation were toluene at a temperature of 105 °C and the amount of $NH_2OH{\cdot}HCl$ varied according to the substrate. The reaction was successful with a wide range of functional groups, including halogens, alkenes, alkynes, free phenolic hydroxyl groups, and heterocycles. Other important observations in the application of this methodology were that the Boc protecting group was unaffected under the typical reaction conditions (Scheme 12a) and that amino acid esters can be *N*-acylated without loss of the ester functionality (Scheme 12b). The highest conversion was generally seen with aliphatic amides; in some cases, only 10 mol % of $NH_2OH{\cdot}HCl$ was necessary to reach full conversion in short reaction times. However, the transamidation of secondary amides using this methodology is not effective.

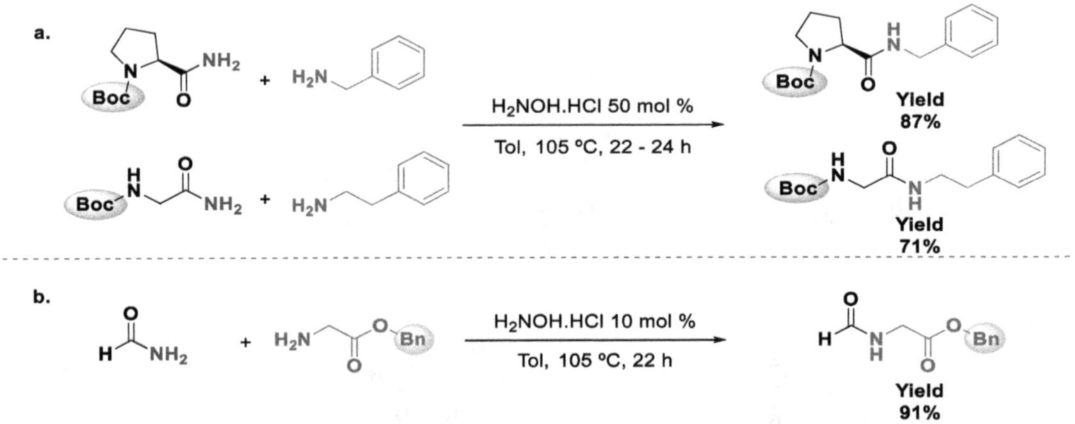

a. $H_2NOH.HCl$ 50 mol %, Tol, 105 °C, 22 - 24 h — Yield 87%; Yield 71%

b. $H_2NOH.HCl$ 10 mol %, Tol, 105 °C, 22 h — Yield 91%

**Scheme 12.** Hydroxylamine hydrochloride as a catalyst of the transamidation reaction of (**a**) substrates with a Boc protecting group and (**b**) α-amino esters with a benzyl group.

Hypervalent iodine compounds have undergone impressive developments, and their uses are increasing in many applications in organic synthesis. This progress served as inspiration to Singh and coworkers [45] to propose the first iodine (III)-catalyzed transamidation reaction. The best results were obtained when the reaction was done in the presence of 5 mol % of (diacetoxyiodo)benzene (DIB) under microwave heating (60–150 °C) without any solvent. Gratifyingly, this methodology showed a very good substrate scope with good to excellent yields. A careful analysis of the crude reaction mixtures was performed since the catalyst used is an oxidant and products derived from Hoffman rearrangements were observed. In addition, the use of stoichiometric amounts of catalyst promoted nitrene formation and increased the number of byproducts, but by using catalytic amounts, the transamidation reaction was favored.

As mentioned earlier in this manuscript, the use of organocatalysts has also been described in transamidation reactions. The study reported by Adimurthy [23], where the process was carried out in the presence of L-proline as an inexpensive catalyst (10 mol %) under solvent-free conditions, suffers from some weaknesses, principally the need for the temperature to be higher than 80 °C and the complete inactivity other amino acid catalysts.

Continuing with the revision of inexpensive catalysts in transamidation reactions, benzoic acid was recently used by Xu and coworkers [46], showing very good activity in transamidations. The reaction, performed in xylene at 130 °C, showed very good functional group tolerance, and many substrates demonstrated good reactivity. Among them, the use of heterocyclic amides (nicotinamide, furan-2- and thiophene-2-carboxamide) with heterocyclic amines should be noted (see Scheme 13a). Unfortunately, this method is not effective with aromatic amines, with which it only afforded moderate yields. The main advantages of this methodology are the availability of the catalysts and the excellent selectivity observed. The authors performed a couple of experiments using mixtures of benzamide with 3-aminobenzylamine and benzamide with tryptamine, showing that only one product was formed in each case and thus making this a potential method for the protection of primary amines (Scheme 13b).

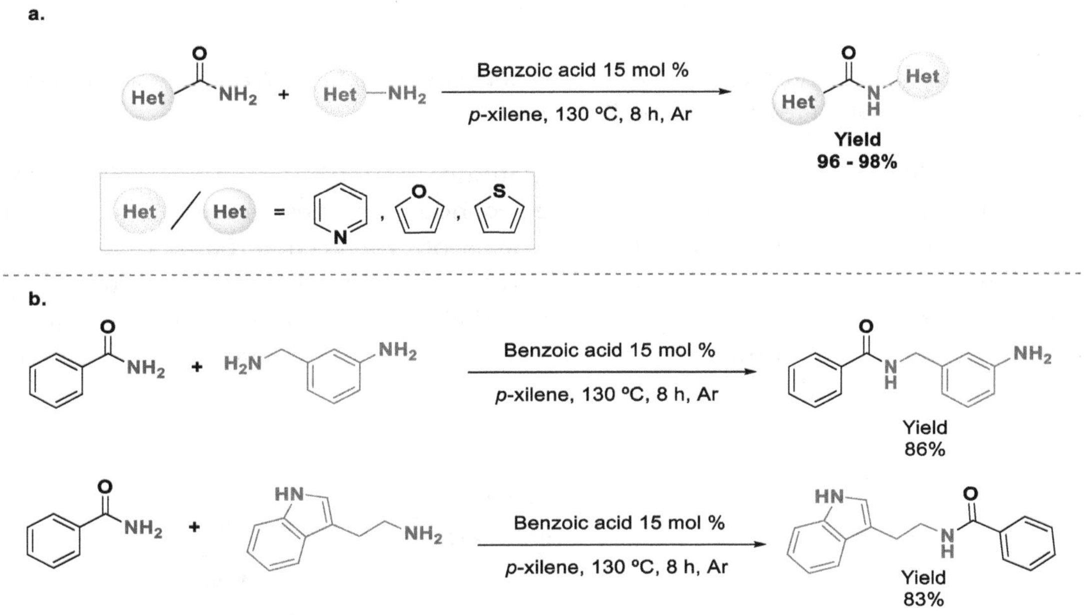

**Scheme 13.** (**a**) Benzoic acid as a catalyst of the transamidation reactions of heterocyclic amides and heterocyclic amines. (**b**) Selectivity of the transamidation reaction using benzoic acid as a catalyst.

In addition to these metal-free strategies for transamidation, the use of $K_2S_2O_8$ in aqueous media was recently described [47]. The reaction can be performed under the influence of microwaves or conventional heating. The study included the screening of other peroxides, such as $H_2O_2$, TBHP, *m*CPBA and oxone, finding that $K_2S_2O_8$ was the best reactant for the transamidation reaction. Unfortunately, stoichiometric amounts of the promoter are required to achieve complete conversions. This method was applied successfully to L-phenylalanine methyl ester hydrochloride with DMF as the formylating reagent, obtaining the *N*-formyl amino acid ester in 95% yield without any change in the configuration and optical purity, making this method possibly advantageous with chiral substrates (Scheme 14a). The authors showed an application in the synthesis of some very important molecules, such as phenacetin, paracetamol, lidocaine and piperine (Scheme 14b).

Recently, Das and coworkers [48] developed a new $H_2SO_4$-$SiO_2$ catalytic system for the direct transamidation of carboxamides, describing a new methodology which complements those previously outlined in this article. Due to the easier ability to manipulate the catalysts, it is described as ecofriendly and low cost. The reaction is performed using 5 mol % $H_2SO_4$-$SiO_2$ at 70 °C under solvent-free conditions, and the products were obtained without any purification.

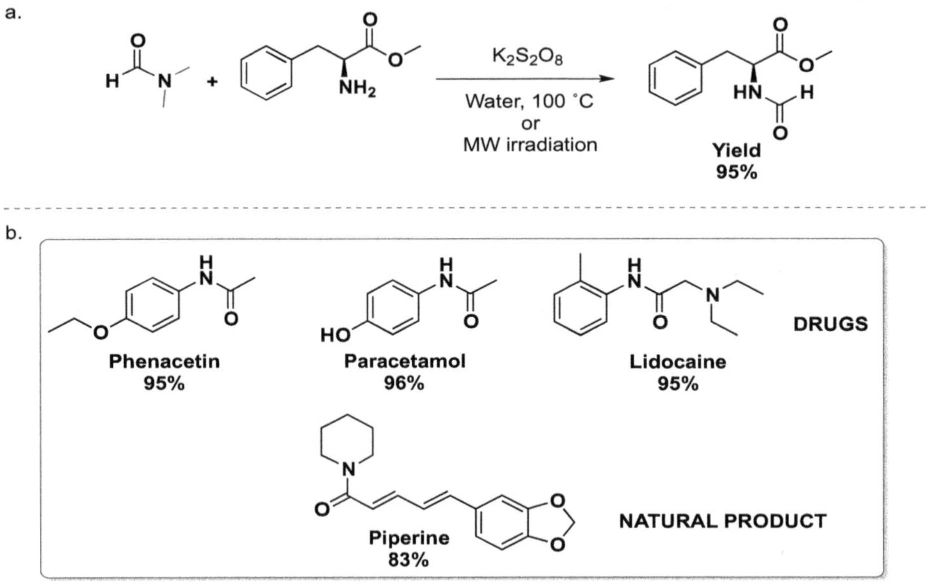

Scheme 14. (a) *N*-Formylation of α-amino esters promoted by $K_2S_2O_8$. (b) Natural products and drugs obtained by transamidation reactions using $K_2S_2O_8$.

The catalytic system was explored using tertiary amides with substituted aromatic, heteroaromatic, and aliphatic/alicyclic primary amines and some secondary amines as well. The most important application of this methodology was in the synthesis of *N*-aryl/heteroaryl pivalamides, which are an important kind of compound in organic synthesis due to their use as directing groups in many transition-metal catalyzed reactions (Scheme 15a). Furthermore, following this optimized transamidation protocol, the authors reported the synthesis of procainamide, a drug used for the treatment of cardiac arrhythmias (Scheme 15c).

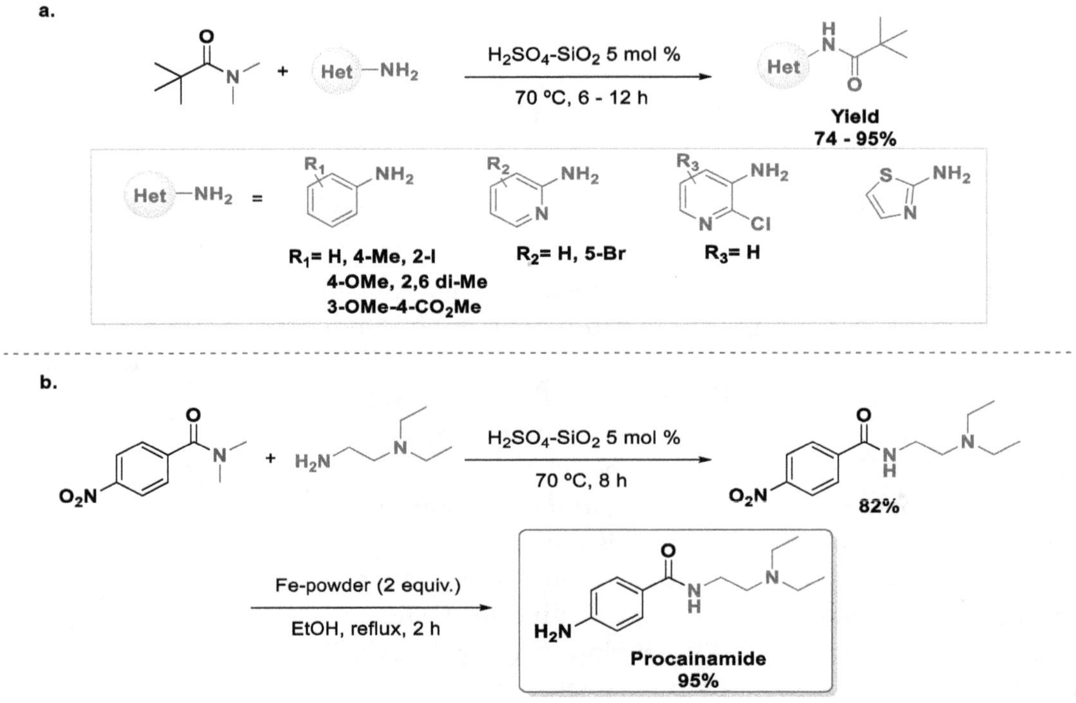

Scheme 15. (a) Synthesis of *N*-aryl/heteroaryl pivalamides via transamidation catalyzed by $H_2SO_4$-$SiO_2$. (b) Synthesis of procainamide.

The examples cited in this section make it clear that the transamidation reaction under metal-free conditions is an attractive alternative because of its environmentally friendly and inexpensive nature. Other less relevant reports for metal-free transamidations are also described in the literature. Among them, new catalytic systems using ionic liquids [49], OSU-6 (a modified silica) [50], $Fe_3O_4$-guanidine acetic acid nanoparticles [51], chitosan [52], graphene oxide [53] and nanosized zeolite beta [54] have shown good results and a good substrate scope. However, they do not have any special consideration that make a more detailed discussion necessary.

## 5. Conclusions

The amide moiety is one of the most important functional groups in organic, pharmaceutical and biological chemistry. Its synthesis has been the focus of many researchers; alternative methods to obtain amides are still needed, and this field is in continuous growth. Currently, alternative activation methods for amides are among the most innovative and recent methods; however, the simple activation of the C–N bond in amides by coordination with metals or by interaction with small molecules is an important alternative in amide bond synthesis. Herein, we surveyed the direct activation of amides in transamidation reactions and presented selected examples in this field. We hope this overview will not only that help researchers and serve as inspiration in their future work, but also that they will find it useful in understanding some recently described transamidation processes.

**Author Contributions:** P.A.-G. wrote more than 70% of the manuscript and contributed to editing the schemes and researching cited papers; A.M.-G. drew the schemes and contributed to the editing and research of the cited papers, as well as the discussion; D.G.-S. wrote approximately 30% of the manuscript and made major revisions and edits.

**Acknowledgments:** P.A.-G. and A.M.-G. acknowledge the Universidad de los Andes and particularly the Chemistry Department for providing fellowships. D.G.-S. kindly acknowledges the Chemistry Department of Universidad de los Andes for logistical support.

## References

1. Humphrey, J.M.; Chamberlin, A.R. Chemical synthesis of natural product peptides: Coupling methods for the incorporation of noncoded amino acids into peptides. *Chem. Rev.* **1997**, *97*, 2243–2266. [CrossRef] [PubMed]
2. Greenberg, A.; Breneman, C.M.; Liebman, J.F. *The Amide Linkage: Structural Significance in Chemistry, Biochemistry, and Materials Science*; Wiley-Interscience: New York, NY, USA, 2000.
3. Ghose, A.K.; Viswanadhan, V.N.; Wendoloski, J.J. A knowledge-based approach in designing combinatorial or medicinal chemistry libraries for drug discovery. 1. A qualitative and quantitative characterization of known drug databases. *J. Comb. Chem.* **1999**, *1*, 55–68. [CrossRef] [PubMed]
4. Carey, J.S.; Laffan, D.; Thomson, C.; Williams, M.T. Analysis of the reactions used for the preparation of drug candidate molecules. *Org. Biomol. Chem.* **2006**, *4*, 2337–2347. [CrossRef] [PubMed]
5. Roughley, S.D.; Jordan, A.M. The medicinal chemist's toolbox: An analysis of reactions used in the pursuit of drug candidates. *J. Med. Chem.* **2011**, *54*, 3451–3479. [CrossRef] [PubMed]
6. Constable, D.J.C.; Dunn, P.J.; Hayler, J.D.; Humphrey, G.R.; Leazer, J.J.L.; Linderman, R.J.; Lorenz, K.; Manley, J.; Pearlman, B.A.; Wells, A.; et al. Key green chemistry research areas-a perspective from pharmaceutical manufacturers. *Green Chem.* **2007**, *9*, 411–420. [CrossRef]
7. Lundberg, H.; Tinnis, F.; Selander, N.; Adolfsson, H. Catalytic amide formation from non-activated carboxylic acids and amines. *Chem. Soc. Rev.* **2014**, *43*, 2714–2742. [CrossRef] [PubMed]
8. Lanigan, R.M.; Sheppard, T.D. Recent developments in amide synthesis: Direct amidation of carboxylic acids and transamidation reactions. *Eur. J. Org. Chem.* **2013**, *2013*, 7453–7465. [CrossRef]
9. Ojeda-Porras, A.; Gamba-Sánchez, D. Recent developments in amide synthesis using nonactivated starting materials. *J. Org. Chem.* **2016**, *81*, 11548–11555. [CrossRef] [PubMed]

10. de Figueiredo, R.M.; Suppo, J.-S.; Campagne, J.-M. Nonclassical routes for amide bond formation. *Chem. Rev.* **2016**, *116*, 12029–12122. [CrossRef] [PubMed]
11. Fleischer, A. Ueber bildung von azoverbindungen. *Ber. Bunsen-Ges. Phys. Chem.* **1876**, *9*, 992–995. [CrossRef]
12. Galat, A.; Elion, G. The interaction of amides with amines: A general method of acylation. *J. Am. Chem. Soc.* **1943**, *65*, 1566–1567. [CrossRef]
13. Pettit, G.; Kalnins, M.; Liu, T.; Thomas, E.; Parent, K. Notes-potential cancerocidal agents. III. Formanilides. *J. Org. Chem.* **1961**, *26*, 2563–2566. [CrossRef]
14. Yoshio, O.; Noboru, M.; Eiji, I. Transacylation from acid amides to amines catalyzed by carbon dioxide. *Bull. Chem. Soc. Jpn.* **1968**, *41*, 1485.
15. Kraus, M.A. The formylation of aliphatic amines by dimethylformamide. *Synthesis* **1973**, *1973*, 361–362. [CrossRef]
16. Bon, E.; Bigg, D.C.H.; Bertrand, G. Aluminum chloride-promoted transamidation reactions. *J. Org. Chem.* **1994**, *59*, 4035–4036. [CrossRef]
17. Thomson, R.K.; Zahariev, F.E.; Zhang, Z.; Patrick, B.O.; Wang, Y.A.; Schafer, L.L. Structure, bonding, and reactivity of Ti and Zr amidate complexes: DFT and X-ray crystallographic studies. *Inorg. Chem.* **2005**, *44*, 8680–8689. [CrossRef] [PubMed]
18. Hoerter, J.M.; Otte, K.M.; Gellman, S.H.; Stahl, S.S. Mechanism of Al(III)-catalyzed transamidation of unactivated secondary carboxamides. *J. Am. Chem. Soc.* **2006**, *128*, 5177–5183. [CrossRef] [PubMed]
19. Kissounko, D.A.; Hoerter, J.M.; Guzei, I.A.; Cui, Q.; Gellman, S.H.; Stahl, S.S. $Ti^{IV}$-mediated reactions between primary amines and secondary carboxamides: Amidine formation versus transamidation. *J. Am. Chem. Soc.* **2007**, *129*, 1776–1783. [CrossRef] [PubMed]
20. Kissounko, D.A.; Guzei, I.A.; Gellman, S.H.; Stahl, S.S. Titanium(IV)-mediated conversion of carboxamides to amidines and implications for catalytic transamidation. *Organometallics* **2005**, *24*, 5208–5210. [CrossRef]
21. Hoerter, J.M.; Otte, K.M.; Gellman, S.H.; Cui, Q.; Stahl, S.S. Discovery and mechanistic study of Al(III)-catalyzed transamidation of tertiary amides. *J. Am. Chem. Soc.* **2008**, *130*, 647–654. [CrossRef] [PubMed]
22. Stephenson, N.A.; Zhu, J.; Gellman, S.H.; Stahl, S.S. Catalytic transamidation reactions compatible with tertiary amide metathesis under ambient conditions. *J. Am. Chem. Soc.* **2009**, *131*, 10003–10008. [CrossRef] [PubMed]
23. Rao, S.N.; Mohan, D.C.; Adimurthy, S. L-proline: An efficient catalyst for transamidation of carboxamides with amines. *Org. Lett.* **2013**, *15*, 1496–1499. [CrossRef] [PubMed]
24. Yang, X.; Fan, L.; Xue, Y. Mechanistic insights into L-proline-catalyzed transamidation of carboxamide with benzylamine from density functional theory calculations. *RSC Adv.* **2014**, *4*, 30108–30117. [CrossRef]
25. Ayub Ali, M.; Hakim Siddiki, S.M.A.; Kon, K.; Shimizu, K. $Fe^{3+}$-exchanged clay catalyzed transamidation of amides with amines under solvent-free condition. *Tetrahedron Lett.* **2014**, *55*, 1316–1319. [CrossRef]
26. Becerra-Figueroa, L.; Ojeda-Porras, A.; Gamba-Sánchez, D. Transamidation of carboxamides catalyzed by Fe(III) and water. *J. Org. Chem.* **2014**, *79*, 4544–4552. [CrossRef] [PubMed]
27. Thale, P.B.; Borase, P.N.; Shankarling, G.S. Transamidation catalysed by a magnetically separable $Fe_3O_4$ nano catalyst under solvent-free conditions. *RSC Adv.* **2016**, *6*, 52724–52728. [CrossRef]
28. Arefi, M.; Heydari, A. Transamidation of primary carboxamides, phthalimide, urea and thiourea with amines using $Fe(OH)_3@Fe_3O_4$ magnetic nanoparticles as an efficient recyclable catalyst. *RSC Adv.* **2016**, *6*, 24684–24689. [CrossRef]
29. Eldred, S.E.; Stone, D.A.; Gellman, S.H.; Stahl, S.S. Catalytic transamidation under moderate conditions. *J. Am. Chem. Soc.* **2003**, *125*, 3422–3423. [CrossRef] [PubMed]
30. Zhang, M.; Imm, S.; Bahn, S.; Neubert, L.; Neumann, H.; Beller, M. Efficient copper(II)-catalyzed transamidation of non-activated primary carboxamides and ureas with amines. *Angew. Chem. Int. Ed.* **2012**, *51*, 3905–3909. [CrossRef] [PubMed]
31. Pathare, S.P.; Jain, A.K.H.; Akamanchi, K.G. Sulfated tungstate: A highly efficient catalyst for transamidation of carboxamides with amines. *RSC Adv.* **2013**, *3*, 7697–7703. [CrossRef]
32. Gu, D.-W.; Guo, X.-X. Synthesis of *N*-arylcarboxamides by the efficient transamidation of DMF and derivatives with anilines. *Tetrahedron* **2015**, *71*, 9117–9122. [CrossRef]
33. Cheung, C.W.; Ma, J.A.; Hu, X. Manganese-mediated reductive transamidation of tertiary amide. *J. Am. Chem. Soc.* **2018**, *140*, 6789–6792. [CrossRef] [PubMed]

34. Yedage, S.L.; D'Silva, D.S.; Bhanage, B.M. $MnO_2$ catalyzed formylation of amines and transamidation of amides under solvent-free conditions. *RSC Adv.* **2015**, *5*, 80441–80449. [CrossRef]
35. Sheng, H.; Zeng, R.; Wang, W.; Luo, S.; Feng, Y.; Liu, J.; Chen, W.; Zhu, M.; Guo, Q. An efficient heterobimetallic lanthanide alkoxide catalyst for transamidation of amides under solvent-free conditions. *Adv. Synth. Catal.* **2017**, *359*, 302–313. [CrossRef]
36. Zarei, Z.; Akhlaghinia, B. Ce(III) immobilised on aminated epichlorohydrin-activated agarose matrix—"green" and efficient catalyst for transamidation of carboxamides. *Chem. Pap.* **2015**, *69*, 1421–1437. [CrossRef]
37. Ghosh, S.C.; Li, C.C.; Zeng, H.C.; Ngiam, J.S.; Seayad, A.M.; Chen, A. Mesoporous niobium oxide spheres as an effective catalyst for the transamidation of primary amides with amines. *Adv. Synth. Catal.* **2014**, *356*, 475–484. [CrossRef]
38. Nirmala, M.; Prakash, G.; Viswanathamurthi, P.; Malecki, J.G. An attractive route to transamidation catalysis: Facile synthesis of new *o*-aryloxide-*N*-heterocyclic carbene ruthenium(II) complexes containing trans triphenylphosphine donors. *J. Mol. Catal. A Chem.* **2015**, *403*, 15–26. [CrossRef]
39. Ma, J; Zhang, F.; Zhang, J; Gong, H. Cobalt(II)-catalyzed *N*-acylation of amines through a transamidation reaction. *Eur. J. Org. Chem.* **2018**. [CrossRef]
40. Shi, M.; Cui, S.C. Transamidation catalyzed by a recoverable and reusable PolyDMAP-based hafnium chloride and montmorillonite KSF. *Synth. Commun.* **2006**, *35*, 2847–2858. [CrossRef]
41. Starkov, P.; Sheppard, T.D. Borate esters as convenient reagents for direct amidation of carboxylic acids and transamidation of primary amides. *Org. Biomol. Chem.* **2011**, *9*, 1320–1323. [CrossRef] [PubMed]
42. Nguyen, T.B.; Sorres, J.; Tran, M.Q.; Ermolenko, L.; Al-Mourabit, A. Boric acid: A highly efficient catalyst for transamidation of carboxamides with amines. *Org. Lett.* **2012**, *14*, 3202–3205. [CrossRef] [PubMed]
43. El Dine, T.M.; Evans, D.; Rouden, J.; Blanchet, J. Formamide synthesis through borinic acid catalysed transamidation under mild conditions. *Chem. Eur. J.* **2016**, *22*, 5894–5898. [CrossRef] [PubMed]
44. Allen, C.L.; Atkinson, B.N.; Williams, J.M. Transamidation of primary amides with amines using hydroxylamine hydrochloride as an inorganic catalyst. *Angew. Chem. Int. Ed.* **2012**, *51*, 1383–1386. [CrossRef] [PubMed]
45. Vanjari, R.; Kumar Allam, B.; Nand Singh, K. Hypervalent iodine catalyzed transamidation of carboxamides with amines. *RSC Adv.* **2013**, *3*, 1691–1694. [CrossRef]
46. Wu, J.-W.; Wu, Y.-D.; Dai, J.-J.; Xu, H.-J. Benzoic acid-catalyzed transamidation reactions of carboxamides, phthalimide, ureas and thioamide with amines. *Adv. Synth. Catal.* **2014**, *356*, 2429–2436. [CrossRef]
47. Srinivas, M.; Hudwekar, A.D.; Venkateswarlu, V.; Reddy, G.L.; Kumar, K.A.A.; Vishwakarma, R.A.; Sawant, S.D. A metal-free approach for transamidation of amides with amines in aqueous media. *Tetrahedron Lett.* **2015**, *56*, 4775–4779. [CrossRef]
48. Rasheed, S.; Rao, D.N.; Reddy, A.S.; Shankar, R.; Das, P. Sulphuric acid immobilized on silica gel ($H_2SO_4$-$SiO_2$) as an eco-friendly catalyst for transamidation. *RSC Adv.* **2015**, *5*, 10567–10574. [CrossRef]
49. Fu, R.; Yang, Y.; Chen, Z.; Lai, W.; Ma, Y.; Wang, Q.; Yuan, R. Microwave-assisted heteropolyanion-based ionic liquids catalyzed transamidation of non-activated carboxamides with amines under solvent-free conditions. *Tetrahedron* **2014**, *70*, 9492–9499. [CrossRef]
50. Nammalwar, B.; Muddala, N.P.; Watts, F.M.; Bunce, R.A. Efficient conversion of acids and esters to amides and transamidation of primary amides using OSU-6. *Tetrahedron* **2015**, *71*, 9101–9111. [CrossRef]
51. Maryam, K.M.; Marzban, A.; Elahe, Y.; Sepideh, A.; Meghdad, K.; Kobra, A.; Akbar, H. Guanidine acetic acid functionalized magnetic nanoparticles: Recoverable green catalyst for transamidation. *ChemistrySelect* **2016**, *1*, 6328–6333.
52. Rao, S.; Mohan, D.; Adimurthy, S. Chitosan: An fficiente recyclable catalyst for transamidation of carboxamdes with amines under neat condition. *Green Chem.* **2014**, *16*, 4122–4126.
53. Bhattacharya, S.; Ghoh, P.; Basu, B. Graphene oxide (GO) catalyzed transamidations of aliphatic amides: An efficient metal-free procedure. *Tetrahedron Lett.* **2018**, *59*, 899–903. [CrossRef]
54. Durgaiah, C.; Naresh, M.; Swamy, P.; Srujana, K.; Rammurthy, B.; Narender, N. Transamidation of carboxamides with amines over nanosized zeolite beta under solvent-free conditions. *Catal. Commun.* **2016**, *81*, 29–32. [CrossRef]

# 12

# Ruthenium-Based Catalytic Systems Incorporating a Labile Cyclooctadiene Ligand with *N*-Heterocyclic Carbene Precursors for the Atom-Economic Alcohol Amidation using Amines

**Cheng Chen [1], Yang Miao [2], Kimmy De Winter [3], Hua-Jing Wang [4], Patrick Demeyere [3], Ye Yuan [1,*] and Francis Verpoort [1,2,5,6,*]**

1 State Key Laboratory of Advanced Technology for Materials Synthesis and Processing, Wuhan University of Technology, 122 Luoshi Road, Wuhan 430070, China; chengchen@whut.edu.cn
2 School of Materials Science and Engineering, Wuhan University of Technology, 122 Luoshi Road, Wuhan 430070, China; miaoyang2018@126.com
3 Odisee/KU Leuven Technology Campus, Gebroeders de Smetstraat 1, 9000 Ghent, Belgium; kimmydewinter@hotmail.com (K.D.W.); patrick.demeyere@odisee.be (P.D.)
4 School of Chemistry, Chemical Engineering and Life Sciences, Wuhan University of Technology, 122 Luoshi Road, Wuhan 430070, China; 18772101949@163.com
5 National Research Tomsk Polytechnic University, Lenin Avenue 30, Tomsk 634050, Russian
6 Ghent University Global Campus, 119 Songdomunhwa-Ro, Yeonsu-Gu, Incheon 21985, Korea
* Correspondence: fyyuanye@whut.edu.cn (Y.Y.); francis.verpoort@ghent.ac.kr (F.V.)

**Abstract:** Transition-metal-catalyzed amide-bond formation from alcohols and amines is an atom-economic and eco-friendly route. Herein, we identified a highly active in situ *N*-heterocyclic carbene (NHC)/ruthenium (Ru) catalytic system for this amide synthesis. Various substrates, including sterically hindered ones, could be directly transformed into the corresponding amides with the catalyst loading as low as 0.25 mol.%. In this system, we replaced the *p*-cymene ligand of the Ru source with a relatively labile cyclooctadiene (cod) ligand so as to more efficiently obtain the corresponding poly-carbene Ru species. Expectedly, the weaker cod ligand could be more easily substituted with multiple mono-NHC ligands. Further high-resolution mass spectrometry (HRMS) analyses revealed that two tetra-carbene complexes were probably generated from the in situ catalytic system.

**Keywords:** ruthenium (Ru); *N*-heterocyclic carbenes (NHCs); homogeneous catalysis; in situ; amide bonds; synthesis

---

## 1. Introduction

Amides are a series of fundamental functional structures in nature and biological systems, as well as crucial building blocks for organic synthesis [1–6]. As of late, numerous synthetic methods were reported for the construction of amide bonds. However, they generally suffer from the usage of various stoichiometric additives and the production of unfavorable equimolar byproducts [7–14]. Therefore, green and eco-friendly strategies are highly required for amide synthesis [15]. Recently, a methodology employing transition-metal-based catalytic systems for direct amide synthesis from alcohols and amines was proven to be far more atom-economic and environmentally friendly as the only byproduct is hydrogen [16–22]. Throughout this research, ruthenium (Ru) was most extensively studied [23]. Initially, the Murahashi [24] and Milstein [25] groups pioneered Ru-catalyzed amide synthesis in intramolecular and intermolecular manners, respectively. Notably, the Milstein catalyst, a Ru complex bearing a PNN-type pincer ligand, was highly active for this reaction. With a catalyst loading of 0.1 mol.%, various amides could be synthesized from alcohols and amines [25]. Later, great progress was achieved by the Milstein [26–28], Madsen [29–31], Williams [32,33], Hong [34–43], Crabtree [44,45], Albrecht [46], Guan [47,48], Glorius [49], Möller [50,51], Bera [52], Huynh [53], Viswanathamurthi [54–56], Mashima [57], Verpoort [58,59], and Kundu [60] groups. In particular, Ru combined with *N*-heterocyclic carbenes (NHCs) attracted more and more interest due to the flexible tunability of the electronic and steric properties of NHCs, which may easily access the optimal structures of the corresponding NHC/Ru complexes [61–63]. Accordingly, a multitude of efficient NHC/Ru catalytic systems were discovered for this reaction. Furthermore, considering the merits of the in situ catalytic systems, such as easy operation and convenient investigation of electronically and sterically distinct NHCs, a number of versatile and potent in situ NHC/Ru catalytic systems recently emerged. However, satisfactory yields could only be attained by these reported systems if relatively high Ru loadings of 2.0–5.0 mol.% were employed [29,34,36,37,49]. Therefore, the development of more efficient in situ NHC/Ru catalytic systems which can accomplish the formation of amide linkage are urgently required.

In our previous work, the development of various in situ generated (*p*-cymene)/Ru catalytic systems, which contain benzimidazole-based NHC precursors bearing different electronic and steric properties, was accomplished [58]. Further experiments revealed that two mono-NHC/Ru complexes were observed as major species and two poly-carbene complexes were detected as only minor species (as depicted in Figure 1a) [59]. Herein, we envisioned that replacing the *p*-cymene ligand of the Ru center with a relatively labile cyclooctadiene (cod) ligand could possibly give rise to poly-carbene complexes as a major species (as shown in Figure 1b). Expectedly, the weaker cod ligand could be more easily substituted with multiple mono-NHC ligands. Based on this, an efficient in situ NHC/Ru catalytic system was developed through extensive screening of various conditions. Notably, this system demonstrated excellent catalytic activity for amide synthesis with the applied catalyst loading as low as 0.25 mol.%. Various amides, including sterically congested ones, were directly synthesized from alcohols and amines in moderate to excellent yields. Furthermore, high-resolution mass spectrometry (HRMS) analyses suggested several Ru species bearing multiple NHC ligands as major species, which was in accordance with our prospection.

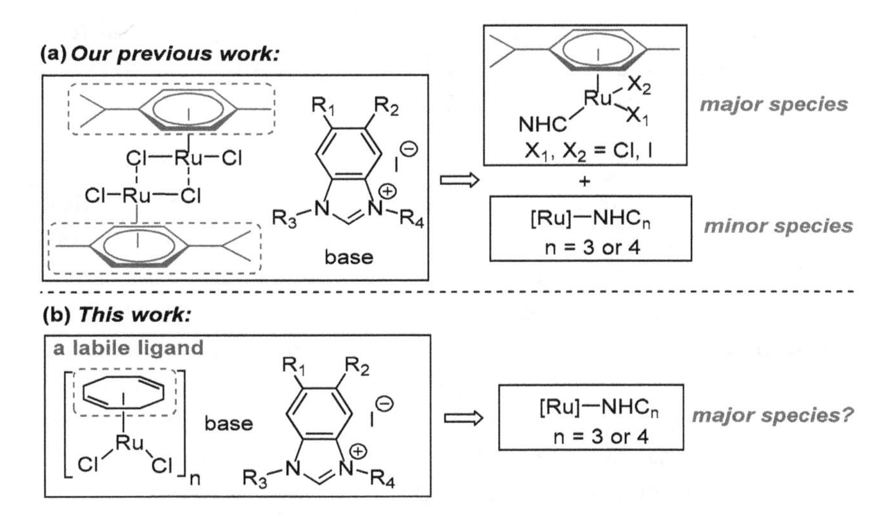


**Figure 1.** The design strategy of this work.

## 2. Results and Discussion

The reaction of benzyl alcohol (**1a**) and benzylamine (**2a**) was selected as a model reaction for the optimization of the reaction conditions. Based on our previous work [59], 0.5 mol.% of $[RuCl_2(cod)]_n$, 2.00 mol.% of an NHC precursor, 3.50 mol.% of NaH, 0.5 h of catalyst generation time, and 16 h of reaction time were originally applied (as listed in Table 1). In the beginning, NHC precursors **L1**–**L6** with different backbone and wingtip substituents were prepared (entries 1–6, Table 1). The first and foremost, 62% of amide **3a** and 15% of imine **4a**, were obtained with 18% of **1a** remaining if **L1** was used (entry 1). Electron-deficient precursor **L2** gave rise to lower amide content in the product distribution, demonstrating its disadvantage for amide formation (entry 2 vs. entry 1). In the case of an electron-rich NHC precursor (**L3**), a similar result was obtained compared with **L1** (entry 3 vs. entry 1). Moreover, the substituents on the *N*-terminus of the NHC precursors were adjusted (entries 1, 4–6). With retaining Me as the substituent for one *N*-terminus, different groups including Et, *n*Pr, and *i*Pr were introduced for the other terminus. The result was indicative that Et was the optimized group for this reaction (entry 4 vs. entries 1, 5, and 6). After establishing the ideal NHC precursor (**L4**), we continued the optimization by screening other reaction conditions. It was found that the catalyst generation time was crucial for the catalysis (entries 4, 7–11); 57% of the amide product could be detected if every substance was added simultaneously (entry 5). As we elongated the period for the in situ catalyst generation from 0 h to 2.0 h, the yields of **3a** gradually increased (entries 4, 7–10). A further increment of the time led to a similar yield (entry 11 vs. entry 10). Therefore, the ideal duration for the catalyst generation was finalized as 2 h. Next, the ratio of [Ru]: **L4**:NaH was varied (entries 12–17). It is worth emphasizing that the amount of both **L4** and NaH changed so as to ascertain three additional equivalents of NaH to activate $[RuCl_2(cod)]_n$ for all cases. Without **L4**, no amide was formed (entry 12). As the ratio increased from 1:0:3 to 1:5:8, gradually higher yields of **3a** were observed (entries 10, 12–16). However, a higher ratio prompted a reduced yield of **3a** (entry 17 vs. entry 16). Thus, the ratio of 1:5:8 was recognized as the best one (entry 16), and further increasing the reaction time from 16 h to 36 h produced **3a** in 93% yield (entry 18).

In order to identify a more active catalytic system, a reduced Ru loading of 0.25 mol.% was attempted (as listed in Table 2). At the outset, 65% of **3a** was afforded if the loading of the above-optimized catalytic system was directly reduced to 0.25 mol.% (entry 1). In addition, different bases including potassium bis(trimethylsilyl)amide (KHMDS), KO*t*Bu, and $Cs_2CO_3$ were exploited instead of NaH (entries 2–4). Interestingly, compared with NaH, the milder $Cs_2CO_3$ led to an increased yield of **3a** (entry 4 vs. entry 1). It was also noticed that the volume of toluene was crucial for the reaction (entries 4–8). Either a more concentrated or diluted solution triggered a lower amide/imine selectivity (entry 5–8 vs. entry 4). Furthermore, the adjustment of the base amounts influenced the reaction (entries 4, 9–12), and 1.75 mol.% of $Cs_2CO_3$ was found to be optimal for the selective amide formation (entry 10). Therefore, the optimized reaction conditions were identified as **1** (5.00 mmol), **2** (5.50 mmol), $[RuCl_2(cod)]_n$ (0.0125 mmol), **L4** (0.0625 mmol), $Cs_2CO_3$ (0.075 mmol), toluene (1.50 mL), reflux, and 36 h unless otherwise noted.

With the optimized reaction conditions at hand, the substrate scope and limitations of this strategy were further investigated (as depicted in Figure 2). For the sterically non-hindered substrates (**1a**–**1e**), the corresponding amides could be obtained in good to excellent yields. If a secondary amine (**1f**) was employed, tertiary amide **3f** was also given in 80% yield with 0.5 mol.% of [Ru]. Expectedly, lactam **3g** was efficiently afforded from amino alcohol **1f** in an intramolecular pattern. On the other hand, the reactions of benzyl alcohol with substituted benzylamines were evaluated. It seemed that these substituents had no obvious influence on the reactivity, and amides **3h**–**3k** were synthesized in 75–85% yields. In the case of coupling benzylamine with various benzyl alcohols, a substituent at either the *para* or *meta* position resulted in good yields of amides **3l**–**3n**. However, an *ortho* group gave amide **3o** in a moderate yield. Apparently, aromatic amines were less reactive, and aniline (**2p**) produced amide **3p** in only 25% yield. To our delight, this newly developed catalytic system was not as sensitive to steric bulks as our previous systems [58,59]. With an Ru loading of 0.5 mol.%, several sterically hindered substrates could be efficiently transformed into amides **3q**–**3t**.

**Table 1.** Optimization of reaction conditions with a catalyst loading of 0.5 mol.% [a].

Ph⌒OH + Ph⌒NH₂ → [RuCl₂(cod)]ₙ (0.50 mol%), L (x mol%), NaH (y mol%), toluene, reflux, n h → reflux, 16 h → **3a** + **4a**

**1a** (1.00 equiv.) **2a** (1.10 equiv.)

**L1** **L2** **L3** **L4** **L5** **L6**

| Entry | L | x | y | n | Yields (%) [b] | | |
|---|---|---|---|---|---|---|---|
| | | | | | 3a | 4a | Unreacted 1a |
| 1 | **L1** | 2.00 | 3.50 | 0.5 | 62 | 15 | 18 |
| 2 | **L2** | 2.00 | 3.50 | 0.5 | 28 | 30 | 39 |
| 3 | **L3** | 2.00 | 3.50 | 0.5 | 63 | 15 | 16 |
| 4 | **L4** | 2.00 | 3.50 | 0.5 | 78 | 10 | 8 |
| 5 | **L5** | 2.00 | 3.50 | 0.5 | 72 | 12 | 8 |
| 6 | **L6** | 2.00 | 3.50 | 0.5 | 28 | 30 | 39 |
| 7 | **L4** | 2.00 | 3.50 | 0.0 | 57 | 10 | 31 |
| 8 | **L4** | 2.00 | 3.50 | 1.0 | 79 | 6 | 8 |
| 9 | **L4** | 2.00 | 3.50 | 1.5 | 81 | 5 | 7 |
| 10 | **L4** | 2.00 | 3.50 | 2.0 | 83 | 4 | 5 |
| 11 | **L4** | 2.00 | 3.50 | 2.5 | 82 | 4 | 6 |
| 12 | **L4** | 0.00 | 1.50 | 2.0 | 0 | 19 | 76 |
| 13 | **L4** | 0.50 | 2.00 | 2.0 | 37 | 10 | 51 |
| 14 | **L4** | 1.00 | 2.50 | 2.0 | 60 | 11 | 28 |
| 15 | **L4** | 1.50 | 3.00 | 2.0 | 75 | 7 | 16 |
| 16 | **L4** | 2.50 | 4.00 | 2.0 | 86 | 4 | 9 |
| 17 | **L4** | 3.00 | 4.50 | 2.0 | 81 | 6 | 3 |
| 18 [c] | **L4** | **2.50** | **4.00** | **2.0** | **93** | **5** | **0** |

[a] **1a** (2.50 mmol), **2a** (2.75 mmol), $[RuCl_2(cod)]_n$ (0.50 mol.%), **L** (x mol.%), NaH (y mol.%), toluene (1.25 mL), 120 °C, n h of catalyst generation time, and 16 h of reaction time; [b] NMR yields (average of two consistent runs) using 1,3,5-trimethoxybenzene as an internal standard; [c] 36 h of reaction time.

Concerning the in situ catalytic systems, it is crucial to explore the possible structures of the generated Ru species. As a result, HRMS analyses were performed to clarify this matter (as shown in Figure 3). In accordance with our speculation, no mono-carbene complexes were detected. Instead, two poly-carbene Ru species were observed from the spectrum. **[Ru]-1** (corresponding to an isotopic peak at $m/z = 812.24209$), consistent with an Ru species comprising four-fold NHC ligands, was observed as a major species. Furthermore, another tetra-carbene Ru species, assigned as **[Ru]-2** with the isotopic peak at $m/z = 793.26709$, was also found as a minor species. Presumably, during exposure to air and/or the HRMS measurements, the Ru centers in **[Ru]-1** and **[Ru]-2** were oxidized to +3 and +4, respectively. Unfortunately, attempts to isolate these tetra-carbene complexes were unsuccessful, probably due to the complexity of the in situ catalyst generation. Therefore, it was still unclear whether the high activity of the current catalytic system was attributed to the observed tetra-carbene Ru species or other species.

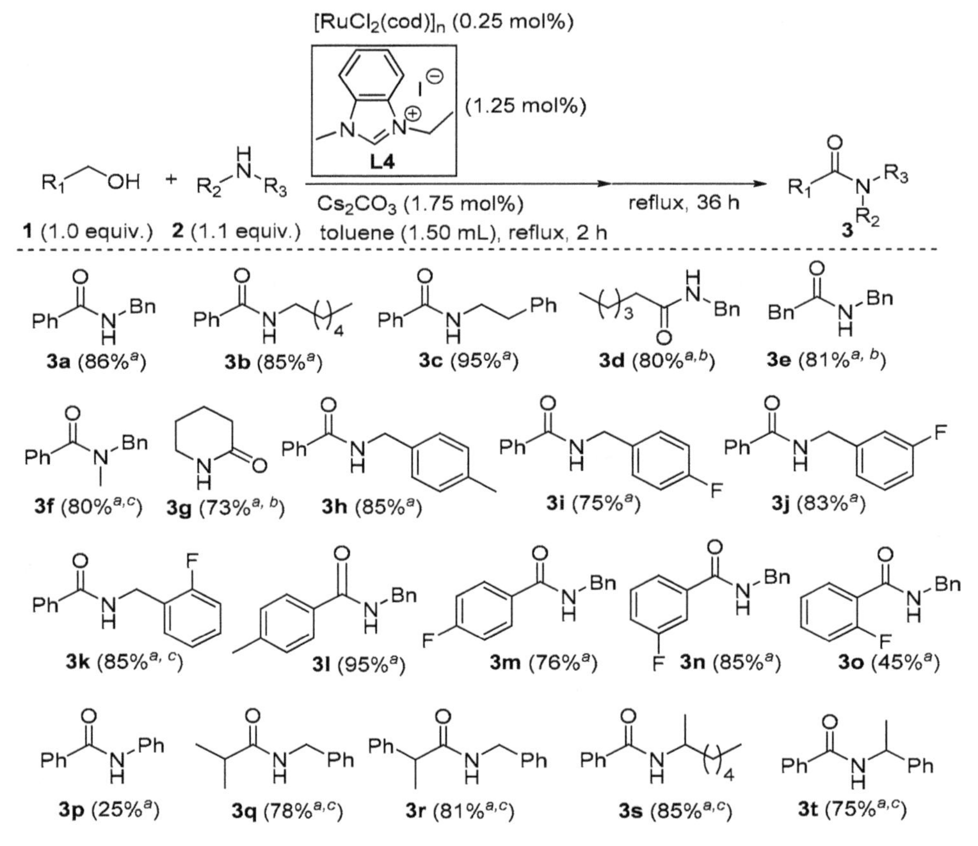


**Figure 2.** Amide synthesis from various alcohols and amines; [a] isolated yields (averages of two consistent runs); [b] in *m*-xylene at reflux; [c] 0.5 mol.% of [Ru].

**Table 2.** Optimization of reaction conditions with a catalyst loading of 0.25 mol.% [a].

[RuCl2(cod)]n (0.25 mol%), L4 (1.25 mol%), base (x mol%), toluene (y mL), reflux, 2 h; reflux, 36 h

1a (1.00 equiv.) 2a (1.10 equiv.) → 3a + 4a

| Entry | Base | x | y | Yields (%) [b] | | |
|---|---|---|---|---|---|---|
| | | | | 3a | 4a | Unreacted 1a |
| 1 | NaH | 2.00 | 1.50 | 65 | 7 | 24 |
| 2 | KHMDS | 2.00 | 1.50 | 27 | 11 | 57 |
| 3 | KO*t*Bu | 2.00 | 1.50 | 45 | 15 | 32 |
| 4 | $Cs_2CO_3$ | 2.00 | 1.50 | 86 | 7 | 5 |
| 5 | $Cs_2CO_3$ | 2.00 | 0.50 | 57 | 18 | 22 |
| 6 | $Cs_2CO_3$ | 2.00 | 1.00 | 71 | 13 | 12 |
| 7 | $Cs_2CO_3$ | 2.00 | 2.00 | 69 | 16 | 13 |
| 8 | $Cs_2CO_3$ | 2.00 | 2.50 | 45 | 38 | 15 |
| 9 | $Cs_2CO_3$ | 1.50 | 1.50 | 66 | 15 | 12 |
| **10** | **$Cs_2CO_3$** | **1.75** | **1.50** | **90** | **7** | **2** |
| 11 | $Cs_2CO_3$ | 2.25 | 1.50 | 81 | 10 | 8 |
| 12 | $Cs_2CO_3$ | 2.50 | 1.50 | 72 | 12 | 15 |

[a] **1a** (5.00 mmol), **2a** (5.50 mmol), $[RuCl_2(cod)]_n$ (0.25 mol.%), **L4** (1.25 mol.%), base (x mol.%), toluene (y mL), 120 °C, 2 h of catalyst generation time, and 36 h of reaction time; [b] NMR yields (average of two consistent runs) using 1,3,5-trimethoxybenzene as an internal standard.

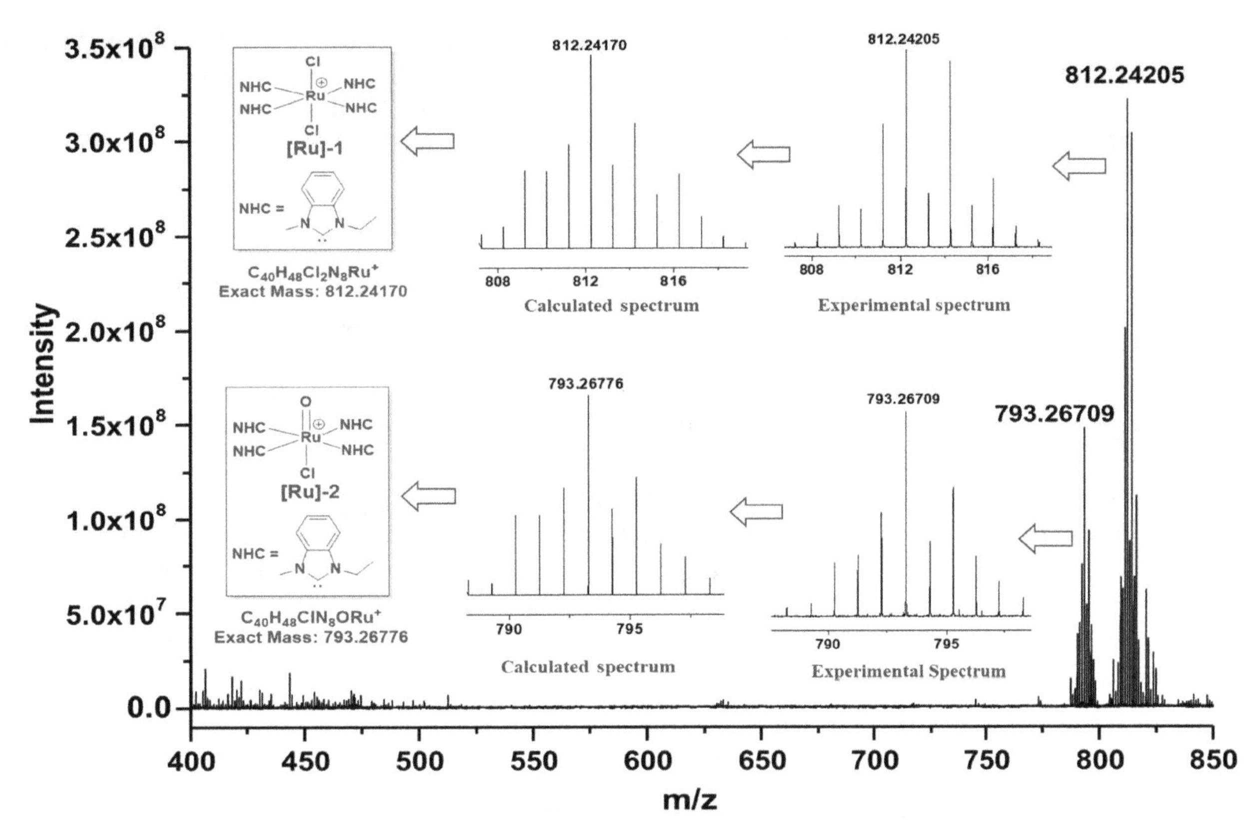


**Figure 3.** The high-resolution mass spectrometry (HRMS) analyses for the identification of the possible Ru species.

## 3. Experimental

### *3.1. General Considerations*

All reactions were carried out using standard Schlenk techniques or in an argon-filled glove box unless otherwise mentioned. All the substrates and solvents were obtained from commercial suppliers and used as received without further purification. $^1$H-NMR spectra were recorded on a Bruker Avance 500 spectrometer (Billerica, MA, USA) in $CDCl_3$ or DMSO-$d_6$ with TMS as the internal reference, and $^{13}$C-NMR spectra were recorded in $CDCl_3$ or DMSO-$d_6$ on a Bruker Avance 500 (126 MHz) spectrometer. The following abbreviations were used to designate multiplicities: s = singlet, brs = broad singlet, d = doublet, t = triplet, dd = doublet of doublets, dq = doublet of quartets, td = triplet of doublets, ddd = doublet of doublets of doublets, and m = multiplet. Melting points were taken on a Buchi M-560 melting point apparatus (Flawil, Switzerland) and were uncorrected. HRMS analyses were done with a Bruker Daltonics microTOF-QII instrument (Billerica, MA, USA). NHC precursors **L1–L6** were prepared according to a previous publication [58,59], and all the amide products were identified by spectral comparison with the literature data [58,59]. $^1$H-NMR, $^{13}$C-NMR data and original spectra of amides **3a–3t** could be found in the Supplementary Materials.

### *3.2. General Procedure for the Amide Synthesis*

Inside an argon-filled glove box, $[Ru(cod)Cl_2]_n$ (3.5 mg, 0.0125 mmol), **L4** (18.0 mg, 0.0625 mmol), $Cs_2CO_3$ (28.6 mg, 0.0875 mmol), and dry toluene (1.50 mL) were added to an oven-dried 25-mL Schlenk flask. The tube was taken out of the glove box and heated to reflux under argon for 2 h. Then, an alcohol (5.00 mmol) and an amine (5.50 mmol) were added, and the mixture was stirred at a refluxing temperature for 36 h. The procedures for calculating the NMR yields were as follows: when the reaction was complete, 1,3,5-trimethoxybenzene (0.5 mmol, 84.0 mg) and $CHCl_3$ (1.0 mL) were added to the reaction mixture. Afterward, to an NMR tube was added 0.1 mL of the above solution and

0.4 mL of $CDCl_3$. The NMR yields were obtained based on the exact amount of 1,3,5-trimethoxybenzene. In order to obtain the isolated yields of the amides, the reaction mixture was cooled down to room temperature, and the solvent was removed under reduced pressure. Finally, the residue was purified by silica-gel flash column chromatography to afford the amides.

## 4. Conclusions

In summary, based on the assumption that the relatively labile cod ligand could be replaced by multiple NHC ligands to obtain versatile and active catalytic systems, we prepared several NHC precursors with distinct electronic and steric properties, then combined them with $[RuCl_2(cod)]_n$ and a mild $Cs_2CO_3$ to obtain a series of in situ NHC/Ru catalytic systems. Through extensive screening of these systems and other conditions, the **L4**-based NHC/Ru catalytic system exhibited optimal activity for the dehydrogenative amidation of alcohols and amines. Various amides, especially sterically hindered ones, could be afforded in an efficient manner. Notably, the applied catalyst loading was as low as 0.25 mol.%. Further experiments revealed that the higher amount of **L4** compared to Ru probably facilitated the formation of two tetra-carbene species (**[Ru]-1** and **[Ru]-2**), which were observed from HRMS analyses. However, since the in situ catalytic system was relatively complicated, it is still uncertain whether these tetra-carbene Ru species or other species were key catalytic intermediates for this reaction.

**Author Contributions:** C.C., Y.Y and F.V. discussed and designed the whole project together. C.C., Y.M., K.D.W., and H.W. performed the experiments. C.C. and Y.M. wrote the manuscript. Y.Y., F.V. and P.D. revised the manuscript. All authors read and approved the final manuscript.

## References

1. Humphrey, J.M.; Chamberlin, A.R. Chemical synthesis of natural product peptides: Coupling methods for the incorporation of noncoded amino acids into peptides. *Chem. Rev.* **1997**, *97*, 2243–2266. [CrossRef] [PubMed]
2. Bode, J.W. Emerging methods in amide- and peptide-bond formation. *Curr. Opin. Drug Discov. Dev.* **2006**, *9*, 765–775. [CrossRef]
3. Cupido, T.; Tulla-Puche, J.; Spengler, J.; Albericio, F. The synthesis of naturally occurring peptides and their analogs. *Curr. Opin. Drug Discov. Dev.* **2007**, *10*, 768–783.
4. Valeur, E.; Bradley, M. Amide bond formation: Beyond the myth of coupling reagents. *Chem. Soc. Rev.* **2009**, *38*, 606–631. [CrossRef] [PubMed]
5. Ali, M.A.; Punniyamurthy, T. Palladium-catalyzed one-pot conversion of aldehydes to amides. *Adv. Synth. Catal.* **2010**, *352*, 288–292. [CrossRef]
6. Pattabiraman, V.R.; Bode, J.W. Rethinking amide bond synthesis. *Nature* **2011**, *480*, 471–479. [CrossRef] [PubMed]
7. Han, S.Y.; Kim, Y.A. Recent development of peptide coupling reagents in organic synthesis. *Tetrahedron* **2004**, *60*, 2447–2467. [CrossRef]
8. Kohn, M.; Breinbauer, R. The Staudinger ligation-A gift to chemical biology. *Angew. Chem. Int. Ed.* **2004**, *43*, 3106–3116. [CrossRef] [PubMed]
9. Montalbetti, C.A.G.N.; Falque, V. Amide bond formation and peptide coupling. *Tetrahedron* **2005**, *61*, 10827–10852. [CrossRef]
10. Kolakowski, R.V.; Shangguan, N.; Sauers, R.R.; Williams, L.J. Mechanism of thio acid/azide amidation. *J. Am. Chem. Soc.* **2006**, *128*, 5695–5702. [CrossRef] [PubMed]
11. Lang, S.; Murphy, J.A. Azide rearrangements in electron-deficient systems. *Chem. Soc. Rev.* **2006**, *35*, 146–156. [CrossRef] [PubMed]
12. Martinelli, J.R.; Clark, T.P.; Watson, D.A.; Munday, R.H.; Buchwald, S.L. Palladium-catalyzed aminocarbonylation of aryl chlorides at atmospheric pressure: The dual role of sodium phenoxide. *Angew. Chem. Int. Ed.* **2007**, *46*, 8460–8463. [CrossRef] [PubMed]

13. Owston, N.A.; Parker, A.J.; Williams, J.M.J. Iridium-catalyzed conversion of alcohols into amides via oximes. *Org. Lett.* **2007**, *9*, 73–75. [CrossRef] [PubMed]
14. Chang, J.W.W.; Chan, P.W.H. Highly efficient ruthenium (II) porphyrin catalyzed amidation of aldehydes. *Angew. Chem. Int. Ed.* **2008**, *47*, 1138–1140. [CrossRef] [PubMed]
15. Constable, D.J.C.; Dunn, P.J.; Hayler, J.D.; Humphrey, G.R.; Leazer, J.L., Jr.; Linderman, R.J.; Lorenz, K.; Manley, J.; Pearlman, B.A.; Wells, A.; et al. Key green chemistry research areas-a perspective from pharmaceutical manufacturers. *Green Chem.* **2007**, *9*, 411–420. [CrossRef]
16. Allen, C.L.; Williams, J.M.J. Metal-catalysed approaches to amide bond formation. *Chem. Soc. Rev.* **2011**, *40*, 3405–3415. [CrossRef] [PubMed]
17. Chen, C.; Hong, S.H. Oxidative amide synthesis directly from alcohols with amines. *Org. Biomol. Chem.* **2011**, *9*, 20–26. [CrossRef] [PubMed]
18. Gunanathan, C.; Milstein, D. Applications of acceptorless dehydrogenation and related transformations in chemical synthesis. *Science* **2013**. [CrossRef] [PubMed]
19. Gunanathan, C.; Milstein, D. Bond activation and catalysis by ruthenium pincer complexes. *Chem. Rev.* **2014**, *114*, 12024–12087. [CrossRef] [PubMed]
20. de Figueiredo, R.M.; Suppo, J.S.; Campagne, J.M. Nonclassical routes for amide bond formation. *Chem. Rev.* **2016**, *116*, 12029–12122. [CrossRef] [PubMed]
21. Xiong, X.Q.; Fan, G.M.; Zhu, R.J.; Shi, L.; Xiao, S.Y.; Bi, C. Highly efficient synthesis of amides. *Prog. Chem.* **2016**, *28*, 497–506.
22. Gusey, D.G. Rethinking the dehydrogenative amide synthesis. *ACS Catal.* **2017**, *7*, 6656–6662.
23. Chen, C.; Verpoort, F.; Wu, Q.Y. Atom-economic dehydrogenative amide synthesis via ruthenium catalysis. *RSC Adv.* **2016**, *6*, 55599–55607. [CrossRef]
24. Naota, T.; Murahashi, S.I. Ruthenium-catalyzed transformations of amino-alcohols to lactams. *Synlett* **1991**, *10*, 693–694. [CrossRef]
25. Gunanathan, C.; Ben-David, Y.; Milstein, D. Direct synthesis of amides from alcohols and amines with liberation of H-2. *Science* **2007**, *317*, 790–792. [CrossRef] [PubMed]
26. Gnanaprakasam, B.; Balaraman, E.; Ben-David, Y.; Milstein, D. Synthesis of peptides and pyrazines from β-Amino alcohols through extrusion of $H_2$ catalyzed by ruthenium pincer complexes: Ligand-controlled selectivity. *Angew. Chem. Int. Ed.* **2011**, *50*, 12240–12244. [CrossRef] [PubMed]
27. Gnanaprakasam, B.; Balaraman, E.; Gunanathan, C.; Milstein, D. Synthesis of polyamides from diols and diamines with liberation of $H_2$. *J. Polym. Sci. Part A Polym. Chem.* **2012**, *50*, 1755–1765. [CrossRef]
28. Srimani, D.; Balaraman, E.; Hu, P.; Ben-David, Y.; Milstein, D. Formation of tertiary amides and dihydrogen by dehydrogenative coupling of primary alcohols with secondary amines catalyzed by ruthenium bipyridine-based pincer complexes. *Adv. Synth. Catal.* **2013**, *355*, 2525–2530. [CrossRef]
29. Nordstrøm, L.U.; Vogt, H.; Madsen, R. Amide synthesis from alcohols and amines by the extrusion of dihydrogen. *J. Am. Chem. Soc.* **2008**, *130*, 17672–17673. [CrossRef] [PubMed]
30. Dam, J.H.; Osztrovszky, G.; Nordstrøm, L.U.; Madsen, R. Amide synthesis from alcohols and amines catalyzed by ruthenium *N*-heterocyclic carbene complexes. *Chem. Eur. J.* **2010**, *16*, 6820–6827. [CrossRef] [PubMed]
31. Makarov, I.S.; Fristrup, P.; Madsen, R. Mechanistic investigation of the ruthenium-*N*-heterocyclic-carbene-catalyzed amidation of amines with alcohols. *Chem. Eur. J.* **2012**, *18*, 15683–15692. [CrossRef] [PubMed]
32. Watson, A.J.A.; Maxwell, A.C.; Williams, J.M.J. Ruthenium-catalyzed oxidation of alcohols into amides. *Org. Lett.* **2009**, *11*, 2667–2670. [CrossRef] [PubMed]
33. Watson, A.J.A.; Wakeham, R.J.; Maxwell, A.C.; Williams, J.M.J. Ruthenium-catalysed oxidation of alcohols to amides using a hydrogen acceptor. *Tetrahedron* **2014**, *70*, 3683–3690. [CrossRef]
34. Ghosh, S.C.; Muthaiah, S.; Zhang, Y.; Xu, X.Y.; Hong, S.H. Direct amide synthesis from alcohols and amines by phosphine-free ruthenium catalyst systems. *Adv. Synth. Catal.* **2009**, *351*, 2643–2649. [CrossRef]
35. Zhang, Y.; Chen, C.; Ghosh, S.C.; Li, Y.X.; Hong, S.H. Well-defined *N*-heterocyclic carbene based ruthenium catalysts for direct amide synthesis from alcohols and amines. *Organometallics* **2010**, *29*, 1374–1378. [CrossRef]
36. Muthaiah, S.; Ghosh, S.C.; Jee, J.E.; Chen, C.; Zhang, J.; Hong, S.H. Direct amide synthesis from either alcohols or aldehydes with amines: Activity of Ru (II) hydride and Ru (0) complexes. *J. Org. Chem.* **2010**, *75*, 3002–3006. [CrossRef] [PubMed]

37. Ghosh, S.C.; Hong, S.H. Simple $RuCl_3$-catalyzed amide synthesis from alcohols and amines. *Eur. J. Org. Chem.* **2010**, 4266–4270. [CrossRef]
38. Zhang, J.; Senthilkumar, M.; Ghosh, S.C.; Hong, S.H. Synthesis of cyclic imides from simple diols. *Angew. Chem. Int. Ed.* **2010**, *49*, 6391–6395. [CrossRef] [PubMed]
39. Chen, C.; Zhang, Y.; Hong, S.H. *N*-heterocyclic carbene based ruthenium-catalyzed direct amide synthesis from alcohols and secondary amines: Involvement of esters. *J. Org. Chem.* **2011**, *76*, 10005–10010. [CrossRef] [PubMed]
40. Chen, C.; Hong, S.H. Selective catalytic $sp^3$ C-O bond cleavage with C-N bond formation in 3-alkoxy-1-propanols. *Org. Lett.* **2012**, *14*, 2992–2995. [CrossRef] [PubMed]
41. Kim, K.; Kang, B.; Hong, S.H. *N*-Heterocyclic carbene-based well-defined ruthenium hydride complexes for direct amide synthesis from alcohols and amines under base-free conditions. *Tetrahedron* **2015**, *71*, 4565–4569. [CrossRef]
42. Kang, B.; Hong, S.H. Hydrogen acceptor- and base-free *N*-formylation of nitriles and amines using methanol as C-1 Source. *Adv. Synth. Catal.* **2015**, *357*, 834–840. [CrossRef]
43. Kim, S.H.; Hong, S.H. Ruthenium-catalyzed urea synthesis using methanol as the C1 source. *Org. Lett.* **2016**, *18*, 212–215. [CrossRef] [PubMed]
44. Nova, A.; Balcells, D.; Schley, N.D.; Dobereiner, G.E.; Crabtree, R.H.; Eisenstein, O. An experimental-theoretical study of the factors that affect the switch between ruthenium-catalyzed dehydrogenative amide formation versus amine alkylation. *Organometallics* **2010**, *29*, 6548–6558. [CrossRef]
45. Schley, N.D.; Dobereiner, G.E.; Crabtree, R.H. Oxidative synthesis of amides and pyrroles via dehydrogenative alcohol oxidation by ruthenium diphosphine diamine complexes. *Organometallics* **2011**, *30*, 4174–4179. [CrossRef]
46. Prades, A.; Peris, E.; Albrecht, M. Oxidations and oxidative couplings catalyzed by triazolylidene ruthenium complexes. *Organometallics* **2011**, *30*, 1162–1167. [CrossRef]
47. Zeng, H.; Guan, Z. Direct synthesis of polyamides via catalytic dehydrogenation of diols and diamines. *J. Am. Chem. Soc.* **2011**, *133*, 1159–1161. [CrossRef] [PubMed]
48. Oldenhuis, N.J.; Dong, V.M.; Guan, Z. Catalytic acceptorless dehydrogenations: Ru-Macho catalyzed construction of amides and imines. *Tetrahedron* **2014**, *70*, 4213–4218. [CrossRef] [PubMed]
49. Ortega, N.; Richter, C.; Glorius, F. *N*-formylation of amines by methanol activation. *Org. Lett.* **2013**, *15*, 1776–1779. [CrossRef] [PubMed]
50. Malineni, J.; Merkens, C.; Keul, H.; Möller, M. An efficient *N*-heterocyclic carbene based ruthenium-catalyst: Application towards the synthesis of esters and amides. *Catal. Commun.* **2013**, *40*, 80–83. [CrossRef]
51. Malineni, J.; Keul, H.; Möller, M. An efficient *N*-heterocyclic carbene-ruthenium complex: Application towards the synthesis of polyesters and polyamides. *Macromol. Rapid Commun.* **2015**, *36*, 547–552. [CrossRef] [PubMed]
52. Saha, B.; Sengupta, G.; Sarbajna, A.; Dutta, I.; Bera, J.K. Amide synthesis from alcohols and amines catalyzed by a Ru-II-*N*-heterocyclic carbene (NHC)-carbonyl complex. *J. Organomet. Chem.* **2014**, *771*, 124–130. [CrossRef]
53. Xie, X.K.; Huynh, H.V. Tunable dehydrogenative amidation versus amination using a single ruthenium-NHC catalyst. *ACS Catal.* **2015**, *5*, 4143–4151. [CrossRef]
54. Nirmala, M.; Viswanathamurthi, P. Design and synthesis of ruthenium (II) OCO pincer type NHC complexes and their catalytic role towards the synthesis of amides. *J. Chem. Sci.* **2016**, *128*, 9–21. [CrossRef]
55. Selvamurugan, S.; Ramachandran, R.; Prakash, G.; Viswanathamurthi, P.; Malecki, J.G.; Endo, A. Ruthenium (II) carbonyl complexes containing bidentate 2-oxo-1,2-dihydroquinoline-3-carbaldehyde hydrazone ligands as efficient catalysts for catalytic amidation reaction. *J. Organomet. Chem.* **2016**, *803*, 119–127. [CrossRef]
56. Selvamurugan, S.; Ramachandran, R.; Prakash, G.; Nirmala, M.; Viswanathamurthi, P.; Fujiwara, S.; Endo, A. Ruthenium (II) complexes encompassing 2-oxo-1,2-dihydroquinoline-3-carbaldehyde thiosemicarbazone hybrid ligand: A new versatile potential catalyst for dehydrogenative amide synthesis. *Inorg. Chim. Acta* **2017**, *454*, 46–53. [CrossRef]
57. Higuchi, T.; Tagawa, R.; Iimuro, A.; Akiyama, S.; Nagae, H.; Mashima, K. Tunable ligand effects on ruthenium catalyst activity for selectively preparing imines or amides by dehydrogenative coupling reactions of alcohols and amines. *Chem. Eur. J.* **2017**, *23*, 12795–12804. [CrossRef] [PubMed]

58. Cheng, H.; Xiong, M.Q.; Cheng, C.X.; Wang, H.J.; Lu, Q.; Liu, H.F.; Yao, F.B.; Chen, C.; Verpoort, F. In situ generated ruthenium catalyst systems bearing diverse *N*-heterocyclic carbene precursors for atom-economic amide synthesis from alcohols and amines. *Chem. Asian J.* **2018**, *13*, 440–448. [CrossRef] [PubMed]
59. Cheng, H.; Xiong, M.Q.; Zhang, N.; Wang, H.J.; Miao, Y.; Su, W.; Yuan, Y.; Chen, C.; Verpoort, F. Efficient *N*-heterocyclic carbene/ruthenium catalytic systems for the alcohol amidation with amines: Involvement of poly-carbene complexes? *ChemCatChem* **2018**. [CrossRef]
60. Maji, M.; Chakrabarti, K.; Paul, B.; Roy, B.C.; Kundu, S. Ruthenium(II)-NNN-pincer-complex-catalyzed reactions between various alcohols and amines for sustainable C-N and C-C bond formation. *Adv. Synth. Catal.* **2018**, *360*, 722–729. [CrossRef]
61. Huynh, H.V.; Han, Y.; Jothibasu, R.; Yang, J.A. $^{13}$C-NMR spectroscopic determination of ligand donor strengths using *N*-heterocyclic carbene complexes of palladium (II). *Organometallics* **2009**, *28*, 5395–5404. [CrossRef]
62. Chen, C.; Kim, M.H.; Hong, S.H. *N*-heterocyclic carbene-based ruthenium-catalyzed direct amidation of aldehydes with amines. *Org. Chem. Front.* **2015**, *2*, 241–247. [CrossRef]
63. Kaufhold, S.; Petermann, L.; Staehle, R.; Rau, S. Transition metal complexes with *N*-heterocyclic carbene ligands: From organometallic hydrogenation reactions toward water splitting. *Coord. Chem. Rev.* **2015**, *304*, 73–87. [CrossRef]

# Permissions

All chapters in this book were first published in MDPI; hereby published with permission under the Creative Commons Attribution License or equivalent. Every chapter published in this book has been scrutinized by our experts. Their significance has been extensively debated. The topics covered herein carry significant findings which will fuel the growth of the discipline. They may even be implemented as practical applications or may be referred to as a beginning point for another development.

The contributors of this book come from diverse backgrounds, making this book a truly international effort. This book will bring forth new frontiers with its revolutionizing research information and detailed analysis of the nascent developments around the world.

We would like to thank all the contributing authors for lending their expertise to make the book truly unique. They have played a crucial role in the development of this book. Without their invaluable contributions this book wouldn't have been possible. They have made vital efforts to compile up to date information on the varied aspects of this subject to make this book a valuable addition to the collection of many professionals and students.

This book was conceptualized with the vision of imparting up-to-date information and advanced data in this field. To ensure the same, a matchless editorial board was set up. Every individual on the board went through rigorous rounds of assessment to prove their worth. After which they invested a large part of their time researching and compiling the most relevant data for our readers.

The editorial board has been involved in producing this book since its inception. They have spent rigorous hours researching and exploring the diverse topics which have resulted in the successful publishing of this book. They have passed on their knowledge of decades through this book. To expedite this challenging task, the publisher supported the team at every step. A small team of assistant editors was also appointed to further simplify the editing procedure and attain best results for the readers.

Apart from the editorial board, the designing team has also invested a significant amount of their time in understanding the subject and creating the most relevant covers. They scrutinized every image to scout for the most suitable representation of the subject and create an appropriate cover for the book.

The publishing team has been an ardent support to the editorial, designing and production team. Their endless efforts to recruit the best for this project, has resulted in the accomplishment of this book. They are a veteran in the field of academics and their pool of knowledge is as vast as their experience in printing. Their expertise and guidance has proved useful at every step. Their uncompromising quality standards have made this book an exceptional effort. Their encouragement from time to time has been an inspiration for everyone.

The publisher and the editorial board hope that this book will prove to be a valuable piece of knowledge for researchers, students, practitioners and scholars across the globe.

# List of Contributors

**Lisa-Maria Rečnik and Thomas L. Mindt**
Ludwig Boltzmann Institute Applied Diagnostics, General Hospital Vienna, 1090 Vienna, Austria
Institute of Inorganic Chemistry, Faculty of Chemistry, University of Vienna, 1090 Vienna, Austria
Department of Biomedical Imaging and Image Guided Therapy, Division of Nuclear Medicine, Medical University of Vienna, 1090 Vienna, Austria

**Wolfgang Kandioller**
Institute of Inorganic Chemistry, Faculty of Chemistry, University of Vienna, 1090 Vienna, Austria

**Eri Seitoku, Shuhei Hoshika, Takatsumi Ikeda, Toru Tanaka and Hidehiko Sano**
Faculty of Dental Medicine, Hokkaido University, Sapporo 060-8586, Japan

**Shigeaki Abe**
Graduate School of Biomedical Sciences, Nagasaki University, Nagasaki 852-8102, Japan

**Sungwoo Hong and Hoi-Yun Jung**
Center for Catalytic Hydrocarbon Functionalization Institute for Basic Science (IBS), Daejeon 34141, Korea
Department and of Chemistry, Korea Advanced Institute of Science and Technology (KAIST), Daejeon 34141, Korea

**Zhenghuan Fang and Soon-Sun Hong**
Department of Biomedical Sciences, College of Medicine, Inha University, Incheon 22212, Korea

**Jin-Ha Yoon and Han-Joo Maeng**
College of Pharmacy, Gachon University, Incheon 21936, Korea

**Diego Antonio Ocampo Gutiérrez de Velasco, Aoze Su, Luhan Zhai, Yuko Otani and Tomohiko Ohwada**
Laboratory of Organic and Medicinal Chemistry, Graduate School of Pharmaceutical Sciences, University of Tokyo, 7-3-1 Hongo, Bunkyo-ku, Tokyo 113-0033, Japan

**Satowa Kinoshita**
Laboratory of Organic and Medicinal Chemistry, Graduate School of Pharmaceutical Sciences, University of Tokyo, 7-3-1 Hongo, Bunkyo-ku, Tokyo 113-0033, Japan
Department of Chemistry, St John's College, University of Cambridge, St John's Street, Cambridge CB2 1TP, UK

**Sandeep R. Vemula, Michael R. Chhoun and Gregory R. Cook**
Department of Chemistry and Biochemistry, North Dakota State University, Fargo, ND 58108–6050, USA

**Jean Le Bras and Jacques Muzart**
Institut de Chimie Moléculaire de Reims, CNRS—Université de Reims Champagne-Ardenne, B.P. 1039, 51687 Reims CEDEX 2, France

**Yongjun Bian, Xingyu Qu, Yongqiang Chen, Jun Li and Leng Liu**
College of Chemistry and Chemical Engineering, Jinzhong University, Yuci 030619, China

**Giovanni La Penna**
Istituto di Chimica dei Composti Organometallici (ICCOM), Consiglio Nazionale delle Ricerche (CNR), via Madonna Del Piano 10, I-50019 Sesto Fiorentino, Firenze, Italy

**Fabrizio Machetti**
Istituto di Chimica dei Composti Organometallici (ICCOM), Consiglio Nazionale delle Ricerche (CNR), c/o Dipartimento di Chimica "Ugo Schiff" via Della Lastruccia 13, I-50019 Sesto Fiorentino, Firenze, Italy

**Stephen A. Glover and Adam A. Rosser**
Department of Chemistry, School of Science and Technology, University of New England, Armidale, NSW 2351, Australia

**Jagadeesh Kalepu and Lukasz T. Pilarski**
Department of Chemistry-BMC, Uppsala University, 75-123 Uppsala, Sweden

**Paola Acosta-Guzmán, Alejandra Mateus-Gómez and Diego Gamba-Sánchez**
Laboratory of Organic Synthesis Bio- and Organocatalysis, Chemistry Department, Universidad de los Andes, Cra. 1 No 18A-12 Q:305, Bogotá 111711, Colombia

**Cheng Chen and Ye Yuan**
State Key Laboratory of Advanced Technology for Materials Synthesis and Processing, Wuhan University of Technology, 122 Luoshi Road, Wuhan 430070, China

**Yang Miao**
School of Materials Science and Engineering, Wuhan University of Technology, 122 Luoshi Road, Wuhan 430070, China

**Kimmy De Winter and Patrick Demeyere**
Odisee/KU Leuven Technology Campus, Gebroeders de Smetstraat 1, 9000 Ghent, Belgium

**Hua-Jing Wang**
School of Chemistry, Chemical Engineering and Life Sciences,Wuhan University of Technology, 122 Luoshi Road, Wuhan 430070, China

**Francis Verpoort**
State Key Laboratory of Advanced Technology for Materials Synthesis and Processing, Wuhan University of Technology, 122 Luoshi Road, Wuhan 430070, China
School of Materials Science and Engineering, Wuhan University of Technology, 122 Luoshi Road, Wuhan 430070, China
National Research Tomsk Polytechnic University, Lenin Avenue 30, Tomsk 634050, Russian
Ghent University Global Campus, 119 Songdomunhwa-Ro, Yeonsu-Gu, Incheon 21985, Korea

# Index

Printed in the USA
CPSIA information can be obtained
at www.ICGtesting.com
JSHW051406091023
49903JS00006B/304

9 781647 285203